A Practical Approach to Motor Vehicle Engineering and Maintenance

A Practical Approach to Motor Vehicle Engineering and Maintenance

Third Edition

Allan Bonnick

Derek Newbold

AMSTERDAM • BOSTON • HEIDELBERG • LONDON • NEW YORK • OXFORD
PARIS • SAN DIEGO • SAN FRANCISCO • SINGAPORE • SYDNEY • TOKYO

Butterworth-Heinemann is an imprint of Elsevier

Butterworth-Heinemann is an imprint of Elsevier
The Boulevard, Langford Lane, Kidlington, Oxford, OX5 1GB, UK
225 Wyman Street, Waltham, MA 02451, USA

First published by Arnold 2000
Reprinted 2002, 2003, 2004
Second edition 2005
Third edition 2011

British Library Cataloguing in Publication Data
A catalogue record for this book is available from the British Library

Library of Congress Control Number: 2011924728

ISBN: 978-0-08-096998-5

For information on all Butterworth-Heinemann publications
visit our website at www.elsevierdirect.com

Printed and bound in the UK

11 12 13 14 10 9 8 7 6 5 4 3 2 1

Working together to grow
libraries in developing countries

www.elsevier.com | www.bookaid.org | www.sabre.org

ELSEVIER   BOOK AID
             International   Sabre Foundation

Contents

Foreword

In this third edition some of the older technology has been removed and modern technology to include Level 3 topics has been included. Much emphasis has been placed on fault diagnosis because it is such an important part of a modern automotive technician's work.

The careful consideration of fault diagnosis that is undertaken in a range of examples in this edition shows how a methodical approach to fault diagnosis, based on good knowledge of automotive systems, can lead to success.

As an example, the ECU may report the fault code P0301 that tells us that there is a misfire in number 1 cylinder. The misfire may be caused by one or more factors, including:

- Ignition problem, defective spark plug, defective HT cables, defective direct ignition coil, etc.
- Poor compression, badly seating valve, leaking gasket, worn or broken piston rings.
- Incorrect mixture in that cylinder, air leak, defective petrol injector.

It is clear that good ignition in a single cylinder is dependent on many factors, which must be taken into account when considering what action to take. Faults in some of these other areas may also be recorded in fault codes, but others will not. To locate causes of faults in these other areas may require use of specialist equipment, and reference to manuals and circuit diagrams — these factors are covered in detail in several of the examples in this book. The methodical six-step approach is highlighted as a sound basis for successful fault diagnosis.

Outline of the diagnostic process

1. Understand the problem.
    - Receive the customer and listen carefully to their description of the problem.
    - Understand the symptoms and check to ensure that they are understood before proceeding.
    - Make notes of details.
    - Carry out checks and make a visual examination. Perform a road test if thought necessary.
2. Analyse the evidence — firm up the diagnosis.
3. Verify the problem — carry out tests to confirm the diagnosis.
4. Find the cause — use test equipment.
5. Perform the necessary repair — consult the customer if there are alternatives.
6. Carry out thorough tests to ensure that the fault is cured. Ensure that the vehicle is properly prepared for return to the customer.

Questions/exercises

I have included a number of questions at the end of most chapters. In most cases they are designed to tease out certain points and I hope that readers will use them as the basis of discussion with fellow students and tutors. Where appropriate I have included my suggested answers.

Acknowledgements

Thanks are due to the following companies who supplied information and in many cases permission to reproduce photographs and diagrams:

- Audi UK Ltd
- Bowers Metrology Group (Moore & Wright)
- Champion Spark Plugs Ltd
- Crypton Technology Group
- Cummins Engine Company Inc.
- Delmar Publishers Inc.
- Dunlop Holdings Ltd
- Ford Motor Company UK Ltd
- Haynes Publications
- Honda UK Ltd
- KIA Cars Ltd
- Lucas Automotive Ltd
- Rover Group
- Sealey Group Ltd
- TEXA Ltd
- Toyota GB Ltd
- Vauxhall Motors Ltd
- Volvo Group (UK)
- Zahnradfabrik Friedrichshafen AG.

Special thanks are due to the following for supplying much useful information, and with permission to reproduce pictures and diagrams from:

The Automotive Chassis: Engineering principles, 2nd edition, J. Reimpell & H. Stoll (Vogel-Buchverlag, Würzburg, 1995).

If we have used information or mentioned a company name in the text, not listed here, our apologies and acknowledgements.

Introduction to the retail motor industry and Health and Safety at Work

Topics covered in this introduction

- Details of the motor vehicle maintenance and repair industry.
- An introduction to Health and Safety at Work.
- Organization of the firm.
- Customers.
- Workshop activities.

The motor vehicle maintenance and repair industry — the garage industry

By taking a few examples of aspects of modern life, it is possible to gain an insight into areas of activity where motor vehicle maintenance and repair plays an important part (Table A.1.1). One has only to consider the effect of a vehicle breakdown in any of these areas to gain an appreciation of the part that motor vehicle technicians play in the day-to-day operation of society when they maintain and repair vehicles.

In order for these activities to take place, the vehicles must be serviced and maintained at regular intervals and, in the event of a breakdown, action must be taken to clear the road and repair the vehicle so as to restore it to good working condition, as quickly as possible. In the UK it is the motor vehicle repair and maintenance industry that performs the bulk of vehicle maintenance and repair work. It is for trainees and students, preparing for work in this industry, that this book is designed.

Some details about the type of work involved in repair and maintenance of vehicles

There were approximately 30 000 000 vehicles of various types in use in the UK in 2004 and a large vehicle repair and maintenance industry exists to provide the necessary services. The majority of motor cars are repaired and maintained in retail garages and businesses vary in size, from large-sized vehicle dealerships employing several people in a range of occupations to small one-person type businesses. Buses tend to be cared for in specialized workshops operated by Local Authorities and specialized bus companies. The repair and maintenance of heavy goods vehicles is often carried out in garage workshops that are owned by transport companies.

Table A.1.1   Areas of activity and types of vehicle

Area of activity	Type of activity	Type of vehicle
Personal transport	Getting to work. Taking children to school. Going on holiday. Visiting friends. Going shopping	Cars, people carriers, motorcycles, scooters, and mopeds
Public transport	Activities as for personal transport, but much more important in towns and cities where traffic congestion causes delays	Buses, coaches
Goods vehicles	Movement of food, fuel, materials for industry. Bringing food and materials into the country. Exporting manufactured goods, etc.	Trucks, vans, tankers, articulated vehicles
Emergency services	Fire and rescue services. Ambulances. Doctors. Movement of blood supplies and human organs. AA, RAC, and other forms of roadside assistance	Cars, vans, fire engines, rescue vehicles, mobile cranes
Armed forces	Defence of the realm. Action overseas. Maintenance of services in times of need	Trucks, armoured vehicles, tanks, fuel tankers, tank transporters
Postal and other delivery services	Communications for business. Private correspondence. Delivery of mail-order goods	Vans, trucks
Fuel deliveries	Fuel deliveries to service stations and fuel depots. Domestic fuel supplies	Tankers of various sizes

The garage industry employs several hundred thousand people in a range of occupations. The work is interesting, often demanding — both physically and mentally. There are many opportunities for job satisfaction. For example, it is most rewarding to restore a vehicle to full working condition after it has suffered some form of failure. There are opportunities for promotion, technicians often progressing to service managers and general managers, or to run their own companies. The Office of Fair Trading (OFT) report for the year 2000 shows that more than 50% of cars were more than 5 years old. These cars require annual MOT inspections and there is a tendency for them to be serviced and repaired in the independent sector of the industry. The OFT report shows that there are approximately 16 000 independent garages and 6500 garages that are franchised to one or more motor manufacturers. Evidently there are plenty of opportunities for employment.

Health and safety

It is the responsibility of every person involved in work to protect their own safety and that of any other persons who may be affected by their activities. At the basic level this means that everyone must work in a safe manner and know how to react in an emergency. Personal cleanliness issues such as regular washing, removal of substances from skin, use of barrier creams, use of protective clothing and goggles and gloves, etc. are factors that contribute to one's well being. Behaving in an orderly way, not indulging in horseplay, learning how to employ safe working practices, and helping to keep the workplace clean and tidy are all ways in which an individual can contribute to their own and others' health and safety at work.

There are various laws and regulations that govern working practice in the motor vehicle repair industry, in the UK — the main ones are:

- Health & Safety at Work Act 1974
- The Factories Act 1961
- The Offices, Shops and Railway Premises Act 1963.

Some of the other regulations that relate to safety in motor vehicle repair and maintenance are:

- The Control of Substances Hazardous to Health (COSHH) Regulations 1988
- Regulations about the storage and handling of flammable liquids
- The Grinding Wheel Regulations.

Health and safety laws are enforced by a factory inspector from the Health and Safety Executive (HSE) or an environmental health officer from the local council.

Safety policy

As stated at the beginning of this section, each individual has a responsibility to work safely and to avoid causing danger to anyone else. This means that each individual must know how to perform their work in a safe way and

how to react with other people in the event of an accident or emergency. In any establishment, however small, safety planning must be performed by a competent person, who must then familiarize all others engaged in the enterprise with the plans that have been devised, to ensure that all health and safety issues are properly covered. These plans are a set of rules and guidelines for achieving health and safety standards, in effect a policy.

Every motor vehicle repair business that employs five or more people must write down their policy for health and safety and have it to hand for inspection by the HSE.

Motor vehicle repair and maintenance health and safety topics

Readers are advised to purchase a copy of the publication *Health and Safety in Motor Vehicle Repair and Associate Industries*, HSG261 (2009) available from HMSO for approximately £13. The HSE website, www.hse.gov.uk, is also a valuable source of information. Several pages are devoted to motor vehicle repair and the home page offers the user the opportunity to select 'your industry', a click on 'Motor vehicle repair' brings up many opportunities to study a variety of health and safety topics.

This publication (HSG261) states that:

'Most accidents in Motor Vehicle Repair (MVR) involve slips, trips and falls or occur during lifting and handling, and often cause serious injury. Crushing incidents involving the movement or collapse of vehicles under repair result in serious injuries and deaths every year. Petrol-related work is a common cause of serious burns and fires, some fatal.

There is also widespread potential for work-related ill health in MVR. Many of the substances used require careful storage, handling and control. Isocyanate-containing paints have been the biggest cause of occupational asthma in the UK for many years and MVR is also in the top ten industries for cases of disabling dermatitis. Use of power tools can cause vibration white finger.

This guidance has been developed in consultation with representatives from the MVR industry and describes good practice. Following it should help you reduce the likelihood of accidents or damage to health. The book is divided into two main sections, one has guidance for specific industry sectors and the other provides extensive advice on common MVR issues.'

From this you might think that motor vehicle work is hazardous but, if work is carried out properly and safety factors are always considered, it is possible to work without injury to anyone. Safety training and related skills training must figure prominently in any course of education and training that aims to prepare people for work in motor vehicle maintenance and repair. The following descriptions are intended to highlight some of the everyday safety issues.

The information contained in this book does not attempt to provide full coverage of all safety and health issues. The information provided merely draws attention to some safety issues, and is not intended to cover all areas of safety. Readers should ensure that any education course, training scheme, apprenticeship or other training arrangement does contain all necessary safety training.

Lifting equipment

The types of lifting equipment that are commonly used in vehicle repair workshops are: jacks of various types, axle stands, vehicle hoists (lifts), floor cranes, and vehicle recovery equipment. Hydraulic jacks are used to raise the vehicle. For work to be done underneath the vehicle, the vehicle must be on level ground, and the wheels that are remaining on the ground must be chocked or have the handbrake applied. Axle stands must be placed in the correct positions before any attempt is made to get under the vehicle. Figure A.1.1 shows the hydraulic jack placed at a suitable jacking point. This is important: (i) to ensure that the jacking point is secure from the safety point of view; and (ii) to prevent damage to the vehicle.

Hydraulic jacks must be maintained in good condition. The safe working load must be clearly marked on the jack. Axle stands must be of good quality and of a load-carrying capacity that is correct for the vehicle being supported on them. The proper pins that allow for height adjustment should be attached to the stands and the stands must be kept in good condition.

Vehicle hoists

The two-post vehicle hoist shown in Fig. A.1.2 is an example of a type that is widely used in the garage industry. The HSE guide (HSG261) states:

'Two-post lifts typically consist of two upright columns: a master or powered column. plus an auxiliary or 'slave', both fitted with a pair of (vehicle) carrying arms. These arms are pivoted at the column and their lengths are adjustable, usually by telescopic means, though some are articulated. At the free end of each carrying arm there is an adjustable pick-up plate fitted with a rubber mounting pad. A two-post lift achieves 'wheel-free lifting' by aligning the pick-up plates to four jacking points on the underside of the chassis of the vehicles. It is important that:

- *you follow the vehicle manufacturer's recommendations regarding vehicle lifting. In particular, you need to ensure the equipment is suitable when lifting larger SUV and 'light' freight and commercial vans (LCVs);*
- *vehicle chassis and chassis jack points are identifiable and in a satisfactory condition. Vehicle jacking points are usually identified by a symbol on the vehicle sill, if in doubt always consult the car handbook;*
- *the support-arm rubber mounting pads are in serviceable condition and, where necessary, for example to avoid the lift arm fouling the car bodywork, are set at the correct height before the vehicle is raised;*
- *the lifting arms are carefully positioned before each lift, in accordance with the manufacturer's instructions. This is to ensure correct weight distribution and proper contact with load-bearing points so the vehicle is stable. Also take account of the weight distribution of the vehicle, e.g. front/rear*

(a)

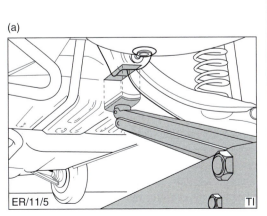

(b)

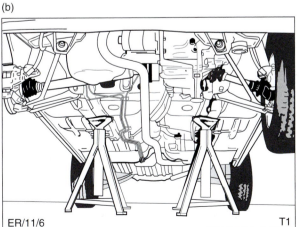

Fig. A.1.1 (a) Hydraulic jack positioned at jacking point. (b) Axle stands in position prior to working under the vehicle. *(Reproduced with the kind permission of Ford Motor Company Limited.)*

Fig. A.1.2 Wheel-free post hoist. *(Reproduced with the kind permission of Rover Motor Company Limited.)*

engined, loads within vehicle/boot, and absence of major components such as engine or gearbox;
- *you consider the effect on the stability of the vehicle caused by the removal of major components or by the application of forces, via tools etc;*
- *the lift arms are fitted with an effective automatic mechanical arm locking system to maintain the angle between them. These should be checked regularly — typically daily.*
- *It is essential that two-post lifts are installed correctly into a suitable base. This is particularly important where the posts are free-standing (with no bridging support). Specialist advice may be required to ensure the structural integrity of the floor and fixings and the lift supplier or specialist garage equipment maintenance company should be consulted.'*

Comment

Hoists, in common with other equipment, should not be used until the method of operation is properly understood. For trainees this means that proper training must be provided and trainees should not attempt to use hoists until they have been trained and given permission to do so. Part of that training will be a demonstration of the procedure for locating jacking points and positioning the support arm pads. This ensures that the vehicle will be stable when lifted and the support arm pads prevent damage to the vehicle structure. The weight distribution of the vehicle on the hoist will be affected according to the work being performed. For example, if the hoist is being used during a transmission unit change, the removal of the transmission unit will greatly affect weight distribution and this must be allowed for as the work progresses.

Regarding four-post hoists, the HSE guide (HSG261) says that:

'Four-post hoists should have effective "dead man's" controls, toe protection and automatic chocking. Toe traps should also be avoided when body straightening jigs are fitted. Raised platforms should never be used as working areas unless proper working balconies or platforms with barrier rails are provided.'

Greater detail is given on pages 45–46 of HSG261.

Comment

The dead man's control is designed to avoid mishaps during the raising and lowering of the hoist. Toe protection is to prevent damage to the feet when the hoist is lowered to the ground, and the automatic chocking is to prevent the vehicle from being accidentally rolled off the raised hoist. Body straightening jigs are used in vehicle body repair shops and their use requires special training. The point about working above floor level and the use of balconies should be noted.

In general, hoists make work on the underside of a vehicle less demanding than lying on one's back. A point to note is that hard hats should be worn when working under a vehicle on a hoist. The area around the hoist must be kept clear and a check should always be made when lowering or raising a hoist to ensure that no person, vehicle, or other object is likely to be harmed by the operation. Hoists must be frequently examined to ensure that they are maintained in safe working order.

Electrical safety

Safety in the use of electrical equipment must figure prominently at all stages of training.

> *Learning task*
> As an exercise to develop knowledge in this area you should now read pages 72–79 of HSG261 and compare that with the training and tuition that you are receiving. Also visit the HSE motor vehicle repair website and study what it has to say about electrical safety. Check with your supervisor to make sure that you know what to do in the event of an electrical emergency in your present circumstances.

Compressed air equipment

Compressed air is used in a number of vehicle repair operations such as tyre inflation, pneumatic tools, greasing machines, oil dispensers, etc. The compressed air is supplied from a compressor and air cylinder, this is equipment that must be inspected regularly by a competent person. The usual procedure is for the insurance company to perform this work. Air lines, hoses, tyre inflaters, pressure gauges, couplings, etc. should be inspected at regular intervals to ensure that they are kept in working order. Inflation of the split rim type of commercial vehicle tyres requires special attention and the tyres should only be inflated behind a specially designed guard. The HSE publication HSG261 covers many aspects of health and safety for those engaged in tyre and exhaust fitting work.

> *Learning task*
> Read pages 55–60 of HSG261 and make notes about the main points. Discuss these points with your tutor and class mates.

Petrol fires

Referring back to the statement at the beginning of this section we have:

> *'Petrol-related work is a common cause of serious burns and fires, some fatal.'*

Comment

Petrol gives off a flammable vapour that is heavier than air; this means that it will settle at a low level and spread over a wide area. Petrol vapour is also invisible. The vapour can be ignited by a flame or spark at some distance from any visible sign of the liquid. Great care must be exercised when dealing with petrol and the appropriate fire extinguisher should be close to hand when performing any task involving petrol or any other flammable substance. Petrol may only be stored in containers specified in the Petroleum Spirit Regulations. Advice on this is readily available from the local Fire Authority. In the event of spillage on clothes, the clothes should be removed for cleaning because petrol vapour can accumulate inside the clothing, which can result in serious burns in the event of ignition.

> *Learning task*
> Read page 66 of HSG261. What does HSG261 say about the fuel gauge sender unit? Discuss this with your tutor/trainer, or supervisor.

Brake and clutch linings

> *'There is also widespread potential for work related ill-health in garages. Many of the substances used require careful storage, handling and control.'*
>
> (HSG261)

Some brake and clutch linings contain asbestos. To guard against harmful effects special measures must be employed. These measures include wearing a mask, using an appropriate vacuum cleaner to remove dust and preventing asbestos dust from getting into the surrounding air.

> *Learning task*
> Read page 10 of HSG261 and then describe to your tutor/trainer the type of vacuum cleaner that should be used to remove brake lining or clutch lining dust. Also explain the purpose of wetting any dust that may have been left on the floor.

Oils and lubricants

Oils and lubricants used in vehicles contain chemicals to change their properties and make them suitable for use in many vehicle applications. Many of the chemicals used are harmful to health if not handled properly.

Used oils must only be disposed of in approved ways. Special containers are available for receiving oil as it is drained from a vehicle. Vehicle repair workshops must have facilities to store used oil prior to disposing of it through approved channels. Used oil has a market value and some companies collect it for reprocessing.

Learning task
Visit the website http://ehsni.gov.uk. Study what this says about disposal of used oil from garages and private homes. Make a note of the main points.

'Accidents involving vehicles are very frequent and cause serious injuries and deaths every year.'
(HSG261)

Comment

Work that involves movement of vehicles both inside and outside of the workshop can be dangerous if not handled properly. Vehicles should only be moved by authorized persons and great care must be taken when manoeuvring them.

Learning task
Study the sections of HSG261 that deal with rolling road dynamometers and brake testers, moving and road testing vehicles. Visit the HSE website and note the entries that deal with this topic. Discuss your findings with your colleagues, fellow students, and workmates.

Other topics

The publication HSG261 draws attention to many topics relating to health and safety in motor vehicle repair and maintenance. Two of these topics are vehicle valeting, and the use of steam cleaners and water pressure cleaners.

Comment

Vehicle valeting tasks, such as cleaning the exterior and interior of the vehicle, removing stains from upholstery,

etc., often involve the use of chemical substances. Cleaning of components during servicing and repair procedures also entails the use of chemicals, and steam and high-pressure water cleaners.

Learning task
Read section that starts on page 14 of HSG261. Note the points about COSHH. Note down the procedures that are used in your workplace for dealing with COSHH. Make a note of the types of steam and water pressure cleaning apparatus and any other cleaning tanks that are used in any place where you are involved in practical work. Note any special cleaning substances that are used. Visit the HSE website and study the motor vehicle repair section that deals with COSHH.

The intention in this section has been to draw attention to some aspects of Health and Safety. Health and Safety are aspects that one must be constantly aware of throughout working life. Many aspects of work, such as keeping a workplace clean and tidy, and working methodically, contribute to health and safety and also form part of an orderly approach to tasks that assist one in the performance of complex tasks such as fault diagnosis.

Organization of the firm

A company, or firm as it is often called, will often display its structure in the form of an organization chart. For any group of people who are engaged in some joint activity it is necessary to have someone in charge and for all members of the 'team' to know what their role is and which person they should speak to when they need advice. Just as a football team is organized and each person is given a position on the field, so a business is organized so that people can work as a team.

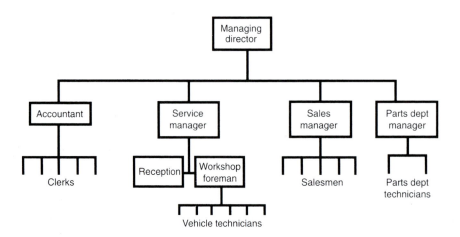

Fig. A.1.3 A garage business line organization chart

The 'line chart' (Fig. A.1.3) shows how the work is divided up into manageable units, or areas of work, and how the personnel in those areas relate to each other. In this example the managing director is the 'boss'. Under the managing director come the accountant, the service manager, the parts department manager, and the sales manager; these are known as 'line' managers. Below these line managers, in the hierarchical structure that is used in many businesses, are supervisors (foremen/women) and then come the technicians, clerks, etc.

Policy

A policy is a set of rules and guidelines that should be written down so that everyone in an organization knows what they are trying to achieve. Policy is decided by the people at the top of an organization, in consultation with whoever else they see fit. It is important for each employee to know how policy affects them and it should be suitably covered in a contract of employment. For example, every firm must have a 'safety policy'. It is the duty of every employee to know about safety and firms are subject to inspection to ensure that they are complying with the laws that relate to health and safety at work, and similar legislation.

Discipline

Standards of workmanship, hours of work, relations with other employees, relations with customers, and many other factors need to be overseen by members of the firm who have some authority because, from time to time, it may be necessary to take steps to change some aspect. For example, it may be the case that a particular technician is starting to arrive late for work; it will be necessary for someone to deal with this situation before it gets out of hand and this is where the question of 'authority' arises.

Authority

Authority is power deriving from position. The organization chart illustrates who has authority (power) over whom. The vertical position on the chart of a member of staff indicates that they are in charge of, and have power over, the employees lower down the chart. The service manager in a garage has authority in relation to the operation of the workshop; in addition they are responsible for the satisfactory performance of the workshop side of the business.

Accountability

We all have to account for actions that we take. This means having to explain why we took a particular line of action. We are all responsible for the actions that we take. This means that we did certain things that led to some outcome. By having to explain why something 'went wrong', i.e. by having to account for our behaviour, we are made to examine our responsibility. We are 'responsible' for seeing that we get to work on time, for making sure that we do a job properly and safely, and for helping to promote good working relationships.

Delegating

It is often said that you can delegate (pass down) authority but not responsibility. So, if you are a skilled technician who has a trainee working with you and you have been given a major service to do, it will be your responsibility to ensure that the job is done properly. You will have had some authority given to you to instruct the trainee, but the responsibility for the quality of the work remains with you.

Having introduced some ideas about business organization and the working relationships that are necessary to ensure success, we can turn our attention to the business activities that are the immediate concern of the service technician. In order to place some structure on this section it is probably a good idea to start with 'reception'.

Reception

Reception in a garage is the main point of contact between customers and the services that the workshop provides. It is vitally important that the communication between the customers and reception and between reception and the workshop is good. It is at reception that the customer will hand over the vehicle and it is at reception that it will be handed back to the customer. The reception engineer is a vital link in the transaction between the customer and the firm, and it is at this point that friction may occur if something has gone wrong. Staff in reception need to be cool headed, able to think on their feet, and equipped with a vocabulary that enables them to cope with any situation that may 'crop up'. From time to time they may need to call on the assistance of the service manager and it is vitally important that there are good relations and clear lines of communication in this area.

In many garages there will be an area set aside for reception and the staff placed there will have set duties to perform. In the case of small garages employing, perhaps, one or two people, reception may take place in the general office and the person who is free at the time may perform the function. In either case, certain principles apply and it is these that are now addressed.

Customer categories

Because a good deal of skill and tact is required when dealing with customers it is useful to categorize them as follows:

- informed
- non-informed
- new
- regular.

Informed customers

Informed customers are those who are knowledgeable about their vehicle and who probably know the whole procedure that the vehicle will follow before it is returned to them. This means that the transaction that takes place between them and the receptionist will largely consist of taking down details of the customer's requirements and agreeing a time for the completion of the work and the collection of the vehicle. Depending on the status of the customer, i.e. new or regular, it will be necessary to make arrangements for payment. It may also be necessary to agree details for contacting the customer should some unforeseen problem arise while work on the vehicle is in progress, so that details can be discussed and additional work and a new completion time agreed.

Non-informed customers

The non-informed customer probably does not know much about cars or the working procedures of garages. This type of customer will need to be treated in quite a different way. The non-informed customer will — depending on the nature of the work they require for their vehicle — need to have more time spent on them. A proportion of this extra time will be devoted to explaining what the garage is going to do with their vehicle and also what the customer has to do while the vehicle is in the care of the garage. The latter part will largely consist of making arrangements for getting the customer to some chosen destination and for collecting the vehicle when the work is completed. On handing the car back to the 'non-informed' customer it will probably be necessary to describe, in a non-patronizing and not too technical way, the work that has been done and what it is that they are being asked to pay for.

New customers

A 'new' customer may be informed or non-informed and this is information that should be 'teased' out during the initial discussions with them. Courtesy and tact are key words in dealing with customers and they are factors that should be uppermost in one's mind at the new-customer-introduction stage. The extent of the introductory interview with the new customer will depend on what it is that they want done and the amount of time that they and the firm have to give to the exercise. If it is a garage with several departments it may be advisable to introduce the new customer to relevant personnel in the departments that are most likely to be concerned with them. Customers may wish to know that their vehicle will be handled by qualified staff and it is common practice for firms to display samples of staff qualification certificates in the reception area. At some stage it will be necessary to broach the subject of payment for services and this may be aided by the firm having a clearly stated policy. Again, it is not uncommon for a notice to be displayed which states the 'terms of business'; drawing a client's attention to this is relatively easy.

Regular customers

Regular customers are valuable to a business and they should always be treated with respect. It hardly needs saying that many of the steps that are needed for the new customer introduction will not be needed when dealing with a regular customer. However, customers will only remain loyal (regular) if they are properly looked after. It is vitally important to listen carefully to their requests and to ensure that the work is done properly and on time, and that the vehicle is returned to them in a clean condition.

Workshop activities

Reception will have recorded the details of the work that is to be done on the vehicle, and they will have agreed details of completion time, etc. with the workshop. At some point in the process a 'job card' will have been generated. The instructions about the work to be done must be clear and unambiguous. In some cases this may be quite brief — for example, 10 000 mile service. The details of the work to be performed will be contained in the service manual for the particular vehicle model. In other cases it may be rather general — for example, 'Attend to noisy wheel bearing. Near side front.' Describing exactly what work is required may entail further investigation by the technician. It may be that the noise is caused by the final drive. The whole thing thus becomes much more complicated and it may be necessary to conduct a preliminary examination and test of the vehicle before the final arrangements are made for carrying out the work.

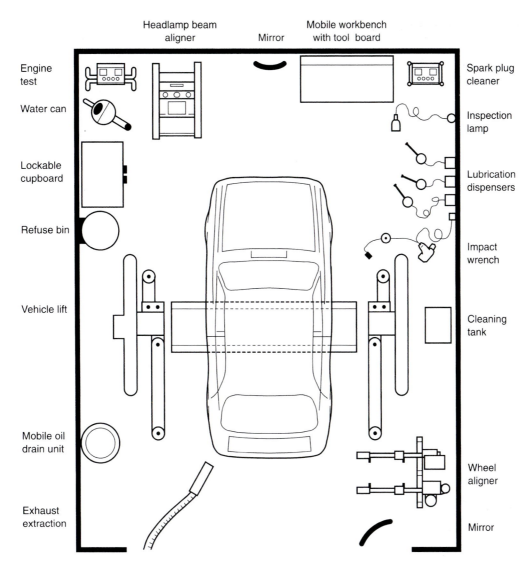

Fig. A.I.4 A service bay

Once the vehicle has been handed over to the workshop with a clear set of instructions about the work to be done, it becomes the responsibility of the technician entrusted with the job and their colleagues to get the work done efficiently and safely, and to make sure that the vehicle is not damaged. This means that the workshop must have all necessary interior and exterior protection for the vehicle such as wing and seat covers, etc. Figure A.1.4 shows a suggested layout for a service bay.

Records

As the work proceeds a record must be made of materials used and time spent because this will be needed when making out the invoice. In large organizations the workshop records will be linked to stock control in the parts department and to other departments, such as accounts, through the company's information system, which will probably be computer based.

Quality control

Vehicle technicians are expected to produce work of high quality and various systems of checking work are deployed. One aspect of checking quality that usually excites attention is the 'road test'. It is evident that this can only be performed by licensed drivers and it is usually restricted to experienced technicians. A road test is an important part of many jobs because it is probably the only way to ensure that the vehicle is functioning correctly. It is vitally important that it is conducted in a responsible way. Figure A.1.5 illustrates the point that there must be a good clear road in front of and behind the vehicle.

Fig. A.1.5   The front and rear view of the road when road testing

Returning the vehicle to the customer

On satisfactory completion of the work, the vehicle will be taken to the customer's collection point; the covers and finger marks, and other small blemishes should be removed. In the meantime the accounts department will, if it is a cash customer, have prepared the account so that everything is ready for completion of the transaction when the customer collects the vehicle.

1
Vehicle layouts and some simple vehicle structures

Topics covered in this chapter

The types of forces to which a vehicle structure is subjected
Unitary construction
Chassis-built vehicles
Front- and rear-wheel drive and the layout of these systems
Passenger protection — crumple zones and side impact bars
Vehicle shape — air resistance
Pros and cons of front- and rear-wheel-drive systems
End-of-life vehicles and methods of disposal

Light vehicles

The term light vehicle is generally taken to mean vehicles weighing less than 3 tonnes. These are vehicles such as cars, vans, and light commercial vehicles. Various types of light vehicles are shown in Fig. 1.1. Most of these vehicles are propelled by an internal combustion engine but increasing concern about atmospheric pollution is causing greater use of electric propulsion. Vehicles that incorporate an internal combustion engine and an electric motor are called hybrid vehicles.

Vehicle structure

Figure 1.2 shows some of the forces that act on a vehicle structure. The passengers and other effects cause a downward force that is resisted by the upward forces at the axles. The vehicle structure acts like a simple beam where the upper surface is in compression and the lower one in tension. When the loading from side to side of the vehicle is unequal the vehicle body is subject to a twisting effect and the vehicle structure is designed to have torsional stiffness that resists distortion through twisting.

There are basically two types of vehicle construction: one uses a frame on to which the vehicle is built; the other is called unitary construction where the body and frame are built as a unit to which subframes are added to support the suspension and other components.

The frame is normally made from low-carbon steel that is formed into shapes to provide maximum strength; box and channel sections are frequently used for this purpose. The frame shown in Fig. 1.3 has a deeper section in the centre area of the side members because this is where the bending stress is greatest. In areas where additional strength is required, such as where suspension members are attached, special strengthening supports are fitted.

Unitary construction

Most of the vehicle structure is made from steel sections that are welded together to provide a rigid structure which is able to cope with the stresses and strains that occur when the vehicle is in use. In most cases the material used is a deep-drawing mild steel that can be readily pressed into the required shapes. An example of unitary construction is shown in Fig. 1.4. In some cases, the outer panels are made from plastics such as Kevlar (Fig. 1.5), which has excellent strength and resistance to corrosion.

Body panels are normally lined with sound-deadening material which is either sprayed, or glued, on to the inside. In order to protect body components against rust and corrosion, they may be galvanized, or treated with some other form of protection.

Passenger protection

In addition to providing the strength required for normal motoring conditions, vehicles are designed to protect the occupants in the event of a collision. Two areas of vehicle construction are particularly related to this problem:

1. Crumple zones
2. Side impact protection.

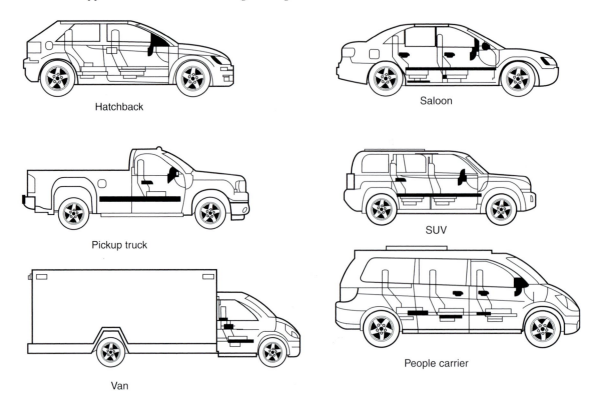

Fig. 1.1   Types of light vehicles

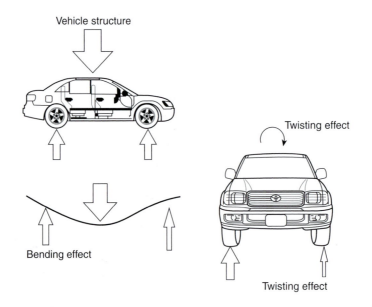

Fig. 1.2   Forces acting on vehicle structure

Crumple zones

A vehicle in motion possesses kinetic energy. Because energy cannot be destroyed, some means has to be found to change its form. Under braking conditions, the kinetic energy is converted into heat by the friction in the brakes – this heat then passes into the atmosphere. When involved in a collision the kinetic energy is absorbed in distorting the vehicle structure – if this distortion can take place outside the passenger compartment, a degree of protection for the vehicle occupants can be provided. The front and rear ends of motor cars are designed to collapse in the event of a collision; the areas of the bodywork that are designed for this purpose are known as 'crumple zones' (Fig. 1.6). At the design

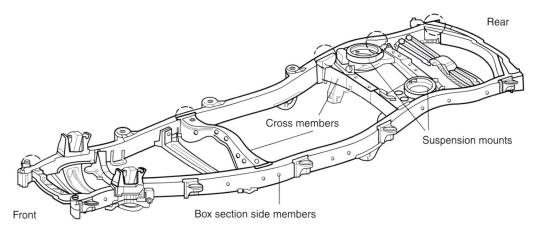

Fig. 1.3 A typical vehicle frame

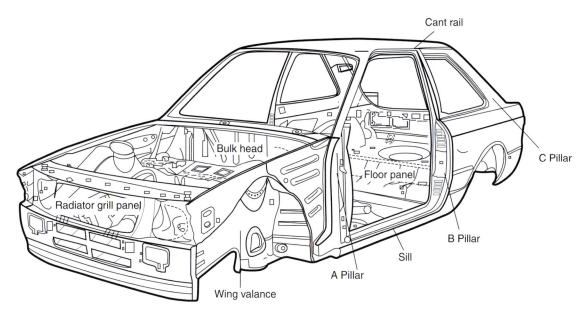

Fig. 1.4 Features of a unitary construction body

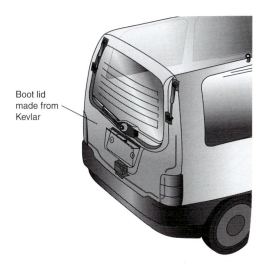

Fig. 1.5 Kevlar body panel

Fig. 1.6 Crumple zone — impact test at the British Transport Research Laboratory (TRL)

Image Courtesy of StaraBlazkova at the Czech language Wikipedia

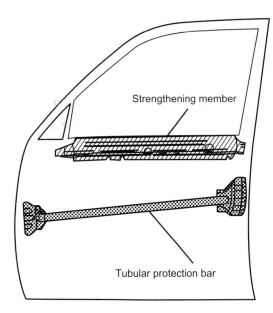

Fig. 1.7   Side impact protection (Toyota)

and development stages, and prior to introduction into general use, samples of vehicles are subjected to rigorous tests to ensure that they comply with the standards that are set by governmental bodies.

Side impact protection

In the event of side impact, a degree of protection to occupants is provided by the bars that are fitted inside the doors (Fig. 1.7).

Vehicle shape

A considerable amount of engine power is consumed in driving a vehicle against the air resistance that is caused by vehicle motion. The air resistance is affected by a factor known as the 'drag coefficient' and it is dependent on the shape of the vehicle.

Air resistance = $C_d \times A \times V^2$, where C_d is the drag coefficient, A is the frontal area of the vehicle, and V is the velocity of the vehicle relative to the wind speed. The way in which engine power is absorbed in overcoming air resistance is shown in the graph in Fig. 1.8.

Streamlining

The efficiency of the streamlining of a vehicle body is a major factor in reducing the drag coefficient; other factors, such as recessing door handles and shaping of exterior mirrors, also contribute to a lowering of drag. Another factor that contributes to the lowering of drag is the air dam that is fitted to the front of a vehicle (Fig. 1.9); this reduces under-body turbulence.

End-of-Life Vehicles Directive

The End-of-Life Vehicles (ELV) Directive aims to reduce the amount of waste produced from vehicles when they are scrapped. Around two million vehicles reach the end of their life in the UK each year. These vehicles are classed as hazardous waste until they have been fully treated.

What does the directive mean?

The directive requires ELV treatment sites to meet stricter environmental standards.

The last owner of a vehicle must be issued with a Certificate of Destruction for their vehicle and they must be able to dispose of their vehicle free of charge. Vehicle manufacturers and importers must cover all or most of the cost of the free take-back system.

It also sets higher reuse, recycling and recovery targets, and limits the use of hazardous substances in both new vehicles and replacement vehicle parts.

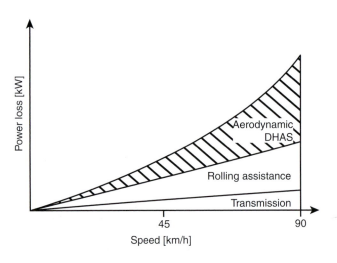

Fig. 1.8   Power used against air resistance

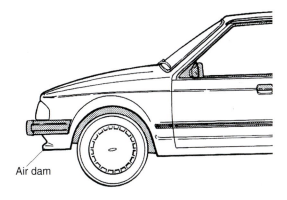

Air dam

Fig. 1.9 An air dam (DuPont)

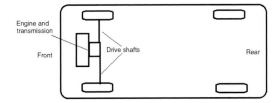

Fig. 1.10 Typical front-wheel-drive arrangement

Who implements the directive?

In the UK, the directive is implemented through ELV Regulations issued in 2003 and 2005, and through the Environmental Permitting (EP) Regulations 2007.

The 2003 regulations deal with information requirements, certificate of destruction requirements, and restricting the use of hazardous substances in new vehicles. The 2005 regulations cover recycling targets and free take-back for ELVs.

The 2007 regulations extended the treatment requirements in the UK to **all** waste motor vehicles (including coaches, buses, motor cycles, goods vehicles, etc).

Authorized Treatment Facilities

Authorized Treatment Facilities (ATFs) are permitted facilities accepting waste motor vehicles, which are able to comply with the requirements of the End-of-Life Vehicle (ELV) and Environmental Permitting (EP) regulations.

In the UK, most local authorities have vehicle recycling companies that are authorized to deal with ELVs:

- Vehicle recyclers must dismantle vehicles in an environmentally responsible manner and achieve between 75% and 85% recycling targets.
- The vehicle owner must dispose of their unwanted vehicle in a legal and responsible manner by using an Authorized Treatment Facility.

Layout of engine and driveline

Front-wheel drive

The majority of light vehicles have the engine at the front of the vehicle with the driving power being transmitted to the front wheels.

In the arrangement shown in Fig. 1.10 the engine and transmission units are placed transversely at the front of the vehicle, which means that they are at right angles to the main axis of the vehicle.

Some of the advantages claimed for front-wheel drive are:

- Because the engine and transmission system are placed over the front wheels the road holding is improved, especially in wet and slippery conditions.
- Good steering stability is achieved because the driving force at the wheels is in the direction that the vehicle is being steered. There is also a tendency for front-wheel drive vehicles to understeer, which can improve driveability when cornering.
- Passenger and cargo space are good because there is no need for a transmission shaft to the rear axle.

Possible disadvantages are:

- Complicated drive shafts are needed for constant-velocity joints.
- Acceleration is affected because load transfer to the rear of the vehicle lightens the load on the drive axle at the front.
- The turning circle radius is limited by the angle through which a constant-velocity joint can function.

Front-engine rear-wheel drive

Until reliable mass-produced constant-velocity joints became available, the front engine and rear drive axle arrangement shown in Fig. 1.11 was used in most light vehicles.

In the layout shown in Fig. 1.11 the engine is mounted in-line with the main axis of the vehicle. The gearbox is at the rear of the engine and power is transmitted through the propeller shaft to the drive axle at the rear. The gearbox, propeller shaft, and rear axle make up what is known as the driveline of the vehicle.

The advantages of a front-engine rear-wheel drive arrangement are:

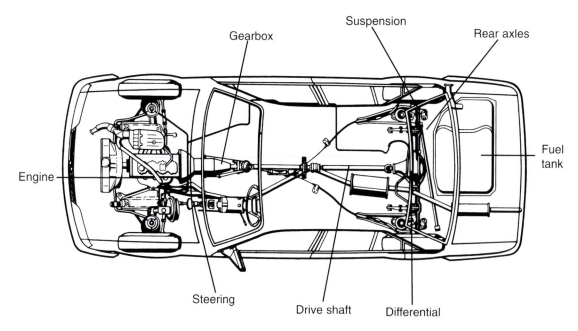

Fig. 1.11   Front-engine rear-wheel drive

- The front axle is relatively simple.
- Acceleration and hill climbing are aided because load transfer to the rear of the vehicle retains traction at the driving wheels.

    Possible disadvantages are:

- Reduced space for driver and front passenger because of the bulge in the floor panel that is required to accommodate the gearbox and clutch housing.
- The raised section known as the propeller shaft affects available space throughout the length of the passenger compartment.
- Long propeller shafts can cause vibration problems.

Rear-engine rear-wheel drive

Figure 1.12 shows an arrangement where the engine is mounted transversely at the rear with the drive being transmitted to the rear axle.

The advantages claimed for the rear-engine layout are:

- Short driveline because the engine, gearbox, and final drive can be built into a single unit.
- A preponderance of weight at the rear of the vehicle gives improved traction during hill climbing and acceleration.

    Possible disadvantages are:

- A tendency to oversteer.
- Difficulty accommodating liquid cooling of the engine.
- Difficulty accommodating the fuel tank in a safe zone of the vehicle.

- Space for luggage is reduced.
- Difficulty steering in slippery conditions.

Four-wheel drive

In this system the engine power is transmitted to all four wheels of a light vehicle. In the arrangement shown in Fig. 1.13(a), the engine is placed at the front of the vehicle. Power to the front wheels is provided through the gearbox to the front axle and from the gearbox to the rear axle via the propeller shaft.

Permanent drive to all four wheels (Fig. 1.13(a) and (b)) poses certain difficulties with braking and steering that require the use of sophisticated electronically controlled devices. A common approach to four-wheel drive makes use of an additional gearbox that is known as the transfer gearbox. This additional unit allows the driver to select four-wheel drive when driving conditions make it beneficial and for cross-country work the transfer gearbox provides an additional range of lower gear ratios.

The advantages claimed for four-wheel drive are:

- Better traction in all conditions.
- Wear of tyres and other driveline components is more evenly shared.

    Possible disadvantages are:

- Increased weight and initial vehicle cost.
- Increased maintenance due to the complexity of transmission systems.
- Increased fuel consumption.
- Possibly difficult to accommodate anti-lock braking systems.

Vehicle layouts and some simple vehicle structures

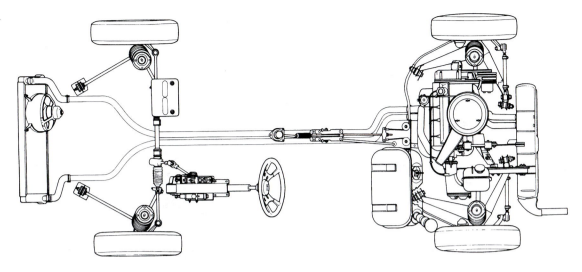

Fig. 1.12 Rear-engine rear-wheel-drive layout

(a)

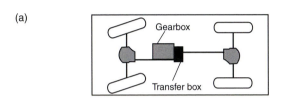

(b)

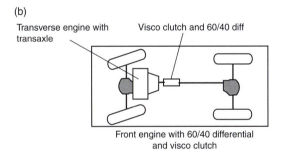

Fig. 1.13 (a) Optional four-wheel drive. (b) Permanent four-wheel drive

---

Self-assessment questions

1. What happens to a motor vehicle that has reached the end of its useful life?
2. Find out the names of the Authorized Treatment Facilities in your area.
3. What percentage of a vehicle is recycled? What happens to the steel that is reclaimed when a vehicle is scrapped?
4. Examine a manual for a vehicle that you work on and describe the features that the design incorporates to protect the occupants in the event of a collision.
5. Write a few notes to describe why you think that four-wheel drive vehicles are now popular for use as family cars.
6. What is meant by the term 'oversteer'. Why do you think that a rear-engine vehicle may be more prone to oversteer than a front-engine vehicle?
7. What is the purpose of an air dam?
8. What measures should be taken to protect paintwork when a vehicle is being worked on?
9. In which position on vehicle panels is soundproofing applied?
10. If the speed of a vehicle is doubled, by what factor is the air resistance increased?
11. What features of vehicle design affect the drag coefficient?
12. Which part of a four-wheel drive transmission system permits the four-wheel drive to be engaged?
13. In which positions on a vehicle are the side impact protection bars fitted?
14. Make a list of the external parts of a motor car that are made from plastic.
15. What methods of joining body panels are used in modern vehicle construction?
16. What materials and methods are used to prevent water entering the interior of a car?
17. Describe, with the aid of a diagram, the type of engine mounting that is used to attach an engine to a vehicle frame.

2
Engine principles

Topics covered in this chapter

The Otto cycle
Compression ratio
The two-stroke cycle
The Wankel rotary engine
The Atkinson cycle as adapted for use in hybrid vehicles
Valve and ignition timing
Variable valve lift and valve timing

The engine is the device that converts the chemical energy contained in the fuel into the mechanical energy that propels the vehicle. The energy in the fuel is converted into heat energy by burning the fuel in a process known as combustion, which is why vehicle engines are often referred to as **internal combustion engines**.

The fuel is burned inside the engine cylinders in the presence of air; when the air is heated its pressure rises and generates the force that operates the engine. Most engines used in motor vehicles make use of the piston and crank mechanism that converts linear motion into rotary motion; the piston moves to and fro in the cylinder in a reciprocating fashion — because of this the engines are frequently called **reciprocating engines**.

The component parts of the simple engine shown in Fig. 2.1 are:

1. The piston, which receives the gas pressure.
2. The cylinder, in which the piston moves to and fro.
3. The connecting rod that transmits force from the piston to the crank.
4. The crank that converts the reciprocating movement of the piston into rotary movement.
5. The flywheel that rotates and stores energy to drive the piston when gas force is not acting on it.

Engine details

Example of calculating swept volume

A single-cylinder engine of the type shown in Fig. 2.2 has a bore of 90 mm and a stroke of 100 mm. Calculate the swept volume in cm^3.

Working in centimetres, bore diameter $D = 9$ cm, stroke length $L = 10$ cm.

The swept volume = area of piston crown × stroke length.

The piston crown is a circle and its area

$$= \frac{\pi D^2}{4} = \frac{3.142 \times 9 \times 9}{4} = 63.6 \, cm^2.$$

The swept volume $= 63.6 \times 10 = 636 \, cm^3$.

A cycle of operations

In order for the engine to function it goes through a sequence of events:

1. Getting air into the cylinder.
2. Getting fuel into the cylinder and igniting the fuel.
3. Expanding the high-pressure air to produce useful work.
4. Getting rid of the spent gas so that the sequence can be repeated.

This sequence of events is called a cycle.

The four-stroke Otto cycle

A large proportion of light vehicle engines use petrol as a fuel and they operate on the Otto cycle. The Otto cycle is named after Dr A. Otto, who developed the first commercially successful engines, in Germany, in the 1860s. Otto cycle engines are also called four-stroke engines because the Otto cycle takes four strokes of the piston for its completion.

The basic engine

A four-stroke engine (Fig. 2.3) has one end of the cylinder sealed — this end of the cylinder is called the cylinder head. In the cylinder head are two valves and a spark plug that supplies the spark that ignites the fuel. One valve is called the inlet valve and it is opened when air and fuel are required; the other valve is the exhaust valve and this is opened when the spent gas is removed from the cylinder.

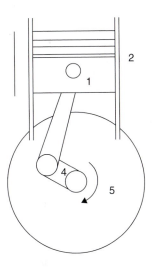

Fig. 2.1 Simple engine mechanism

The four strokes (see Fig. 2.4)

- **First stroke — Induction.** The inlet valve is open and the exhaust valve is closed. The piston is pulled down the cylinder by the action of the crank and connecting rod. As the piston descends it creates

a partial vacuum in the cylinder and this causes the atmospheric air pressure to force a mixture of air and fuel that is supplied by a carburettor or fuel injection system into the cylinder.
- **Second stroke — Compression.** Both valves are now closed and the piston is pushed up the cylinder by the action of the flywheel, crank, and connecting rod. The mixture of air and fuel in the cylinder is now compressed to a high pressure. A high pressure is required to extract the maximum amount of energy from the fuel.
- **Third stroke — Power.** Both valves are closed and the spark ignites the fuel. This causes the pressure in the cylinder to rise and the action pushes the piston down the cylinder to rotate the crankshaft and deliver power to the flywheel.
- **Fourth stroke — Exhaust.** The exhaust valve is open and the inlet valve is closed, the action of the flywheel and crank pushes the piston up the cylinder to expel the spent gas. The cycle is now complete and the engine is ready to start the next cycle.

The four strokes are completed in two revolutions of the crankshaft, which is equivalent to an angular movement of $720°$.

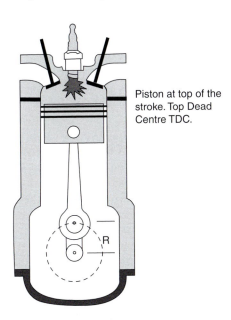

Piston at top of the stroke. Top Dead Centre TDC.

D = diameter of the cylinder and piston. Normally referred to as the cylinder bore.

TDC

Swept Volume

Stroke

BDC

Piston at the bottom of the stroke. Bottom dead centre BDC.

R = radius of crank. Also called the crank throw.

Length of stroke = 2 × crank radius

The swept volume is the space that is created in the cylinder when the piston moves from TDC to BDC. It is also called the cylinder capactiy.

Swept volume = cross sectional area of the pistion × length of the stroke.

Fig. 2.2 Single-cylinder engine dimensions (Renault)

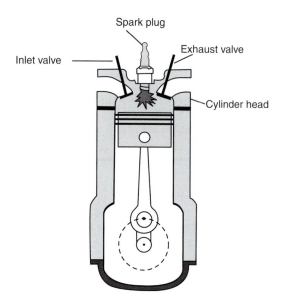

Fig. 2.3   A four-stroke cycle engine

Compression ratio

To a certain extent, the more that the mixture of fuel and air is compressed the greater the amount of power that can be extracted from the fuel. The amount of compression that takes place in an engine is determined by the **compression ratio** of the engine (Fig. 2.5).

Compression ratio is the total volume inside the cylinder when the piston is at bottom dead centre (BDC) divided by the total volume inside the cylinder when the piston is at top dead centre (TDC).

The total volume inside the cylinder when the piston is at bottom dead centre is the clearance volume plus the swept volume. The swept volume is the volume swept by the piston when it moves from TDC to BDC.

The total volume inside the cylinder when the piston is at TDC is the clearance volume, or combustion space.

The formula for compression ratio is:

$$\text{Compression ratio} = \frac{V_s + V_c}{V_c},$$

where V_s = swept volume and V_c = clearance volume.

Example

A certain engine has a swept volume of 400 cm^3 and a clearance volume of 50 cm^3. Calculate the compression ratio.

$$\text{Compression ratio} = \frac{V_s + V_c}{V_c} = \frac{400 + 50}{50} = \frac{450}{50} = 9:1.$$

Valve timing

In the four-stroke cycle the valves are required to open and close at the correct point in the cycle. The inlet valve normally opens a few degrees before the piston reaches TDC on the exhaust stroke and closes again several degrees after BDC on the induction stroke.

The exhaust valve normally opens several degrees before BDC on the power stroke and closes a few degrees after TDC on the exhaust stroke.

These events can be shown on a circular display called a timing diagram. A typical timing diagram is shown in Fig. 2.6.

Valve timing diagrams display details about valve operation in terms of degrees of crankshaft rotation, which also indicates the position of the piston in the cylinder. In the timing diagram shown in Fig. 2.6 the following details apply:

- The inlet valve opens when the crank is 4° before TDC and it remains open down the induction stroke and for 48°, part of the way up the compression stroke.
- The number of degrees for which the valve remains open is called the valve period — in this case, the inlet valve period is 4° + 180° + 48° = 232°.
- The exhaust valve opens 48° before BDC on the power stroke and it remains open up the entire exhaust stroke and for 4° on the induction stroke. The exhaust valve period is 48° + 180° + 4° = 232°.
- The number of degrees around TDC for which both valves are open together is called **valve overlap**.
- The number of degrees that the exhaust valve opens before BDC is called **exhaust valve lead**. Early opening of the exhaust valve while there is still some pressure left in the gas allows gas to escape into the exhaust system and thus reduces the pressure that the piston works against on the exhaust stroke. This improves the efficiency of the engine.
- The number of degrees that the inlet valve remains open after BDC is called **inlet valve lag**. Closing the inlet valve after BDC allows the momentum of the air entering the cylinder to overcome the increasing pressure in the cylinder as the piston moves up the cylinder on the compression stroke. In this way the engine is made more efficient.
- Valve timing varies from engine to engine and the actual details are determined by the type of use that the vehicle is intended for.

The motion of valves is determined by the shape of the camshaft (cam) and this is designed to open and close the valves as quickly as possible without causing undue stress on components.

Setting the valve timing

When reassembling an engine after repair it is necessary to ensure that the camshaft is set in the correct position relative to the crankshaft. This process is called setting the valve timing, and most engines carry marks like those shown in Fig. 2.7 to assist in the process. The marks are carefully aligned prior to fitting the chain.

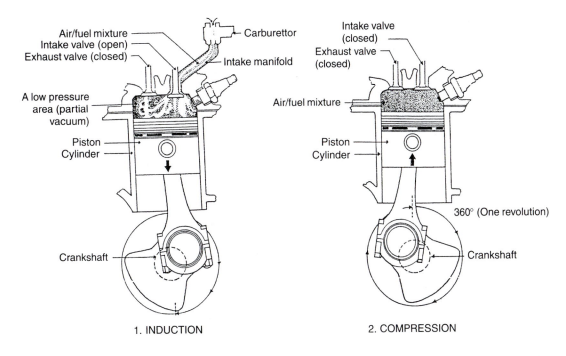

1. INDUCTION

2. COMPRESSION

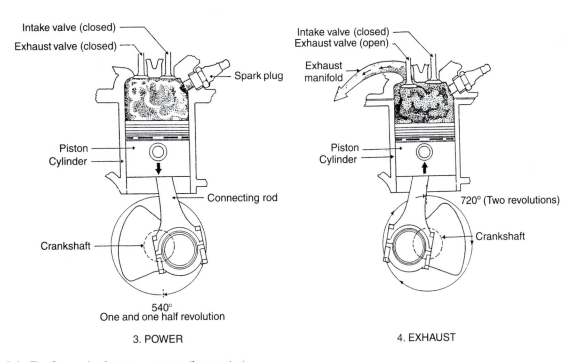

3. POWER

4. EXHAUST

Fig. 2.4 The Otto cycle of engine operations (four strokes)

Engines that use gear or belt drives on the camshaft have similar marks.

Valve timing and emissions

When the engine is operating at low speed the overlap that occurs when the inlet valves and exhaust valves are open simultaneously is a cause of harmful emissions and various forms of variable valve control are used to overcome the problem. Two forms of valve control that are used are:

1. Different amounts of valve lift for low and high engine speeds.
2. Automatically changing the valve timing while the engine is running.

The Honda valve system that is outlined here (Fig. 2.8) is used on engines that have four valves per

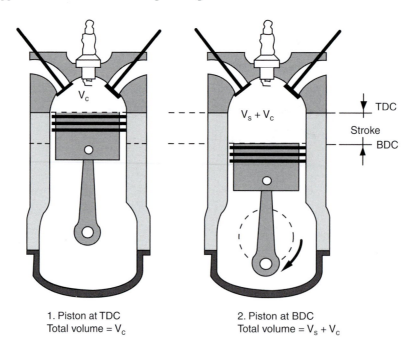

1. Piston at TDC
Total volume = V_c

2. Piston at BDC
Total volume = $V_s + V_c$

Fig. 2.5 Compression ratio (Renault)

cylinder − two inlet and two exhaust − and it provides variable valve lift as well as variable valve timing.

Variable valve lift

There are three cams for each pair of valves − two of the cams provide the low-speed features and the third one that is placed between the other two provides the lift and period for high-speed. At low speed the high-lift cam freewheels until it is required at higher speed. When the high lift is required the cam is brought into operation by the movement of the locking pins. These locking pins are operated by hydraulic pressure from the engine lubricating system under the control of the engine computer. The two low-speed inlet cams that are called the primary and secondary cams have slightly

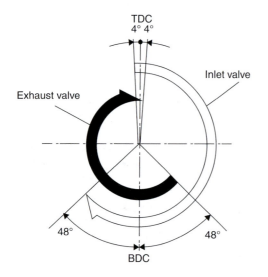

Fig. 2.6 A timing diagram

different profiles and are designed to produce turbulence in the combustion chamber. Details of the method for obtaining variations in valve movement are shown in Fig. 2.8.

Variable valve timing

The actuator on the inlet camshaft (Fig. 2.10) is a hydraulically operated device that advances the opening of the inlet valve at high engine speed to take advantage of the momentum of the inflowing air and to maximize volumetric efficiency.

Details of valve lift are given in Table 2.1 and the effect of valve timing is illustrated in Fig. 2.11.

Figure 2.11 shows how, by opening the inlet valve early, overlap is increased with the effect that the Honda system varies the amount of overlap, and consequently the intake closure moment. This strongly influences engine characteristics: minimum overlap − for smooth idling and cruising, and excellent fuel economy through stable combustion; maximum overlap − for power, by exploiting gas flow inertia to improved cylinder filling.

Ignition timing

The spark at the spark plug is arranged to take place slightly before TDC on the compression stroke so that maximum gas pressure is reached at the beginning of the power stroke. The number of degrees before TDC that the spark is initiated is called the angle of advance. In vehicle repair work the action of setting the ignition timing is called 'setting the timing' and it requires the piston to be in the correct position when the device that triggers the spark is also in the correct position. On most engines there are timing marks on the

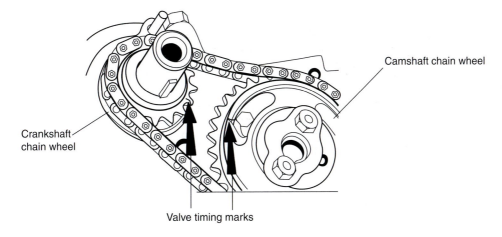

Camshaft chain wheel

Crankshaft chain wheel

Valve timing marks

Fig. 2.7 Valve timing marks

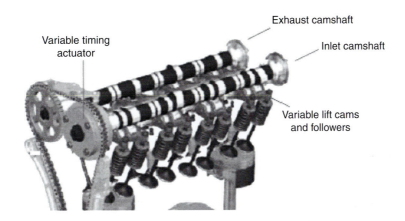

Variable timing actuator

Exhaust camshaft

Inlet camshaft

Variable lift cams and followers

Fig. 2.8 Variable valve lift and timing (Honda)

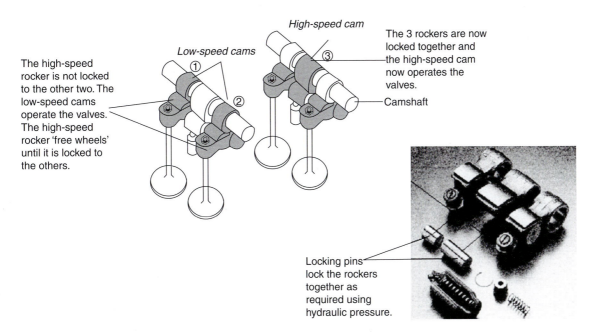

High-speed cam

The 3 rockers are now locked together and the high-speed cam now operates the valves.

Camshaft

Low-speed cams

The high-speed rocker is not locked to the other two. The low-speed cams operate the valves. The high-speed rocker 'free wheels' until it is locked to the others.

Locking pins lock the rockers together as required using hydraulic pressure.

Fig. 2.9 Variable valve lift (Honda)

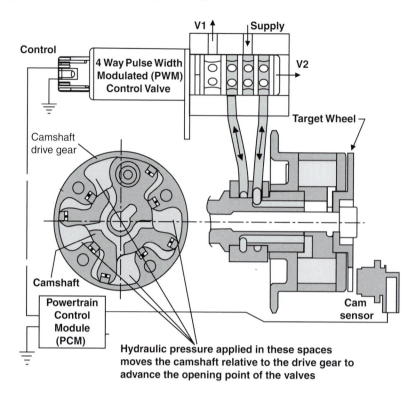

Fig. 2.10   Valve timing actuator (Delphi)

Table 2.1   Valve lift

Operation	Inlet valve lift	Exhaust valve lift
Low-speed	Primary cam 7.2 mm	Primary cam 6.9 mm
	Secondary cam 7.0 mm	Secondary cam 7.1 mm
High-speed	All cams 12.0 mm	All cams 10.7 mm

crankshaft pulley, like those shown in Fig. 2.12, that are used in checking and setting ignition timing.

The two-stroke cycle

In its simplest form the two-stroke cycle offers the following advantages over the four-stroke cycle:

- No valves are used because the piston covers and uncovers ports through which air and fuel enter the engine and exhaust products are expelled.

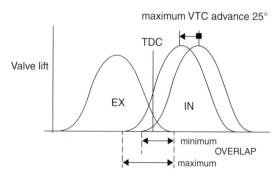

Fig. 2.11   The effect of valve timing on valve overlap

- A power stroke occurs once for each revolution of the crank — in theory this makes a two-stroke engine of a given size twice as powerful as a four-stroke engine.

Two-stroke engines (Fig. 2.13) have been used in some light cars and vans from time to time but their main use has been in motorcycles and mopeds.

The crankcase is sealed because it is used to hold the air—fuel mixture at a stage of the cycle of operations. By using both the top and underside of the piston the four phases of the cycle (induction, compression, power, and exhaust) are completed in two strokes of the piston and one revolution of the crankshaft.

When considering how this type of engine works it is advantageous to consider events above and below the piston separately.

First stroke (piston moving down the cylinder)

Events above the piston

The expanding gases that have been ignited by the spark plug force the piston down the cylinder. About two-thirds of the way down the cylinder the **exhaust port** is uncovered by the piston and the exhaust gases leave the cylinder. As the piston moves further downwards the **transfer port** is uncovered and this allows a fresh charge of fuel and air from the crankcase to enter the cylinder above the piston.

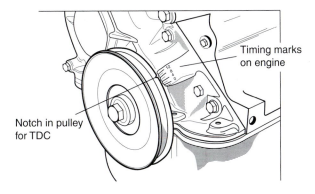

Fig. 2.12 Ignition timing marks

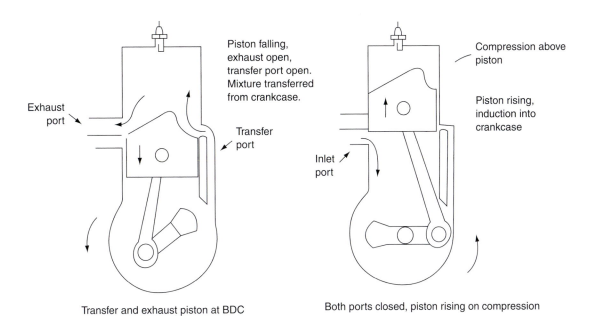

Fig. 2.13 The two-stroke engine

Events below the piston

The descending piston covers the **inlet port** and compresses the air and fuel mixture in the crankcase.

Second stroke (piston moving up the cylinder)

Events above the piston

Compressed fuel and air is forced into the cylinder from the crankcase, through the **transfer port.** With the aid of the deflector on top of the piston the incoming charge of fuel and air helps to drive exhaust gas out. When both the transfer port and the exhaust ports are closed the piston continues to rise and compress the fuel and air mixture. The spark occurs at the end of this stroke and the engine begins the next power stroke.

Events below the piston

As the piston moves upwards the partial vacuum in the crankcase now draws in fuel and air through the **inlet port** as it is uncovered by the bottom of the piston.

Because the piston is used to control the opening and closing of the ports the power stroke is effectively shortened and this reduces the power output of the simple two-stroke engine.

Two-stroke engine with valves

The engine shown in Fig. 2.14 makes use of poppet valves and direct injection of petrol into the cylinder. It is equipped with a supercharger that pumps air into the cylinder rather than relying on crankcase induction and compression as used in the simple engine. Similar

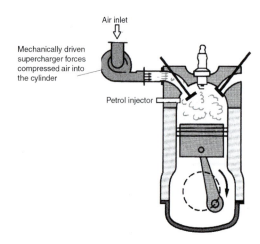

Fig. 2.14 Toyota two-stroke engine

types of two-stroke engines operating on the diesel principle are used in some large vehicles.

Rotary engines

The rotor in this type of engine (Fig. 2.15) replaces the piston and crank of the reciprocating engine. The housing in which the rotor moves has a shape which is called an epitrochoid and it permits the four steps of the Otto cycle to be completed in one revolution of the rotor. On the inside of the rotor is a gear that engages with a smaller gear on the output shaft and this is the medium through which the energy from combustion is transmitted to the engine flywheel.

The Atkinson engine cycle

The theoretical Akinson cycle is shown in the pressure–volume diagram of Fig. 2.16. There are four processes. The first starts at point 3 on the diagram, where a mass of air is compressed up to point 4. At point 4 the air is heated and the pressure rises while the volume remains constant. At point 1 the hot air expands on the power stroke. The power stroke ends at point 2 and the gas is exhausted at constant pressure up to point 3, where the cycle starts again.

A point to note is that the power stroke is longer than the compression stroke because this is the feature that makes the engine more fuel efficient than the Otto engine. The original Atkinson engines were made to produce the four processes in one revolution of the crankshaft. In order to achieve this it was necessary to use a complicated toggle mechanism that proved unreliable due to excessive wear and the engine fell out of use.

In recent years the attraction of more efficient use of fuel and better miles per gallon has led to renewed

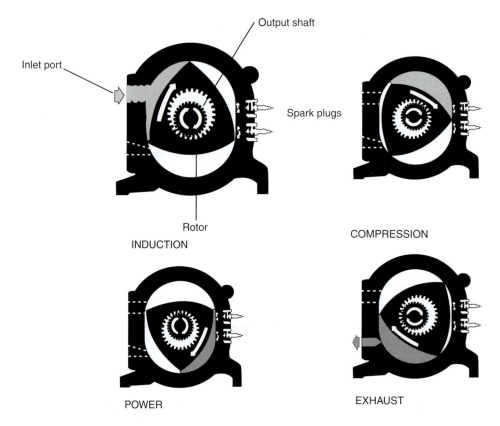

Fig. 2.15 The Wankel-type rotary engine

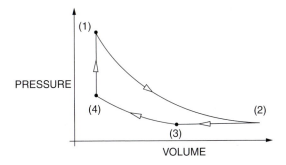

Fig. 2.16 Pressure–volume diagram for the ideal Atkinson cycle

interest in the Atkinson engine. The normal four-stroke engine has been adapted so that it runs on a cycle that bears a resemblance to the Atkinson. This has been achieved by keeping the inlet valve open so that some charge from the induction stroke is pushed back to the induction system where it is used in other cylinders. This effectively shortens the compression stroke because compression does not start until the piston is someway up the compression stroke. This provides a power stroke that is of longer duration than the power stroke and thus provides the feature that produces the greater fuel efficiency. Unfortunately this type of engine is only efficient at a fairly narrow range of speeds and it is necessary to equip vehicles with transmission systems to counteract this problem.

The theoretical thermal efficiency of the Atkinson cycle is given by the following equation:

$$\text{efficiency} = 1 - \gamma\left(\frac{r - \alpha}{r^\gamma - \alpha^\gamma}\right),$$

where r is the expansion ratio, α is the compression ratio, and γ is a constant for air.

If we assume a compression ratio of 8:1 and an expansion ratio of 13:1 we can put some figures in this equation to arrive at a value for thermal efficiency, which we can then compare with an Otto engine with a compression ratio of 8:1.

Atkinson efficiency

γ for air is approximately 1.4, $r = 13$, and $\alpha = 8$. Putting these numbers into the equation in place of the symbols gives the Atkinson thermal efficiency as:

$$1 - 1.4\left(\frac{13 - 8}{13^{1.4} - 8^{1.4}}\right) = 0.61 \text{ or } 61\%.$$

The equivalent theoretical efficiency for an Otto engine with a compression ratio of 8:1 is:

$$1 - \frac{1}{r^{\gamma - 1}} = 1 - \frac{1}{8^{0.4}} = 56.5\%.$$

A value of 61% compared with 56.5% seems a relatively small advantage for the Atkinson cycle over the

Otto cycle, but at a time when emissions and fuel use are so important some manufacturers consider it worthwhile to make use of the Atkinson principle.

> ### Learning task
>
> See if you can find out how the Toyota Prius Hybrid vehicle transmission system overcomes the disadvantages of the four-stroke Atkinson engine.

Self-assessment questions

1. Which valve opens near the end of the power stroke in a four-stroke engine?
2. In a certain engine the cross-sectional area of the piston crown is 80 cm² and the stroke length is 120 mm. The swept volume is:
   (a) 9600 cm³
   (b) 120 cm³
   (c) 960 cm³
   (d) 960 cm².
3. The valve timing details for an engine are:
   - Inlet valve opens 6° before TDC and closes 38° after BDC
   - Exhaust valve opens 35° before BDC and closes 5° after TDC.
   Calculate in degrees:
   (a) The valve overlap
   (b) The exhaust valve lead
   (c) The inlet valve lag
   (d) The period of: (i) the inlet valve, (ii) the exhaust valve.
4. In a simple two-stroke engine the air–fuel mixture is drawn into the crankcase. What is the name of the port that is used to get the mixture into the combustion space above the piston?
5. At the end of which stroke in the four-stroke cycle does the spark occur?
6. How many degrees of crank rotation does it take to complete the four-stroke cycle?
7. What is the reason for starting to open the inlet valve before TDC is reached in a four-stroke engine?
8. An engine has a bore of 79 mm and a stroke of 100 mm. Calculate the compression ratio given that the clearance volume is 50 cm³.
9. If an engine has a stroke of 120 mm what is the radius of the crank throw?
10. An engine has a bore diameter of 98 mm and a stroke length of 90 mm. Calculate its swept volume.
11. Give a short explanation of the reasons for opening the exhaust valve before BDC is reached in a four-stroke cycle engine.

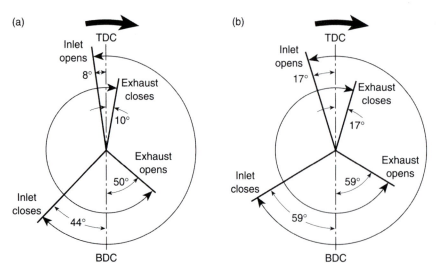

Fig. 2.17 Typical value timing diagrams

12. Figure 2.17 shows two valve timing diagrams. Which of these would be suitable for a high-speed engine?

Questions 13—16 relate to the Honda system.

13. 25° of camshaft advance is equal to:
    (a) 50° of crankshaft rotation
    (b) 100° of crankshaft rotation
    (c) 12.5° of crankshaft rotation
    (d) 25° of crankshaft rotation.
14. What effect on the closing point of the inlet valve is brought about by opening it 25° early?
15. The maximum lift of the inlet valve is:
    (a) 7.2 mm
    (b) 6.9 mm
    (c) 12.0 mm
    (d) 10.7 mm.

16. The difference in lift between the inlet primary and secondary cams at low speed is:
    (a) 4.8 mm
    (b) 5 mm
    (c) 0.20 mm
    (d) 0.30 mm.
17. Discuss with other students vehicles (other than the Trabant and motor cycles) that have been fitted with two-stoke engines in recent years.
18. With the aid of diagrams describe how the clearance volume of an engine can be measured and then describe how knowledge of the bore and stroke would enable you to work out the compression ratio.
19. Make a list of the vehicles equipped with a Wankel-type rotary engine that are currently available in the UK.

3
The main moving parts of an engine

Topics covered in this chapter

Valves

Valves of the type shown in Fig. 3.1 are called poppet valves, or mushroom valves because of their shape. Valves are made from high-quality steel that is hard wearing and, in the case of the exhaust valve, heat resistant because the exhaust valve may run at temperatures in excess of $700°C$.

The valve shown in this figure is typical of the type used in vehicle engines and the pertinent details are:

(A) Valve collets, also known as cotters, secure the valve stem to the spring retainer, which is the part that allows the spring to pull the valve on to its seat (Fig. 3.2).

(B) This is the spring retainer.
(C) The valve spring that keeps the valve under the control of the cam and holds it tight on its seat when closed.
(D) The valve stem seal prevents oil being drawn through the valve guide into the combustion chamber.
(E) Part of the cylinder head that contains the valve seat and the guide that holds the valve in position.
(F) The valve itself. The valve seat is normally cut at an angle of $44.5°$ on the head of the valve; the mating valve seat in the cylinder head is cut at an angle of $45°$. This provides a gas-tight seal when the valve is closed.

The valve guide shown in Fig. 3.3 is a hollow cylinder which is inserted into the cylinder head and its purpose is to ensure that the valve remains concentric with the valve seat. Valve guides are normally made from cast iron or phosphor bronze and they are made to be an interference fit in the cylinder head. In an alternative arrangement, that is sometimes used in cast iron cylinder heads, the valve guide is an integral part of the cylinder head. In operation the valve guide takes heat away from the valve and conducts it into the engine cooling system.

Cam and follower

Followers, or tappets, are used to transmit motion from the cam to the valve. The simple follower shown in Fig. 3.4(a) shows the cam and valve in different positions. The clearance is required to allow for thermal expansion.

The main details of a typical cam are shown in Fig. 3.4(b). This type of cam is called a circular arc cam because the nose circle, base circle, and flanks are formed from arcs of circles. The motion that this type of cam imposes on the follower and valve is shown in Fig. 3.4(c). The initial lift happens rapidly as shown by the high acceleration; this gets the valve open as quickly as possible. This first phase is followed by a drop in acceleration as the valve approaches the fully open position.

The return movement to the closed position follows a similar pattern except that there is a deceleration to slow the valve down as it reaches the seat.

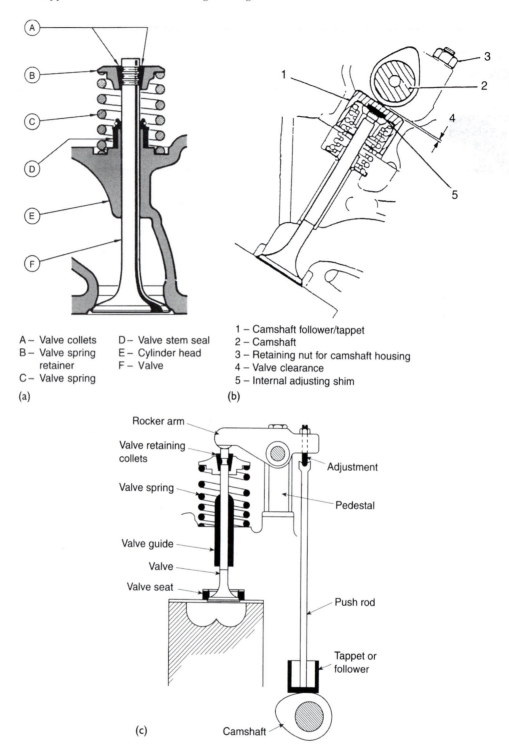

A – Valve collets   D – Valve stem seal
B – Valve spring   E – Cylinder head
      retainer   F – Valve
C – Valve spring

(a)

1 – Camshaft follower/tappet
2 – Camshaft
3 – Retaining nut for camshaft housing
4 – Valve clearance
5 – Internal adjusting shim

(b)

Rocker arm
Valve retaining collets
Valve spring
Valve guide
Valve
Valve seat
Adjustment
Pedestal
Push rod
Tappet or follower
Camshaft

(c)

Fig. 3.1 Valve details

0.5 to 1 degree difference
valve head 44.5, valve seat 45
degrees

Fig. 3.2 Valve seat details

Valve springs

The purpose of the valve spring is to keep the valve under the control of the cam at all times and hold it tightly on the seat to provide a gas-tight seal when it is closed. Valve springs of the type shown in Fig. 3.5 are normally made from high-carbon or alloy steels and they are designed to last for the lifetime of the engine. In practice,

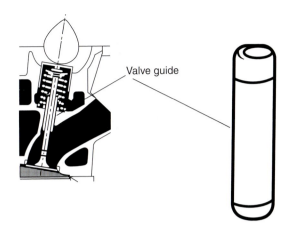

Fig. 3.3 The valve guide

after long hard use, the springs may lose their strength and replacements may be required.

Single coil spring

The single coil spring is the type that is widely used in vehicle engines. Springs will vibrate when compressed and released; the frequency at which they vibrate is called the natural frequency of vibration. If the natural frequency coincides with the frequency at which the valve is being opened and closed a condition known as resonance may occur. When this happens the valve motion is no longer under control of the cam — this condition is called **valve surge**, or **valve bounce**, and it can lead to serious engine damage. It is most likely to happen if the engine speed exceeds its designed maximum or if the valve springs become weak.

Variable rate spring

This spring has its coil closer together at the cylinder head end, which has the effect of changing the spring stiffness as the valve opens and presses the coils closer together. This alters the frequency of vibration and lowers the chances of surge occurring.

Double springs

These springs are wound in opposite directions; the inner spring is of smaller diameter and it fits inside the larger one. The springs have different frequencies

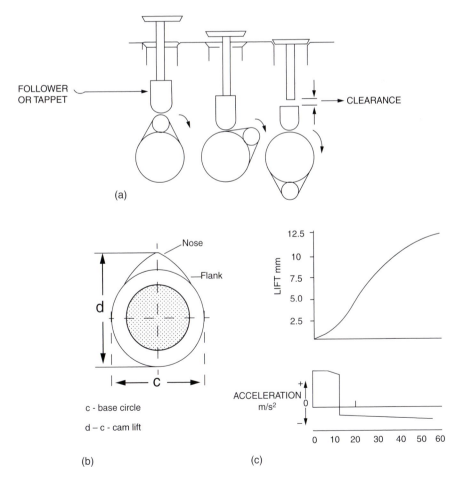

Fig. 3.4 Valve operations

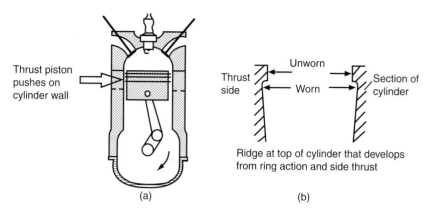

Fig. 3.18   Piston force against cylinder wall — thrust

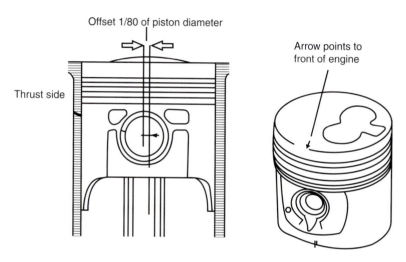

Fig. 3.19   Gudgeon pin offset

the ring groove next to the piston to ensure maximum effectiveness of the scraping action. Small holes in the ring groove allow oil that is scraped from the cylinder wall to be returned to the crankcase.

Figure 3.20(a) shows the main details of a rectangular section compression ring. The radial thickness and the width determine the clearance between the ring and its groove. These are critical dimensions because they affect the way that the ring operates and they should always be checked when replacement rings are being fitted.

The ring gap is another critical dimension because it is there to allow for expansion; if the gap is too small the ring will probably break and if the gap is too large it will allow gas to escape past the piston.

Figure 3.20(b) shows how a feeler gauge can be used to check the ring gap. When performing this operation a piston should be used to push the ring into the ring cylinder to ensure that the ring is at right angles to the cylinder axis. The recommended ring gap is:

- Water-cooled engine, ring gap 0.003 × cylinder diameter.

- Air-cooled engine, ring gap 0.0035 × cylinder diameter.

Because the top ring is closest to the combustion chamber its clearance may be slightly larger than the ring gaps further down the piston to allow for the higher temperature in that region. For example, the manual that I am referring to gives the following figures for an engine with 92.0 mm bore:

- Top compression ring gap 0.35—0.50 mm.
- Second compression ring gap 0.25—0.40 mm.
- Oil control ring gap 0.25—0.55 mm.

The stepped piston ring shown in Fig. 3.20(c, ii) is sometimes called a 'ridge dodger', its purpose being to avoid coming into contact with the wear ridge that develops at the top of the cylinder. This type of ring is often used when slightly worn engines are being reconditioned.

The section of oil control ring shown in Fig. 3.20(c, iii) is shaped so that it can scrape excess oil from the cylinder wall and return it to the crankcase via the hole in the piston.

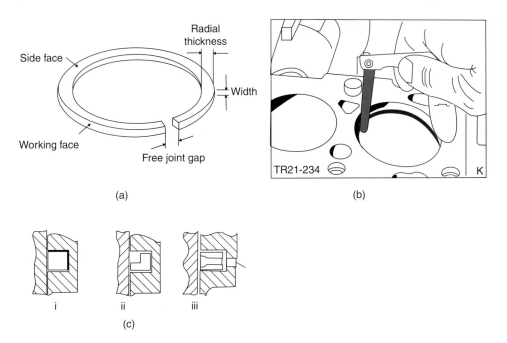

(a)   (b)

(c)

Fig. 3.20 Piston ring details

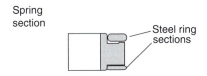

Fig. 3.21 A composite oil control piston ring

Oil control rings such as that shown in Fig. 3.21 are made from several steel sections and a crimped steel spring section that is placed in the ring groove next to the piston to ensure maximum effectiveness of the scraping action.

Condition checks

Piston diameter

The major piston diameter is at right angles to the gudgeon pin bosses and it is measured as shown in Fig. 3.22. It is this diameter that is subtracted from the cylinder bore diameter to determine the clearance. If this clearance is too small, the piston-to-cylinder wall surface will not be lubricated and damage of the form shown in Fig. 3.23 may occur.

In extreme cases complete engine failure can also occur if the piston seizes in the bore.

Piston ring clearance in groove

The side clearance of the piston ring allows the piston ring to exert the pressure that provides an effective gas seal (Fig. 3.24). If the clearance is too small the ring may

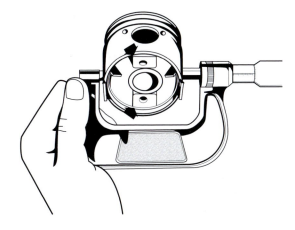

Fig. 3.22 Measuring piston diameter

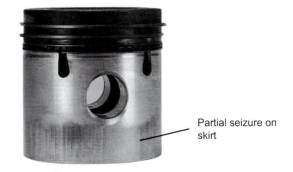

Partial seizure on skirt

Fig. 3.23 Damage at piston skirt

Fig. 3.24   Piston ring side clearance in the piston

seize in the groove and not provide the seal, or it may cause damage to the cylinder and piston. Too large a clearance may cause the ring to vibrate and eventually fracture.

The crankshaft and flywheel assembly

The force from the piston is applied through the connecting rod to the crankpin. This, in turn, creates a turning effect (torque) that rotates the flywheel to provide the power output from the engine. The flywheel is a heavy mass that stores kinetic energy that keeps the engine running smoothly and carries it over the strokes when no power is being produced. The counterweights on the crankshaft help to balance out inertia forces caused by piston motion but, as shown in Fig. 3.25, they cannot achieve this completely.

The crank pin and main journals are hardened at the surface to provide resistance to wear but the core of the shaft and the webs are heat treated to provide maximum strength and toughness.

The connecting rod

The connecting rod transmits the full gas force from the piston to the crank. Connecting rods are normally made of forged steel and have an I-section shape, which gives them maximum strength and a relatively small weight. The small end is connected to the piston via the gudgeon pin and the small-end bearing. Two types of small-end bearing are used:

1. The semi-floating small end, where there is a bush in the connecting rod eye and the gudgeon pin is free to rotate and slide both in the small end and the piston. The gudgeon pin is prevented from sliding out of the piston by means of a circlip that fits into a groove in the gudgeon pin hole, as shown in Fig. 3.26.
2. The other type of small end makes use of a clamping bolt to secure the gudgeon pin — in this case all motion occurs between the gudgeon pin and the piston, as shown in Fig. 3.26.

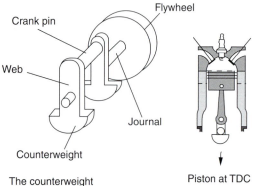
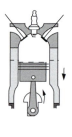
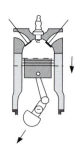

The counterweight creates centrifugal force as the crankshaft rotates. This sets up a force that partly cancels out the force created by acceleration and deceleration of the piston.

Piston at TDC centrifugal force on counterweight acts in opposite direction to inertia force on piston.

Piston at BDC centrifugal force on counterweight acts in opposite direction to inertia force on piston.

Piston is part way along the stroke. The force on the counterweight is no longer directly in line with the stroke of the piston and this causes a certain lack of balance in the engine.

Fig. 3.25   Crankshaft and flywheel

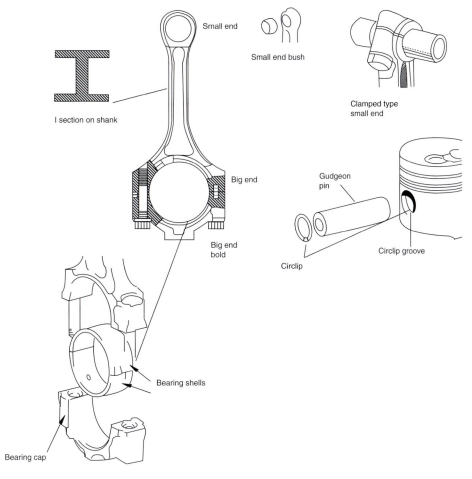

Fig. 3.26 Connecting rod and bearings

The big end of the connecting rod contains the bearing that connects the rod to the crankshaft. The big-end bearing is split to permit assembly to the crankshaft. The split bearings are referred to as shell bearings because they are fabricated from thin sections of steel that are coated with a layer of bearing metal. There is a small gap between the bearing and the shaft to allow for lubrication. The bearing metals that are used for the coating of the shells are various alloys such as those containing lead and copper or tin and lead, for example.

Condition checks

Alignment – straightness and twist

The connecting rods should be visually inspected for signs of obvious damage such as extreme distortion. Less obvious damage can only be detected by thorough examination of the type shown in Fig. 3.27. The operation shown here requires a surface plate and a set of feeler gauges. When the rod has been cleaned it is pressed down on to the surface plate and feeler gauges inserted as shown. If the rod is not distorted it should

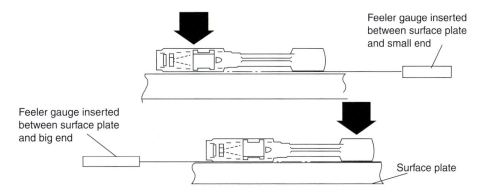

Fig. 3.27 Checking alignment of connecting rod

not be possible to insert a feeler gauge. The acceptable limits will depend on the manufacturer's specification. An example in the manual I have to hand gives the following specification:

- Maximum permissible bend = 0.5 mm.
- Maximum permissible twist = 0.5 mm.

Dimensions of big-end eye

To check the internal diameter of the big-end eye it is necessary to remove the bearing shells. When this is done the rod should be wiped clean and the bearing gap replaced. Each bolt should be lubricated and then tightened to the recommended torque. An internal micrometer (Fig. 3.28) is then used to take readings at three different points and these dimensions should not vary more than a very small amount – the manual I am referring to gives the following figures, which are representative:

- Big-end eye diameter 57.563–57.582 mm.

Gudgeon pin and small-end bush

The gudgeon pin should be checked for signs of damage such as partial seizure caused by overheating or lack of lubrication. In addition the diameter should be checked at three or more points along its length, as shown in Fig. 3.29. The amount of wear should not exceed approximately 0.05 mm. In common with other checks, the figures used should be those that apply to the engine being worked on.

Weight

If a replacement connecting rod is being fitted its weight should be compared with that of the other connecting rods. The weights should not vary by more than a few grams because of the effect on engine balance – a figure of 5–10 grams is suggested for this tolerance.

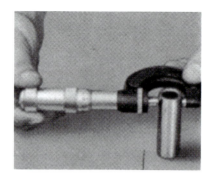

Fig. 3.29 Checking diameter of gudgeon pin

Self-assessment questions

1. It is recommended that the side clearance between the piston ring and its groove should be 0.04 mm. Figure 3.30 shows this clearance being checked with the aid of a feeler gauge. If the ring width is 6 mm and the feeler gauge that just fills the space between the ring and the groove is 0.03 mm, the width of the ring gap is:
   (a) 6.03 mm
   (b) 0.03 mm
   (c) 5.97 mm
   (d) 0.04 mm.

2. Figure 3.31 shows a machine for testing valve springs. The force that is exerted on the spring by means of the handle is recorded on the dial. At the side of the vertical member that presses on the spring is a calibrated scale that shows the amount by

Fig. 3.28 Checking diameter of big-end eye

Feeler gauge

Fig. 3.30 Checking side clearance of piston ring (Fiat)

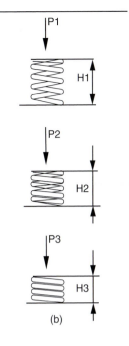

(a)

(b)

Fig. 3.31 Testing valve springs

which the spring is compressed for a given load on it. Figure 3.31(b) shows three degrees of compression and the corresponding heights H1, H2, and H3.

The workshop manual gives the following figures for a certain engine:

Test load on spring (kg)	Height of spring (mm)	Position of valve
0	H1 = 43.20	Free height, spring removed from valve
31–37	H2 = 37.0	Valve closed
88–94	H3 = 26.60	Valve fully open

The following result was obtained during an actual test on a valve spring:

Test load on spring	Height of spring
35.0 kg	H2 = 34 mm

Do you think that something is wrong with this spring? If so, what remedial action would you recommend?

3. Figure 3.32 shows a vernier height gauge being used to take a piston dimension. The compression height of a piston is the distance from the centre of the gudgeon pin to the piston crown. In this case the compression height is:
   (a) 40 mm
   (b) 52 mm
   (c) 60 mm
   (d) 50 mm.

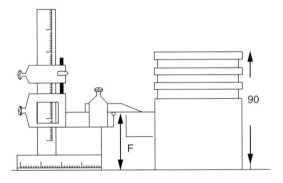

Fig. 3.32 Piston dimensions. F = 50 mm, gudgeon pin diameter = 24 mm

4. What is the purpose of the 1° difference in angle between the valve and its seat (Fig. 3.33)?
5. The crankshaft gear in Fig. 3.34 has 20 teeth. This means that:
   (a) The camshaft gear is 1.5 times larger in diameter than the crankshaft gear

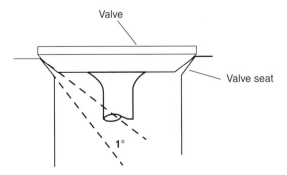

Fig. 3.33 Valve seat details

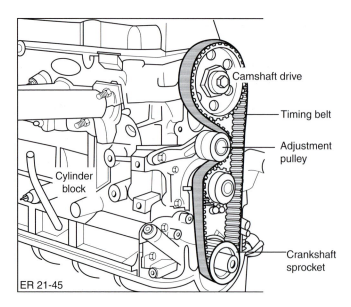

Fig. 3.34   A belt drive camshaft (Ford)

(b) There are 35 teeth on the camshaft gear
(c) There are 40 teeth on the camshaft gear
(d) The camshaft rotates twice as fast as the crankshaft.

6. An aluminium alloy piston has a maximum diameter of 100 mm at a temperature of 20°C. Given that thermal expansion = 0.000023 × diameter × temperature change, the expansion

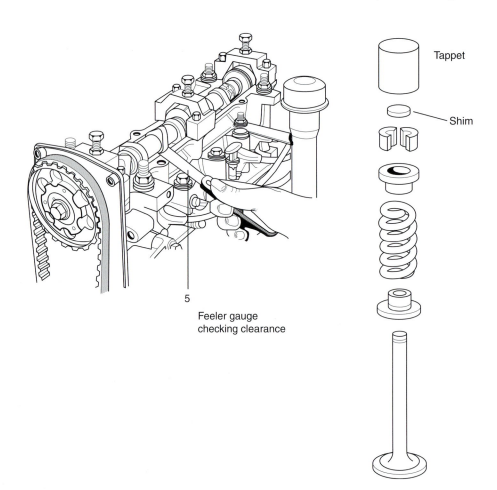

Fig. 3.35   Checking and resetting valve clearance

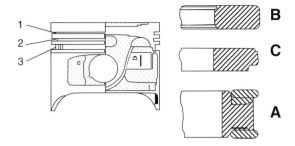

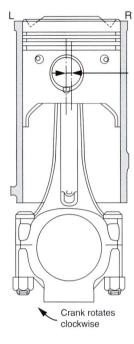

Fig. 3.36 Piston ring details

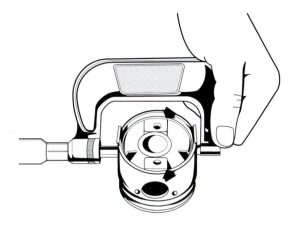

Fig. 3.38 Measuring piston diameter

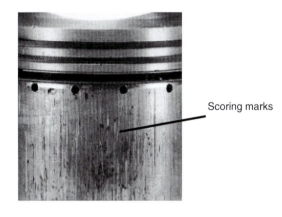

Scoring marks

Fig. 3.39 A damaged piston

L R

Crank rotates
clockwise

Fig. 3.37 Offset gudgeon pin

on this piston when it is at a temperature of 280°C is:
(a) 5.98 mm
(b) 0.598 mm
(c) 0.0598 mm
(d) 59.8 mm.

7. An engine of the type shown in Fig. 3.35 should have a valve clearance of 0.40 mm. During a valve clearance check the valve is found to have a clearance of 0.56 mm. What additional thickness of shim will be required to restore the correct valve clearance?

8. The metal used for an aluminium alloy piston contains 15% silicon. How many grams of silicon are there in a piston that weighs 0.360 kg:
(a) 5.4 g
(b) 54 g
(c) 0.54 g
(d) 0.054 g.

9. Valve period is the number of degrees of crank rotation between the point of valve opening and the point at which the valve closes. How will the length of the valve period be affected if the valve clearance is too great?

10. What effect on mechanical noise from the engine is excessive valve clearance likely to have?

11. Why is valve clearance necessary?

12. What effect on engine performance is excessive valve clearance likely to have?

13. In which position must the valve be in order to check the valve clearance?

14. What should be the temperature of the engine when carrying out a check on valve clearances?

15. On some engines the exhaust valves have a larger clearance than the inlet valves — why is this so?

16. Describe a procedure for checking the straightness of a connecting rod.

17. Oil ring question. What type of piston ring is shown at A in Fig. 3.36? In which ring groove is this piston ring fitted? Which piston ring should be fitted in ring groove 1?

18. With the aid of a sketch, describe the devices that are fitted at the front and rear ends of a crankshaft to prevent oil leaks.

19. The gudgeon pin centre in Fig. 3.37 is offset to the left of the piston centre. Which side of the diagram, L or R, is the thrust side of the cylinder?
20. Why is it important to measure the maximum piston diameter at the position shown in Fig. 3.38? What other engine dimension will be measured in order to determine the clearance between the piston and the cylinder?
21. In an exercise to determine piston clearance in the engine cylinder, the following dimensions were recorded:

- Maximum piston diameter = 73.920 mm
- Cylinder bore diameter = 73.960 mm.
  What is the size of the clearance between the piston and the cylinder?
22. The piston shown in Fig. 3.39 has been damaged by a lack of lubrication between the piston and the cylinder wall. Describe how such a lubrication failure may have arisen. What effect would the clearance between the piston and the cylinder have on lubrication?

4
Multi-cylinder engines

Topics covered in this chapter

Condition checks on cylinder head
Valve guides
Checking valve stem clearance in the valve guide
Valve oil seals
Damaged cylinder head gasket
Cracks and other flaws
Factors that influence combustion chamber design
Pre-ignition
Detonation
Need for high compression ratio
Scavenging
Types of combustion chambers
Details of cylinder block and crankcase
Cylinder liners
Cylinder block condition checks
Cylinder bore wear

During the 120 or so years that motor vehicles have been in use the four-stroke reciprocating engine has become established as the main means of propulsion. Early motor cars often used a single-cylinder engine that was not very powerful. As motor vehicles developed, the need for more engine power became apparent. A single-cylinder engine can be made more powerful by increasing its size, but it still leaves a problem: there is only one power stroke for every two revolutions of the crankshaft — this uneven application of power to the transmission of the vehicle is not very satisfactory. A solution to this problem is to have an engine with a number of cylinders. Because each cylinder has a power stroke every two revolutions of the crankshaft, a four-cylinder engine will produce four power strokes in two revolutions of the crankshaft. This provides a much smoother supply of power. The period between power strokes is often expressed in degrees of crankshaft rotation as: angle between power strokes $= 720/N$, where 720 is the number of degrees in two revolutions of the crankshaft and N is the number of cylinders that the engine has.

A four-cylinder engine has $180°$ of crankshaft rotation between power strokes, a six-cylinder one has $120°$, and an eight-cylinder one has $90°$. In general, the more cylinders there are in an engine, the smoother will be its application of power to the transmission system. Figure 4.1 shows the three basic engine layouts that are commonly used.

Cylinder arrangements

In-line engine

In this type of engine the cylinders are arranged along the length of the engine. Four or six cylinders are the types most often used. Eight-cylinder engines have been used but their greater length requires a very long bonnet and restricts the amount of space that is available for use in carrying passengers or goods.

The 'V' engine

In these engines the cylinders are arranged on each side of the crankshaft as shown in Fig. 4.1 and this results in a shorter engine length. The banks of cylinders are placed at an angle which, in some engines, is $90°$.

The horizontally opposed engine

In this engine the cylinders are horizontal and placed on each side of the crankshaft. This produces an engine that sits low down in the vehicle structure and this makes it suitable for use in rear-engine vehicles.

Engine structure — the main fixed parts

Figure 4.2 shows the layout of the most evident structural features of a four-cylinder in-line engine. The components listed are to be found in all engines, although their form may vary considerably.

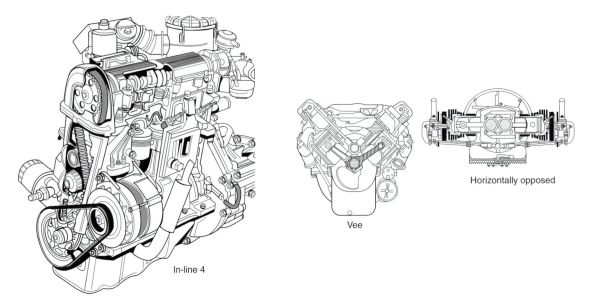

In-line 4

Vee

Horizontally opposed

Fig. 4.1   Multi-cylinder arrangements

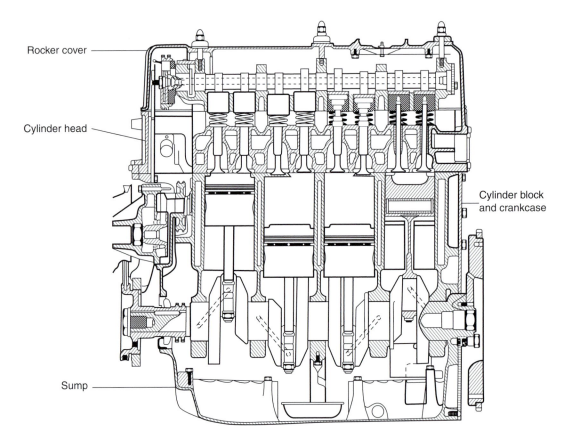

Rocker cover

Cylinder head

Cylinder block
and crankcase

Sump

Fig. 4.2   Layout of a typical four-cylinder in-line engine

The rocker cover

The rocker cover is an oil-tight cover that ensures that the oil that is required for lubrication of the camshaft is readily available. The rocker cover is normally constructed from cast aluminium alloy or a steel pressing. The type shown in Fig. 4.3 contains the oil filler cap.

When access to the valve gear is required for purposes such as checking valve clearance the cover must be removed; this means that the gasket is disturbed. When refitting the rocker cover it is advisable to fit a new gasket. Care should be taken not to overtighten the securing bolts because this may distort or break the cover. In common with other bolt-tightening procedures, the

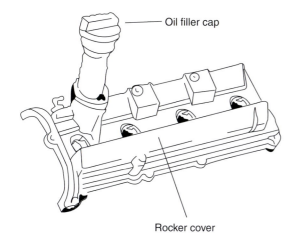

Fig. 4.3 A rocker cover

manufacturers recommended torque wrench settings should always be used.

The cylinder head

Cylinder heads are cast from cast iron or aluminium alloy. The cylinder head shown in Fig. 4.4(a) has provision for the camshaft and holes at the side that connect with the combustion chamber, which is on the underside of the head — these holes accommodate the fuel injectors on a diesel engine and the spark plugs on a petrol engine. Figure 4.4(b) shows a cross-section where the waterways can be seen. The region around the combustion chambers reaches a high temperature when the engine is operating and the waterways play an important part in preventing overheating. The valve guides and the valve seats are contained in the cylinder head. In aluminium alloy cylinder heads valve inserts are used to provide a hard-wearing, heat-resistant valve seat that can be replaced if they become worn or damaged. Valve-seat inserts (Fig. 4.5) are an interference fit and they are normally shrunk by cooling them in liquid oxygen before being inserted in the cylinder head; this

is a specialized process which is normally undertaken by firms that specialize in engine reconditioning.

The cylinder head is connected to the cylinder block by means of the cylinder head studs, or bolts. The joint between the cylinder head and the cylinder block is sealed by the cylinder head gasket (Fig. 4.6), which is made from heat- and pressure-resistant materials because it is in a high-temperature and high-pressure area of the engine. When a cylinder head is removed for any purpose the gasket should be replaced and when the head bolts are being tightened it is important that the correct tightening sequence and torque setting are used.

Example of cylinder head bolt-tightening sequence

The cylinder head bolts should be tightened in stages, starting from the centre and working outwards, as shown in Fig. 4.7. The torque should be applied in stages — the actual torque wrench setting will vary from engine to engine and it should always be checked against the manufacturer's recommended figures.

Condition checks on cylinder head

Valve guides

Prior to performing the check shown in Fig. 4.8 the valve stem should be checked for wear and if it shows signs of wear it should not be used for this test; instead a new unworn valve should be used. When performing the check the valve should be raised so that it is just clear of the cylinder head surface. The valve head is then pushed sideways to determine the amount of clearance in the guide, which should not exceed 0.50 mm. Should the clearance be found excessive the remedial action may take one of two forms:

1. Where the guides are part of the cylinder head casting they can be bored out with a reamer to a recommended oversize. New valves with slightly larger stem diameters are then required.

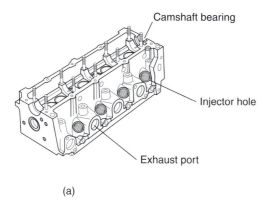

(a)

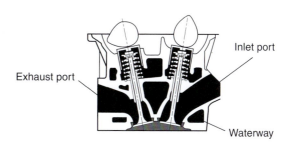

(b)

Fig. 4.4 A cylinder head

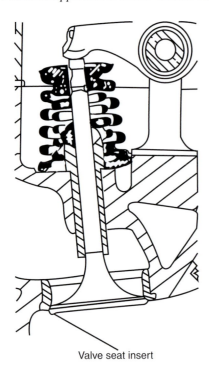

Valve seat insert

Fig. 4.5 Valve-seat insert

2. Where the guides are an interference fit in the cylinder head they may be pressed out and replaced by new ones.

Valve oil seals

Valve stem oil seals of the type shown in Fig. 4.9 play an important part in preventing oil from entering the combustion chamber. If they are defective they can affect engine emissions and oil consumption.

Cylinder head gasket

In the event of cylinder head gasket failure the gasket itself would be examined because this should help to reveal the cause of failure. There are several probable causes of such failure, but the one shown in Fig. 4.10 is believed to have been caused by material trapped under the gasket when it was first fitted − some other

gasket failures are discussed in the chapter on engine faults (Chapter 17).

In this example of gasket failure, combustion gas has entered the engine cooling system and coolant has entered the combustion chamber, both events causing serious problems that are covered in detail in the chapter on engine cooling systems (Chapter 13).

Cracks and other flaws

Cracks between valve seats and between a valve seat and a waterway sometimes arise when an engine has suffered a cooling system failure. This type of crack is shown in Fig. 4.11. Special penetrating dyes have been developed to search for cracks. When the affected part is thoroughly clean the dye is painted on and allowed to dry. After a short period the dye is wiped off and the crack will show up clearly because the dye has entered into it. Various proprietary brands of this treatment are available. Other methods of crack detection rely on principles of magnetism. In this procedure the object being tested is magnetized in special apparatus − iron filings are then applied to the affected surface and cracks will show up as the filings line up along the crack.

Valve seats

Valve seats should be examined to ensure that the valves are seating correctly. If a valve seat is worn it may be re-cut with a special tool of the type shown in Fig. 4.12.

In cases of extreme wear the valve seat may need renewing by fitting a valve-seat insert. The inserts are an interference fit in the cylinder head as shown in Fig. 4.13. In some cases the inserts are pressed in and in other cases they are cooled in liquid nitrogen to shrink them prior to fitting to the cylinder head.

Combustion chambers

The combustion chamber is the space formed by the chamber in the cylinder head and the piston crown. It contains the valve heads and valve seats, and the sparking plug. These features are accommodated in a relatively small space, which is determined by the cylinder bore diameter and the compression ratio.

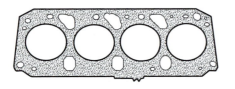

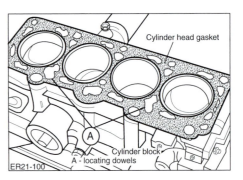

Fig. 4.6 Cylinder head gasket and dowels that locate it

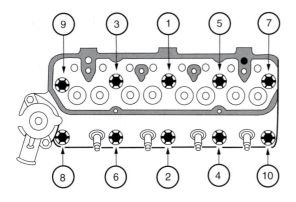

Fig. 4.7 Cylinder head tightening sequence. Start at centre and move outwards

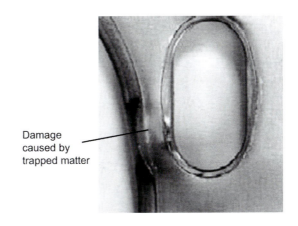

Fig. 4.10 Damaged cylinder head gasket

Damage caused by trapped matter

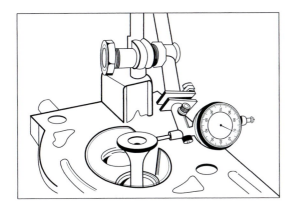

Fig. 4.8 Checking valve stem clearance in the valve guide

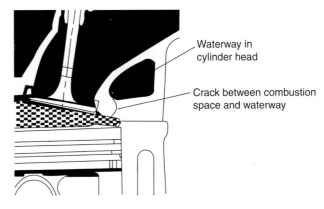

Waterway in cylinder head

Crack between combustion space and waterway

Fig. 4.11 A crack in the cylinder head

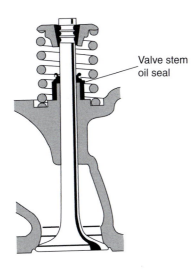

Valve stem oil seal

Fig. 4.9 Valve stem oil seal

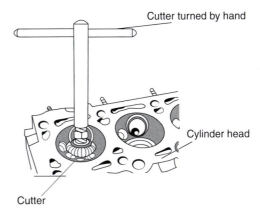

Cutter turned by hand

Cylinder head

Cutter

Fig. 4.12 Re-cutting a valve seat

Factors that influence combustion chamber design

- **The need to get the maximum amount of air into the engine in a short space of time.** Each gram of fuel requires approximately 15 grams of air for combustion. Air is very light and a large volume is drawn into the engine every second. In

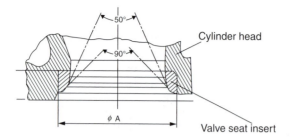

The diameter A is slightly greater than the hole into which it is pressed. This provides the interference fit that keeps the seat securely in place.

Fig. 4.13 Valve-seat insert

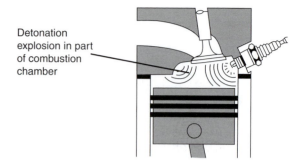

Fig. 4.14 Detonation

unsupercharged engines the air for combustion is pushed into the cylinders by the pressure difference between the atmosphere and that inside the cylinder. The amount of air that can be drawn into the cylinders is affected by the size of the valves and the amount of valve lift.

- **The need to ensure that each particle of fuel is surrounded by enough air to ensure complete combustion.** To obtain the maximum amount of energy the fuel must burn quickly and completely and this process is aided, in the combustion chamber, by making the air circulate in a process called turbulence. Turbulence is generated by various processes, some of which are discussed in the consideration of various combustion chamber designs.

- **The need to avoid heat loss.** The aim of combustion is to convert chemical energy into heat to raise the pressure of the gas in the cylinder and cause it to expand on the power stroke. For this process to be efficient there must be as little heat loss as possible through the walls of the combustion chamber. To achieve this result the ratio of the surface area of the combustion chamber to its volume must be as small as possible. It is this factor that has considerable influence on the shape of the chamber.

- **The need to avoid combustion problems.**
  - *Pre-ignition.* Pre-ignition happens when the charge of fuel and air is ignited before the spark occurs and it is caused by parts of the engine that remain hot from the previous power stroke. These hot spots are caused by small, sharp projections and combustion chambers are designed to avoid them.
  - *Detonation.* Detonation occurs when a portion of the charge is trapped in a small area of the combustion chamber (Fig. 4.14). The spreading flame front causes a region of high pressure to be created in this small area — this leads to

spontaneous combustion and a loud knocking sound with a risk of mechanical damage.
- **Need for high compression ratio.** The efficiency with which the chemical energy in the fuel is converted into power at the flywheel is called the thermal efficiency of the engine. The graph in Fig. 4.15 shows how theoretical thermal efficiency varies with compression ratio — it also shows that above compression ratios of approximately 8:1, the gains in thermal efficiency are quite small. The volume of the combustion chamber in relation to the bore and stroke of the engine determines the compression ratio; the higher the compression ratio, the smaller will be the volume of the combustion chamber.

Scavenging

The process of clearing burnt gas from the cylinder and replacing it with the new air fuel charge is called scavenging. Scavenging is affected by several factors, including the position of the valve ports.

Types of combustion chambers

Some of the various types of combustion chambers that are used in petrol engines are shown in Fig. 4.16 and this

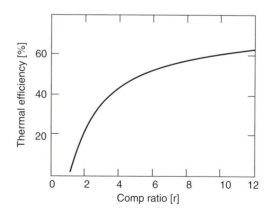

Fig. 4.15 Compression ratio vs. thermal efficiency

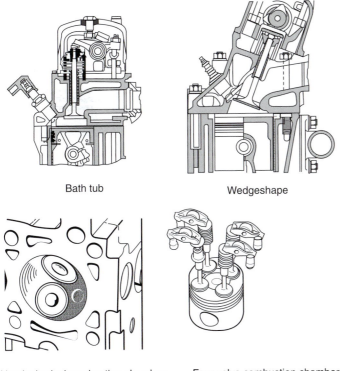

Bath tub   Wedgeshape

Pent roof combustion chamber

Hemispherical combustion chamber   Four valve combustion chamber
Tow valves

Fig. 4.16   Various combustion chamber shapes

section now assesses them in relation to the factors that influence design.

The bath tub chamber

The name derives from the shape, which is like that of an upturned bath tub. This type of chamber is widely used in overhead valve engines. The valves are placed side by side with the inlet and exhaust ports on one side of the cylinder head. The spark plug is placed on one side of the chamber, which allows for reasonably even flame spread across all regions of the chamber and reduces the chances of detonation occurring. In some designs the use of 'squish' created turbulence is used to ensure good combustion.

Squish

As the piston nears the end of the compression stroke the air–fuel mixture is forced out into the main body of the combustion chamber, creating turbulence that aids the spread of flame through the charge (Fig. 4.17).

The wedge-shaped chamber

This type of chamber widens the angle of the inlet and exhaust ports and thus allows relatively unrestricted flow of gas into and out of the cylinder, which aids scavenging and cylinder filling. The large area near the spark plug provides a wide flame front that aids combustion

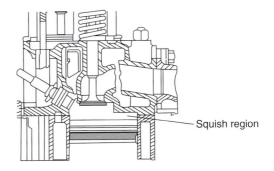

Squish region

Fig. 4.17   Squish turbulence

and squish turbulence is provided in the same way as it is in the bath tub design.

The pent roof combustion chamber

The spark plug can be located in the centre of the chamber between the valves and this should ensure that the flame spreads evenly through the charge, thus avoiding detonation. The design lends itself readily to the twin overhead camshaft arrangement and the use of four valves per cylinder, which provides the maximum area through which the incoming charge can enter and the outgoing exhaust gas can leave. This improves scavenging and volumetric efficiency; these topics are dealt with in the chapter on air supply intake and exhaust systems (Chapter 11).

The hemispherical combustion chamber

In this type of chamber the space in the cylinder head is formed from a section of a sphere. A hemisphere itself has a low surface area to volume ratio, but it does not allow a high compression ratio to be used. For this reason a section of the hemisphere is used because it has the advantage of the low surface area to volume ratio. The spark plug can be situated close to the centre of the chamber, thus reducing the chances of detonation occurring.

Cylinder block and crankcase

Figure 4.18 shows the main features of a cylinder block and crankcase. In the example shown, the cylinder block and crankcase are made from a single casting. An alternative to this is to use separate castings for the block and the crankcase. Cast iron is often used to make the

cylinder block and the cylinder bores are made in the casting.

Cylinder liners

In some cases the cylinder block cast iron may not meet the designer's requirements and dry cylinder liners are pressed into the block, as shown in Fig. 4.19(a). Most dry liners are an interference fit in the cylinder and they are machined after being pressed in to ensure that the correct dimensions are obtained. In some cases a type of liner known as a slip fit is used; these are pushed into place by light pressure and then secured by means of a flange at the top of the liner that is clamped down by the cylinder head.

Wet liners (Fig. 4.19(b)) are so named because the exterior surface of the line is in direct contact with the engine coolant. They have a disadvantage that is caused by the extra sealing that is required in order to prevent

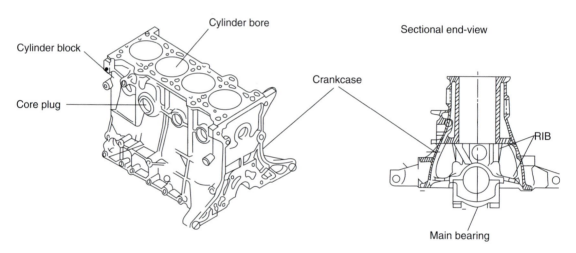

Fig. 4.18   Details of cylinder block and crankcase

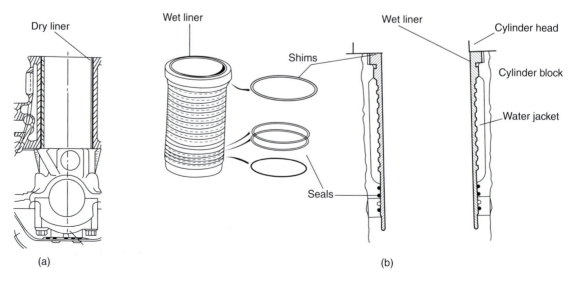

Fig. 4.19   Cylinder liners

coolant leaking into the lubricating oil. However, they have the advantage that they can be replaced without major machining work.

When aluminium alloy is used for the cylinder block, cylinder liners of the wet or dry type become a necessity, which partly offsets the advantage gained by the lighter weight of the aluminium alloy casting.

Cylinder block condition checks

Because piston rings do not reach the top edge of the cylinder bore, when the piston is at the top of its stroke, an unworn ridge develops at the top of the cylinder and further cylinder bore wear occurs on the side where most thrust is exerted. This means that the cylinders of an engine wear unevenly from top to bottom of the stroke. In order to assess cylinder bore wear it is necessary to measure the diameter at several points; the Mercer gauge has been developed for this purpose (Fig. 4.20). The gauge can be zeroed by setting it to the diameter of the unworn section at the top of the cylinder bore, or more accurately by means of an outside micrometer. Once zeroed, the bar gauge is inserted into the cylinder, as shown in Fig. 4.20(a). Measurements are taken in line

with, and at right angles to, the crankshaft. The maximum difference between the standard size and gauge reading is the wear. The difference between the in-line reading and the one at right angles is the ovality. The measurements are repeated at points in the cylinder, as shown in Fig. 4.20(c), and differences between the diameter at the top and bottom of the cylinder are referred to as taper.

The measurements taken may then be compared with the quoted bore size and excessive wear can be remedied by reboring or, in the case of minor wear, new piston rings.

Wet liner protrusion for clamping purposes

The type of wet liner shown in Fig. 4.19(b) is secured in place by the clamping force that is provided by the cylinder bolts. In order to obtain the requisite amount of clamping force the rim of the liner is made to protrude by a small amount above the top of the cylinder block. Figure 4.21 shows an arrangement for checking liner protrusion. The liner is securely clamped down as shown and the protrusion is measured by the dial gauge – in this example the protrusion should be 0.05 mm.

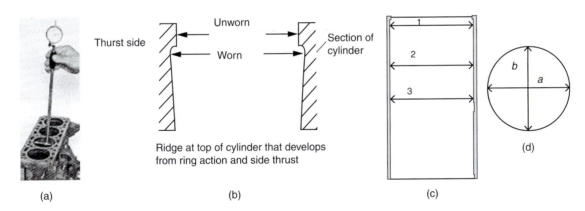

(a)   (b)   (c)   (d)

Fig. 4.20   Cylinder bore wear

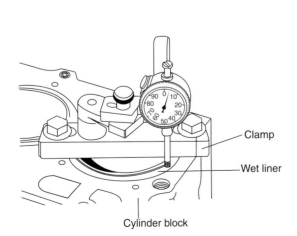

Fig. 4.21   Checking wet liner protrusion

Fig. 4.22   Liner damaged by incorrect fitting

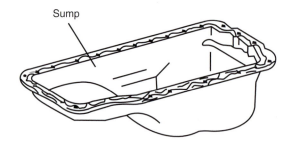

Fig. 4.23   The sump

Figure 4.22 shows the type of failure that may occur if the liner is not fitted correctly. The effect of such a failure upon engine performance is likely to be severe and may include:

- Loss of power
- Coolant in the oil
- Combustion gas in the cooling system
- Coolant inside the cylinder
- Problems with combustion and emissions controls.

The crankcase

The crankcase carries the main bearing supports and in some cases it is made as a separate casting from a light alloy, which reduces overall engine weight.

Sump

The sump (Fig. 4.23), or oil pan, is normally made as a steel pressing and its main function is to act as a reservoir for the oil that lubricates the engine. It is separated from the crankcase by a gasket and this, in common with other gaskets, must be positioned accurately prior to reassembling the sump to the crankcase. When correctly positioned, the set screws that hold it in place must be tightened gradually, in sequence, and finally to the correct torque.

Self-assessment questions

1. Figure 4.20 shows the positions in an engine cylinder at which measurements of diameter are taken in order to assess wear. The measurements are taken in line and at right angles to the crankshaft, at three levels as shown. The third measurement is taken at the lowest extent of piston ring travel. The terms ovality and taper are used to describe the condition of the cylinder bore. Ovality is the difference betweeen a and b, and taper is the difference between the largest diameter at the first measurement and the smallest diameter

at the third measurement. In an example, the following dimensions were recorded:
   - First measurement, $a = 74.22$ mm, $b = 73.95$ mm
   - Second measurement, $a = 74.19$ mm, $b = 74.00$ mm
   - Third measurement, $a = 74.00$ mm, $b = 73.97$ mm.

   The maximum ovality is:
   (a) 0.22 mm
   (b) 0.19 mm
   (c) 0.27 mm
   (d) 0.03 mm.
2. The taper is:
   (a) 0.25 mm
   (b) 0.03 mm
   (c) 0.19 mm
   (d) 0.27 mm.
3. Describe the repairs that can be undertaken if excessive oil consumption is caused by worn valve guides.
4. How many degrees of crank rotation are there between firing strokes on a four-cylinder, four-stroke, in-line engine?
5. What steps are necessary when refitting an aluminium alloy cylinder head?
6. Why are wet cylinder liners normally fitted to aluminium alloy cylinder blocks? How are wet liners secured in a cylinder block?
7. Describe the features of engine design that produce 'squish' turbulence.
8. What is the reason for producing turbulence in a combustion chamber?
9. What is the probable effect of a 'blown' cylinder head gasket on an engine cooling system?
10. Describe the procedure for checking piston ring gap. How can the ring gap be increased if it is too small?
11. How does compression ratio affect the thermal efficiency of an engine?
12. Describe how a cylinder head can be examined for probable cracks?
13. With the aid of a sketch, describe the types of valve seats that are used in aluminium alloy cylinder heads.
14. What is the difference between pre-ignition and detonation?
15. Describe some of the possible causes of pre-ignition? What are the symptoms of pre-ignition?
16. Why is an eight-cylinder engine generally smoother running than a four-cylinder one?
17. When cylinders and pistons are worn, combustion gas enters the crankcase. What happens to this gas and how can it affect the lubricating oil in the sump?
18. How can a badly pitted valve seat be repaired?

5
Firing orders and engine balance

Topics covered in this chapter

Firing orders and associated crank arrangements
Engine balance
Secondary force (harmonic) balancer
Dual mass flywheel

Firing orders

The firing order of an engine is the sequence in which the power strokes take place and it is determined by the design of the crankshaft. Each time a firing impulse occurs the resulting torque on the crankshaft tends to twist it and the effect is noticeable on long crankshafts. In order to minimize the torsional twist on the crankshaft the cylinders are fired first at the front of the engine and then at the rear — this process is followed, as far as possible, on all engines.

Crankshaft layout

The crankshaft shown in Fig. 5.1 causes the pistons to reach top dead centre (TDC) in pairs, namely 1 and 4, and then 2 and 3. Table 5.1 shows how the two possible firing orders for a four-cylinder in-line engine arise.

The reasoning is as follows:

- When piston number 1 is going down the power stroke, number 4 piston is also going down the stroke and the only possibility is that it is on the induction stroke.
- When pistons 1 and 4 are moving down the stroke, pistons 2 and 3 are moving up their strokes. This gives the two possibilities shown: either it is compression or power. If it is decided to make number 2 cylinder fire next, the firing order will be 1 followed by 2. This means the next cylinder to fire is number 4 and that is followed by number 3.
- The resulting firing order is 1—2—4—3.

An alternative firing order for a four-cylinder in-line engine is 1—3—4—2.

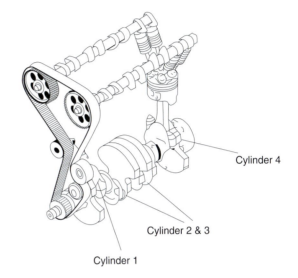

Fig. 5.1 A four-cylinder in-line engine crankshaft (KIA)

Table 5.1 Firing orders, four-cylinder in-line engine

	Cylinder number			
	1	**2**	**3**	**4**
First stroke	Power	Compression or exhaust	Compression or exhaust	Induction
Second stroke	Exhaust	Induction or power	Induction or power	Compression
Third stroke	Induction	Exhaust or compression	Compression or exhaust	Power
Fourth stroke	Compression	Induction or power	Induction or power	Exhaust

A six-cylinder in-line crankshaft and firing order

The crank throws are arranged in pairs, as shown in Fig. 5.2. In the crank arrangement at the top, pistons 1 and 6 are at TDC together; when cylinder 1 fires it can be followed by cylinders 3 or 4 firing. The most common arrangement is to fire number 4 cylinder and this leads to the firing order of 1—4—2—6—3—5.

In the lower of the two diagrams, the crank throws are arranged so that pistons 1 and 6 are at TDC together, followed by the other pistons as shown. This leads to

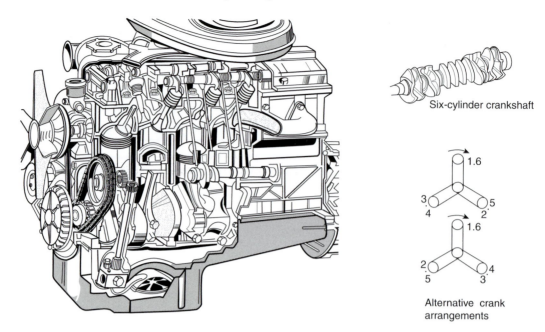

Six-cylinder crankshaft

Alternative crank arrangements

Fig. 5.2 A six-cylinder in-line engine (Toyota)

the other firing order commonly used in six-cylinder in-line engines, i.e. 1−5−3−6−2−4.

Firing first in the front half of the crankshaft and then as far as possible in the back half minimizes the twisting of the crankshaft that is the cause of torsional vibrations.

A V8 crankshaft and firing order

The use of the V8 layout permits the advantages of many cylinders to be accommodated in a short space.

The crankshaft has four crankpins, each of which serves two connecting rods as shown in Fig. 5.3.

Vibrations and engine balance

If an engine is not properly balanced it will develop vibrations, which cause noise and discomfort to vehicle occupants and, in time, cause damage to components.

There are two main areas of the engine where vibrations are a problem:

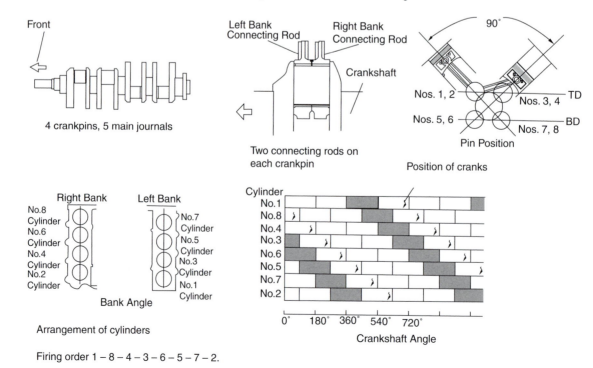

Front

4 crankpins, 5 main journals

Two connecting rods on each crankpin

Position of cranks

Arrangement of cylinders

Firing order 1 − 8 − 4 − 3 − 6 − 5 − 7 − 2.

Crankshaft Angle

Fig. 5.3 A V8 engine (Toyota)

- Pistons and connecting rods
- Crankshaft and flywheel.

The reciprocating movement of the pistons and connecting rods gives rise to two types of unbalanced forces:

1. Primary forces
2. Secondary forces.

Pistons and connecting rods

Primary and secondary forces

The acceleration of the piston is at its maximum at each end of the stroke, as shown in Fig. 5.4.

Analysis of piston and crank motion shows that at any point during the stroke a very close approximation of the piston acceleration (a) is given by the formula:

$$a = \omega^2 r\{(\cos \theta) + (\cos 2\theta)/n\},$$

where ω = angular velocity of crank, θ = angle that crank has turned through, r = radius of the crank throw and n = connecting rod to crank ratio,

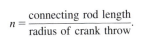

$$n = \frac{\text{connecting rod length}}{\text{radius of crank throw}}.$$

The force on the piston is:

$$F_p = m\omega^2 r\{(\cos \theta) + (\cos 2\theta)In\},$$

where m = piston mass. For engine balance purposes the force F_p is normally considered in two parts, called the **primary force** and the **secondary force** respectively.

The primary force is given by the first part of the equation, e.g. primary force:

$$F_p = m\omega^2 r(\cos \theta).$$

The secondary force is given by the second part of the equation, e.g. secondary force:

$$F_s = m\omega^2 r(\cos 2\theta)/n.$$

Figure 5.5 shows how the primary and secondary forces vary during one complete revolution of the crankshaft; the plus and minus signs indicate the direction of the force along the line of the piston stroke. The graph shows that the primary force is at a maximum at TDC and bottom dead centre (BDC), i.e. **twice per revolution of the crankshaft**.

The graph also shows that the secondary force is at a maximum at TDC, $90°$ after TDC, BDC and $270°$ of crank rotation, i.e. **four times per revolution of the crankshaft**.

Four cylinder in-line engine balance

Figure 5.6 shows a layout that is widely used in the construction of four-cylinder in-line engines. The reciprocating parts are identical for each cylinder and the cranks are arranged to provide uniform firing intervals and, in most cases, the distances between the cylinders are equal. With this design of engine only

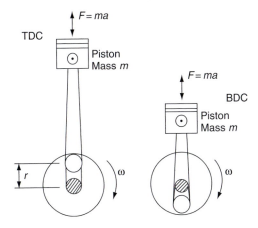

Fig. 5.4 Maximum force at end of stroke

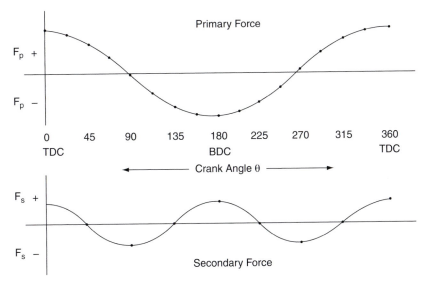

Fig. 5.5 Primary and secondary force variation

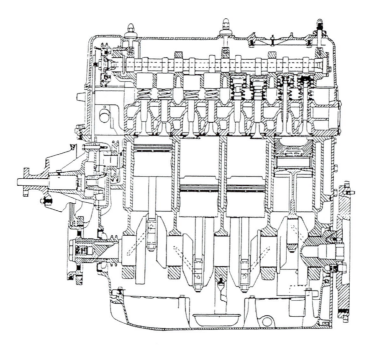

Fig. 5.6 Four-cylinder in-line engine

the primary forces are balanced; the secondary forces and other harmonic forces are unbalanced. In engines operating at high speed, the secondary forces can affect the operation of the vehicle and secondary balancers may be incorporated into the engine design to counter any ill effects.

Secondary force balancer

The balance weights are attached to shafts that are driven from the crankshaft (see Fig. 5.7). One shaft rotates in a clockwise direction, the other rotates in an anticlockwise direction; both shafts are driven at twice engine speed.

The rotation of the balance weights is timed so that they are effective at the points required, i.e. TDC, 45°, 90°, and 135° after TDC (ATDC), of crankshaft rotation. The diagrams in Fig. 5.8 show how the weights are effective when required and how the forces they generate cancel each other out when the secondary force is zero.

Harmonics

Vibrations caused by unbalanced forces in an engine occur at certain engine speeds; the lowest speed at which the vibrations are a problem is called the **fundamental frequency**. Further vibrations will occur at higher speeds, which are multiples of the fundamental frequency; these higher speeds are called **harmonics** and the harmonic balancer is designed to counteract the effects of the vibrations at these speeds. That is why the secondary force balancer is often referred to as a harmonic balancer.

Balance of rotating parts of the single-cylinder engine

For the purposes of this explanation the rotating parts considered are the crankpin and the crankshaft, and part of the mass of the connecting rod. In multi-cylinder engines the counterweights that are part of the crankshaft provide the necessary balancing forces, as shown in Fig. 5.9.

Couples and distance between crank throws

Figure 5.10(a) shows a twin-cylinder in-line engine where the cranks are 180° apart. The primary forces caused by the reciprocating parts produce a turning effect (couple) that causes the engine to rock on its mountings. Figure 5.10(b) shows a four-cylinder engine where the cranks are arranged as shown. The cranks are equal distances apart and the couples on each half of the crankshaft oppose each other. The couples balance out inside the engine and no effect is felt at the engine mountings.

Balance of individual components

During manufacture great care is taken to ensure that the moving parts of the engine are manufactured to fine tolerances. In the case of pistons and connecting rods this also applies to their weight. If, in the course of repair, it becomes necessary to replace these units they should be checked to ensure that their weights are identical within approved limits.

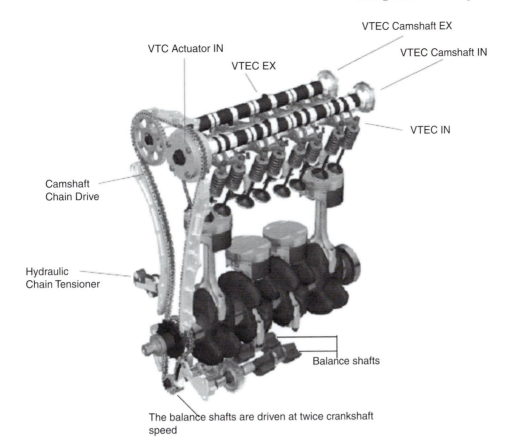

Fig. 5.7   A Honda VTEC engine with harmonic balancer

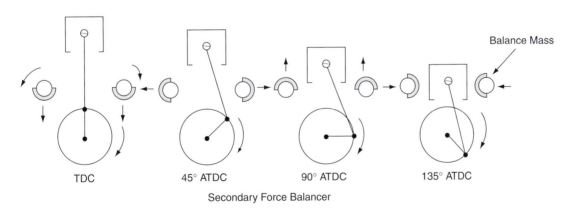

Fig. 5.8   Principle of secondary force balancer

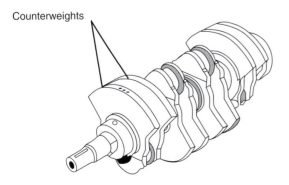

Fig. 5.9   Crankshaft counterweights

The crankshaft and flywheel are usually balanced as a unit and balance is achieved by drilling holes in appropriate positions; these can often be seen in the crank web or flywheel. To ensure that the balance is not upset crankshafts are fitted with devices such as dowels that ensure that the flywheel is always correctly fitted to it, as shown in Fig. 5.11.

Torsional vibration

The vibrations associated with the crankshaft and flywheel are caused by the interaction between them

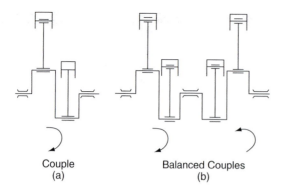

Fig. 5.10   Couples

Couple (a)   Balanced Couples (b)

and the effect is most evident in engines with a long crankshaft. When a cylinder at the front end of a crankshaft fires, the torque exerted on the shaft accelerates the flywheel and in doing so the crankshaft is twisted slightly. The periodic application and removal of crankshaft torque as each cylinder fires can set up a vibration called torsional vibration. The effect is minimized by firing cylinders at alternate ends of the engine, but in many cases the vibrations are a problem because they can fracture the crankshaft or cause damage in the transmission system.

Various types of vibration dampers are fitted to engines to lessen the effect.

Crankshaft damper

The crankshaft torsional vibration damper shown in Fig. 5.12 makes use of a layer of rubber between the centre of the pulley and the mass of the drive-belt rim. The damping effect is provided by the rubber, which distorts when vibration attempts to accelerate the drive-belt pulley mass; the distortion of the rubber converts vibrational energy into heat, which is then passed into the atmosphere and it is this action that provides the damping.

Crankshaft damper

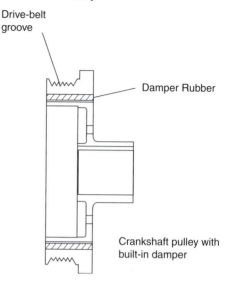

Fig. 5.12   Crankshaft torsional vibration damper

The vibration damper in the clutch plate

In a clutch plate of the type shown in Fig. 5.13, the damping effect is provided by the coil springs and the small friction surface at the hub of the clutch plate. The springs permit slight movement of the part of the plate that holds the linings relative to the hub. This action compresses the springs and converts energy into heat at the small friction surface to provide the damping action.

The dual mass flywheel

Dual mass flywheels are constructed from a number of separate parts and they are designed to overcome low-speed torsional vibration that is a problem in some diesel engine vehicles. They are called dual mass flywheels because they have two major mass components. One is the mass that affects the crankshaft and another mass

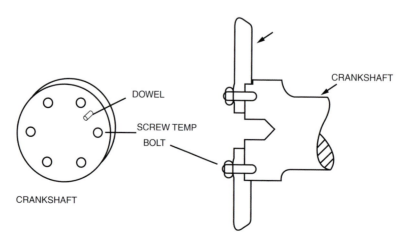

Fig. 5.11   Flywheel fixing to crankshaft

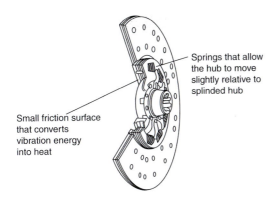

Fig. 5.13 Vibration damper at centre of clutch plate

Springs that allow the hub to move slightly relative to splinded hub

Small friction surface that converts vibration energy into heat

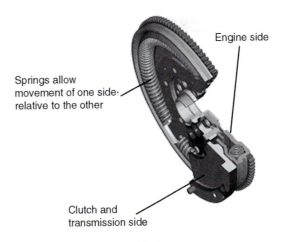

Engine side

Springs allow movement of one side relative to the other

Clutch and transmission side

Fig. 5.14 A dual mass flywheel (LuK)

that affects the transmission. The two masses rotate as a unit, but one mass — the engine side — can rotate by a small angle relative to the gearbox side mass by virtue of the coil springs. The action of the flywheel removes vibrational energy from the system and protects the engine and transmission from mechanical damage, and also reduces noise transmission into the vehicle structure.

Examination of Fig. 5.14 will give an indication of the complexity of the device as compared with the relative simplicity of the standard steel disc type of flywheel.

For further information you should look at the LuK website (www.luk.com) — or just do a general search for dual mass flywheel and you will come across a number of interesting items.

Self-assessment questions

1. A four-cylinder, in-line, four-stroke engine has a firing order 1—3—4—2. When number 1 cylinder is on its power stroke, what stroke is number 3 cylinder on?
2. Make a list of the vehicles that are fitted with dual mass flywheels. What type of vehicles are most likely to have dual mass flywheels?
3. How fast, in relation to crankshaft speed, do secondary balancer shafts rotate?
4. What is the purpose of the counterweights that form part of the crankshaft webs?
5. What is the angle between firing impulses in a six-cylinder, four-stroke, in-line engine?
6. Sketch and describe a crankshaft vibration damper that makes use of a rubber insert.
7. What is the purpose of the vibration damper that is fitted at the centre of some clutch plates?
8. A certain four-cylinder engine has suffered a piston failure and the cylinder has been scored. In an attempt to remedy this failure, a cylinder has been bored out and an oversize piston fitted. What is the likely effect of this repair on engine balance?
9. What is the reason for firing cylinders at alternate ends of the crankshaft, as far as that is possible?
10. Provided that the valve timing is correctly set, how can observation of the order in which inlet valves open help to determine the firing order of an engine?

6
Crankshaft, camshaft, and valve operating mechanisms

Topics covered in this chapter

Heat treatment
Nitriding
Crankshaft bearings
Bearing metals
Crankshaft condition checks
Checking crankshaft end float
Crankpin — taper and ovality
The flywheel and its mounting
Crankshaft oil seals
Multi-cylinder engine camshaft
Condition checks on camshaft
Valve train
Overhead valve engines (OHV)
Checking valve clearance

Crankshaft

The crankshaft is part of the engine where reciprocating motion of the pistons is converted into rotary motion.

The main features that are shown in Fig. 6.1 are:

- The crankpin, which is the part of the crankshaft to which the big end of the connecting rod is connected.

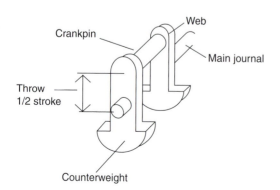

Fig. 6.1 Details of a simple crankshaft

- The webs are the parts that connect the crankpin to the main journal, which are the parts of the shaft on which the whole shaft rotates.
- The main journals are also the parts where the main bearings support the crankshaft in the engine frame.
- The throw is the radius between the centre of the main journal and the centre of the crankpin; it is equal to half the stroke of the engine.
- The counterweights provide the balance that is described in Chapter 5.

Crankshafts are normally made from high-quality steel by a process known as drop forging. This process preserves the grain flow in the finished product, which is believed to improve its strength. After forging, the journals and the crankpins are machined and in the process the fillet radii are carefully produced because they play an important part in the ability of the shaft to resist fracture. Internal drillings transfer oil from the main bearings to the crankpins and big-end bearings. As shown in Fig. 6.2.

Heat treatment

The bearing surfaces at the journals and crankpins are hardened to provide good resistance to wear — other parts are heat treated to remove stresses and produce the necessary strength.

In carbon steel shafts the hardening is produced by a process called case hardening, or carburizing. In this process the shaft is placed at high temperature in the presence of carbon-rich material — this results in a layer of metal that can be made very hard while the remainder of the shaft remains softer and tough. After hardening the final machining is performed to produce journals and crankpins of the correct size.

Nitriding

Alloy steels containing elements such as chromium, molybdenum and others that are sometimes used in the manufacture of crankshafts, can be hardened by a process called nitriding. In this process the machined

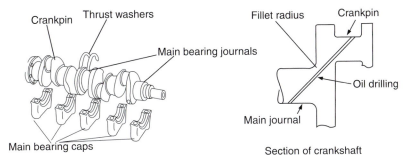

Section of crankshaft

Fig. 6.2 Details of four-cylinder engine crankshaft

shaft is heated under pressure in a special container that contains ammonia. The inside of the container is heated to about 500°C for 5–25 hours, during which time a thin layer of hard metal approximately 0.2 mm thick is produced on the bearing surfaces.

Crankshaft bearings

Because multi-cylinder engine crankshafts are normally made in one piece it is necessary to use split bearings (Fig. 6.3). The two halves of the split bearing are often referred to as the bearing shells. The actual shell is made from steel sheet approximately 2.5 mm thick and this is coated with a 0.15 mm layer of bearing metal.

Bearing metals

White metal or Babbit metal is named after Isaac Babbit, who developed it in the early nineteenth century. The original Babbit metal was made from 50 parts of tin, 5 of antimony and 1 of copper. Later alloys contain some lead and the proportions of the alloying elements are varied according to the use to which the metal is to be put. It is quite soft and has low friction when used with a hard steel shaft. The soft metal permits small abrasive particles to become embedded below the surface and this reduces shaft wear. Because of the high stresses imposed on big-end and main bearings, Babbit bearings are normally used for camshafts.

Tin–aluminium

Alloys of tin and aluminium are commonly used for bearing metal because they are resistant to fatigue and they have low friction when used with a steel shaft. A thin layer of the tin–aluminium alloy is applied to the steel shell as shown in Fig. 6.3.

Copper–lead

The copper–lead alloys are approximately 70% copper and 30% tin. They are harder than aluminium–tin alloys and they have high fatigue resistance but they are less able to allow the embedding of small particles, which means that the journals and crankpins are harder.

Lead–bronze

These are alloys consisting of copper and lead with a small amount of tin. They are used in turbo-charged and other heavy-duty engines because of their load-bearing capacity. However, they suffer the disadvantage that they are susceptible to corrosive attack from acids found in some oils. To overcome this problem they are coated with a thin layer of alloy of lead, tin, and indium approximately 0.02 mm thick.

The bearing shell in the bearing housing

The bearing shells are an interference fit in the housing (Fig. 6.4). When assembling new bearings the shell will protrude slightly from the bearing housing; as the securing bolts are tightened the shells are pressed into the housing to provide the tight fit that is required.

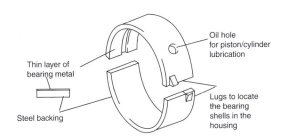

Fig. 6.3 Crankshaft bearing

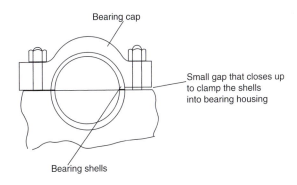

Fig. 6.4 Bearing shell fit in housing

F21-119

A. Filament (plastigage) applied
B. Shape of filament
   after bearing cap is
   removed
C. The measuring
   gauge

Fig. 6.5   Main bearing clearance

Crankshaft condition checks

Main bearings

Main bearing journals should be checked for signs of damage and wear. The journals can be checked by means of an external micrometer and the readings obtained then checked against figures in the workshop manual. As an example, the manual that I have to hand gives these figures: 57.98−58.00 mm.

If the diameter is smaller than the lower figure the crankshaft would be reground so that the journals become slightly smaller. This major repair then necessitates the use of undersize bearing shells, which are normally supplied in a range of sizes. On reassembly the clearance between the main journals and the bearing shell may be checked with the aid of a plastic filament called plastigage.

When the crankshaft is resting in the crankcase, the plastigage material is placed on the journal as shown in Fig. 6.5. The bearing gap is then fitted and tightened down. After tightening the bearing gap is removed and the small gauge is applied. The width as shown on the gauge gives the bearing clearance.

End float

The thrust washers that are placed at the sides of the centre main bearing are needed to take the end thrust on the crankshaft that arises from forces such as those produced by the clutch release operation. A small amount of space, called 'end float', of approximately 0.20 mm between the crankshaft and the thrust washer is required to provide lubrication. End float may be checked as shown in Fig. 6.6 and then checked against manufacturer's specification.

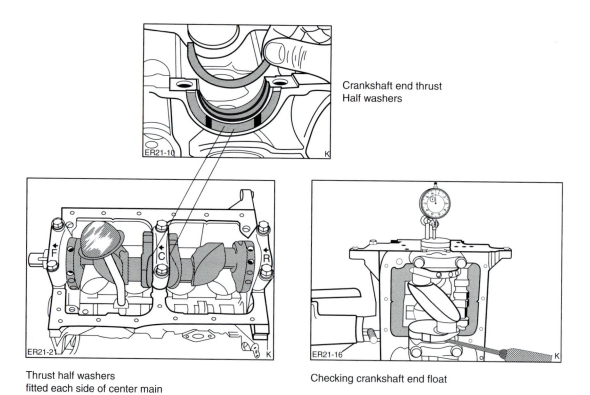

Crankshaft end thrust
Half washers

Thrust half washers
fitted each side of center main
bearing

Checking crankshaft end float

Fig. 6.6   Checking crankshaft end float

Crankpins and big-end bearings

The part of the crankshaft that connects it to the connecting rod is called the crankpin. Crankpins are subjected to forces that cause them to wear unevenly and a range of measurements are required to ascertain their condition.

Crankpin dimensions

The fluctuating inertia forces that the big end is subject to cause the crankpin to wear unevenly and if these forces are excessive they may distort the connecting rod bore.

Clearance between the crankpin and the bearing shells can be determined by the use of a plastigage as shown in Fig. 6.5. Prior to checking this clearance the diameter of the crankpin should be measured at various points as shown in Fig. 6.7.

Example

A crankpin is checked for wear as shown in Fig. 6.7 and the following results are recorded:

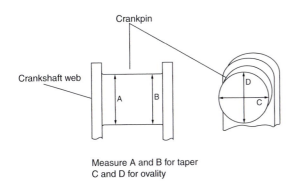

Measure A and B for taper
C and D for ovality

Fig. 6.7 Crankpin — taper and ovality

- A = 42.99 mm
- B = 42.88 mm
- C = 42.99 mm
- D = 42.90 mm.

The taper on the crankpin = A − B = 42.99 − 42.88 = 0.11 mm.

The ovality on the crankpin = C − D = 42.99 − 42.90 = 0.09 mm.

Although these variations in diameter of the crankpin are quite small they will, when added to the normal bearing clearance, cause a gap that could lead to a loss of oil pressure and bearing failure.

Connecting rod eye

With the bearing shell removed, the bearing cap is assembled to the rod and the bolts are tightened to the correct torque. An internal micrometer is then used to record the internal diameter of the connecting rod eye at three different points to ascertain that it is not distorted. The readings should then be checked against the workshop manual figures.

Flywheel

A primary function of the flywheel is to store the energy that helps the engine to run smoothly. On most engines the drive to the transmission takes place via the flywheel and the starter ring gear provides the means by which the starter motor starts the engine when required.

The rear end of the crankshaft is formed into a flange to which the flywheel is attached. At the centre of this flange is situated a small bearing called the spigot bearing, which supports the first motion shaft of the gearbox. Where a friction clutch is used, the gearbox side of the flywheel is machined to act as a friction surface for the clutch. Figure 6.8 shows a clutch friction

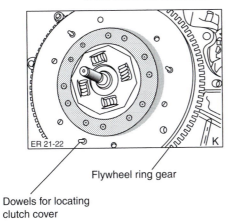

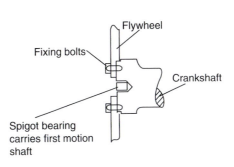

Fig. 6.8 The flywheel and its mounting

plate being aligned to the flywheel with the aid of a mandrel. These mandrels are often referred to as 'dummy first motion shafts'.

Oil seals

As the front end of the crankshaft protrudes outside the engine to accommodate a drive pulley and at the rear to accommodate the flywheel, it is necessary to prevent oil escaping from the sump and this is achieved by the use of various types of oil seals. Two types of crankshaft oil seals are shown in Fig. 6.9.

Camshaft

Camshafts are made from high-grade steel, either by drop forging or casting. The bearing surface and the wearing surfaces of the cams are heat treated to produce the requisite amount of strength and resistance to wear. End thrust on the camshaft is controlled by a semicircular thrust washer made from bearing metal that fits in a groove in the bearing housing and a corresponding one in the camshaft journal. Effective lip-type oil seals are normally situated at each end of the camshaft to retain the oil that is necessary for the lubrication in overhead camshaft engines. The camshaft drive gear shown in Fig. 6.10 is secured to the camshaft by means of the bolt and a dowel that protrudes from the camshaft and provides a tight fit in the camshaft drive gear to prevent the gear from moving relative to the shaft.

Camshaft bearings

Overhead camshaft

On overhead camshaft engines the camshaft is situated in the cylinder head. Three types of bearing arrangement are used to support the camshaft:

1. A plain cylindrical bearing that is inserted into the cylinder head as shown in Fig. 6.11(a). The internal diameter of the bearing is slightly larger than the maximum diameter of the crankshaft to allow the camshaft to be removed from the cylinder head.
2. No detachable bearing — the camshaft runs in the holes machined in the cylinder head material.
3. Split shell-type bearings where the shells are held in place by a bearing cap (Fig. 6.11(b)).

Overhead valve

On an overhead valve engine the camshaft is normally situated at the side of the engine (Fig. 6.12); in this case it will run in bearings machined into the cylinder block, or cylindrical-type bearings of phosphor bronze or other bearing material.

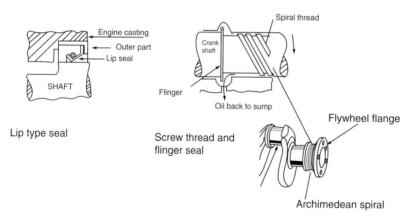

Fig. 6.9 Crankshaft oil seals

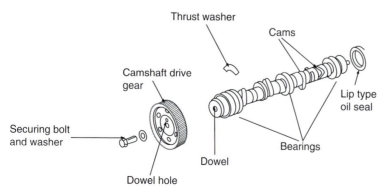

Fig. 6.10 Multi-cylinder engine camshaft

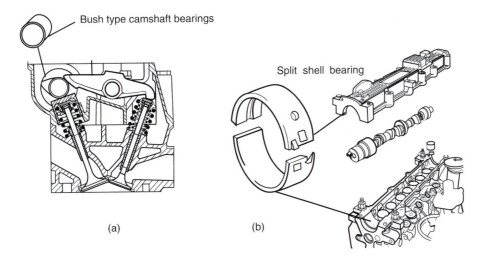

Bush type camshaft bearings

Split shell bearing

(a)

(b)

Fig. 6.11 Alternative camshaft bearings

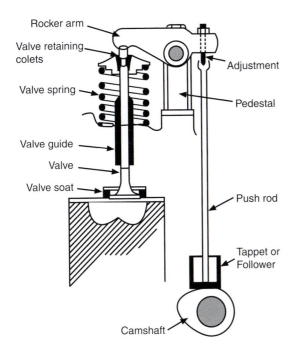

Rocker arm

Valve retaining colets

Adjustment

Valve spring

Pedestal

Valve guide

Valve

Valve soat

Push rod

Tappet or Follower

Camshaft

Fig. 6.12 Overhead valve engine with side camshaft

Condition checks on camshaft

After cleaning, cams and journals should be visually inspected for signs of wear or damage. The extent of wear on cams and journals can be determined by measuring with an external micrometer. An example of the relevant dimensions is shown in Fig. 6.13. To check for distortion the camshaft is mounted between centres on a lathe. A dial test gauge is then mounted so that the stylus rests on the centre bearing of the camshaft; when the camshaft is rotated any distortion will be recorded on the gauge. A representative figure for maximum distortion is approximately 0.05 mm, but this should always be checked against an individual manufacturer's figures.

Valve train

Overhead valve engines (OHV)

The valve train is the mechanism that is used to operate the valves. In an overhead valve engine of the type shown in Fig. 6.12, the valve train consists of the cam follower, which is also called a tappet, the push rod, the rocker shaft, and the rocker arm.

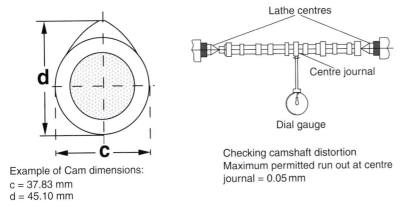

Lathe centres

Centre journal

Dial gauge

d

c

Example of Cam dimensions:
c = 37.83 mm
d = 45.10 mm

Maximum permitted variation = 0.05 mm

Checking camshaft distortion
Maximum permitted run out at centre journal = 0.05 mm

Fig. 6.13 Camshaft dimension checks

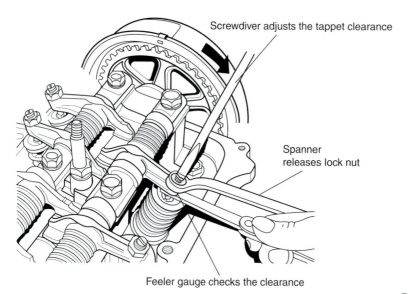

(a)

1. Inlet valve
2. Exhaust valve
3. Camshaft
4. Rocker arms

1. Inlet valve
2. Adjustment shims
3. Single camshaft
4. Rocker arm
5. Rocker shaft
6. Adjustment shims
7. Exhaust valve

(b)

(c)

Fig. 6.14 Valve gear layout, overhead camshaft engines

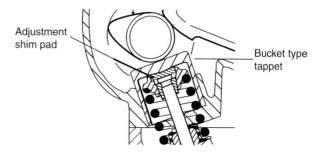

Fig. 6.15 Valve clearance adjustment

Overhead camshaft engines

The valve train in these engines is shorter than in the OHV engine because the camshaft is situated close to the valves, as shown in Fig. 6.14.

In engine Fig. 6.14(a) the valves are operated by rocker arms that are mounted on a pair of rocker shafts. The valve clearance is adjusted by means of a screw and lock nut.

In engine Fig. 6.14(b) a single camshaft operates the inlet valve directly via a bucket-type tappet, and the exhaust valve indirectly via a rocker arm. In both cases the valve clearance is adjusted by means of the shims that are located at the end of the valve stem.

Tappet rotation

In order to minimize wear, bucket tappets are placed slightly off-centre from the cam; this causes a partial rotation of the tappet as the valve is operated. On some engines the valve is made to rotate by the use of

Fig. 6.16 Valve clearance shims

specially designed cotters and a grooved valve stem; this prevents particles of carbon becoming trapped in a single spot on the valve seat and thus reduces the possibility of damaged valves.

In engine Fig. 6.14(c) the valves are operated by twin overhead camshafts and bucket-type tappets, the valve clearances are adjusted by means of shims, as shown in Fig. 6.16. In some engines these shims may be

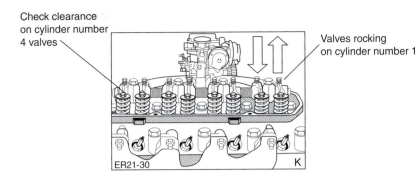

Check clearance on cylinder number 4 valves

Valves rocking on cylinder number 1

ER21-30

Fig. 6.17 Checking valve clearance

changed with the aid of special pliers, or a magnetic device.

Checking valve clearance

Valve clearances are checked at some routine service intervals, or when engine repairs have been carried out. The procedure for checking valve clearance varies according to engine design and the workshop manual should always be used as a reference. Whichever method is used the essential point is that the cam follower must be resting on the base circle of the cam. Figure 6.15 shows the general procedure for checking valve clearance. Feeler gauges are selected and inserted between the rocker arm and the valve stem. When the selection of feeler gauges is moved in and out of the gap, the resistance to movement should just be felt. Should the gap be too large, the lock nut should be reduced and the adjustment made by means of the screwdriver. When the correct clearance is achieved, the lock nut is tightened and the correct clearance again checked.

The following examples show the general procedure for checking valve clearance. For a four-cylinder, in-line, single-camshaft engine:

- Cylinder number 4 rocking − check cylinder number 1.
- Cylinder number 3 rocking − check cylinder number 2.
- Cylinder number 1 rocking − check cylinder number 4.
- Cylinder number 2 rocking − check cylinder number 3.

An alternative to this procedure that can be used on engines of this type is called the rule of 9. Figure 6.17 makes the point about valves rocking. The valves are numbered from the front of the engine so that the first valve is number 1 and the last one is number 8. The procedure is:

- When number 1 valve is fully open, check number 8 clearance: 1 + 8 = 9.
- When number 2 valve is fully open, check number 7 clearance: 2 + 7 = 9.
- When number 3 valve is fully open, check number 6 clearance: 3 + 6 = 9.
- When number 4 valve is fully open, check number 5 clearance: 4 + 5 = 9.
- And so on until valve number 8 is fully open.

For other engines the procedure is often more complicated, as shown in this example of a six-cylinder engine:

- No. 1 cylinder − inlet valve open, check cylinder 6 inlet valve and cylinder 4 exhaust.
- No. 2 cylinder − inlet open, check cylinder number 5 inlet and 6 exhaust.
- No. 3 cylinder − inlet open, check cylinder number 4 inlet and cylinder 5 exhaust.
- No. 4 cylinder − inlet open, check cylinder number 3 inlet and cylinder 2 exhaust.
- No. 5 cylinder − inlet open, check cylinder number 2 inlet and cylinder 1 exhaust.
- No. 6 cylinder − inlet open, check cylinder number 1 inlet and cylinder 3 exhaust.

These two examples show how widely different some of these procedures can be and how important it is to check the workshop manual.

Valve head broken off

Fig. 6.18 Broken valve

Self-assessment questions

1. Figure 6.18 shows the head of a valve that has broken away from the stem. It is believed that the damage was caused by the valve hitting the piston. Describe:
   (a) The valve spring defects that may have caused the problem.
   (b) The extent of damage that may have been caused to other engine components.

2. Name the parts marked A, B and C in Fig. 6.19. What is the purpose of the hole at D?
3. What is the purpose of the seals shown in Fig. 6.20? Why is the exhaust valve seal different to the inlet valve seal?
4. What steps are necessary to ensure that the piston ring is placed at right angles to the cylinder when checking the ring gap (Fig. 6.21)? It is recommended that the ring gap should be 0.003 × bore diameter. If the bore diameter is 80 mm what should be the ring gap?

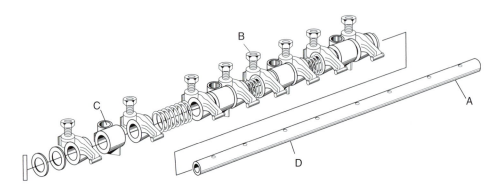

Fig. 6.19   Engine components

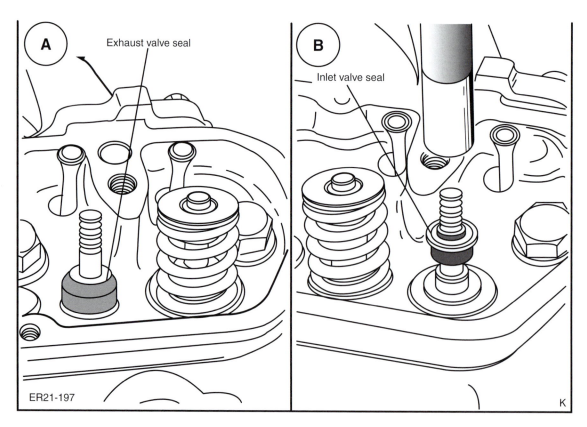

Fig. 6.20   Valve stem seals

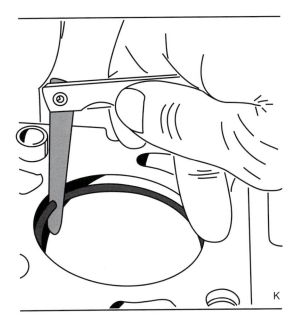

Fig. 6.21 Checking piston ring gap

Fig. 6.23 Engine assembly procedure

5. In which position on an engine are the half washers (Fig. 6.22) fitted? What problem arises if these washers are too thin?

6. Describe the operation that is shown in Fig. 6.23. What is the name of the device at A?

7. Name the device shown at A. How is the valve prevented from moving during the operation shown in Fig. 6.24?

8. Name each of the components marked 1—6 in Fig. 6.25 and decribe their function.

9. A cam belt tensioning device of the type shown in Fig. 6.26 is applied at position 1. The dial on the gauge records the tension as a number — in the example shown, the tension of a new cam belt should give a reading of 7 on the gauge. How many newtons does this represent? Write a short explanation of the reasons why the correct belt tension is an important setting.

10. The following details relate to a 1600 cc engine:

   - Crankpin diameter, 47.890—47.910 mm
   - Inside diameter of big-end bearing when fitted, 47.916—47.950 mm.

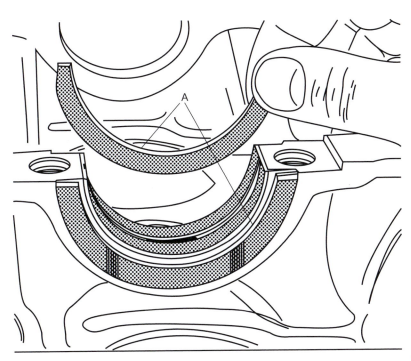

Fig. 6.22 Half washers

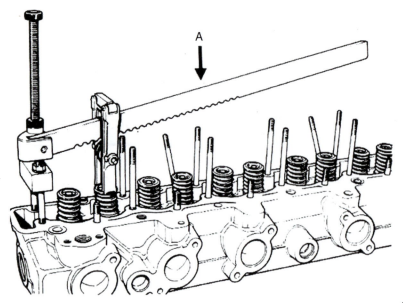

Fig. 6.24   Reassembling a cylinder head

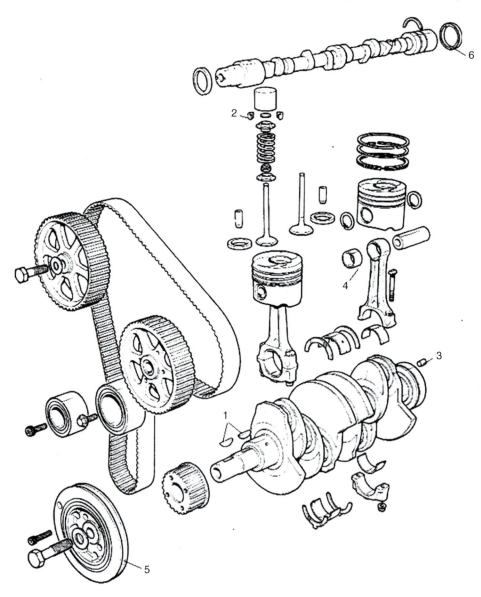

Fig. 6.25   Engine components

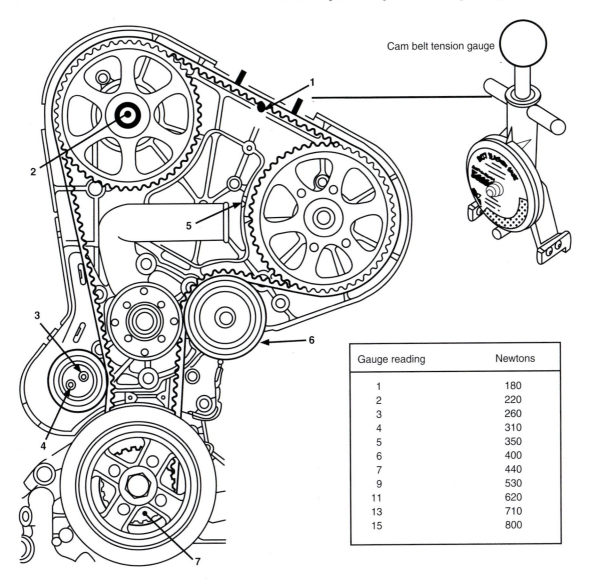

Cam belt tension gauge

Gauge reading	Newtons
1	180
2	220
3	260
4	310
5	350
6	400
7	440
9	530
11	620
13	710
15	800

Fig. 6.26 Cam belt tension

How big is the bearing clearance when the shaft is at its smallest diameter and the inside diameter of the bearing is at its maximum value:

(a) 0.060 mm
(b) 0.006 mm
(c) 0.040 mm
(d) 0.400 mm?

11. With the aid of Fig. 6.27, describe how oil reaches the big-end bearings.

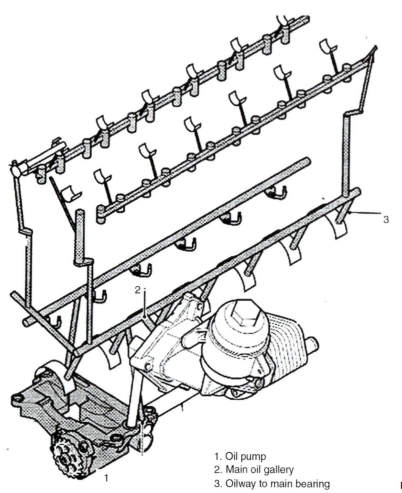

1. Oil pump
2. Main oil gallery
3. Oilway to main bearing

Fig. 6.27   Lubrication circuit for a six-cylinder Land Rover engine

7

A petrol fuel system

Topics covered in this chapter

The fuel tank
Other fuel tank details
Pump details
Fuel return
Mechanical fuel pump
Pipes
Filters
Carbon canister
Inertia switch
Working with petrol
Flammability of petrol and the proper precautions
to be taken when working with it
Fuel retrievers
Maintenance and repair

The purpose of the fuel system is to hold a supply of fuel that is readily available at the engine when it is required. The principal elements of a modern petrol fuel system are shown in Fig. 7.1.

The fuel tank

The fuel tank on front-engine light vehicles is normally situated at the rear of the vehicle in a position where it is reasonably well protected against damage in the event of a rear-end collision. The tank may be made from steel, aluminium, or plastic:

- Steel fuel tank. A steel sheet is pressed into the desired shape. The inside of the tank is normally coated with tin to minimize the risk of corrosion.
- Aluminium alloy fuel tank. These are corrosion resistant and of lighter weight than the steel ones.
- Plastic fuel tank. These are moulded from various proprietary plastic materials. They may be protected against impact damage by the use of external steel plates.

Other fuel tank details

The filler cap is designed to allow ease of access for refuelling and to prevent the escape of fuel when the vehicle is in use. The neck of the fuel tank contains a valve that is opened when the fuel supply hose is inserted – the purpose of this valve is to prevent fuel 'blowing back' during the refuelling process.

In, or close to, the neck is placed a pipe that connects to the carbon canister so that vapour can be taken away from the tank and processed in the engine.

The fuel tank sender unit is the sensor that transmits an electrical signal to the fuel gauge to provide a reasonably accurate measure of the amount of fuel in the tank.

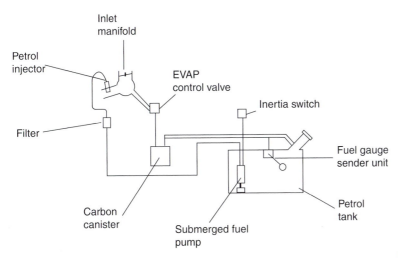

Fig. 7.1 Petrol fuel system

Internal baffle plates are fitted to prevent fuel from moving away from the pump during acceleration, braking, and other vehicle motions.

An electrically operated fuel pump is fitted inside the tank and this pumps fuel to the engine through pipes and filters; an alternative is to mount the pump outside the tank.

An inertia switch is fitted outside the tank so that, in the event of a severe collision, the fuel pump is switched off to prevent the escape of fuel.

The tank is attached to the vehicle by means of brackets built on to the tank, or by means of straps. Both of these methods are used in the example shown in Fig. 7.2.

Pump details

On most modern vehicles the fuel is transferred from the fuel tank to the engine by means of an electric pump of the type shown in Fig. 7.3. In this example the pump is housed in the fuel tank, inside a swirl pot. The swirl pot holds an amount of fuel that ensures that the pick-up filter on the pump inlet is always immersed in fuel. The pump may also be situated outside the fuel tank.

The pumping element consists of an impeller rotor that has grooves on its periphery. When the electric motor is energized it causes the impeller to rotate; the action of the grooves creates a low pressure at the pump inlet and fuel is drawn into the pump. The space

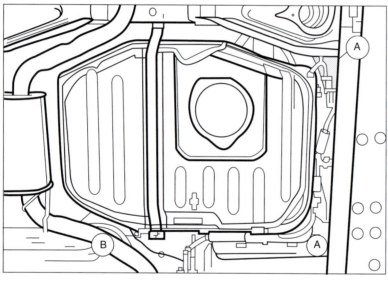

Tank securing bolts and strap.
A - Tank bolts
B - Strap securing bolts

Fig. 7.2 Fuel tank fixings (Ford), viewed from underside of vehicle

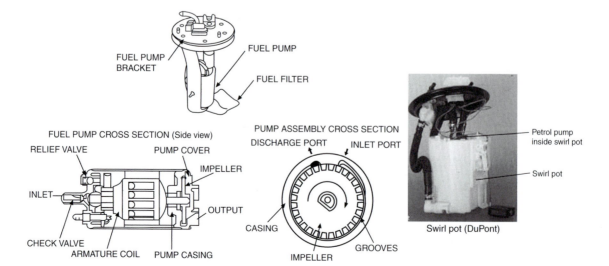

Fig. 7.3 An electric fuel pump (ARG) and swirl pot (DuPont)

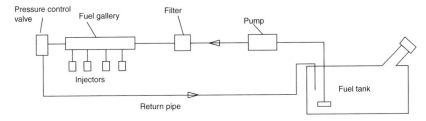

Fig. 7.4 Fuel return

between the impeller and pump casing is larger at the inlet than at the outlet and this results in a region of higher pressure, which is used to convey the fuel through the system. The check valve at the pump inlet prevents fuel from draining back when the pump is not energized and the relief valve protects the pump from excessive pressure.

Fuel return

Figure 7.4 shows the layout of a system that is used for petrol injection systems. In order to maintain an adequate amount of fuel at the injectors, fuel is held under pressure in a pipe called a fuel gallery. The gallery is kept full by the pump and the pressure valve, which allows excess fuel to be returned to the fuel tank.

Mechanical fuel pump

Mechanical pumps of the type shown in Fig. 7.5 were widely used in engines that were equipped with a carburettor fuel system. The pump is driven by a special cam that is mounted on the engine camshaft; unlike an electric pump there is no pumping action when the engine is not running and mechanical pumps are normally equipped with a lever that allows them to be primed manually.

The operating mechanism causes the diaphragm rod to move up and down, and this causes the diaphragm to deflect against the control of the return spring. The downward movement of the diaphragm increases the space above it and petrol is drawn into the space, through the spring-loaded inlet valve. On upward movement of the diaphragm, petrol is pushed out through the outlet valve into the delivery pipe and on to the carburettor. In the space under the cap there is a filter through which the incoming fuel is filtered prior to entering the diaphragm chamber. These pumps require occasional checks to make sure that the filter is clean, and pressure checks on the pump outlet may be required if lack of fuel pressure is suspected.

Pipes

Where there is relative movement between the engine and the chassis it is normal practice to use flexible pipes

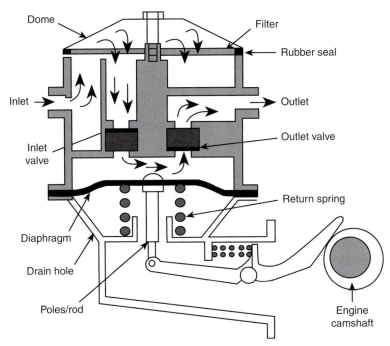

Fig. 7.5 Mechanical fuel pump

that can absorb the movement. Plastic materials such as Teflon may be used for this purpose. The other pipes that transfer the fuel along the vehicle are normally made from steel tube that is coated with tin, or a plastic material, to prevent corrosion. The pipes are secured to the vehicle frame by means of clips that are suitably placed so as to prevent:

- vibration that may cause them to fracture;
- contact with high-temperature parts such as the exhaust system;
- contact with moving parts of the suspension or other parts of the vehicle system.

Filters

Filters are required to prevent unwanted materials from causing blockages and damage to the pump and injectors. The primary filter in the fuel tank is normally made from a fine wire or plastic mesh. The filter on the pressure side may be made from fine nylon mesh, or other material, contained in a cylindrical capsule.

Carbon canister

Petrol is volatile, producing a vapour of hydrocarbons which can have several effects, including the following:

- Because the volume of fuel changes as the tank empties and is refilled, the pressure in the tank will vary. If the pressure is low the tank may implode, if the pressure is too high it may burst. To prevent this occurring the evaporative purge control system contains a valve that maintains the fuel tank pressure at a constant level.
- Hydrocarbon vapour must not be allowed to escape into the atmosphere because it is harmful to human health and the environment.

The carbon canister (Fig. 7.6) is the device through which the fuel tank is vented.

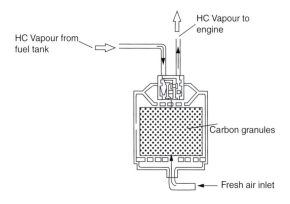

Fig. 7.6 A carbon canister

The carbon canister contains granulated carbon (charcoal) that acts like a sponge to absorb the hydrocarbon vapour when the engine is not running. When the engine is running and conditions are suitable, the hydrocarbon vapour is drawn into the engine to be burned along with the fuel — the process of taking the vapour into the engine is called purging the canister. On vehicles that have engine management the purging is controlled by the engine computer; on other engines the purging is controlled by a valve that responds to manifold vacuum.

Inertia switch

The inertia switch is a safety device that switches off the fuel pump to prevent the escape of fuel. It is activated by the deceleration of the vehicle when it is involved in a collision. The switch is equipped with a reset button to enable resumption of pumping when the situation warrants it.

Working with petrol

Petrol is highly flammable and proper precautions must be taken when working with it. The following information, taken from the Health and Safety Executive (HSE) book entitled *Health and Safety in Motor Vehicle Repair and Associated Industries*, is a good general guide. When working on vehicles some special processes may be necessary and the manufacturer's advice should be followed in addition to those recommended by the HSE.

Safe use of petrol

Petrol forms a highly flammable vapour that is easily ignited. Even low-voltage inspection lamps (if damaged) and static electricity (e.g. generated when petrol flows through pipework and into containers) can cause ignition. Vapour will arise from evaporation of spillages or from a petrol container as it is filled. Small leaks or spills of petrol can escalate into a major incident with fatal or major injuries and extensive property damage (1 litre of spilt petrol can produce up to 15 000 litres of a flammable gas mixture).

Most petrol spillages and fires directly associated with vehicle fuel systems occur during fuel draining operations rather than repair work on the fuel system.

Emergency exits should be well signposted and clear of obstructions.

Removal of petrol from all or part of the fuel system is often required to carry out repairs on it or possibly other mechanical or bodywork repairs. The risk also arises during the removal of contaminated fuel from tanks following misfuelling, e.g. where a diesel vehicle has been filled with petrol and vice versa. It is estimated that misfuelling happens over 120 000 times a year, often requiring fuel draining and replacement. This may be carried out off-site using specially adapted fuel-recovery vehicles. Alternatively, vehicles may be transported to fixed sites for repair.

Fuel retrievers

Using a proprietary fuel retriever correctly will eliminate spillage, minimize escaping petrol vapour, and provide a suitable and stable container in which to collect petrol. The petrol can then be transferred back to the vehicle or to another suitable container (e.g. a United Nations approved container for the carriage of petrol) for disposal or storage. Proprietary fuel retrievers use earthing straps to eliminate dangerous static discharge and some have vapour recovery pipework (see Fig. 7.7).

Fuel retrievers can and should be used in most circumstances when transferring petrol:

- Always work outdoors or in well-ventilated areas, well away from pits or other openings in the ground.
- Follow the manufacturer's instructions, paying particular attention to the correct use of vapour recovery pipework.
- Ensure the vehicle chassis and the retriever are both earthed.
- Ensure that other sources of ignition have been excluded (including disconnecting the vehicle battery).
- Alert other people that fuel draining is taking place, e.g. by using signs.

Where rollover or anti-siphon devices are fitted, use the appropriate fuel line connector, following the supplier's and vehicle manufacturer's advice. In exceptional circumstances it may be necessary to gain access

Fig. 7.7 Fuel retriever equipment (HSE). Note metal collection tank and earthing straps

by removing the tank sender unit. If this is located on the side of the tank, do not try to remove it, unless you are sure the fuel level is below the level of the sender unit.

Draining without a fuel retriever

Draining petrol without a retriever is particularly hazardous — only do this as a last resort. A hand-operated siphon or independent manual pump (i.e. not electrically operated) may be acceptable, provided that transfer pipework is securely positioned at both ends and the vehicle chassis and container are grounded by earthing straps. Following the precautions given for fuel retrievers, petrol should be drained into a suitable metal container large enough to hold the contents of the fuel tank, which can be securely closed after use. The container should have the appropriate hazard labels to show its contents and be stable or held within a stable framework so it cannot be easily knocked over.

Other precautions

- Only a competent person who has been shown how to use the equipment and understands the hazards of the operation should carry out fuel removal. Keep a foam or dry-powder fire extinguisher nearby and ensure operators are trained in its use.
- Do not dispose of unwanted petrol by adding it to the waste oil tank, or by burning it — any contaminated petrol or petrol/diesel mixtures should be consigned to waste, giving a clear description of the nature of the material.
- Do not store drained or contaminated fuel in the workplace unless it is to be returned to the vehicle immediately. Store fuel retrievers or containers in a clearly designated area in the workshop, from which ignition sources and combustible materials are excluded. It should not be kept on an escape route and a suitable fire extinguisher should be close at hand. At the end of the working day, move the fuel retriever or petrol container to a petrol or flammable liquids store.
- Soak up any spills immediately, using absorbent granules or similar material, and dispose of the material safely.

Maintenance and repair

EVAP system

Because the EVAP (evaporative purge system) system is related to emissions, its operation is constantly monitored by the ECU (electronic control unit). Any fault

Table 7.1   Standard diagnostic codes

OBD fault code	Brief description of fault
PO 441	EVAP purge valve leak
PO 443	Purge solenoid electrical fault
PO 464	Fuel level sensor circuit – intermittent electrical defect
PO 456	Leak in EVAP system
PO 462	Fuel level sender voltage low
PO 455	Large leak in vacuum system

will cause the engine malfunction indicator lamp to be illuminated; at the same time, a fault code will be flagged up in the ECU memory. This code can then be read out, by a code reader, through the diagnostic port. The selection of fault codes in Table 7.1 gives an indication of the extent of the fault detection provided.

In common with other fault codes, they point to an area in which a fault lies. Full diagnosis requires further tests, some of which are outlined in the following questions.

Self-assessment questions

1. By referring to a modern workshop manual, study how an EVAP system can be tested. When you are satisfied that you understand the procedure, discuss it with fellow students to see how it compares with the tests on the system that they studied.

2. What is the function of the item marked 5 in the fuel pump circuit shown in Fig. 7.8? Where is this switch normally placed on a vehicle? How can this switch be reset if it is accidentally brought into operation?

3. Figure 7.9 shows how a pressure gauge is attached to a petrol fuel system in order to test the fuel-line pressure. Describe the safety precautions that must be observed when releasing the line pressure prior to inserting the test gauge into the system.

4. Figure 7.10 shows some detail of a petrol gauge and its sender unit. Describe how the level of fuel in the tank affects the bimetallic strip so that the gauge needle shows the fuel level.

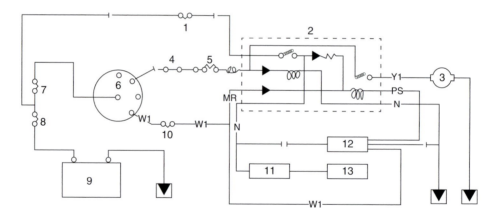

1 - Fuse   2 - Main relay   3 - Fuel pump   4 - Fuse   5 - Inertia switch

Fig. 7.8   Fuel pump circuit

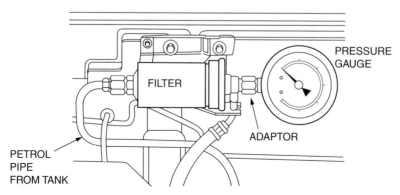

PRESSURE GAUGE

FILTER

ADAPTOR

PETROL PIPE FROM TANK

Fig. 7.9   Pressure test on fuel system

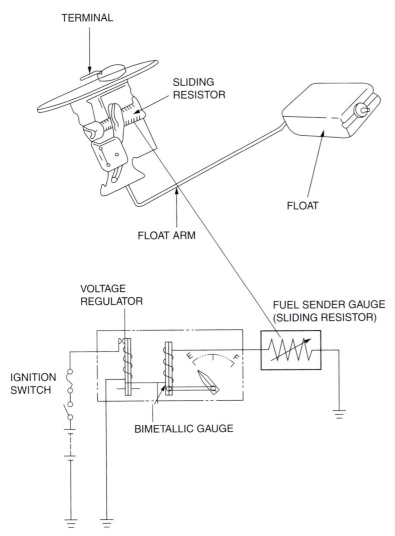

TERMINAL

SLIDING
RESISTOR

FLOAT

FLOAT ARM

VOLTAGE
REGULATOR

FUEL SENDER GAUGE
(SLIDING RESISTOR)

IGNITION
SWITCH

BIMETALLIC GAUGE

Fig. 7.10 Petrol gauge and sender unit

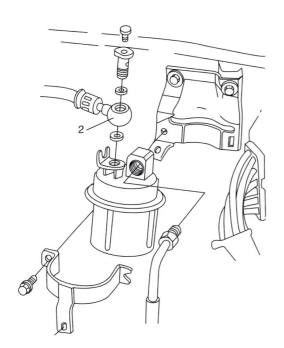

2

Fig. 7.11 Petrol filter

5. Figure 7.11 shows some detail of an in-line petrol filter. What type of pipe connection is shown at position 2 on the diagram? What type of washer is placed on each side of this connector to ensure that there are no leaks? Check a service schedule that you are familiar with and write down the recommended mileage at which the fuel filter should be renewed.
6. Name some of the positions in the fuel system where flexible pipes are used. What type of material is used for flexible pipes? How are flexible pipes secured to the metal pipes?
7. The fuel tank of a petrol engine vehicle is accidentally topped up with diesel fuel. Describe the procedure for dealing with this incident. How should the contaminated fuel be disposed of?

8
Motor fuels

Topics covered in this chapter

Calorific value
Combustion
Products of combustion and emissions
Hydrocarbon fuels
Alternative sources of energy
Catalysts and exhaust after treatment
Exhaust gas re-circulation
Zero emissions vehicles

Most motor fuels and lubricating oils are derived from crude oil that comes from oil wells. Crude oil is a finite resource and various estimates exist about the probable number of years that supplies will last. This, coupled with concern about atmospheric pollution, is a major driving force behind many of the technological changes in vehicle design. Changes in vehicle design and the ever increasing demands from legislators to conserve oil supplies and reduce emissions are challenges that automotive service technicians are expected to meet on a daily basis.

Fuels

Calorific value

The calorific value of a fuel is the amount of energy that is released by the combustion of 1 kg of the fuel. Two figures for the calorific value of hydrocarbon fuels are normally quoted: these are the higher or gross calorific value, and the lower calorific value. These two values arise from the steam that arises from the combustion of hydrogen. The higher calorific value includes the heat of the steam, whereas the lower calorific value assumes that the heat of the steam is not available to do useful work. The calorific value that is quoted for motor fuels is that which is used in calculations associated with engine and vehicle performance. For example, the calorific value of petrol is approximately 44 MJ/kg. The approximate values of the properties of other fuels are given in Table 8.1.

Table 8.1 Approximate values of fuel properties

Property	LPG	Petrol	Diesel oil	Methanol
Relative density	0.55	0.84	0.84	0.8
Usable calorific value (MJ/kg)	48	44	42.5	20
Octane rating	100	90	–	105
Chemical composition	82% C, 18% H_2	85% C, 15% H_2	88% C, 11.5% H_2, 0.5% S	38% C, 12% H_2, 50% O_2
Air–fuel ratio	15.5	14.8	14.9	6.5
Mass of CO_2 per kg of fuel used (kg/kg)	3.04	3.2	3.2	1.4

Combustion

Fuels for motor vehicles, such as petrol and diesel fuel, are principally hydrocarbons. That is, they consist largely of the two elements hydrogen and carbon. The proportions of these elements in the fuel vary, but a reasonably accurate average figure is that motor fuels such as petrol and diesel fuel are approximately 85% carbon and 15% hydrogen.

Products of combustion

When considering products of combustion it is useful to take account of some simple chemistry relating to the combustion equations for carbon and hydrogen. An adequate supply of oxygen is required to ensure complete combustion and this is obtained from the atmospheric air.

The relevant combustion equations are:

$$\text{For carbon, } C + O_2 = CO_2.$$

Because of the relative molecular masses of oxygen and carbon, this may be interpreted as: 1 kg of carbon requires 2.68 kg of oxygen and produces 3.68 kg of carbon dioxide when combustion is complete.

$$\text{For hydrogen, } 2H_2 + O_2 = 2H_2O.$$

Again, because of the relative molecular masses of H and O_2, this may be interpreted as: 1 kg of hydrogen requires 8 kg of oxygen and produces 9 kg of H_2O when combustion is complete.

Air—fuel ratio

Petrol has an approximate composition of 15% hydrogen and 85% carbon. The oxygen for combustion is contained in the air supply and approximately 15 kg of air contains the amount of oxygen that will ensure complete combustion of 1 kg of petrol. This means that the air—fuel ratio for complete combustion of petrol is approximately 15:1; a more precise figure is 14.8:1.

Petrol engine combustion

Combustion in spark ignition engines such as the petrol engine is initiated by the spark at the sparking plug and the burning process is aided by factors such as combustion chamber design, temperature in the cylinder, mixture strength, etc.

Because petrol is volatile, each element of the fuel is readily supplied with sufficient oxygen from the induced air to ensure complete combustion when the spark occurs. Petrol engine combustion chambers are designed so that the combustion that is initiated by the spark at the sparking plug is able to spread uniformly throughout the combustion chamber.

For normal operation of a petrol engine, a range of mixture strengths (air—fuel ratios) are required from slightly weak mixtures — say, 20 parts of air to 1 part of petrol for economy cruising, to 10 parts of air to 1 part of petrol for cold starting. During normal motoring, a variety of mixture strengths within this range will occur — for example, acceleration requires approximately 12 parts of air to 1 part of petrol. These varying conditions plus other factors such as atmospheric conditions that affect engine performance lead to variations in combustion efficiency and undesirable combustion products known as exhaust emissions are produced. Exhaust emissions and engine performance are affected by conditions in the combustion chamber. Two effects that are associated with combustion in petrol engines are (1) detonation and (2) pre-ignition.

Detonation

Detonation is characterized by a knocking and loss of engine performance. The knocking arises after the spark has occurred and it is caused by regions of high pressure that arise when the flame spread throughout the charge in the cylinder is uneven. Uneven flame spread leads to pockets of high pressure and temperature that cause elements of the charge to burn more rapidly than the main body of the charge. Detonation is influenced by engine design factors such as turbulence, heat flow, combustion chamber shape, etc. Quality of the fuel, including octane rating, also has an effect. Detonation

may lead to increased emissions of CO and NO_x (oxides of nitrogen), and hydrocarbons result.

Pre-ignition

Pre-ignition is characterized by a high pitched 'pinking' sound that is emitted when combustion prior to the spark occurs and it is caused by regions of high temperature. These regions of high temperature may be caused by sparking plug electrodes overheating, sharp or rough edges in the combustion region, carbon deposits, and other factors. In addition to loss of power and mechanical damage that may be caused by the high pressures generated by pre-ignition, combustion may be affected and this will cause harmful exhaust emissions.

Octane rating

The octane rating of a fuel is a measure of the fuel's resistance to knock. A high octane number indicates a high knock resistance. Octane ratings are determined by standard tests in a single-cylinder, variable compression ratio engine. The research octane number (RON) of a fuel is determined by running the test engine at a steady 600 rpm while the compression ratio is increased until knock occurs. The motor octane number (MON) is determined by a similar test, but the engine is operated at higher speed. The RON is usually higher than the MON and fuel suppliers often quote the RON on their fuel pumps. An alternative rating that is sometimes used gives a figure which is the average of RON and MON.

Factors affecting exhaust emissions

During normal operation, the engine of a road vehicle is required to operate in a number of quite different modes as follows:

- Idling — slow running
- Coasting
- Deceleration — overrun braking
- Acceleration
- Maximum power.

These various modes of operation give rise to variations in pressure, temperature, and mixture strength in the engine cylinder with the result that exhaust pollutants are produced.

Hydrocarbons (HC)

These appear in the exhaust as a gas arising from incomplete combustion due to a lack of oxygen. The answer to this might seem to be to increase the amount of oxygen to weaken the mixture. However, weakening the mixture gives rise to slow burning; combustion will be incomplete as the exhaust valve opens and unburnt HC will appear in the exhaust gases.

Idling

When the engine is idling the quantity of fuel present is small. Some dilution of the charge occurs because, owing to valve overlap and low engine speed, scavenging is poor.

The temperature in the cylinders tends to be lower during idling and this leads to poor vaporization of the fuel and HC in the exhaust.

Coasting, overrun braking

Under these conditions the throttle valve is normally closed. The result is that no, or very little, air is drawn into the cylinders. Fuel may be drawn in from the idling system. The closed throttle leads to low compression pressure and very little air; the shortage of oxygen arising from these conditions causes incomplete combustion of any fuel that enters the cylinders and this results in HC gas in the exhaust.

Acceleration

Examination of torque V-specific fuel consumption for the spark ignition engine reveals that maximum torque occurs when the specific fuel consumption is high. Because the best acceleration is likely to occur at the maximum engine torque, a richer mixture is required in order to produce satisfactory acceleration; fuelling systems provide this temporary enrichment of appropriate increases in mixture strength to meet demands placed on the engine. This leads to a temporary increase in emissions of HC and CO. As the engine speed rises,

combustion speed and temperature increase and this gives rise to increased amounts of NO_x.

High-speed, heavy-load running

Here the engine will be operating at or near maximum power. Examination of the power versus specific fuel consumption shows that maximum power is produced at higher specific fuel consumption figures and richer mixtures. Increased emissions of CO and HC are likely to result.

Cruising speed − light engine load

Under these conditions, where the engine is probably operating in the low specific fuel consumption speed, the mixture strength is likely to be around 10:1 or higher to provide good fuel economy. Other emissions are lower under these conditions.

European emissions standards EURO 4

Emission limits are given in Table 8.2. In addition to these standards, modern vehicles must be equipped with on-board diagnostics (OBD) that notify the driver of the vehicle that a system malfunction is causing the emissions limits to be exceeded. Figure 8.1 shows how the exhaust emissions are affected by air−fuel ratio.

Emissions and their causes

Oxides of nitrogen (NO_x)

Oxides of nitrogen are formed when combustion temperatures rise above 1800 K.

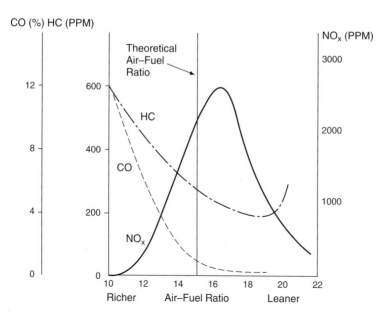

Fig. 8.1 Relationship between air−fuel ratio and exhaust emissions

Table 8.2 Emission limits (g/km)

	CO	HC	HC + NO$_x$	NO$_x$	PM
Petrol	1.0	0.10	—	0.08	—
Diesel	0.50	—	0.30	0.25	0.025

Hydrocarbons (HC)

Unburnt hydrocarbons arise from:

- unburnt fuel remaining near the cylinder walls after incomplete combustion being removed during the exhaust stroke;
- incomplete combustion due to incorrect mixture strength.

Carbon monoxide (CO)

This is produced by incomplete combustion arising from lack of oxygen.

Sulphur dioxide (SO$_2$)

Some diesel fuels contain small amounts of sulphur, which combines with oxygen during combustion. This leads to the production of sulphur dioxide that can, under certain conditions, combine with steam to produce H_2SO_3, which is a corrosive substance.

Particulate matter (PM)

The bulk of particulate matter is soot, which arises from incomplete combustion of carbon. Other particulates arise from lubricating oil on cylinder walls and metallic substances from engine wear. Figure 8.2 gives an indication of the composition of particulate matter.

Carbon dioxide (CO$_2$)

Whilst CO_2 is not treated as a harmful emission, it is thought to be a major contributor to the greenhouse effect and efforts are constantly being made to reduce the amount of CO_2 that is produced. In the UK, the amount of CO_2 that a vehicle produces in a standard test appears in the vehicle specification so that it is possible to make comparisons between vehicles on this score. The figure is presented in grams per kilometre — for example, a small economy vehicle may have a CO_2 figure of 145 g/km and a large saloon car a figure of 240 g/km. Differential car tax rates are applied to provide incentives to users of vehicles that produce smaller amounts of CO_2.

Methods of controlling exhaust emissions

Two approaches to dealing with exhaust emissions suggest themselves:

1. **Design measures to prevent the harmful emissions being produced.** In diesel engines, such measures include improved fuel quality, fuel injection in a number of electronically controlled stages, better atomization of fuel as occurs at higher injection pressures such as those used in common rail fuel injection systems, and exhaust gas recirculation.

   In petrol engines the approach is somewhat similar; lean burn engines employing direct petrol injection are a development that has become reasonably well established. Exhaust gas recirculation is widely used, as is variable valve timing to overcome the combustion problems caused by valve overlap at idling speeds.
2. **Post combustion – exhaust emission control.** In diesel engines this consists of catalytic convertors to oxidize CO and HC, reduction catalysts to reduce NO$_x$, and particulate filters to remove particulate matter.

   Petrol engines utilize three-way catalysts and computer-controlled fuelling systems. Other technologies such as evaporative emissions control to eliminate HC emissions escaping from the fuel system are also used.

Exhaust gas recirculation (EGR)

The air that is used to provide the oxygen for combustion contains approximately 88% nitrogen by mass. When nitrogen is heated above approximately 1800 K (1528°C), in the presence of oxygen, oxides of nitrogen (NO$_x$) are formed. These conditions occur in the combustion chamber when excess oxygen is present, as happens at an air–fuel ratio of approximately 10:1. If the combustion chamber temperature is kept below 1800 K the conditions for the creation of NO$_x$ no longer exist. Exhaust gas recirculation is a method that is used

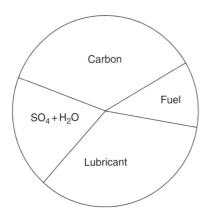

Fig. 8.2 Approximate composition of PM

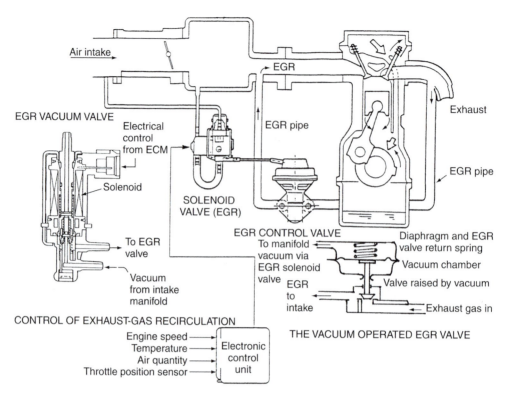

Fig. 8.3 Exhaust gas recirculation system

to keep combustion chamber temperatures below the critical figure. A proportion of exhaust gas is redirected from the exhaust system to the induction system by means of electronically controlled valves. Figure 8.3 shows the layout of an engine management system that incorporates exhaust gas recirculation.

Catalysts

- **Platinum.** Catalysts are materials that assist chemical reactions but are not themselves changed in the process. At the correct temperature, in excess of $300\,°C$, platinum is a catalyst that aids the conversion of CO to CO_2, and HC to H_2O and CO_2. In order for the exhaust system catalytic converter to function correctly, the air–fuel ratio is maintained close to the chemically correct (stoichiometric) air–fuel ratio by means of exhaust gas sensors and electronic control.
- **Rhodium.** Rhodium is a catalyst that reduces NO_x to N_2.

These metals are added to a ceramic honeycomb structure that exposes the exhaust gases to the maximum area of the catalysing material. Figure 8.4 shows the layout of an emission control system for a petrol engine vehicle.

Selective reduction catalysts

The power output of diesel engines is controlled by the quantity of fuel that is injected and the engines operate with excess air over much of the operating range. Excess air and high combustion temperatures give rise to NO_x.

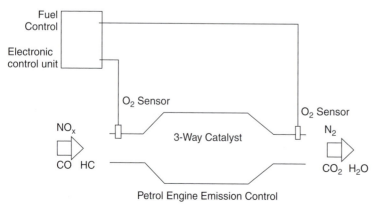

Fig. 8.4 A three-way catalyst emission control system

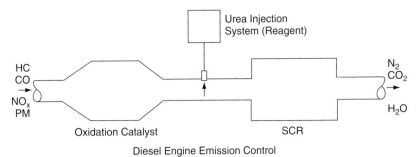
Diesel Engine Emission Control

Fig. 8.5 SCR system

The selective catalyst reduction system (SCR) shown in Fig. 8.5 is a method that is used in heavy vehicles to reduce NO_x emissions.

Diesel particulate filters (DPFs)

Particulate matter that arises from combustion in diesel engines consists mainly of particles of carbon with some absorbed hydrocarbons. Filtration of the exhaust products is a technique that is widely used for the removal of particulate matter from the exhaust gases before they are passed into the atmosphere. Various forms of filtering medium are used to trap the PM and, in common with most filtering processes, cleaning of the filter to avoid blockage is required. The cleaning process in DPFs, which consists primarily of controlled burning to convert the carbon (soot) into carbon dioxide, is known as regeneration. Among the methods of regeneration are:

- **Passive systems.** The heat in the exhaust gas, acting with the materials used in the construction of the particulate filter, produces sufficient temperature to remove the filtered deposits. External sources of heat are not required.
- **Active systems.** Heat is supplied either by injecting fuel into the exhaust stream, or by secondary injection of fuel in the engine. This produces the temperature that is required to burn off the particulate matter that accumulates in the filter.

Bio-fuels

It is now (2011) generally accepted that the world's oil resources are finite and that they are being depleted at a rapid rate. Attention is concentrating on alternative fuels and methods of propulsion for motor vehicles. Alcohols such as methyl alcohol, or methanol as it is commonly known, is produced from vegetable matter. Bio-fuels are said to be environmentally valuable because their products of combustion are, very roughly, water (H_2O) and CO_2. It is argued that the CO_2 from the combustion of these fuels is consumed by the vegetation that is producing the crop that will make the next supply of fuel – this process is referred to as a 'closed carbon cycle'.

Methanol is not strictly a hydrocarbon because it contains some oxygen. The calorific value of methanol is approximately 26 MJ/kg. Methanol has a higher latent heat value than petrol and it has higher resistance to detonation. Whilst the higher latent heat value and relatively high ignition temperature of methanol indicate that higher compression ratios can be used, there is a disadvantage, which is that vaporization at low temperatures is poor and this can lead to poor cold starting ability.

Liquified petroleum gas (LPG)

Petroleum gases such as butane and propane are produced when oil is refined to produce liquid fuels such as derv (diesel) and petrol. Once liquified and stored under pressure, LPG will remain liquid until is exposed to atmospheric pressure. The chemical composition of propane is C_3H_8 and butane is C_4H_{10}; the relative densities at $15°C$ are approximately 0.5 and 0.58 respectively, while the calorific value is slightly higher than petrol at approximately 46 MJ/kg. A considerable industry exists to support the conversion of road fuelling from petrol and derv to LPG.

In some countries, including the UK, favourable tax systems support the use of LPG.

Hydrogen

Compressed hydrogen may be stored on a vehicle and used in an internal combustion engine. Among the advantages claimed for this system are (a) no carbon dioxide emissions and (b) products of combustion that are primarily water. The main future use of hydrogen as a propellant for vehicles is thought to be as a source of energy in fuel cells. The electricity produced in the fuel cell is used as a power source for the electric motor that replaces the internal combustion engine of the vehicle. A simple fuel cell is shown in Fig. 8.6. The fuel cell consists of two electrodes, an anode and a cathode, that are separated by an electrolyte. The hydrogen acts on the anode and oxygen from the

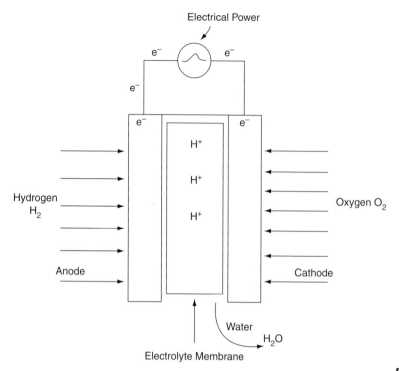

Fig. 8.6 A simple fuel cell

atmosphere acts on the cathode. The catalytic action of the anode causes the hydrogen atom to form a proton and an electron; the proton passes through the polymer electrolyte to the cathode and the electron passes through an external circuit to the cathode. This action provides an electric current in the external circuit. In the process the hydrogen and oxygen combine to produce water, which is the principal emission.

Zero emissions vehicles (ZEVs)

Operation of fuel cells for vehicle propulsion does not involve combustion in an engine and the normal products of combustion and associated pollutants are not produced. The main product of the electrochemical processes in the fuel cell is water; consequently, vehicles propelled by fuel cells and electric motors are known as zero emissions vehicles.

Self-assessment questions

1. Conduct some research to satisfy yourself about the effect on fuelling requirements that arise when considering the merits and demerits of converting a spark ignition engine to run on alcohol fuel. Pay particular attention to the calorific value and general effect on engine performance.

2. A motorist who bought a new car from your garage complains that it is using more petrol than he had anticipated. In the first 2000 miles of use the car has used 59 gallons of petrol. The advertisement for the car gave the fuel consumption as 7.5 litres per 100 km.
   (a) Calculate the car's petrol consumption in mpg for the first 2000 miles.
   (b) Convert the mpg figure to litres/100 km.
   (c) Make a list of the factors that you would discuss with the customer that might help towards an understanding of the difference between the advertised figure for petrol consumption and the actual figure.

3. A large commercial vehicle returns an average fuel consumption of 8 mpg and covers 50 000 miles in a year.
   (a) How many gallons of fuel are used in the year?
   (b) Calculate the annual cost of fuel at £1.20 per litre.

4. The vehicle in question 3 is fitted with a roof deflector that improves fuel consumption to 8.15 mpg. Calculate the annual saving in fuel cost that is gained by the use of the deflector.

9
Carburettors

Topics covered in this chapter

Operating principles of carburettors
Fixed and variable choke

For a large part of the twentieth century the majority of light vehicles used petrol engines that were equipped with a carburettor. Early carburettors were very simple devices but, as the demand for improved performance grew, carburettors became quite complicated. From about 1970 onwards, concerns about fuel reserves and atmospheric pollution grew and fuel injection was seen to offer partial solutions to some of the problems and the use of carburettors declined. However, because many vehicles that are still in use in 2010 are fitted with a carburettor it is thought to be sensible to include a section on carburation.

The simple carburettor

The carburettor is connected to the intake port of the engine. The action of the piston on the induction stroke draws air through the venturi and the open throttle valve. The throttle valve is operated by the accelerator pedal by means of a cable or rod, and the amount of throttle valve opening determines the amount of air and fuel entering the engine and thus the power output (Fig. 9.1).

The venturi is normally called the choke tube and its function is to create a region of low pressure at the narrowest part, which is called its throat.

The jet is placed at the throat of the choke tube and the petrol from the float chamber is held at a level slightly below the jet outlet – this level is controlled by the needle valve and float. When the engine is not running, the pressure at the jet outlet is the same as the pressure on the fuel in the float chamber.

When the engine is running, the pressure at the jet outlet in the choke tube is lower than the pressure in the float chamber and this pressure difference causes petrol to flow out of the jet into the air stream. The action of the air stream on the petrol causes it to vaporize so that the air and fuel are thoroughly mixed. The jet size and the diameter of the choke tube are selected so that the required air–fuel ratio of 14.7:1 is obtained.

Limitations of the simple carburettor

Figure 9.2 shows how, in the simple carburettor, the fuel flow increases at a greater rate than the air flow. At a single speed the air–fuel ratio is correct, but above this speed the fuel flow increases disproportionately; this causes the mixture of air and fuel to contain too much fuel. Below this single speed there is insufficient fuel in the mixture. The result is that the simple carburettor is only suitable for an engine operating at a constant speed and load. Because vehicle engines operate at widely varying speeds and loads, the simple

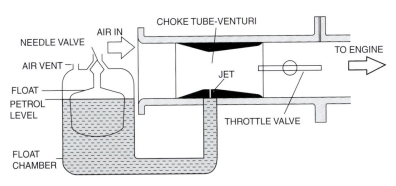

Fig. 9.1 A simple carburettor

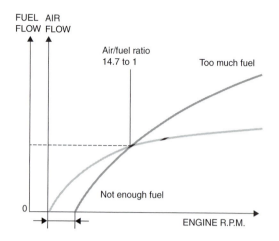

Fig. 9.2   Limitations of the simple carburettor

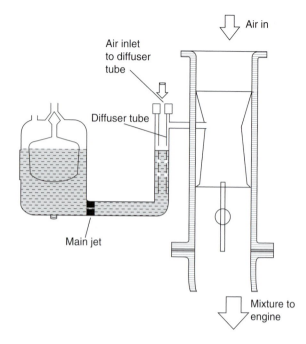

Fig. 9.3   Diffuser tube air bleed compensation

carburettor requires certain additions to make it suitable for use on a vehicle engine.

Rich or weak mixture

The term 'rich mixture' describes an air—fuel mixture that has an air—fuel ratio smaller than 14.7:1. A 'weak mixture' is one in which the air—fuel ratio is greater than 14.7:1.

Keeping the mixture strength constant

Carburettors may be divided into two categories:

1. Constant choke or variable vacuum
2. Constant vacuum, variable choke.

Constant-choke carburettor

In the constant-choke carburettor, the air—fuel ratio can be kept near to 14.7:1 by a process known as air bleed compensation. There are two types of air bleed systems:

1. Diffuser tube
2. Compensating jet and well.

Diffuser tube

The diffuser tube (Fig. 9.3) is a small cylinder with a series of holes down its length and it is situated in a tube that contains petrol up to the level in the float chamber. When the engine is not running the holes in the tube are covered by the petrol. When the engine is started, petrol is drawn into the venturi via the main

jet — as the engine speed increases, the petrol is drawn off faster than it can flow through the main jet and this causes the fuel level to fall in the tube surrounding the diffuser tube. As the fuel level in the tube falls, holes in the diffuser are uncovered and air enters to mix with the fuel; this also reduces the 'suction' on the main jet. By a suitable matching of main jet size and diffuser tube dimensions the air—fuel ratio can be held at 14.7:1 for a range of engine speeds.

Compensating jet and well

This system uses two jets: a main jet and a compensating jet. The main jet connects to the choke tube and is not affected by the air bleed system. The compensating jet feeds into a fuel passage that is open to atmospheric pressure through a tube called a well (Fig. 9.4).

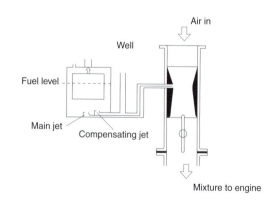

Fig. 9.4   Compensating jet and well

When the engine first starts the well is filled up to the level of the float chamber, through the compensating jet. As the engine speed increases the fuel is drawn down in the well — this allows air to enter with the fuel and reduces the flow through the compensating jet. The air entering in this way plus the reduction in flow caused by the pressure difference across the compensating jet maintains the air–fuel ratio at the required 14.7:1, or thereabouts.

Cold starts, idling, progression, and acceleration

In normal use, motor vehicles are required to operate under a range of conditions and the simple carburettor has been developed to cope with them.

Cold starts

When the engine is cold the fuel in the mixture condenses on the manifold and other parts so that the mixture entering the combustion chamber is too weak to ignite. To overcome this problem the simple carburettor is equipped with a choke, or strangler of the type shown in Fig. 9.5. The purpose of this choke is to temporarily restrict the air flow to the engine so that a richer mixture is provided to overcome the problem of condensation. The spindle of the choke flap is placed off-centre so that the flow of air causes an imbalance to open the choke to prevent the engine from being flooded with fuel once it has started.

Idling

When the engine is idling the throttle valve is almost closed and an orifice placed in the carburettor, just below the throttle valve, allows the requisite amount of mixture to flow to the engine. On engines with fixed valve timing there tends to be an amount of burnt gas in the combustion chamber that dilutes the charge of air and fuel. To overcome this problem the mixture supplied for idling is normally quite rich and this causes problems with the emissions. However, this problem is minimized by making the engine idle at a fairly high speed. In the simple carburettor shown in Fig. 9.6 the idle mixture port is connected to a tube that contains the idling jet and an air jet. These two combined determine the air–fuel ratio of the idle mixture. The idle control screw controls the quantity of mixture that flows to the engine.

Progression

As the engine speed is increased from idling by opening the throttle, the 'suction' over the idling

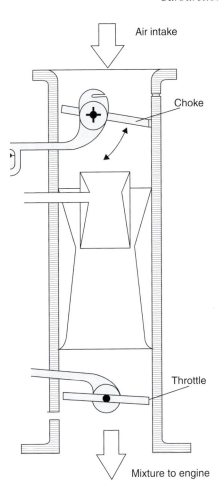

Fig. 9.5 The cold start choke valve

port decreases the mixture flow. To make the transition from idling to higher speed, a progression hole is situated slightly above the throttle valve and this allows extra idling mixture to enter to keep the engine running smoothly until the main jet becomes fully operational.

Acceleration

When the accelerator pedal is pressed promptly there is an inrush of air that raises the pressure in the venturi. This causes a weakening of the mixture that may cause the engine to hesitate. The accelerator pump that is fitted to most carburettors (Fig. 9.7) overcomes this problem and it is designed so that it only works when the throttle is opened rapidly.

Running on

Emission control regulations have caused engines fitted with carburettors to idle at a fairly high speed and to run on weak mixtures, which means that they operate at high temperatures. The combination of high idling speed and high operating temperature brings about

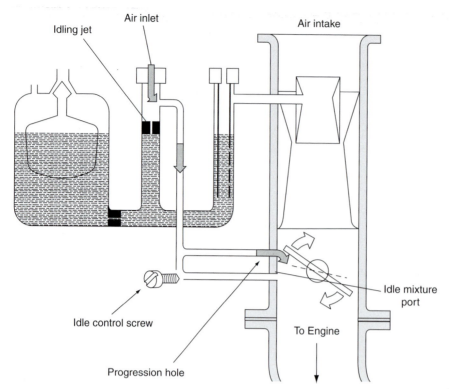

Fig. 9.6   Idling and progression

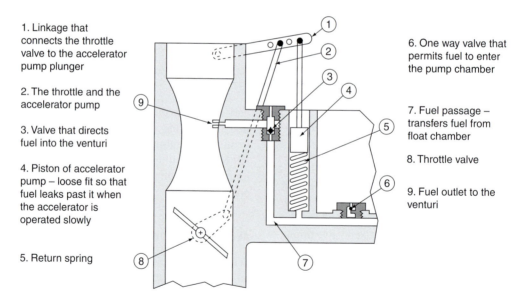

1. Linkage that connects the throttle valve to the accelerator pump plunger

2. The throttle and the accelerator pump

3. Valve that directs fuel into the venturi

4. Piston of accelerator pump – loose fit so that fuel leaks past it when the accelerator is operated slowly

5. Return spring

6. One way valve that permits fuel to enter the pump chamber

7. Fuel passage – transfers fuel from float chamber

8. Throttle valve

9. Fuel outlet to the venturi

Fig. 9.7   An accelerator pump

a tendency for engines to continue running on after the ignition is switched off. This happens because, even with the throttle closed, some air–fuel mixture can enter the combustion chambers, where it is ignited by hot parts of the engine. In order to prevent running on, carburettors are fitted with a solenoid-operated valve that cuts off the idling circuit of the carburettor, as shown in Fig. 9.8.

Constant-vacuum carburettor

In this type of carburettor the size of the venturi varies throughout the engine operating range and the depression (vacuum) over the jet remains constant. This overcomes the need for the compensating system of the fixed-choke carburettor. In addition, the petrol jet size is varied to suit the venturi size and this ensures that

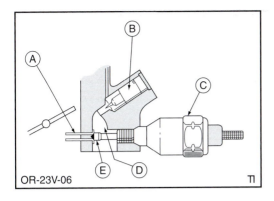

OR-23V-06   TI

Fig. 9.8 Anti-running on valve. A — Idle mixture port. B — Idle mixture control. C — Solenoid. D — Mixture space. E — Mixture cut-off valve

the air–fuel ratio remains accurate at all engine speeds. The carburettor shown in Fig. 9.9 is based on the SU carburettor that has been in use for many years. There are other types, but this design will suffice as an example to explain the principle of operation of a constant-vacuum carburettor.

As the throttle is opened air is drawn out of the dashpot through the depression transfer drilling in the piston. This lowers the pressure on the top of the piston and air at atmospheric pressure entering, through the atmospheric vent, causes the piston to rise against the return spring. As the piston rises it increases the effective area of the venturi opening and ensures that the

depression over the petrol (fuel) jet remains constant. Attached to the lowest part of the piston is a tapered rod called the needle. The needle is positioned centrally in the petrol jet and as it rises and falls with the piston it varies the size of the jet opening that determines the amount of fuel that passes into the venturi. The needles are designed to provide a range of air–fuel ratios to suit the various mixture strengths that are required for normal engine operation.

For cold starting purposes the petrol jet can be lowered away from the needle, by a simple lever system, to increase the petrol flow. The hydraulic damper serves to slow down the lifting of the piston under engine acceleration and causes a temporary slight mixture enrichment that helps to reduce hesitation as the engine speed increases.

Self-assessment questions

1. Why is the spindle of the choke valve offset to one side?
2. Which feature of the constant-vacuum carburettor changes the effective size of the choke tube?
3. Figure 9.10 shows a device that is used to prevent the engine 'running on' after the ignition is switched off. What is the position of the valve E when the engine is switched on?

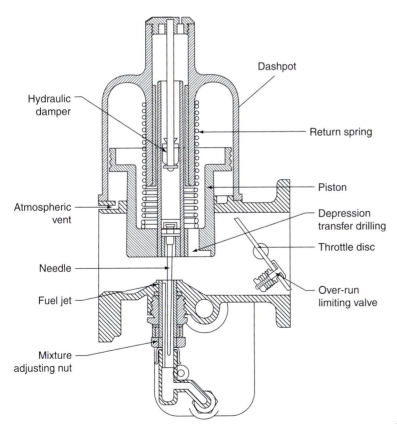

Fig. 9.9 Constant-vacuum carburettor

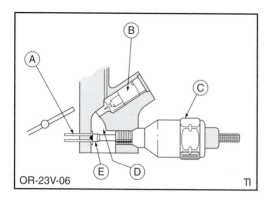

Fig. 9.10 Anti-running on valve

4. What device is incorporated into a fixed-choke carburettor to provide mixture enrichment for acceleration?
5. What effect on the exhaust gas CO content does the idle mixture control screw have? How does this system of idle mixture control vary from the one shown in Fig. 9.11?
6. What is meant by the term 'compensation' in relation to a carburettor? Describe how a carburettor overcomes the problem of the mixture strength becoming too rich as the engine speed increases.
7. What is a carburettor venturi? How does it create a region of low pressure?

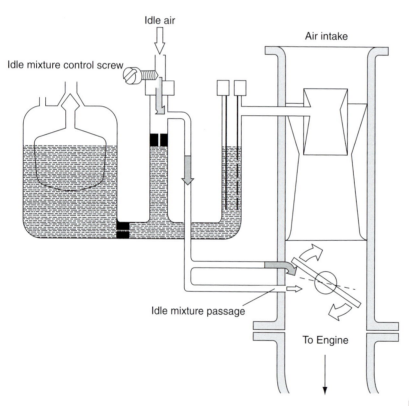

Fig. 9.11 Idle mixture control

10
Petrol injection — introduction

Topics covered in this chapter

Petrol injection systems
Single-point injection
Multi-point injection
Direct petrol injection
Fuel system maintenance and repair

Modern petrol injection systems are designed to ensure that the engine extracts the maximum possible amount of energy from each gram of fuel and produces the minimum possible amount of harmful emissions in the process. In order to achieve these objectives petrol fuelling systems are controlled electronically by a central computer that is called an ECU (electronic control unit).

A principal aim of the fuel and air system is to ensure that each small particle of fuel is surrounded by sufficient air to provide the oxygen that is required for complete combustion, so that the range of air–fuel ratios of approximately 10:1 to 20:1 is provided to meet all engine operating conditions. In petrol injection systems the mixing of air and fuel is achieved by spraying the fuel into the air stream as it enters the induction system, whereas in engines fitted with a carburettor the air and fuel are mixed in the venturi (choke tube) of the carburettor.

Petrol injection systems
The principle

A simple petrol injection system is shown in Fig. 10.1. The main components are as follows:

1. The air-flow meter produces an electrical signal, usually a voltage, that exactly represents the amount of air that is flowing to the engine. This signal is sent to the ECU.
2. The throttle valve controls the amount of air flowing and it also produces an electrical signal that is sent to the ECU.
3. The petrol pump supplies petrol at a constant pressure to the injector.
4. The fuel filter ensures that the petrol is clean when it reaches the injector.
5. The petrol injector produces a fine spray of petrol that is sprayed into the air stream to ensure that air and fuel are mixed.
6. The ECU is a computer that is programmed to hold the injector open for sufficient time to ensure that the amount of fuel injected corresponds to the air flow so that the correct air–fuel ratio is preserved. Part of the program in the computer is a fuelling map — the map is a section of program that tells the computer how long to hold the injector open because the amount of fuel injected is determined by this period of time.

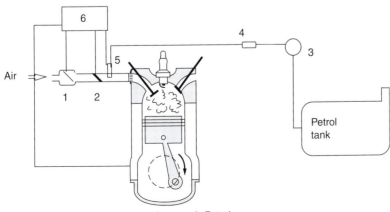

1. Air flow meter.　2. Throttle valve.　3. Petrol pump.
4. Fuel filter.　5. Petrol injector.　6. Electronic control unit.

Fig. 10.1　A simple petrol injection system

Continuous

In continuous injection systems the injection pulses occur on each revolution of the crankshaft, with half the required amount of fuel for each cylinder being delivered in two stages.

Sequential

In sequential petrol injection systems the injection pulses take place in the same sequence as the engine firing order, with the full amount of fuel being delivered on each induction stroke.

The catalyst and oxygen sensor

A key element in the closed-loop control system is the catalytic converter in the exhaust system (Fig. 10.7).

The function of the catalyst is to convert CO, NO_x and HC into CO_2, H_2O and N_2, and it can only do this if the air–fuel ratio λ is held between about 0.107 and 1.03. The oxygen sensor is the device that informs the ECU about the value of λ while the engine is running. The other devices in the petrol injection control system then act together with the ECU to ensure that the air–fuel ratio λ is kept within the limits so that the emissions control system is effective. The catalyst does not function at temperatures below approximately $400\,^{\circ}\text{C}$, which means that the engine only operates in closed-loop control when it is hot enough to do so.

The oxygen sensor

The voltaic oxygen sensor (Fig. 10.8) acts like a small battery. It consists of a pair of platinum electrodes that are separated by a layer of zirconium oxide. The platinum electrodes perform two functions:

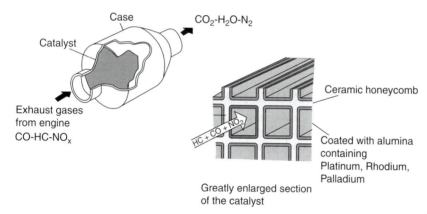

Fig. 10.7 Three-way catalytic converter

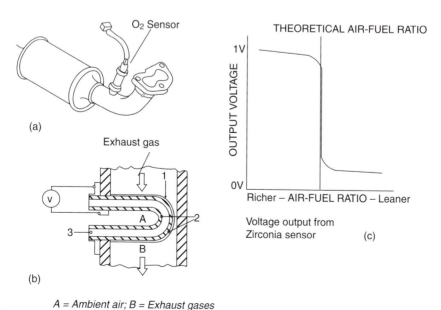

A = Ambient air; B = Exhaust gases

1. Protective ceramic coating
2. Electrodes
3. Zirconium oxide

Fig. 10.8 Voltaic-type oxygen sensor

1. They provide the electron-conducting part of the battery (sensor).
2. They catalyse the reaction between the oxygen in the atmosphere and the exhaust gas.

One side of the sensor is exposed to the atmosphere and the other side to the exhaust gas. Small differences between the oxygen content of the air and the exhaust gas cause the sensor to produce a voltage. The graph in Fig. 10.8(c) shows how there is a sudden change in sensor voltage when the air—fuel ratio is chemically correct. When the mixture is rich the sensor voltage is high — about 800 mV — and when the mixture is weak the sensor voltage is low. The switch from low sensor voltage to high sensor voltage between $\lambda = 0.107$ (rich mixture) and $\lambda = 1.03$ (weak mixture) occurs rapidly and provides the ideal characteristics for accurate control of mixture strength. The voltage from the oxygen sensor is fed back to the ECU, where it is used to adjust the air—fuel ratio.

Increasingly stringent anti-pollution legislation led to the introduction of European on-board diagnostics (EOBD); this system requires the ECU to constantly monitor systems and alert the driver through the engine warning lamp when a component fails to work correctly. Part of the monitoring system consists of a second exhaust gas oxygen sensor that is placed downstream from the catalyst (Fig. 10.9). This second sensor feeds back data to the ECU so that it can compare the readings from the first and second sensors to determine whether the catalytic converter is working correctly.

Determining the amount of fuel required

Because the engine is required to operate in a range of varying conditions, the ECU must be provided with the information that is necessary for it to perform its controlling function — this information is provided by a number of sensors, in addition to the exhaust gas oxygen sensor. The fuelling map shown in Fig. 10.2 shows that there are two variables that affect the air—fuel ratio:

1. Engine load as indicated by the throttle position sensor.
2. Engine speed as indicated by the speed sensor on the crankshaft.

In order to provide the correct air—fuel ratio the ECU must know how much air is entering the engine so that it can determine how much fuel to inject. The amount of air entering the engine is indicated by the air-flow sensor. Under certain conditions such as cold starts and acceleration, when extra rich mixtures are required, the engine operates under open-loop control — the engine coolant temperature sensor and the atmospheric air temperature sensor provide the ECU inputs that are required for these modes of operation.

The throttle position sensor

Throttle position is used to determine engine load and the sensor is normally based on the voltage divider principle (Fig. 10.10). A wiper arm is attached to the throttle valve spindle and as the throttle is opened or closed the wiper moves across a resistor. The resistor is supplied with a constant voltage of 5 volts and the voltage that is transmitted to the ECU is taken from the wiper arm — the position of the wiper arm between the ends of the resistor determines the voltage that is read by the ECU; it is this voltage that is an analogue of engine load. For example, when the wiper arm is halfway between the ends of the resistor the voltage sent to the ECU is 2.5 volts. In practice the throttle

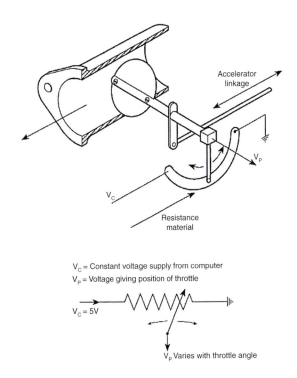

V_c = Constant voltage supply from computer
V_p = Voltage giving position of throttle

Fig. 10.10 Throttle position sensor

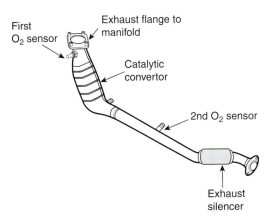

Fig. 10.9 The second exhaust oxygen sensor

position sensor is slightly more complicated because it will contain extra provision to indicate that the throttle is closed.

The air-flow sensor

The vane-type air-flow sensor (Fig. 10.11) is one of several different types of air-flow meters that are used for vehicle control systems. The flap is held in the closed position by a spring. As the air is drawn into the engine, the flap is moved inwards by the force of the incoming air stream; the greater the flow of air, the further the flap will move. The spindle on which the flap moves is attached to the wiper of a voltage divider and the control voltage that represents air flow is read from the wiper. It is this voltage that represents the amount of air entering the engine and it is used by the ECU to determine the amount of fuel to inject.

The engine speed and piston (crank) position sensor

Figure 10.12 shows an electromagnetic sensor that is used to measure engine speed and piston position. In the upper diagram there are two sensors: one generates electrical pulses as the flywheel rotates (these are used by the ECU to represent engine speed); the other generates an electrical pulse as the crank passes the top dead centre (TDC) position.

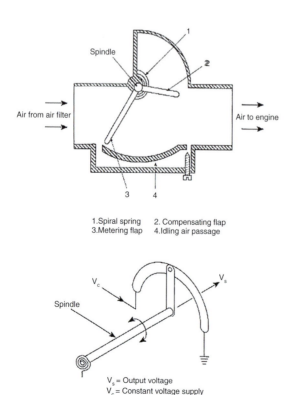

1. Spiral spring   2. Compensating flap
3. Metering flap   4. Idling air passage

V_s = Output voltage
V_c = Constant voltage supply
The voltage Vs represents air flow

Fig. 10.11   Air-flow sensor

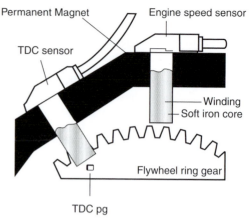

Details of engine speed and crank position sensors

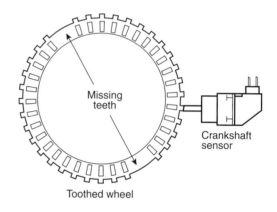

Fig. 10.12   Engine speed sensor

The coolant temperature sensor

The engine coolant temperature sensor makes use of the thermistor principle, which makes use of the fact that certain semiconductor materials produce less electrical resistance as the temperature rises, as shown in the graph in Fig. 10.13. The sensor is supplied with a controlled voltage and the current flowing though the sensing element varies as the temperature and resistance change. The voltage is an accurate representation of the coolant temperature and is used by the ECU to decide when a cold-start procedure is required and also to switch on the electric fan when conditions require it.

Cold start

As with carburettors the petrol injection system is required to provide an extra rich mixture for cold starts in order to overcome condensation on cold engine parts. Under cold-start conditions the emission control catalyst and oxygen sensor are not hot enough for closed-loop control so the system works in open-loop mode, where the extra fuel is injected and the engine speed is controlled by the ECU through the idle speed control (ISC). The ECU constantly monitors the temperature and as it rises

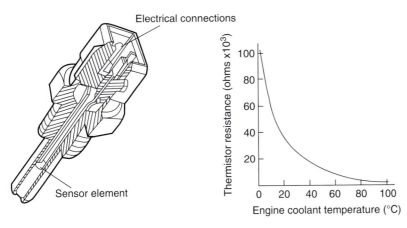

Fig. 10.13 A thermistor-type engine coolant temperature sensor

the air–fuel ratio is gradually weakened until the catalyst and oxygen sensor are operating normally; at this stage the ECU takes the system into closed-loop mode.

Idle speed control

When the engine is idling the throttle valve is virtually closed and the idle speed control valve bypasses the throttle valve to allow sufficient air to keep the engine running smoothly. The idle speed valve shown in Fig. 10.14 is operated by a small electric motor. The partial rotation of the armature of the motor is controlled by the ECU using a process called pulse width modulation. As the armature rotates the screw thread raises or lowers the air valve, under the control of the ECU, to ensure that the correct amount of air for idling reaches the combustion chambers.

A petrol injector

Petrol injectors are operated by an electromagnetic device called a solenoid. The ECU controls a transistor as shown in Fig. 10.15(b) that is used to switch the injector on and off as required. Because the fuel at the

injector is held at constant pressure the amount of fuel injected is determined by the length of time for which the injector is switched on. The nozzle of the injector is designed to ensure that the fuel is injected into the air stream as fine spray. The main parts of a petrol injector are shown in Fig. 10.15(a). The armature and nozzle valve are pulled inside the solenoid coil when the injector is switched on; when the current is switched off the armature and nozzle valve are returned to the closed position by the return spring.

Duty cycle

The maximum time for which the injector can be switched on is denoted by the distance C in Fig. 10.16. The actual time it is switched on is controlled by the ECU so that the amount of fuel injected is correct for the conditions at the time. The switched-on time is denoted by the distance A, which means that the injector is activated for 25% of the available time, i.e. the duty cycle is 25%. For extra fuel the on time is increased, giving an increase in duty cycle.

Exhaust gas recirculation (EGR)

Although the three-way catalyst reduces NO_x to nitrogen (N_2) and CO_2, it is made more effective if the amount of NO_x produced by combustion is reduced. High temperatures in the combustion chamber cause the production of NO_x and when calibrated amounts of exhaust gas are recirculated into the induction system the combustion chamber temperature is reduced, thus reducing the production of NO_x. The exhaust gas recirculation actuator is controlled by the ECU so that measured amounts of exhaust gas are sent back to the induction system when engine operating conditions are suitable — the actual amounts and the conditions under which recirculation takes place are determined by the engine designers.

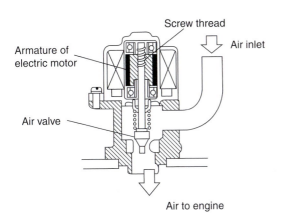

Fig. 10.14 Idle speed control valve (Robert Bosch)

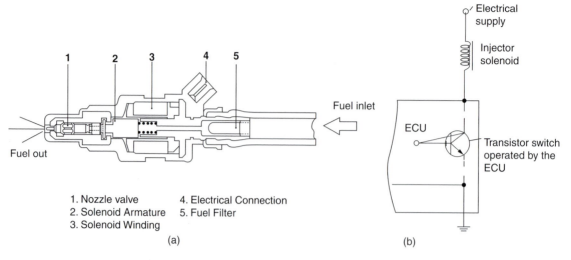

1. Nozzle valve   4. Electrical Connection
2. Solenoid Armature   5. Fuel Filter
3. Solenoid Winding

(a)   (b)

Fig. 10.15 A petrol injector

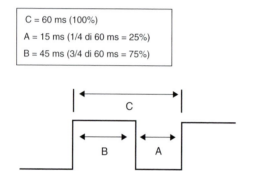

C = 60 ms (100%)
A = 15 ms (1/4 di 60 ms = 25%)
B = 45 ms (3/4 di 60 ms = 75%)

Fig. 10.16 Duty cycle

Direct petrol injection

As efforts to reduce exhaust emissions have intensified, the search for alternative technologies has increased. One such technology is direct petrol injection (see Fig. 10.17), where the petrol is injected directly into the engine cylinders. Two modes of injection and mixture formation are normally associated with direct petrol injection: one takes place on the induction stroke and the other on the compression stroke. The first mode of injection employs homogeneous mixture formation and the second employs a process known as stratified charge formation.

Idling and light-load cruising

Petrol at approximately 40 bar is injected direct into the cylinder a few degrees before TDC on the compression stroke. The specially designed cavity in the piston crown causes the charge of fuel to mix with the air in the cylinder to produce a mixture that is readily ignited.

Under these conditions the air—fuel ratio can be as high as 30:1 or 40:1.

Medium- and full-load conditions

Under medium- and full-load conditions the fuel is injected on the induction stroke shortly after TDC. The air—fuel ratio is approximately 20:1 for medium load and 14:1 for full load.

The ECU is programmed to switch between modes of injection as the operating conditions dictate. The improved fuel economy and lower emissions seem to make the increased complexity of the system a worthwhile development.

Fuel system maintenance and repair

Exhaust emissions

Manufacturing stage exhaust emissions standards

Before a vehicle is used on the road it must comply with the regulations that are in use at the time. In the European Union the regulations are those known as EU4, EU5, etc. The aim of these regulations is to limit atmospheric pollution and harm to health and the environment.

EU5 regulations

Emissions from petrol vehicles or those running on natural gas or LPG are typically as follows:

- carbon monoxide, 1000 mg/km;
- non-methane hydrocarbons, 68 mg/km;
- total hydrocarbons, 100 mg/km;

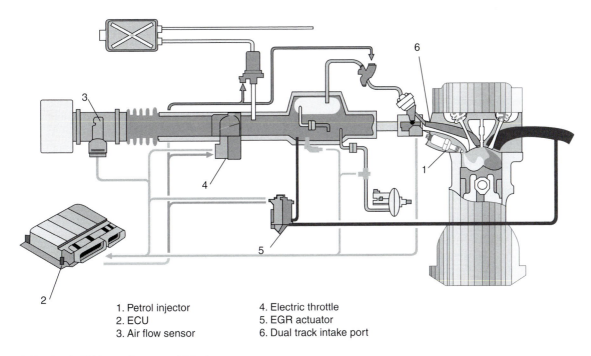

1. Petrol injector
2. ECU
3. Air flow sensor
4. Electric throttle
5. EGR actuator
6. Dual track intake port

Fig. 10.17 Bosch (VW)-type direct petrol injection

- nitrogen oxides (NO_x), 60 mg/km (25% reduction of emissions in comparison to the Euro 4 standard);
- particulates (solely for lean burn direct-injection petrol vehicles), 5 mg/km.

In-service stage

When a light vehicle is 3 years old in the UK it becomes subject to a test known as the MOT; after this first test it must be tested annually. Part of the MOT is a test of exhaust emissions. The test is carried out by means of approved test equipment that must be checked for accuracy at regular intervals. The procedure for performing the exhaust emissions test is described in the MOT testers' manual that is issued by the Vehicle and Operators Service Agency (VOSA) — the manual also contains the figures for emissions limits against which vehicles are tested. Modern computer-based exhaust gas analysers are equipped with a database that contains all of the necessary test data.

The procedures that are prescribed by the VOSA must be rigorously applied because they are a condition of the licence that permits a garage to conduct the tests; frequent inspections are carried out to ensure that standards are being maintained.

The petrol engine emissions that are checked at the MOT are:

- CO — carbon monoxide
- HC — hydrocarbons
- NO_x — nitrogen oxides
- Air—fuel ratio λ.

The permissible figures vary according to the vehicle being tested and it is necessary to refer to current data that is supplied by the VOSA. Gas analyser manufacturers and other organizations such as Autodata also supply the relevant data.

Service checks and repairs

In order for an engine to work properly it must be supplied with the correct amount of clean fuel to meet all operating conditions. The checks that are carried out at servicing intervals are designed to ensure that the fuel system remains in proper working order throughout the life of the vehicle.

Fuel filters

Provided that clean petrol is used the filters should provide relatively trouble-free service; however, as a precautionary measure most service schedules require them to be replaced. An example is to replace every 48 000 miles, or every 4 years.

Should a filter become blocked it will lead to a shortage of fuel, leading to weak mixtures and poor performance.

Fuel pipes

These should be checked to ensure that all securing brackets are in place so that there is no danger of the pipes rubbing against moving parts.

Amount of fuel injected

The amount of fuel that is injected is dependent on fuel-line pressure and the length of time for which the injector is switched on; these two variables can be checked by the following methods.

Safety note

Sparks, hot engines, and petrol are a dangerous combination and the importance of always following safe procedures cannot be overstated. Among other factors, work on petrol systems should never be attempted in a confined space.

Fuel-line pressure

Fuel-line pressure varies according to the type of vehicle (see Fig. 10.18). An example that I have to hand gives a figure for fuel pressure of 2.9—3.1 bar with the engine idling and the pressure regulator disabled.

Fuel delivery as shown by scope pattern of injectors

The two oscilloscope patterns in Fig. 10.19 show how the duty cycle increases to suit engine load. These

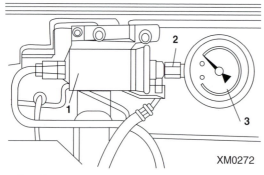

XM0272

1. Fuel filter. 2. Adaptor. 3. Pressure gauge.

Fig. 10.18 Fuel-line pressure

patterns are observed as the engine speed is increased and decreased, and they provide an excellent guide to the condition of the injectors.

Other checks on injector performance

Resistance

The electrical part of a petrol injector consists of a wire coil that has a known resistance. If there is a problem with the injector it may be due to a poor connection or a partial short-circuit; in either case an ohmmeter test of the type shown in Fig. 10.20 will show the condition of the electrical part of the injector. The actual figures obtained should be checked against those given in the manufacturer's data. A figure that I have states that the resistance should be 1.5—2.5 ohms.

Testing on a test bench

If the electrical condition of the injector is found to be good, it may be necessary to conduct further tests. The injector test equipment shown in Fig. 10.21 permits the spray pattern and quantity of fuel delivered to be examined under conditions that are equivalent to those that the injectors work under on the engine.

European on-board diagnostics (EOBD)

The computer part of a modern ECU is required to monitor a range of inputs from sensors and outputs to actuators. Should any of the inputs or outputs not conform to the criteria built in at the design stage, the ECU is programmed to produce a fault code and take remedial action; in some cases the action may be to cause the system to operate in default mode,

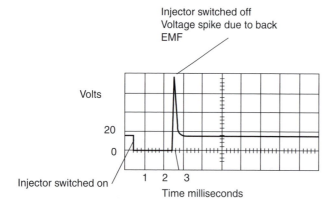

Injector switched off
Voltage spike due to back EMF

Volts

20

0

Injector switched on

1 2 3
Time milliseconds

Low engine load
Injection duration 2 ms

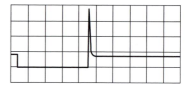

High engine load
Injection duration 4.2 ms

Fig. 10.19 Oscilloscope pattern for petrol injectors

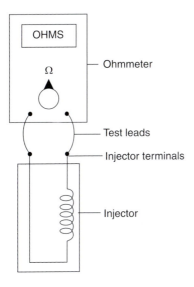

Fig. 10.20 Resistance check on injector

while in other cases it may be to alert the driver by displaying a warning at the engine fault lamp. The fault codes that are produced are a useful aid in dealing with faults in engine fuelling and exhaust emission problems.

Exhaust oxygen sensor faults

EOBD systems incorporate a second oxygen sensor downstream from the first. The second sensor samples at half the frequency of the first one and the ECU compares the two results; if the second sensor detects a different oxygen content from the first it is taken to mean that the catalytic convertor is not functioning correctly.

Exhaust emissions

The exhaust gases are the products of combustion. Under ideal circumstances, the exhaust products would be carbon dioxide, steam (water), and nitrogen. However, owing to the large range of operating conditions that engines experience, exhaust gas contains several other gases and materials such as those in the following list:

- CO — due to rich mixture and incomplete combustion
- NO_x — due to very high temperature
- HC — due to poor combustion
- Particulate matter (PM) — soot and organometallic materials
- SO_2 — arising from combustion of the small amount of sulphur in diesel fuel.

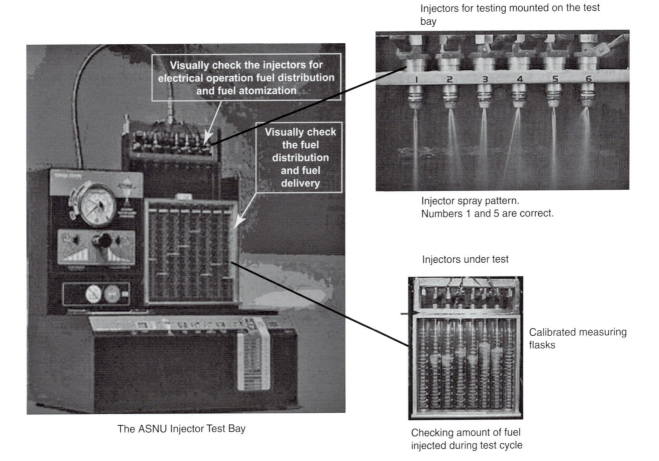

Injectors for testing mounted on the test bay

Visually check the injectors for electrical operation fuel distribution and fuel atomization

Visually check the fuel distribution and fuel delivery

Injector spray pattern.
Numbers 1 and 5 are correct.

Injectors under test

Calibrated measuring flasks

The ASNU Injector Test Bay

Checking amount of fuel injected during test cycle

Figure 10.21 Injector test bench

Self-assessment questions

1. What is the approximate operating voltage of a zirconia-type exhaust oxygen sensor?
2. What is the reason for fitting a second exhaust oxygen sensor downstream from the first one?
3. The catalyst test is part of the MOT test for most spark ignition engined light vehicles, with four or more wheels, first used on and after 1 August 1992. The exhaust emissions test for light vehicles requires the exhaust gas to be sampled and checked at fast idle speed for:
   - Carbon monoxide (CO)
   - Hydrocarbons (HC)
   - Lambda value
   - CO is checked again at idle speed.

   Check on the test data for vehicles that you are familiar with and write down the MOT pass level for CO and HC.
4. What is the effect of the duty cycle on the amount of fuel injected?
5. How many times in each cylinder, in a four-stroke cycle, is fuel injected in a sequential-type petrol injection system? How does this differ from the continuous injection system?
6. If the duty cycle for an injector is 40% and the pulse width is 1.2 milliseconds, for how long is the injector open to allow fuel to pass into the induction system?
7. If the lambda value is 1.05, what does this mean in terms of the richness, or weakness, of the air–fuel mixture?
8. Where, in the engine, is the petrol injector fitted in a direct injection engine?
9. Why is an idle speed control valve required for a petrol injection engine?
10. Name the materials that are used for the three-way catalyst.
11. How does a petrol injection system provide the rich mixture that is required for very cold engine starts?
12. With the aid of sketches, describe a low emissions hybrid vehicle that makes use of a diesel engine and electric motor propulsion.
13. Describe a vehicle that uses hydrogen as a gas to provide the fuel to run an adapted four-stroke engine.

11
Air supply and exhaust systems

Topics covered in this chapter

Engine air supply, air filters, manifolds
Exhaust systems — silencers
Oxygen sensor
Air-flow meters
Turbocharger

Liquid fuels such as petrol and diesel oil contain the energy that powers the engine. **Hydrogen** and **carbon** are the main constituents of these fuels and they need **oxygen** to make them burn and release the energy that they contain. The oxygen that is used for combustion comes from the atmosphere. Atmospheric air contains approximately 77% nitrogen and 23% oxygen by weight. For proper combustion, vehicle fuels require approximately 15 kg of air for every 1 kg of fuel. When the fuel and air are mixed and burned in the engine, chemical changes take place and a mixture of gases is produced. These gases are the exhaust. The principal exhaust gases are **carbon dioxide**, **nitrogen**, and **water** (steam).

Safety note

You should be aware that combustion is often not complete and carbon monoxide is produced in the exhaust. Carbon monoxide is deadly if inhaled. For this reason, engines should never be run in confined spaces. Workshops where engines are operated inside must be equipped with adequate exhaust extraction equipment.

If all conditions were perfect the fuel and air would burn to produce carbon dioxide, superheated steam, and nitrogen (Fig. 11.1). Unfortunately, these 'perfect' conditions are rarely achieved and significant amounts of other harmful gases, namely **carbon monoxide (CO)**, **hydrocarbons (HC)**, **oxides of nitrogen (NO$_x$)** and **solids** (particulates like soot and tiny metallic particles), are produced (Fig. 11.2).

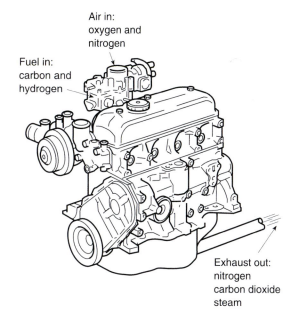

Fig. 11.1 Engine showing fuel/air in and exhaust out

Air in:
oxygen and nitrogen

Fuel in:
carbon and hydrogen

Exhaust out:
nitrogen
carbon dioxide
steam

Most countries have laws that control the amounts of the harmful gases (emissions) that are permitted, and engine air and exhaust systems are designed to comply with these laws.

Engine air supply systems (carburettor)

Figure 11.3 shows the main features of a petrol (spark ignition) engine **air supply system**. Our main concern, at the

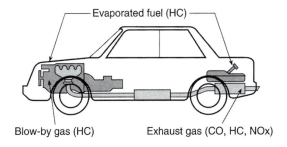

Evaporated fuel (HC)

Blow-by gas (HC)   Exhaust gas (CO, HC, NOx)

Fig. 11.2 Various bad emissions from engine and fuel system

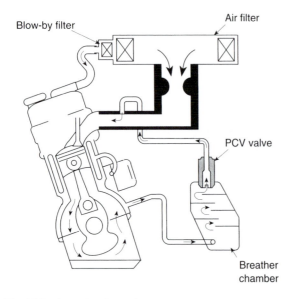

Fig. 11.3 An engine air supply system

moment, is the **air cleaner** and the **intake pipe** (induction tract) leading to the combustion chamber. The **blow-by filter** and the **positive crankcase ventilation** (PCV) system are part of the **emission control system**.

Air filter

The purpose of the air filter (Fig. 11.4) is to remove dust and other particles so that the air reaching the combustion chamber is clean. Pulsations in the air intake generate quite a lot of noise and the air cleaner is designed so that it also acts as an **intake silencer**.

The porous paper element is commonly used as the main filtering medium. Intake air is drawn through the

Fig. 11.4 An air filter/cleaner

paper and this traps dirt on the surface. In time this trapped dirt can block the filter and the air filter element is replaced at regular service intervals. Failure to keep the air filter clean can lead to problems of restricted air supply, which in turn affects combustion and exhaust emissions.

Temperature control of the engine air supply

To improve engine 'warm-up' and to provide as near constant air temperature as possible, the air cleaner may be fitted with a device that heats up the incoming air as and when required. Figure 11.5 shows the layout of a temperature-controlled air intake.

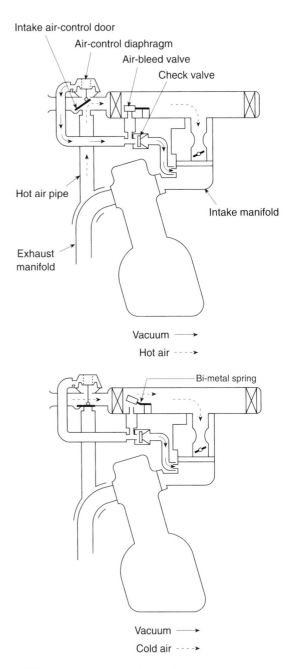

Fig. 11.5 Temperature control of intake air

Figure 11.5 shows that the **manifold vacuum** has lifted the **intake control door** off its seat and air is drawn into the air cleaner over the surface of the **exhaust manifold**. The **bimetallic spring** has pushed the **air bleed valve** on to its seat. When the air in the **air cleaner** reaches a temperature of 25–30°C, the bimetal spring lifts the air bleed valve from its seat. This removes the vacuum from above the **air control door** diaphragm, thus closing the intake control door. The intake air now enters direct

from the atmosphere. During cold weather the temperature control system will be in constant operation.

Checking the operation of the air control valve

Figure 11.6 shows the type of practical test that can be applied to check the operation of the air control diaphragm and the air door. The flap is raised against the spring and a finger is placed over the vacuum pipe connection. The air door valve is then released and the valve should remain up off its seat.

Fuel-injected engine air intake system

Figure 11.7 shows an air intake system for a modern engine fitted with electronic controls. This system has several components that we need to consider briefly.

Air-flow meter

The air intake for a fuel-injected petrol engine includes an **air-flow meter** (Fig. 11.8). The purpose of the

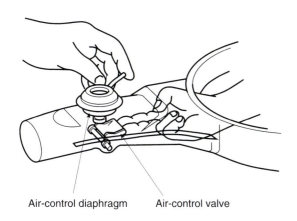

Air-control diaphragm   Air-control valve

Fig. 11.6 Checking the air control valve

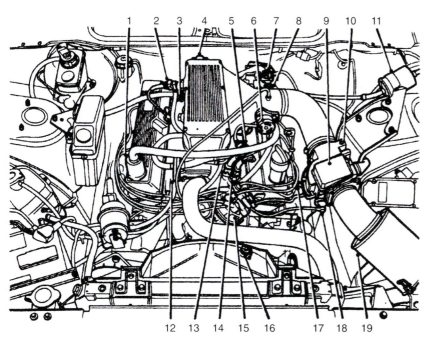

1 – Breather flame trap
2 – Cold start injector
3 – Fuel pressure regulator
4 – Overrun valve
5 – Throttle potentiometer connection
6 – Throttle potentiometer
7 – Engine air breather
8 – Idle speed adjustment screw
9 – Air-flow meter
10 – Idle mixture adjustment screw
11 – Fuel filter
12 – Extra air valve
13 – Coolant temperature switch
14 – Thermo-time switch
15 – Distributor
16 – Diagnostic plug
17 – Spark plug – No.1 cylinder
18 – Ignition coil
19 – Air cleaner

Fig. 11.7 Air supply system – petrol injection engine

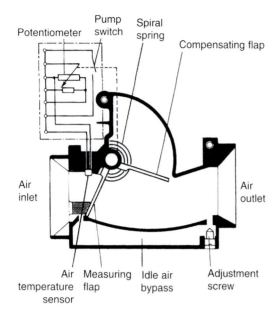

Fig. 11.8   An air-flow meter

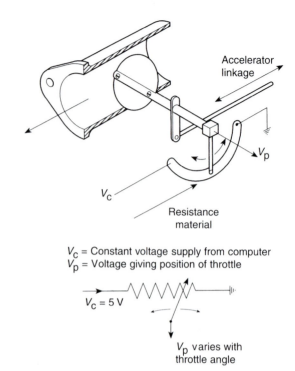

V_c = Constant voltage supply from computer
V_p = Voltage giving position of throttle

$$V_c = 5\text{ V}$$

V_p varies with throttle angle

Fig. 11.9   Throttle position sensor

air-flow meter is to generate an electrical signal that 'tells' the electronic controller how much air is flowing into the combustion chambers. It is also known as an 'air-flow sensor'.

The intake air stream causes a rotating action on the sensor flap, about its pivot. The angular movement of the flap produces a voltage at the sensor that is an accurate representation of the amount of air entering the engine. This voltage signal is conducted to the electronic control unit (**ECU**), where it is used in 'working out' the amount of fuel to be injected.

Throttle position sensor (potentiometer)

This is the sensor (shown as 6 in Fig. 11.7) that 'tells' the ECU the angular position of the throttle butterfly valve. The principle of a throttle position sensor is shown in Fig. 11.9. The voltage from the throttle position sensor is also conducted to the ECU for use in the process of determining the amount of fuel to be injected.

Idle air control

When the driver's foot is removed from the accelerator pedal, the throttle butterfly valve is virtually closed. In order that the engine will continue to run properly (idle), the intake system is provided with a separate system that provides an air supply for idling purposes. Figure 11.10 shows an electronically controlled air valve.

Air valves of the type shown in Fig. 11.10 are controlled by the ECU. The ECU will admit extra air

to provide fast-idle speed so that the engine keeps running even when extra load is imposed with the throttle closed, such as by switching on the headlights.

Air intake manifold

On many engines the intake air is supplied to a central point on the intake manifold, as shown in Fig. 11.11. The manifold is provided with ports that connect to the individual cylinders of the engine. The seal, often a gasket, between the intake manifold and the cylinder head must be secure and it is advisable to check the tightness of the securing nuts and bolts periodically. These nuts and bolts must be tightened in the proper sequence, shown for a particular engine in Fig. 11.12.

Variable-length induction tract

The length of the induction tract (tube) through which the engine's air supply is drawn has an effect on the operating efficiency of the engine. For low to medium engine speeds a long induction tract is beneficial, whereas a shorter induction tract is beneficial at high engine speed. Figure 11.13(a) shows a simplified variable-length induction system. An electronically controlled valve between the throttle valve and the engine switches air flow between the long and short tracts as required by engine speed. At low to medium engine speed this air valve is closed, as shown in

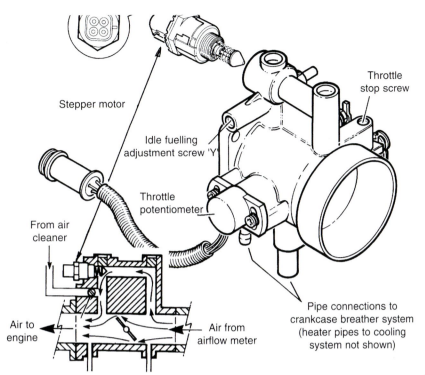

Fig. 11.10   An air control valve

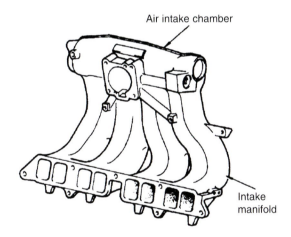

Fig. 11.11   An air intake manifold

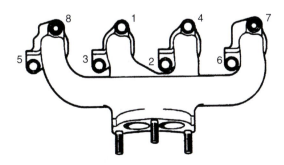

Fig. 11.12   Tightening sequence for exhaust manifold retaining hardware (nuts)

Fig. 11.13(b). As the engine speed rises the electronic control unit causes the air valve to open, as shown in Fig. 11.13(c), and the main mass of air enters the engine by the short route.

Turbocharging

Engines that rely on atmospheric air pressure for their operation are known as naturally aspirated engines. The power output of naturally aspirated engines is limited by the amount of air that can be induced into the cylinder on each induction stroke. Engine power can be increased by forcing air into the cylinders under pressure. The turbocharger is commonly used on modern engines to boost the pressure of the incoming air. The turbocharger consists of two main units, the turbine and the compressor, mounted on a common shaft. The turbine is driven by exhaust gas and utilizes energy that would otherwise be wasted by being expelled through the exhaust system into the atmosphere.

The general principle of turbocharging is shown in Fig. 11.14. Exhaust gas energy is directed to the small turbine, which is connected to the shaft that also drives the compressor. After driving the turbine, the exhaust gas travels through the exhaust system and out through the tail pipe. Intake air is drawn into the compressor through the air filter. The compressor raises the pressure to approximately 0.5 bar above atmospheric

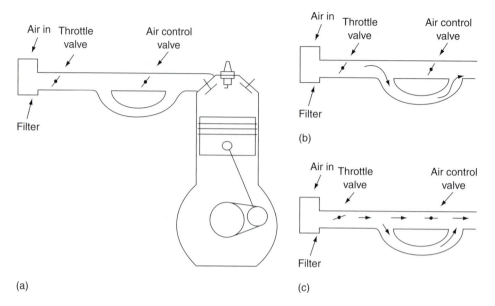

Fig. 11.13 (a) Variable-length induction tract. (b) Idling to medium speed. (c) High speed

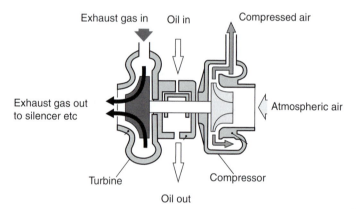

Fig. 11.14 A turbocharger

pressure and the pressure is controlled by the waste gate valve that allows the exhaust gas to bypass the turbine.

The turbocharger therefore consists of a small gas turbine that drives a small compressor that is attached to a common shaft. Although they are used on petrol engines their main use is on diesel engines, where their effect is more beneficial.

The exhaust gas under pressure enters the outer edge of the turbine rotor and flows along radial blades that are shaped to extract the maximum amount of energy from the gas. The speed of rotation of the turbine is very high, of the order of 100 000 rev/min. The energy that is extracted from the exhaust gas would otherwise be wasted by being passed to the atmosphere and although the turbine causes back pressure the gain in power offsets the losses. The turbine has to withstand temperatures in the region of 900°C and the components are made from heat-resistant materials such as

nickel-resist, which is a cast iron that contains 14% nickel, 6% copper, and 2% chromium; the turbine rotor may be made from Nimonic, which is an alloy of approximately 80% nickel and 20% chromium. The shaft that connects the turbine to the compressor is supported on plain bearings that are lubricated from the engine's main lubrication system. The oil that is circulating through the turbocharger serves two purposes:

1. It provides the lubrication of the plain bearings.
2. It acts as a cooling agent to convey heat away from the assembly.

The compressor draws air in at the centre of the rotor and passes it out under pressure at the outer edge of the rotor. The blades on the rotor are designed to provide the maximum pressure increase with the smallest possible use of power. In order to regulate the speed of the assembly and the pressure produced, the turbocharger

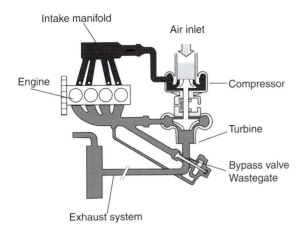

Fig. 11.15 The wastegate bypass valve

system includes a bypass valve that is called a wastegate. The wastegate is a valve that allows exhaust gas to bypass the turbine.

The wastegate

The purpose of the bypass valve (wastegate; see Fig. 11.15) is to control boost pressure by restricting the turbine speed. As soon as the control unit senses that the maximum permitted boost pressure has been reached, it transmits a signal that operates the actuator that opens the wastegate so that the bulk of the exhaust gas bypasses the turbine, thus limiting the turbine speed and boost pressure.

The intercooler

The intercooler (Fig. 11.16) is an air, or liquid, heat exchanger that cools the compressed air as it leaves the compressor. It is necessary to provide the engine with air at a reasonably constant temperature to avoid detonation and possible engine damage.

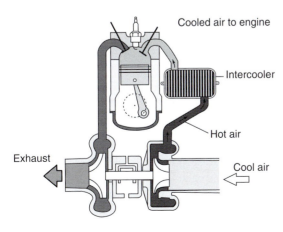

Fig. 11.16 The intercooler

Turbocharger maintenance

Starting the engine

The engine should be allowed to idle for a few seconds after starting to allow the oil flow through the turbo-charger to become established.

Stopping the engine

When the engine first stops the oil no longer circulates but the turbocharger turbine is still rotating. Provided that the engine was idling when it was switched off the residual supply of oil at the turbine shaft bearings will protect them. However, if the engine was running at high speed when it was switched off the turbine rotor will continue rotating at high speed and bearing failure may occur. To prevent this happening it is recommended that the engine should be allowed to idle for 10 seconds or so before switching it off; this also allows the turbo assembly to cool down.

Provided that these precautions are taken the turbo unit requires little maintenance. The main problems arise from:

- Lack of lubricant
- Contamination of the lubricant
- Foreign matter entering the air intake.

Lack of lubricant can be caused by a low oil level in the sump, blocked oil ways, or some other problem such as a defective oil pump – in any event it is likely to lead to bearing failure. It is important to use the grade of oil prescribed by the engine manufacturer because this also affects bearing life.

The oil may be contaminated by products of combustion and small metallic particles from other components; observance of prescribed oil change frequency should avoid damage from contaminated oil.

If the air filter is damaged, or the engine is run without the filter, foreign matter may cause serious damage to the compressor. If this happens pieces of metal may enter the engine, causing major damage.

Practical assignment – air supply and exhaust systems

Objective

The purpose of this assignment is to allow you to develop your knowledge and skill in this important area of motor vehicle work, and to help you to build up your file of evidence for NVQ.

Activity

The nature of this assignment is such that you will need to work under supervision, unless you have previously been trained and are proficient. The assignment is intended to be undertaken during a routine service and MOT test on a petrol-engined vehicle that is equipped with electronically controlled petrol injection. The assignment is in two parts.

Part 1

1. Locate the idle control valve. Describe the purpose of this valve and describe any adjustments that can be made to it.
2. Specify the effects on engine performance and exhaust emissions that may arise if this valve is not working correctly.

Part 2

1. Perform the emissions test part of the MOT test.
2. Record the results and compare them with the legal limits.
3. Write short notes on the test procedure and describe any work that should be carried out before starting the emissions test.
4. Give details of remedial work that can be performed if a vehicle should fail the emissions test.
5. Make a list of the tools and equipment used.

Self-assessment questions

1. A blocked air filter may cause:
   (a) low fuel consumption
   (b) the engine to run rich
   (c) increased cooling system pressure
   (d) excess oxygen in the exhaust.
2. An air leak through the intake manifold flange gasket may cause:
   (a) the engine to run rich
   (b) the malfunction indicator lamp to illuminate
   (c) the positive crankcase ventilation system to fail
   (d) excess NO_x in the exhaust.
3. Turbocharged engines:
   (a) are less efficient than naturally aspirated ones because power is absorbed in driving the turbocharger
   (b) are more efficient than naturally aspirated engines because they make use of energy in the exhaust gas that would otherwise be wasted
   (c) are normally diesel engines
   (d) have high fuel consumption.
4. An air-flow sensor:
   (a) provides an electrical signal that tells the ECU how much air is flowing to the engine
   (b) controls the air flow to the air-conditioning system
   (c) allows the exhaust gases to bypass the turbocharger
   (d) restricts air flow to prevent engine overload.
5. The potentiometer-type throttle position sensor:
   (a) generates electricity that is used as a signal to inform the ECU of the amount of throttle opening
   (b) receives a supply of electricity and sends a signal to the ECU that is an accurate representation of throttle position
   (c) is always situated on the accelerator pedal
   (d) always operates on the variable capacitance principle.
6. Exhaust gas oxygen sensors:
   (a) are used to detect oxygen levels in the exhaust gas
   (b) operate effectively at all temperatures
   (c) always generate electricity
   (d) are only needed on direct injection engines.
7. During combustion, hydrogen in the fuel combines with oxygen from the air:
   (a) and produces water in the exhaust system
   (b) and produces NO_x in the exhaust gas
   (c) and produces carbon monoxide in the exhaust gas
   (d) and causes the ignition to be retarded.
8. When an engine fails the emissions test it means that:
   (a) the catalytic converter is not working
   (b) the emissions control system is not working and this may be caused by a range of engine problems
   (c) the oxygen sensor has failed
   (d) the air filter is blocked.
9. Exhaust systems may be protected against corrosion by:
   (a) coating them inside and outside with a layer of bitumen
   (b) a thin layer of aluminium applied to the inner and outer surfaces
   (c) reducing the temperature of the exhaust gases
   (d) coating them with underseal.
10. Atmospheric air contains approximately 77% nitrogen and 23% oxygen by mass:
   (a) the nitrogen is a hindrance because it is not used in combustion
   (b) the nitrogen helps to form water
   (c) the nitrogen forms part of the gas that expands in the cylinder to make the engine work
   (d) the oxygen and nitrogen form NO_x at very low temperatures.

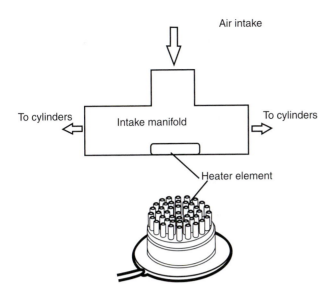

Air intake

To cylinders

Intake manifold

To cylinders

Heater element

Fig. 11.17 Manifold heater

11. Describe the effect that a partially blocked air filter element may have on the carbon monoxide content of the exhaust gas.

12. Figure 11.17 shows a system that provides for pre-heating of the air as it enters the engine. How is the heating element energized and what effect does the pre-heating have on cold starting performance?

13. What effect does the turbocharger have on the maximum pressure after combustion?

14. Why is it necessary to let the engine speed die down before switching off the ignition in a turbocharged engine?

15. What is the approximate temperature of the exhaust gas as it leaves the combustion chamber?

12
Lubricants and engine lubrication

Topics covered in this chapter

Oil properties
Multi-grade oils
Viscosity index
Synthetic oils
Maintenance checks and repair procedures
Working details of engine lubrication systems
Pressure-relief valve
Full-flow and bypass filters
Oil cooler
Oil pumps
Oil pressure checks

Lubricants

Oil is the principal lubricant used in automobiles and its main purpose is to reduce friction and prevent wear. In engines and transmission systems the oil performs an additional function, which is to conduct heat away from moving parts and thus help to keep a system cool.

Friction

When examined under a microscope, machined surfaces are seen to be covered with small peaks and valleys, as shown in greatly magnified form in Fig. 12.1. When one surface moves over another, the peaks cause resistance to motion. This resistance is called friction and it causes two problems:

1. Wear due to tiny fragments of metal being torn off.
2. Energy is used to move the surfaces and heat is generated.

These conditions are a particular problem in shaft bearings and sliding surfaces such as those between the piston and cylinder wall.

Coefficient of friction

The friction force F divided by the force pressing the surfaces together W is called the coefficient of friction, which is denoted by the symbol μ:

$$\mu = \frac{F}{W}.$$

Reducing friction

In most moving parts of a mechanism friction is a problem. Two methods of reducing it are:

1. **Rolling friction** (Fig. 12.2). When a hard steel ball rolls across a hard flat surface the ball and flat surface make contact at a single point. There is no relative sliding at the point of contact, which means that there is no friction. This is the principle of ball and roller bearings that are used extensively in automotive systems.

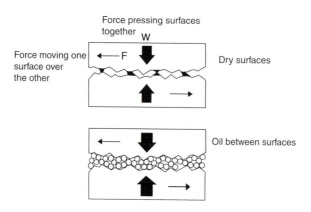

Fig. 12.1   Coefficient of friction

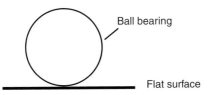

Fig. 12.2 Rolling friction

2. **Lubrication**. A principal purpose of lubrication is to reduce friction. Two important properties of a lubricant are:
   (a) Viscosity
   (b) Oiliness.

Viscosity is a measure of resistance offered to the sliding of one layer of lubricant over an adjacent layer. In automotive use the viscosity is taken to be a measure of the lubricant's resistance to flow. For practical purposes the viscosity of a lubricant is measured by recording the time that it takes for a quantity of the lubricant to flow through an orifice of fixed size at a given temperature. In automotive practice oils are normally identified by a grading system devised by the Society of Automotive Engineers (SAE) of the USA. An oil with a grading of SAE 40 has a higher resistance to flow than one with a grading of SAE 20.

If two lubricants with identical viscosities are smeared on to two pairs of surfaces and the friction between the surfaces is measured and compared, the friction force will be lower for one pair of surfaces than the other. The lubricant that produces the lowest friction force under these circumstances is said to have the greater **oiliness**. Oiliness is the property of an oil to cling to a surface and it is particularly noticeable in vegetable oils and synthetic oils that have been treated so that they possess oiliness.

Boundary lubrication

Boundary lubrication is largely restricted to surfaces where sliding contact occurs in components such as pistons and cylinder walls. In most cases splash lubrication and jets of oil sprayed on the moving parts is the method used to provide the oil film.

Hydrodynamic lubrication

In a plain bearing assembly the shaft is slightly smaller in diameter than the internal diameter of the bearing to provide space for the oil and allow for thermal expansion as shown in Fig. 12.3.

When the shaft is stationary there is direct contact between the shaft and the bearing except for a thin film of oil. The remainder of the crescent-shaped space between the shaft and the bearing is filled with oil. When the shaft begins to rotate, the point of contact between the shaft and the bearing tends to move up the bearing surface against the direction of rotation of the shaft and this results in a wedge of oil building up. Eventually, as the speed increases, the oil fills the entire space around the shaft and the load on the shaft is transferred through the oil to the bearing. In this way there is no metal-to-metal contact and the only friction force that exists is that which shears the oil. This type of lubrication is called hydrodynamic lubrication because it relies on motion for its operation.

Sources of oil

Mineral oil

Mineral oil that is used as the base of most automotive lubricants is obtained from crude petroleum. Petrol and diesel fuel are other crude petroleum products.

Vegetable oil

Whilst vegetable oil is not suitable for engine lubrication because of its tendency to decompose under working conditions, in the twenty-first century it is

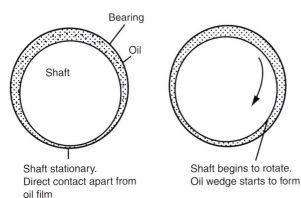

Shaft stationary.
Direct contact apart from
oil film

Shaft begins to rotate.
Oil wedge starts to form

Shaft rotating at high speed. The
shaft is now supported by the layer
of oil between it and the bearing surface

Fig. 12.3 Shaft and bearing lubrication

being increasingly used as a source of fuel. The large areas of bright yellow crops of rape seed that can be seen throughout the countryside in the UK are evidence of this trend.

Types of oil

Viscosity vs. temperature

The viscosity of oil is affected by temperature, as shown in Fig. 12.4. In effect, the oil becomes thinner as its temperature rises and this causes problems in engines that operate across a fairly wide range of temperatures. To address this problem motor oils are designed to have a high viscosity index. The viscosity index is a number that indicates how much an oil thins out as its temperature rises — an oil that is very thick when cold and very thin when hot has a low viscosity index. Most modern engine oils are treated with viscosity index improvers; these are chemicals that restrict the tendency for an oil to get thinner as it is heated and they give the oil a high viscosity index. Oils that are treated in this way are called multi-grade oils because they conform to two SAE methods of classifying them: SAE W for cold temperature performance and the usual SAE number for higher temperatures.

Multi-grade oils

An oil that is graded as 10W/30 is known as a multi-grade oil because it conforms to two viscosity ratings: one that is based on subzero temperatures, which is the reason for the W in the 10W, denoting winter, and the other number, which is based on higher temperatures.

Additives

Motor oil needs to perform a variety of functions under a wide range of engine operating conditions.

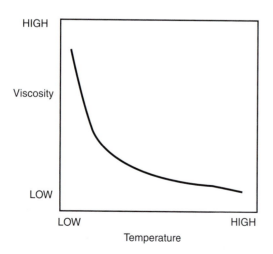

Fig. 12.4 Oil viscosity vs. temperature

Therefore, several additives are incorporated into the formulation:

1. **Detergent/dispersant additives** — used to maintain engine cleanliness, keeping the various contaminants in a fine suspension and preventing them from settling out on vital engine components.
2. **Rust and corrosion inhibitors** — added to protect the engine from water and acids formed as combustion by-products.
3. **Antioxidants** — added to inhibit the oxidation process, which can result in oil thickening and sludge formation.
4. **Anti-wear additives** — form a film on metal surfaces to help prevent metal-to-metal contact.
5. **Viscosity modifiers and pour point depressants** — help improve the flow characteristics of motor oil.

Oil quality

The SAE rating is merely concerned with viscosity; other methods of grading are used as a guide to quality and suitability of an oil for specific purposes. Two of these grading systems are:

1. American Petroleum Institute (API)
2. European Automobile Manufacturers' Association (Association des Constructeurs Européens d'Automobiles; ACEA).

The gradings of these organizations are designed to ensure that oils meet the requirements of modern automobiles and emissions regulations. Examples of these gradings are:

- API SL. For all automotive engines presently in use. Introduced on 30 November 2004, SM oils are designed to provide improved oxidation resistance, improved deposit protection, better wear protection, and better low-temperature performance over the life of the oil.
- ACEA A1/B1. A stable, stay-in-grade oil intended for use at extended drain intervals in gasoline engines, and car and light van diesel engines that are specifically designed to use low-viscosity oils.
- The grade of oil to be used in a specific vehicle is normally stated in the driver's handbook.

Synthetic oil

Modern motor oils are designed to operate with the sophisticated after-treatment systems such as oxygen sensors, three-way catalysts, and particulate filters. These systems are susceptible to damage from sulphur, ash, and phosphorus. Oils known as synthetics, or near synthetics, are examples of such oils.

Synthetic, or man-made, oils are manufactured specifically to stand up to the severe conditions under which conventional oils might falter. They possess viscosity characteristics superior to those of mineral oils. The resulting lubricants have a molecular structure that meets and often exceeds manufacturers' criteria for high-performance engines.

Among the many performance advantages that synthetic oils offer is their ability to remain stable at high temperatures (under which conventional oils begin to break down) and remain fluid at low temperatures (under which conventional oils begin to thicken). This provides optimum lubrication at extreme temperatures, reducing wear for a cleaner, more efficient engine.

Synthetic oil grades

Synthetics are sometimes mixed with conventional mineral oils to produce a cost-effective middle ground between the two, referred to as 'semi-' or 'part-synthetic'. However, while semi- or part-synthetics and conventional mineral oils are both capable lubricants, fully synthetic oils provide the highest level of engine protection.

Handling oil and disposal of waste

Frequent and prolonged contact with used engine oil may cause dermatitis and other skin disorders, including skin cancer, so avoid unnecessary contact. Adopt safe systems of work and wear protective clothing of the type shown in Fig. 12.5 – this clothing

Fig. 12.5 Gloves and protective clothing when working with oil. *From Health and Safety in Motor Vehicle Repair, HSE*

should be cleaned or replaced regularly. Maintain high standards of personal hygiene and cleanliness at all times.

According to the Environment Agency (EA), oil accounts for 25% of all pollution incidents. Do not pour oil into drains as most drains link to watercourses, which become polluted. Recycle oil and filters at an oil bank, or in the case of large quantities waste should be collected by a registered contractor, who will normally buy it.

Lubrication system

There are three main types of lubrication systems in common use on internal combustion engines:

- Wet sump
- Dry sump
- Total loss.

The object of the lubrication system is to feed oil to all the moving parts of the engine to reduce friction and wear, and to dissipate heat. Modern oils also clean the engine by keeping the products of combustion, dirt, etc. in suspension. This makes it essential that oil and filters are changed in accordance with the manufacturers' instructions. It can be seen from this that the oil performs four important functions:

- It keeps friction and wear on the moving parts to a minimum.
- It acts as a coolant and transfers the heat from the moving parts.
- It keeps the moving parts clean and carries the impurities to the oil filter.
- It reduces corrosion and noise in the engine. It also acts as a sealant around the piston and rings.

Learning tasks

1. What special measures should be observed when storing/using oil? How should waste oil be disposed of?
2. Write to an oil manufacturer and ask for details of their current products for the vehicles most commonly serviced in your workshop, e.g. types of oils recommended for the engine, gearbox, final drive, special additives used in the oils, types of grease available, etc.
3. Write to an oil manufacturer and ask for charts showing quantities of oil used in vehicles, viscosity and API ratings, etc. that could be displayed in the motor vehicle workshop.

Main components in the lubrication system

The lubrication system is mostly pressurized and consists of the following main components:

- **Oil pump** — draws the oil from the sump and delivers it under pressure to the engine lubrication system.
- **Relief valve** — limits the maximum pressure of the oil supplied by the pump to the system.
- **Sump** — serves as a reservoir for the oil.
- **Oil galleries** — are channels or drillings through which the oil passes to the different lubrication points in the engine.
- **Oil pressure indicator** — shows whether the oil pressure is being kept within the manufacturers' limits.
- **Oil filter** — filters the oil, removing impurities to keep it clean.

A typical lubrication system works in the following way:

1. Oil is drawn from the sump by the oil pump.
2. The pump pressurizes the oil and passes it through the oil filter into the oil galleries and passages that lead to the crankshaft and camshaft bearings and, in some engines, the rocker shaft and rocker arms.
3. The oil splashed from the crankshaft lubricates the pistons and other internal parts of the engine.
4. After lubricating the moving components, the oil drips back down into the sump. Figure 12.6 shows the oil flow in an engine lubrication system.

Function of the oil

When looking at the working surfaces of, say, the crankshaft (A in Fig. 12.7) and the main bearings in which it runs (B), they appear to be blank and smooth. But when observed under a powerful microscope they will be seen to be uneven and rough. When oil is introduced between these surfaces, it fills up the slight irregularities and forms a thin layer, called an oil film. It is this oil film that separates the surfaces and, when the components are rotating, prevents metal-to-metal contact. When the engine is operating the oil must be strong enough to withstand the heavy loads imposed on all the moving parts. The oil is therefore delivered under pressure to the bearings and, to enable it to enter, a very small clearance between the shaft and bearing is necessary.

The clearance must be sufficient for the oil to enter, but small enough to resist the heavy loadings to which the bearings are subjected. This clearance is approximately 0.05 mm for a shaft of 60–70 mm diameter. When the shaft is not rotating but is resting on the bearings, only a very thin film of residual oil separates the surfaces. As the engine starts the only lubrication for the first revolution is provided by this thin film of oil; as the revolutions increase, the oil pump starts to deliver the oil under pressure to the bearings. The oil is drawn round by the rotating shaft which, together with the pressure, forms an oil wedge that lifts the shaft up from the bearing. The shaft then rotates freely, separated from the bearings by this thin film of oil. A correct bearing clearance is shown in Fig. 12.8.

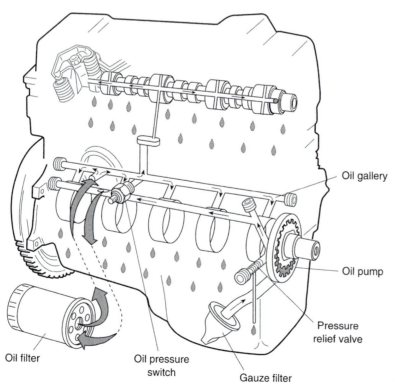

Oil gallery

Oil pump

Pressure relief valve

Oil filter

Oil pressure switch

Gauze filter

Fig. 12.6 Lubrication circuit for an overhead camshaft (OHC) engine

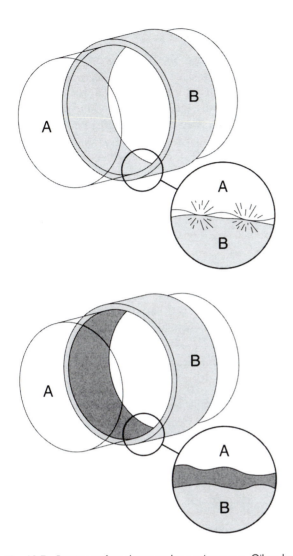

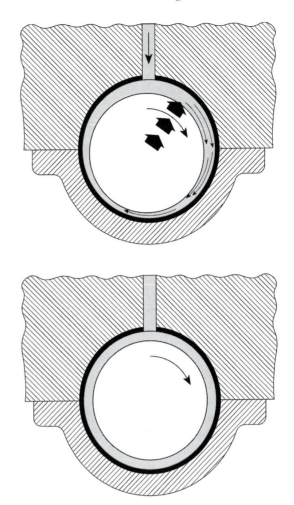

Fig. 12.8 Bearing clearance correct

Fig. 12.7 Bearing surface shown under a microscope. Oil under pressure keeps the surfaces apart

It is important that the bearings have the correct clearance. Too much will cause the oil to escape from the bearing without being able to create the required **oil wedge**; too little will restrict the oil from entering the bearing, causing metal-to-metal contact. In both cases wear will increase.

The oil that is pressed out from the bearings is splashed around and forms an oil mist, which lubricates the cylinder and piston (Fig. 12.9). In some cases a hole is drilled from the side of the connecting rod into the bearing shell to spray oil on to the thrust side of the cylinder wall.

Function of the sump

When the engine is filled with oil, it flows down through the engine into a container called the sump. This is attached to the bottom of the engine block with a series of small bolts, usually with a gasket between the block and the sump. It is commonly formed from sheet steel pressed to a shape that has one end slightly lower to form the oil reservoir. In the bottom of the sump is the drain plug. **Baffle plates** are fitted to prevent the oil from splashing around or surging when the vehicle is accelerating and braking or going round corners. If all the oil is allowed to move to the rear or to the side of the sump, the oil pick-up may become exposed, causing air to be drawn into the lubrication. The sump also acts as an oil cooler because it extends into the air stream under the vehicle. To assist with the cooling process there may also be small fins formed on the outside to increase its surface area. Aluminium is sometimes used to give a more rigid structure to support the crankshaft and crankcase of the engine. Two types of sump are shown in Fig. 12.10.

Oil level indicators

The level in the sump is checked by means of a dipstick on which the maximum and minimum oil levels are indicated. A number of vehicles fit indicators to show the driver the level of oil in the engine without having to lift the bonnet and remove the dipstick manually.

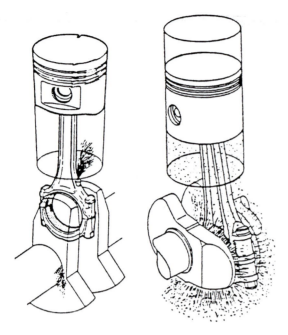

Fig. 12.9 Splash lubrication from the big-end bearing

One of the popular types used is the 'hot wire' dipstick, where a resistance wire is fitted inside the hollow stick between the oil level marks. The current is only supplied to the wire for about 1.5 seconds at the instant the ignition is switched on. If the wire is not in the oil it overheats and an extra electrical resistance is created, which signals the ECU (electronic control unit) to operate the driver's warning light. This is illustrated in Fig. 12.11.

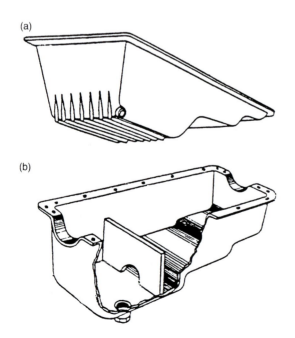

Fig. 12.10 Types of sump. (a) Aluminium cast. (b) Pressed steel

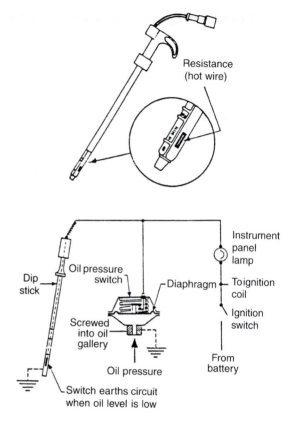

Fig. 12.11 Combined oil pressure and low oil level warning circuit

---

Learning tasks

1. What would be the effects of running an engine with the oil level (a) above the maximum and (b) below the minimum on the dipstick?
2. A vehicle is brought into the workshop with a suspected oil leak from the sump. Suggest one method of checking exactly where the leak is coming from and describe the procedure for correcting the fault.
3. How should a resistance-type oil level indicator be checked for correct operation?

Oil pump

The oil enters the pump via a pipe with a strainer on the end, which is immersed in the oil reservoir in the sump. This strainer prevents larger particles from being sucked into the lubrication system. The oil pump creates the required pressure that forces the oil to the various lubrication points. The quantity of oil delivered by the pump varies greatly from vehicle to vehicle and also depends on engine speed, but will be approximately 120 litres when the speed of the vehicle is 100 km/h.

The most common types of pumps used in the motor vehicle engines are the **gear**, **rotary**, or **vane**.

Gear pump

As shown in Fig. 12.12, the gear pump consists of two gears in a compact housing with an inlet and outlet. The gears can be either **spur** or **helical** in shape (the helical being quieter in operation). The pump drive shaft is mounted in the housing and fixed to this is the driving gear. Oil is drawn via the inlet into the pump. It passes through the pump in the spaces between the gear teeth and pump casing, and out through the outlet at a faster rate than is used by the system. In this way pressure is created in the system until the maximum pressure is reached, at which time the pressure-relief valve will open and release the excess pressure into the sump.

Rotary pump

The main parts of this type of pump (shown in Fig. 12.13) are the **inner rotor**, the **outer rotor**, and the **housing** containing the inlet and outlet ports. The inner rotor, which has four lobes, is fixed to the end of a shaft; the shaft is mounted off-centre in the outer rotor, which has five recesses corresponding to the lobes. When the inner rotor turns, its lobes slide over the

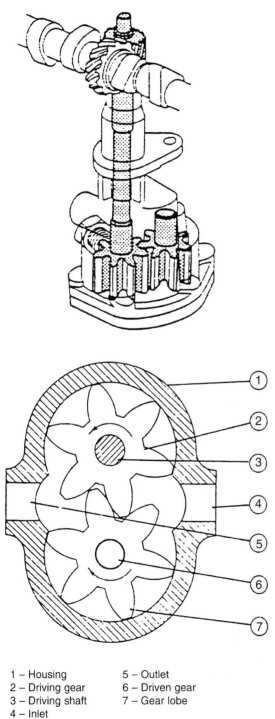

1 – Housing
2 – Driving gear
3 – Driving shaft
4 – Inlet
5 – Outlet
6 – Driven gear
7 – Gear lobe

Fig. 12.12 Gear-type pump

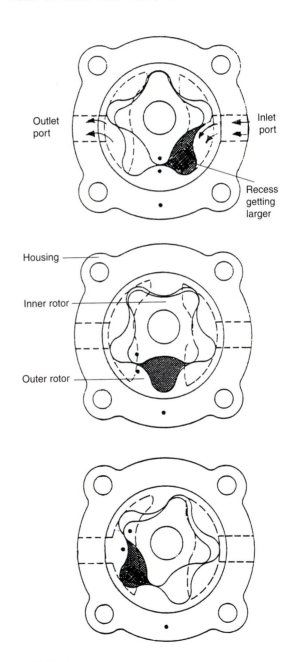

Fig. 12.13 Rotary pump

corresponding recesses in the outer rotor turning it in the pump housing. At the inlet side the recess is small; as the rotor turns the recess increases in size, drawing oil up from the sump into the pump. When the recess is at its largest the inlet port finishes, further movement of the rotor reveals the outlet port, and the recess begins to decrease in size, forcing the oil under pressure through the outlet port.

Vane-type pump

This pump, shown in Fig. 12.14, takes the form of a driven rotor that is eccentrically mounted (mounted offset) inside a circular housing. The rotor is slotted and the **eccentric vanes** are free to slide within the slots, a pair of thrust rings ensuring that the vanes maintain a close clearance with the housing. When in operation the vanes are pressurized outwards by the centrifugal action of the rotor rotating at high speed. As the pump rotates the volume between the vanes at the inlet increases, thus drawing oil from the sump into the pump; this volume decreases as the oil reaches the outlet, pressurizing the oil and delivering it to the oil gallery. This type has the advantage of giving a continuous oil flow rather than the pulsating flow that is rather characteristic of the gear-type pump.

Pressure-relief valve

As engine speed increases, the oil pump produces a higher pressure than is required by the engine lubrication system. A **pressure-relief valve** (see Fig. 12.15) is therefore fitted in the system to take away the excess pressure and maintain it at a level appropriate for the bearings and seals used. It will be seen then that the relief valve performs two important functions: first, it acts as a pressure regulator; second, it acts as a safety device in the lubrication system. The main types in use are the ball valve, the plate, and the plunger or poppet valve. Each is held in the closed position by a spring. As the oil pressure in the oil gallery rises above the setting for the relief valve, the valve opens against

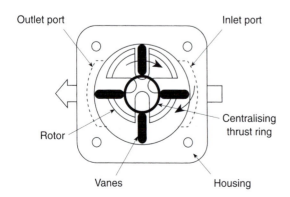

Fig. 12.14 Vane pump

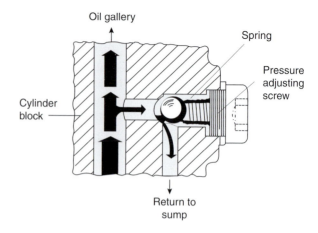

Fig. 12.15 Oil pressure-relief valve

spring pressure, allowing the oil to bypass the system and return back to the sump via the return outlet. The force on the spring determines the oil pressure in the lubrication system.

Learning tasks

1. When next in the workshop, dismantle a vane-, rotor-, and gear-type oil pump. Using the manufacturers' data, measure and record the tolerances for each type and give your recommendations on serviceability.
2. Remove the oil pressure switch from the oil gallery of an engine and attach the appropriate adapter to measure the oil pressure. Record the results at the relevant speeds. Identify any major differences between the recorded results and the manufacturer's data and give your recommendations.
3. Give two reasons for low oil pressure together with recommendations for correcting the fault.

Oil filter

When the oil passes through the engine it becomes contaminated with carbon (the by-product of the combustion process), dust (drawn in from the atmosphere), small metal particles (from components rubbing together), water, and sludge (a combination of all these impurities mixed together). All these will cause engine wear if they remain in the oil, so the engine must be equipped with a filtering system that will remove them and keep the oil as clean as possible. Most modern engines are equipped with a filtering system where all the oil is filtered before it reaches the bearings. This arrangement is called the **full-flow** system. There is another system also in use where only a portion of the oil passes through the filter, called the **bypass** filter system. The two systems are shown in Fig. 12.16.

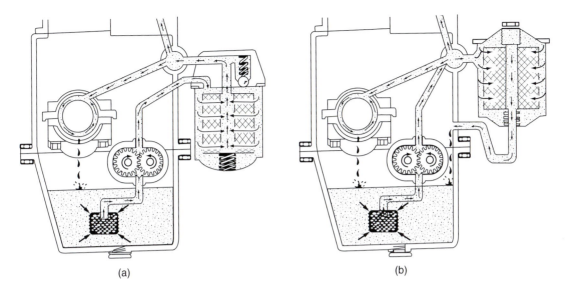

Fig. 12.16 Lubrication systems. (a) Full-flow system. (b) Bypass system

The importance of filtering the oil is shown by the results of an investigation into the wear on the cylinder and piston, using the two filtering systems. It was found that maximum wear (100%) occurs in engines working without an oil filter. When a bypass filter is used, wear is reduced to about 43% on the cylinder and 73% on the piston, which means that the life of the piston and cylinder are almost doubled. Minimum wear occurs when a full-flow filter is used; wear is again reduced by a further 15% on the cylinder and 22% on the piston. This means that the life of the piston and cylinder is four to five times longer than in an engine working without a filter. A good oil filter must be capable of stopping the flow of very small particles without restricting the

flow of oil through the filter. To meet this requirement, different materials are used as the filtering medium. Resin-impregnated paper is widely used, the paper being folded in order to make a large surface area available for the oil to flow through; particles are left on the paper and clean oil is passed to the lubrication system. In this way, when the filter is changed the impurities are removed at the same time. In other types of oil filters, different kinds of fibrous materials are used. The filtering material is enclosed in perforated cylinders, one outer and one inner, to form a filter element (Fig. 12.17).

The oil enters through the perforations in the cylinder, passes through the filtering element and leaves through

Clean oil out   Dirty oil in

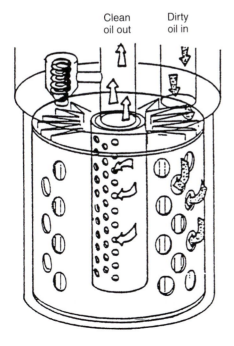

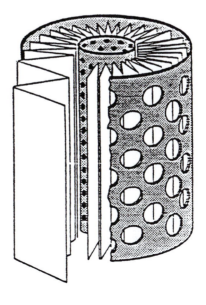

Fig. 12.17 The oil filter and pleated element

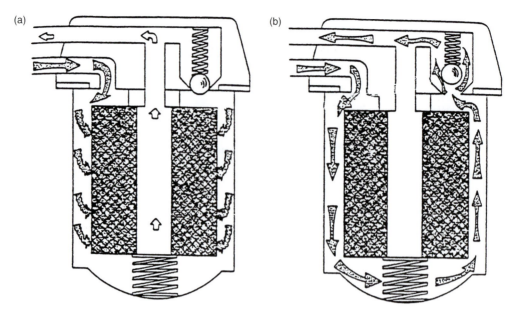

Fig. 12.18 Oil filter operation. (a) Oil being filtered. (b) Oil filter blocked

the central tube outlet. Many modern filters are now the cartridge type, which is removed complete. The advantages of this disposable type are that it cleans the oil very efficiently, it is relatively easy to change, and it is less messy to remove. The filter element can also be located in a removable metal container. With the replaceable-element type it is only the element itself that is changed, the container is thoroughly cleaned, and the 'O' ring replaced.

Full-flow filter

Oil filter operation is shown in Fig. 12.18. The most widely used filtering system is the full-flow filter. The construction of the filter is very efficient because all the oil is passed through the filter before it flows to the bearings. After a certain length of time the element becomes dirty and less efficient, and must therefore be changed. If the element is not changed regularly the impurities will accumulate and the element will become clogged, restricting or preventing the oil from passing through the filter. For this reason a relief valve is fitted into the filter, which opens and allows the oil to bypass the clogged filter element and flow directly to the bearings unrestricted. If the condition is allowed to continue, unfiltered oil will carry abrasive particles to the bearings, causing rapid wear.

Cartridge filter

In the cartridge filter the relief valve is in the filter (shown in the open position in Fig. 12.19). Many of the filters now contain a valve underneath the inlet hole, which opens when oil pressure forces oil into the filter. When the engine stops and the oil flow ceases, the valve closes and the oil is kept within the filter.

This prevents it from draining back into the sump. It also has the advantage of enabling the engine to develop the oil pressure more quickly when starting from cold. Correct operation is shown in Fig. 12.20.

Disc filter

This type of full-flow filter (shown in Fig. 12.21) is used in large diesel engines. The oil is filtered by being forced

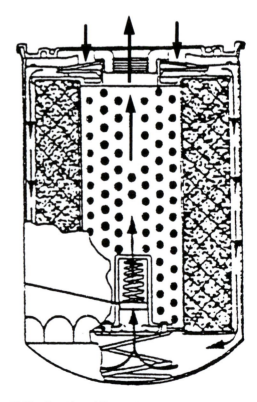

Fig. 12.19 Cartridge oil filter with oil relief valve open

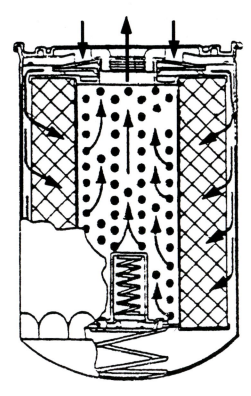

Fig. 12.20 Cartridge oil filter operating correctly

through very narrow gaps (0.05 mm) between thin steel discs, which form an assembly that can be rotated. The narrow gap between the discs prevents impurities in the oil from passing through. The deposits accumulate on the outside of the discs, which are kept clean by scrapers that scrape off the deposits as the disc assembly rotates. In most cases the assembly is connected to the clutch pedal; each time the pedal is operated the disc assembly

is rotated a small amount. The filter must be drained as per manufacturers' recommendations, this being done by removing the drain plug, allowing dirt and some oil to be flushed out.

Centrifugal filter

Again mainly found on larger engines, this consists of a housing with a shaft and rotor inside. The oil is forced through the inlet ports by the pump and fills the rotor through the inlet holes in the rotor shaft, passing down the pipes to the jets. Due to the force of the oil passing through the jets, the rotor rotates at very high speed. Owing to the centrifugal force, the impurities (which are heavier than the oil) accumulate on the walls of the rotor. The filter must be periodically cleaned by dismantling the filter and washing with a suitable cleaning fluid.

> ### Learning tasks
> 1. Draw up a lubrication service schedule for a 12 000-mile major service.
> 2. Complete a lubrication service on a vehicle using the above service schedule. Identify any faults found, such as oil leaks, and complete the report on the workshop job card.

Dry-sump lubrication

This type of system (Fig. 12.22) is fitted to vehicles where the engine is mounted on its side, or where

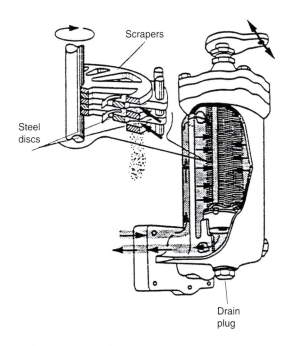

Fig. 12.21 Disc-type oil filter

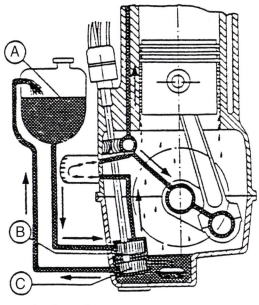

A – Reservoir
B – Pressure pump
C – Scavenge pump

Fig. 12.22 Operation of dry-sump lubrication system

greater ground clearance is required. It is also used for motor cycle engines, cross-country vehicles, and racing engines, where under certain conditions the pick-up pipe could be exposed for a period of time and therefore the oil supply to the engine lubrication system could be interrupted. To overcome this problem, a dry-sump system is often fitted.

The oil is stored in a separate oil tank instead of in the sump. The oil pump takes the oil from the tank and passes it to the lubrication system. The oil then drops down to the crankcase, where a separate scavenge oil pump often running at a higher speed than the pressure pump returns it back to the oil tank. This means that the sump remains almost dry. The faster speed of the scavenge pump is due to the fact that it must be capable of pumping a mixture of air and oil (which has a larger volume than just oil) back to the tank.

Oil coolers (sometimes called heat exchangers)

Two types of oil coolers are fitted where the heat is removed from the oil: one is the oil-to-air, where the heat is passed directly to the air; the other is the oil-to-water, where the heat from the oil is passed to the water cooling system. Both types are shown in Fig. 12.23. In water-cooled engines the oil cooler is normally located in front of, and sometimes combined with, the radiator. The advantage of the water-type heat exchanger is that the oil and water are operating at roughly the same temperature and each is maintained at its most efficient working temperature under most operating conditions. In air-cooled heat exchanger engines the cooler is usually located in the air stream of the cooling fan and is similar in construction to the cooling system radiator. An oil cooler bypass valve is fitted in the system, which allows the oil to heat up more rapidly from cold by initially restricting its circulation to the engine only.

Total loss lubrication

There is one system commonly used that has so far not been mentioned — that is the total loss system. This is where the oil used to lubricate the piston, main, and big-end bearings is burnt during the combustion stroke, and therefore lost through the exhaust system to the atmosphere. One example of this is the two-stroke petrol engine used in the motor cycle.

Learning tasks

1. Remove the pump from a motor cycle engine and identify the type of pump fitted and method of operation. After reassemby, immerse the pump in clean engine oil and test the pumping side for pressure and the scavenge side for suction. Record the results.
2. Give two methods used in the total loss lubrication system of introducing oil into the engine.
3. How should engine oil be removed from the workshop floor?
4. List the main safety precautions that should be observed when working on the lubrication system.

Self-assessment questions

1. Figure 12.24 shows a positive crankcase ventilation system.
   (a) Describe how the PCV valve operates. How does fresh air circulate through the engine?
   (b) Why is such a system fitted to engines?
   (c) What will be the effect on engine idling if the PCV valve is stuck in the open position?
   (d) Examine a service schedule to see how frequently the PCV system should be checked.

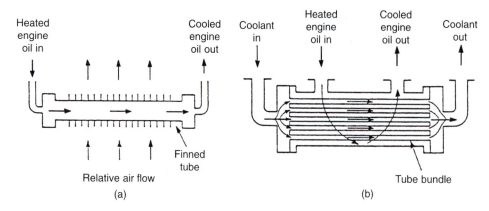

Fig. 12.23 Basic principles of oil coolers. (a) Oil-to-air. (b) Oil-to-water

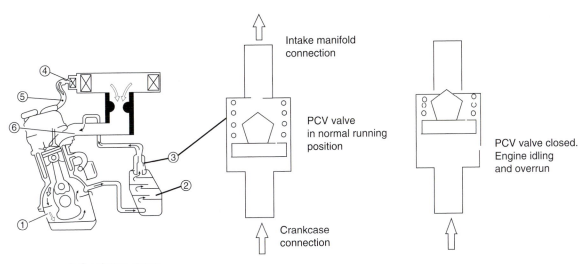

Intake manifold connection

PCV valve in normal running position

Crankcase connection

PCV valve closed. Engine idling and overrun

1. Crankcase gases
2. PCV breather chamber and moisture separator
3. The PCV control valve
4. Blow by filter – permits fresh air into the crankcase

Fig. 12.24   Crankcase ventilation system

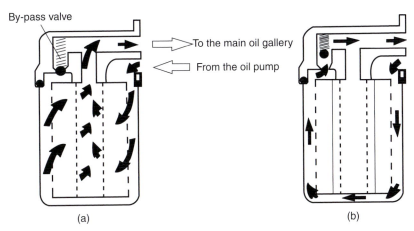

By-pass valve

To the main oil gallery

From the oil pump

(a)

(b)

Fig. 12.25   Full-flow oil filter

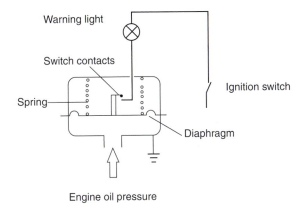

Warning light

Switch contacts

Spring

Ignition switch

Diaphragm

Engine oil pressure

Fig. 12.26   Oil pressure warning light switch

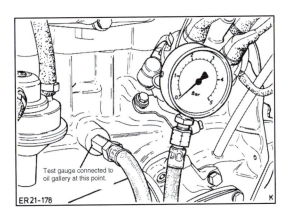

Test gauge connected to oil gallery at this point.

ER 21-178

Fig. 12.27   Oil pressure check

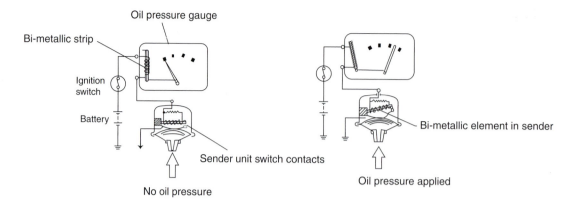

Fig. 12.28 Oil pressure gauge

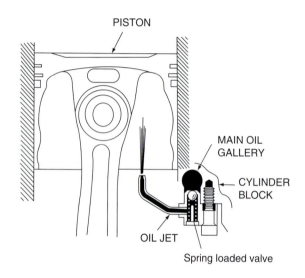

Fig. 12.29 Oil sprayed on underside of piston

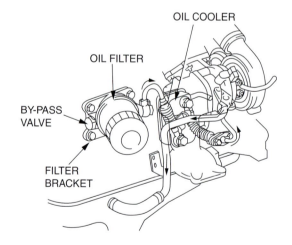

Fig. 12.30 An oil-to-water cooler

2. Figure 12.25 shows two views of a full-flow filter. Describe the function of the bypass valve. What is the condition of the filter element that is shown in B? What routine service operation is performed in order to prevent the occurrence shown in B?

3. In which position on the engine is the oil warning light switch (Fig. 12.26) fitted? In the figure the switch contacts are open; will the warning light be on or off?

4. Figure 12.27 shows a test gauge connected in place of the oil pressure switch in order to make an accurate check on the oil pressure under operating conditions. On the engine shown the pressure should be:
   - a minimum of 0.6 bar at idling speed
   - a minimum of 1.5 bar at 2000 rev/min.

   What should be the engine temperature when conducting the tests? Make a list of the factors that may case the oil pressure to be lower than these figures. What is the probable result of running an engine with oil pressure that is lower than the accepted minimum?

5. The oil pressure gauge circuit shown in Fig. 12.28 contains two bimetallic elements. What causes the sender unit switch contacts to close? How does the bimetallic element cause the gauge needle to move? What prevents the gauge needle from moving completely to the highest position when the sender switch contacts are responding to a low oil pressure?

6. Figure 12.29 shows a system that is used to spray oil at a pressure of 2.5 bar on to the underside of a piston. What is the reason for doing this? What part of the system prevents oil being sprayed when the main oil pressure is less than 2.5 bar?

7. Describe, with the aid of a sketch, how an oil cooler of the type shown in Fig. 12.30 operates. Why is such an oil cooler likely to be found on a turbocharged engine?

8. Why is the filter element in Fig. 12.31 convoluted as shown?

9. If a worn engine oil pump causes low oil pressure:
   (a) the oil pressure can be increased by increasing the pressure exerted by the spring in the oil pressure-relief valve
   (b) increasing the relief valve spring pressure will make no difference

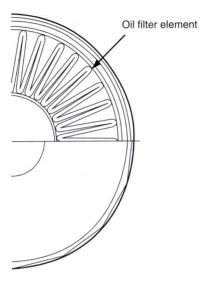

Oil filter element

Fig. 12.31   Section of an oil filter element

(c) the engine oil should be changed to one having a higher viscosity index

(d) the bypass valve in the oil filter should be renewed.

10. A full-flow oil filter is fitted with a bypass valve:
    (a) to prevent excessive oil consumption
    (b) to allow oil to reach the main galleries in the event of a filter blockage
    (c) as a substitute for an oil pressure-relief valve
    (d) to allow excess oil from the gallery to leak back through the oil pump.

11. Big-end bearings are lubricated:
    (a) by oil mist that rises from agitation of oil in the sump
    (b) by splash lubrication
    (c) by oil that is fed to them by oilways in the crankshaft
    (d) by oilways drilled in the connecting rods.

13
Cooling systems

Topics covered in this chapter

Engine cooling systems
Common faults and their diagnosis and repair
Blocked radiator core
Combustion gas leakage into coolant
Coolant loss
Antifreeze and its composition
Vehicle interior heating and air-conditioning systems

During combustion, when the engine is operating at full throttle, the maximum temperature reached by the burning gases may be as high as $1500-2000°C$. The expansion of the gases during the power stroke lowers their temperature considerably, but during the exhaust stroke the gas temperature may still be approximately $800°C$. All the engine components with which these hot gases come into contact will absorb heat from them in proportion to:

- the gas temperature;
- the area of surface exposed to the gas;
- the duration of the exposure.

Engine operating temperatures are shown in Fig. 13.1.

Overheating

For all these reasons the heat will raise the temperature of the engine components. If the temperature of the exhaust gas is above red heat it will be above the melting point of metals such as aluminium, from which the pistons are made. Unless steps are taken to reduce these temperatures a number of serious problems could arise:

- The combustion chamber walls, piston crown, the upper end of the cylinder, and the region of the exhaust port are exposed to the hottest gases and will therefore reach the highest temperatures. This will create distortion, causing a leakage of gas, water, or oil. It may even cause the valve to burn or the

cylinder head to crack and as a consequence there will be a loss of power output.
- The oil film will be burnt, causing excessive carbon to form. The loss of lubrication of the piston and rings will cause excessive wear or the piston to seize in the cylinder.
- Power output will be reduced because the incoming mixture will become heated, so reducing its density. It may also cause **detonation** (this is an uncontrolled explosion in the cylinder), making it necessary to reduce the compression ratio.

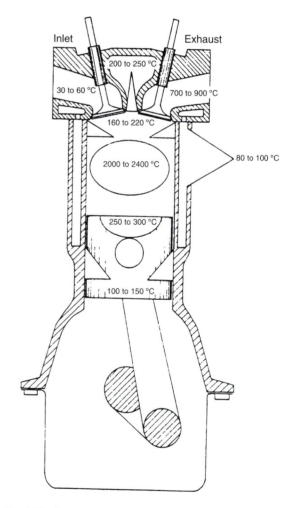

Fig. 13.1 Engine operating temperature ranges

- Some part of the surface of the combustion chamber could become hot enough to ignite the incoming charge before the spark occurs (called **pre-ignition**), which could cause serious damage to the engine if allowed to continue.

For these reasons the engine must be provided with a system of cooling, so that it can be maintained at its most efficient practicable operating temperature. This means that the average temperature of the cylinder walls should not exceed about 250°C, whereas the actual temperature of the gases in the cylinder during combustion may reach 10 times this figure. One of the other things to remember is that the engine should not be run too cool, as this would reduce **thermal efficiency** (this is how good the engine is at converting heat into mechanical power), increase fuel consumption and oil dilution, and cause wear and corrosion of the engine.

Heat transfer

The cooling system works on the principles of **heat transfer**. Heat will always travel from hot to cold (e.g. from a hot object to a cold object; this would be by conduction). This transfer occurs in three different ways:

- Conduction
- Convection
- Radiation.

Conduction is defined as the transfer of heat between two solid objects, e.g. valve stem to valve guide, as shown in Fig. 13.2. Since both objects are solid, heat

is transferred from the hot valve stem to the cool valve guide by conduction and also from the guide to the cylinder head.

Convection is the transfer of heat by the circulation of heated parts of a liquid or gas. When the hot cylinder block transfers heat to the coolant it produces a change in its density and causes the warmer, less dense water to rise, thus setting up convection currents in the cooling system.

Radiation is defined as the transfer of heat by converting it to radiant energy. Radiant heat is emitted by all substances and may be reflected or absorbed by others. This ability will depend upon the colour and nature of the surface of the objects — for example, black rough ones are best for absorption of heat and light polished ones best for reflection of heat.

The cooling system relies on all three of these principles to remove excess heat from the engine.

> *Learning tasks*
>
> 1. Take a look at three different types of radiators and check the colour and texture of the surface finish. Why are they finished like this?
> 2. How is the heat taken away from the top of the piston and spark plug? Which of the three methods named above are used?

Over-cooling

As we have seen, various problems can occur if the engine temperature gets too high, but if the temperature becomes too low then another set of problems can occur:

- Fewer miles per gallon, as the combustion process will be less efficient.
- There will be an increase in the build-up of carbon (as the fuel enters the cylinder it will condense and cause excessive build-up of carbon on the inlet valves).
- There will be an increase in the varnish and sludges formed within the lubrication system. Cooler engines make it easier for these to form.
- A loss of power, because if the combustion process is less efficient the power output will be reduced.
- The fuel not being burned completely, which will cause fuel to dilute the oil and cause excessive engine wear.

The purposes of the cooling system can be summarized as follows:

- To maintain the highest and most efficient operating temperature within the engine
- To remove excess heat from the engine

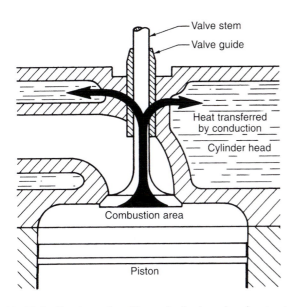

Fig. 13.2 Heat is transferred by conduction from the valve stem to the valve guide. Both objects are solid

- To bring the engine up to operating temperature as quickly as possible — in heavy-duty driving, an engine could theoretically produce enough heat to melt an average 100 kg engine block in 20 minutes.

Types of cooling systems

There are two main types of cooling systems in common use, air and water. Both dissipate (radiate) heat removed from the cylinder into the surrounding air. Air cooling is described immediately below and water cooling later in the chapter.

Air cooling

In this system heat is radiated from the cylinder and head directly into the surrounding air. The rate at which heat is radiated from an object is dependent on:

1. The difference in temperature between the object and the surrounding air.
2. The surface area from which the heat is radiated (since (1) must be limited, the surface area of the cylinder and head exposed to the air must be increased, by forming fins on their external surfaces; Fig. 13.3).
3. The nature of the surface.

It is also necessary to remove the heated air from around the cylinder and deliver a constant supply of cool air around and between the fins. This means that the cylinders must be sufficiently widely spaced to permit a suitable depth of finning all around them, and the engine must be placed where the movement of the vehicle can provide the necessary supply of cool air. A large **fan** is often used and the engine is surrounded by large **cowls** to direct air to where it is required, e.g. around the cylinder head and valve area.

The choice of air or liquid cooling has always been controversial. Air is cheaper, lighter, and more readily obtainable than water — though to remove a given quantity of heat demands four times the weight and 4000 times the volume of air than it does water. It also gives less control of the engine temperature and air-cooled engines tend to be noisier. But air can be collected and rejected, whereas water must be carried on the car, and the jacketing, hoses, pump and radiator of a water-cooled engine will probably weigh more than the substantial fins of an air-cooled engine.

However, the fins force the cylinders of an air-cooled engine to be more widely separated than those of a water-cooled engine, so the crankshaft and crankcase must be longer and therefore heavier. For an engine of many cylinders, this is one of the greatest objections to air cooling.

Engine air-cooling system

Circulation of cooling air

With air cooling the engine structure is directly cooled by forcing air over its high-temperature surfaces. These are finned to present a greater cooling surface area to the air, which in non-motor-cycle applications is forced to circulate over them by means of a powerful fan.

The engine structure is almost entirely enclosed by sheet metal ducting (called a cowl), which incorporates a system of partitions (called **baffles**); these ensure that the air flow is properly directed over the cylinders and cylinder heads. To obtain uniform temperatures the air is forced to circulate around the entire circumference of each cylinder and its cylinder head. The direction of air flow will be along the cooling fins, the greatest number of which are found towards the top of the cylinder (around the exhaust valve), as this is the hottest part of the engine.

The complete system forms what is known as the **plenum chamber**, in which the internal air pressure is higher than that of the atmosphere. The heated air is discharged from the plenum chamber to the atmosphere, or redirected to heat the car interior. Figure 13.4 shows a modern, air-cooled diesel engine. In order to provide the necessary air flow around the cylinders of an enclosed engine, a powerful fan is essential.

Types of air-cooling fans
Axial flow

This is a simple curved blade type in which the direction of air flow is parallel to the axis of the fan spindle (Fig. 13.5).

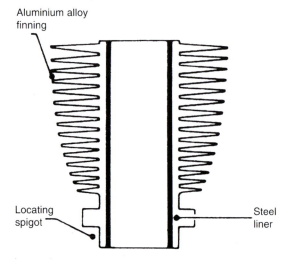

Aluminium alloy finning

Locating spigot

Steel liner

Fig. 13.3   Air-cooled finned cylinder wall method of heat transfer

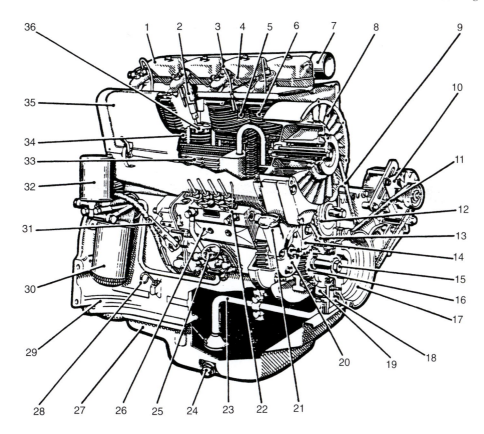

1 – Rocker chamber cover
2 – Injector
3 – Injection line to no.3 cylinder
4 – Back-leakage line
5 – Cylinder head anti-fatigue bolt (four bolts securing each cylinder head with cylinder to crankcase)
6 – Cylinder head (light alloy)
7 – Air intake manifold
8 – Cooling blower (V-belt driven)
9 – Cooling blower V-belt
10 – Generator (dynamo or alternator)
11 – Generator V-belt
12 – Camshaft gear
13 – Oil gallery
14 – Idler gear (driving injection pump and camshaft)
15 – Anti-fatigue bolt (securing V-belt pulley to crankshaft)
16 – Crankshaft gear
17 – V-belt pulley
18 – Vibration damper
19 – Oil pump
20 – Injection pump drive gear with advance/retard unit
21 – Oil filler neck
22 – Overflow line
23 – Oil suction pipe
24 – Oil drain plug
25 – Fuel feed pump
26 – Bosch in-line injection pump with mechanical centrifugal governor
27 – Oil sump (sheet metal or cast iron)
28 – Oil dipstick
29 – Crankcase (cast iron)
30 – Oil filter
31 – Speed control lever
32 – Fuel filter
33 – Integral oil cooler
34 – Finned cylinder (grey cast iron), separately removable
35 – Removable air cowling
36 – Piston

Fig. 13.4   Cut-away view of modern air-cooled diesel engine

Radial flow

Often called a **centrifugal**, this type (Fig. 13.6) is more commonly used because it is more effective and a fan of smaller diameter can be used for a given air flow. This type of fan has a number of curved radial vanes mounted between two discs, one or both having a large central hole. When the fan is rotated, air between the vanes rotates with it and is thrown outwards by centrifugal force.

Figure 13.7 shows a simple air-cooled system for a four-cylinder in-line engine. A centrifugal fan, driven at approximately twice crankshaft speed, is mounted at the front of the engine and takes in air through a central opening (5) in the fan casing. This air is delivered into the cowl (3), where it is directed over the fins of the cylinders (1). Baffles (2) ensure that the air passes between the fins, where it picks up heat, thus cooling the cylinders.

The in-line engine is the most difficult to cool by air. V-type or horizontally opposed engines are easier to cool as the cylinders are spaced further apart to leave room for the crankshaft bearings and this allows more room between the cylinders for a good air flow, whilst at the same time keeping the total engine length fairly short.

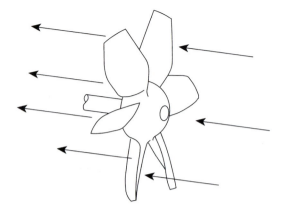

Fig. 13.5   Axial air flow fan

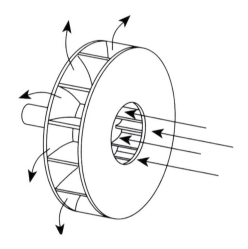

Fig. 13.6   Radial air flow fan

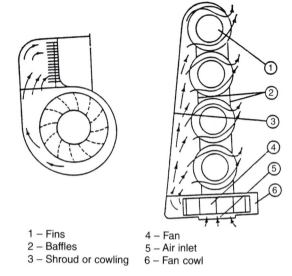

1 – Fins   4 – Fan
2 – Baffles   5 – Air inlet
3 – Shroud or cowling   6 – Fan cowl

Fig. 13.7   Simple in-line air-cooled engine using a radial-type fan

---

Learning tasks

1. What safety precautions should be observed when working on air-cooled systems?
2. What are the main difficulties with working on air-cooled engines?
3. Remove the cowling from an air-cooled engine, clean any excessive dirt, etc. from between the fins of the cylinders. Reassemble the cowling and test the engine for correct operating temperature.

Liquid cooling

In this arrangement the outer surfaces of the cylinder and head are enclosed in a jacket, leaving a space between the cylinder and the jacket through which a suitable liquid is circulated. The liquid generally used is water, which is in many ways the most suitable for this purpose, even though it has a number of drawbacks.

Whilst passing through the jacket the water absorbs heat from the cylinder and head, and it is cooled by being passed through a radiator before being returned to the jacket.

Thermo-syphon system

When heated, the water becomes less dense and therefore lighter than cold water. Thermo-syphon is the action of the water being heated, rising, and setting up convection currents in the water. The thermo-syphon system is no longer used in the modern motor vehicle as it has a number of disadvantages:

- To ensure sufficient circulation, the radiator must be arranged higher than the engine to ensure that the heated coolant will rise into the top of the radiator header tank and the cooled water in the radiator will flow into the bottom of the engine.
- Water circulation will be slow, so a relatively large amount of water must be carried.
- Large water passages must be used to allow an unrestricted flow of water around the system.

This system is usually now confined to small stationary engines such as those used to power narrow boats, small generators, water pumps, etc.

Pump-assisted circulation

Most modern engines use a pump to provide a positive circulation of the coolant. This is shown in Fig. 13.8 and gives the following advantages:

1. A smaller radiator can be used than in the thermo-syphon system.

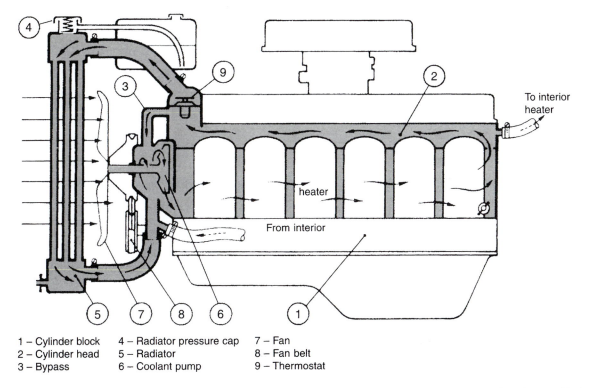

1 – Cylinder block   4 – Radiator pressure cap   7 – Fan
2 – Cylinder head   5 – Radiator   8 – Fan belt
3 – Bypass   6 – Coolant pump   9 – Thermostat

Fig. 13.8 Coolant is pumped from the water pump, through the cylinder block and heads, through the thermostat into the radiator, and back to the water pump

2. Less coolant is carried as the water is circulated faster and therefore the heat is removed more quickly.
3. Smaller passages and hoses are used because of (2) above.
4. The radiator does not need to be above the level of the engine, giving a lower bonnet line; this also has the advantage of less wind resistance, giving a better fuel consumption.
5. Because the water flow is given positive direction, the engine will operate at a more even temperature.

Comparison of air- and water-cooled systems

Advantages of air cooling

- An air-cooled engine is generally lighter than an equivalent water-cooled engine.
- It warms up to its normal running temperature very quickly.
- The engine can operate at a higher temperature than a water-cooled engine.
- The system is free from coolant leakage problems and requires no maintenance.
- There is no risk of damage due to freezing of the coolant in cold weather.

Disadvantages of air cooling

- A fan and suitable cowls are necessary to provide and direct the air flow. The fan can be noisy and absorbs a large amount of engine power. The cowl makes it difficult to get at various parts of the engine when servicing is required.
- The engine is more liable to overheating under difficult conditions than a water-cooled engine.
- Mechanical engine noises tend to be amplified by the fins.
- The cylinders usually have to be made separately to ensure proper formation of the fins. This makes the engine more costly to manufacture.
- Cylinders must be spaced well apart to allow sufficient depth of fins.
- It is more difficult to arrange a satisfactory car-heating system.

Advantages of water cooling

- The temperatures throughout the engine are more uniform, thus keeping distortion to a minimum.
- Cylinders can be placed closer together, making the engine more compact.
- Although a fan is usually fitted to force air through the radiator, it is much smaller than the type required for an air-cooled engine. It therefore absorbs less power and is quieter in operation.

- There is no cowl to obstruct access to the engine.
- The water jacket absorbs some of the mechanical noise, making the running engine quieter.
- The engine is better able to operate under difficult conditions without overheating.

Disadvantages of water cooling

- Weight — not only of the radiator and connections but also of the water; the whole engine installation is likely to be heavier than an equivalent air-cooled engine.
- Because the water has to be heated, it takes longer to warm up after starting from cold.
- If water is used, the maximum temperature is limited to about 85–90°C to avoid the risk of boiling away the water. However, modern cooling systems are pressurized and this permits higher temperatures and better efficiency.
- If the engine is left standing in very cold weather, precautions must be taken to prevent the water freezing in the cylinder jackets and cracking them.
- There is a constant risk of a coolant leakage developing.
- A certain amount of maintenance is necessary — for example, checking water level, anti-frost precautions, cleaning out deposits, etc.

Learning task

Using the above information write a short paragraph to explain the main components and their purpose in the water cooling system.

Radiator and heater matrix

The purpose of the radiator is to provide a cooling area for the water and to expose it to the air stream. A reservoir for the water is included in the construction of the radiator. This is known as the header tank and is made of thin steel or brass sheet and is connected to the bottom tank by brass or copper tubes; these are surrounded by 'fins'. This assembly is known as the **matrix**, **core**, **block**, or **stack**. The more modern radiator uses plastic for the tanks and aluminium for the matrix. Shown in Fig. 13.9(a) is a conventional type of radiator. Figure 13.9(b) shows a typical type of cross-flow radiator that has an integral oil cooler fitted.

The function of the radiator, as we have said, is to transfer heat from the coolant to the air stream. It is designed with a very large surface area combined with a relatively small frontal area, and it forms a container for some of the coolant in the system. The radiators

usually have mounting feet or brackets, a filler cap, an overflow pipe, and sometimes a drain tap is fitted to the lower tank.

A large number of different types of radiator matrix are in common use depending on application, size of engine, etc. Details of the main types in common use are given below.

Film core

The tubes are the full width of the core and are bent to form square spaces through which air can pass. They are sometimes crinkled to extend their length. The top and bottom tanks are secured to side frames with the core located between them and a fan cowl often completes the unit.

Tube and fin

This consists of copper or brass tubes of round, oval, or rectangular cross-section. The tubes pass through a series of thin copper fins with the top and bottom tanks attached to the upper and lower fins respectively. The fins secure the tubes and increase the surface area from which the heat can be dissipated. The tubes are placed edge-on to the air flow for minimum air resistance and they are now produced from strip lock-seaming.

Tube and corrugated film

Sometimes used as an alternative to the tube and fin, the corrugated separator filming is made from copper and laid between the tubes to provide an airway. Each face of the filming is louvred to increase air turbulence as the air passes through. This improves the cooling efficiency of this design. A commonly used form of radiator matrix (core) construction is shown in Fig. 13.10.

Separate tubes

Radiators with separate coolant tubes are occasionally used. They provide a stronger core than the other types but they are more costly to build, heavy, time-consuming to repair, and because of this are mainly confined to commercial vehicle applications. The tanks and side frames are usually bolted together and locate the thick-walled tubes of rectangular or circular cross-section. The tubes are made watertight in the upper and lower tanks with rubber and metal seals, and they have bonded copper fins or a spiral copper wire wound over their complete length to increase their ability to dissipate the heat. Tube removal and refitting may be done by two methods depending on their construction. One is to remove top and bottom tanks from their side

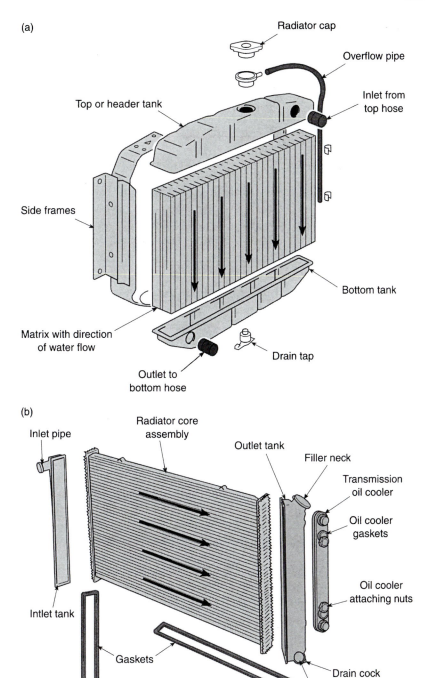

Fig. 13.9 (a) Conventional vertical flow radiator. (b) The parts of a standard cross-flow radiator

frames, the other is to spring the tubes in and out, which is only possible because of their flexibility.

Separate expansion or header tanks

Separate expansion or header tanks are now commonly used. These allow the radiator to be fitted lower than the engine and, on a commercial vehicle, to be fitted in a more accessible position for checking and refilling the coolant. The tank is also used to reduce the risk of **aeration** (this is air bubbles forming in the water) of the coolant when the engine is running. The early types of radiators were all of the conventional type — for example, the coolant flows from the top of the radiator to the bottom (vertically). Many vehicles now use a cross-flow type in which the coolant flows horizontally through the core from the top of one side tank across to the bottom of the other side tank, as shown in Fig. 13.11.

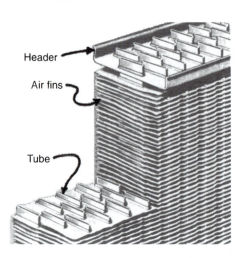 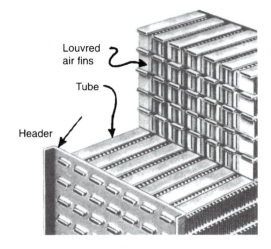

Fig. 13.10   Inside the radiator there are small tubes through which the coolant flows, helping to remove the heat into the air

---

Learning tasks

1. What are the advantages of a cross-flow radiator?
2. How should the radiator be tested for leaks when removed from the vehicle?
3. Why is the radiator painted matt black?

Water pump

The purpose of the water pump is to provide a positive means of circulating the water (it gives a direction to its flow, it does not pressurize the system).

The body of the pump is usually made of cast iron or aluminium. In most cases it is bolted to the front of the engine block and draws cool water from the bottom of the radiator via the bottom hose. This cooler water is directed into the water jacket and over the hottest part of the engine, such as the exhaust valve seats. Two types of pump are used, **axial flow** and **radial flow**.

Axial flow

With axial flow, when the pump is full of water, the rotating impeller carries with it the water contained in the spaces between the impeller blades and the

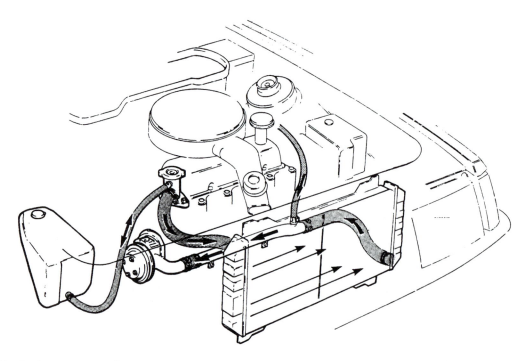

Fig. 13.11   Cooling system water flow

casing. This water is subjected to centrifugal force, which causes it to flow outwards from inlet to outlet. A carbon ring bonded to a rubber sleeve is fitted into the housing and pressed into light contact with a machined face on the impeller by a light spring.

This provides a watertight seal along the shaft. The pump is driven by a belt from the crankshaft via a pulley, which is a press fit on the end of the pump spindle. The fan is then bolted to the pulley, which draws air through the radiator.

Radial flow

The radial flow centrifugal pump operates with a slightly higher flow and therefore the circulation is slightly faster. This type is fitted to commercial vehicles. The construction of such a water pump is shown in Fig. 13.12.

Learning tasks

1. What are the symptoms of the water pump not working? What could be the cause of this fault and how should it be repaired?
2. How should the drive belt to the pump be checked? Remove the belt, check for signs of wear or cracking. Replace and correctly tension the belt.
3. Remove one type of water pump from an engine, check for premature bearing failure, signs of water leaks, and general serviceability. Replace the pump, refill with coolant, and pressure test the system.

Thermostat

This is a temperature-sensitive valve that controls the water flow to the radiator. There are two main reasons for its use:

- To enable the engine to warm up quickly from cold.
- To control the rate of flow and so maintain a constant temperature in the engine.

Two types are in common use, the **wax** type and the **bellows** type.

Wax type

This is used in the pressurized system as it is not sensitive to pressure like the bellows type. A special wax is used, contained in a strong steel cylinder. The reaction pin is surrounded by a rubber sleeve and is positioned inside the cylinder.

As the temperature increases the wax begins to melt, changing from a solid to a liquid, and at the same time it expands. This forces the rubber against the fixed reaction pin, opening the valve against spring pressure, thus allowing the water to circulate through the radiator. There is a small hole in the valve disc to assist in bleeding the system as filling takes place. The 'jiggle pin' closes the hole during engine warm-up. This thermostat is shown in Fig. 13.13.

Bellows type

As shown in Fig. 13.14, the bellows type consists of a flexible metal bellows that is partly filled with a liquid having a boiling point lower than that of water (e.g. alcohol, ether, or acetone). Air is removed from the

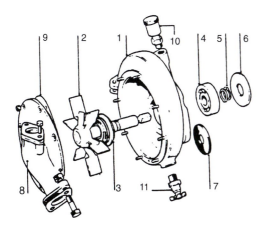

1 – Pump body
2 – Impeller
3 – Unit seal
4 – Ball bearing
5 – Spring clips
6 – Grease retaining end-plate
7 – Synthetic rubber sealing ring
8 – Pump cover
9 – Cover packing
10 – Greaser
11 – Drain cock

Fig. 13.12 Radial flow centrifugal pump

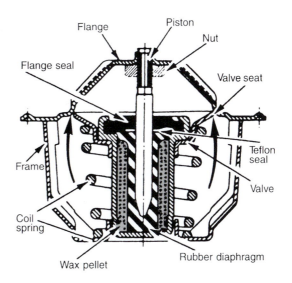

Fig. 13.13 Wax pellet-type thermostat

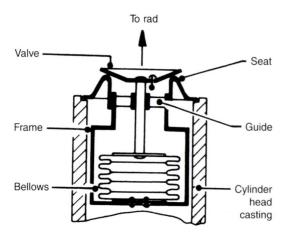

Fig. 13.14   Bellows-type thermostat

bellows, leaving only the liquid and its vapour. The pressure in the bellows is then only due to the vapour pressure of the liquid. This varies with temperature and is equal to atmospheric pressure at the boiling temperature of the liquid, less at lower temperatures and more at higher temperatures. As the temperature of the water increases, the liquid in the bellows begins to turn to a vapour and increases in pressure; this expands the bellows and opens the valve, allowing water to pass to the radiator.

Note

This type is not suitable for pressurized systems as the valves are pressure sensitive. The wax element type does not have this disadvantage.

Testing thermostats

The thermostat cannot be repaired and so must be replaced if found to be faulty. It can be tested by placing the thermostat in a beaker of water and gradually heating it. A thermometer is used to check the temperature of the water (Fig. 13.15). The thermostat should begin

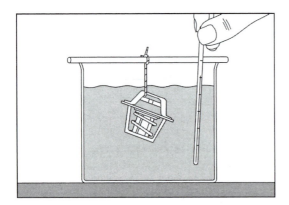

Fig. 13.15   Thermostat suspended in water container

to open when the temperature marked on the valve is reached. An increase of approximately $10-20°C$ will elapse before the valve is fully open.

Learning tasks

1. Write a schedule for removing, testing, and replacing a thermostat that is suspected of being faulty.
2. What is the symptom, fault, cause, and remedial action that should be taken when the driver complains of the heater blowing hot and cold?
3. Remove, test, and replace a thermostat according to your work schedule drawn up in (1) above. State the type of unit fitted, the temperature at which it opened/closed, and check with the manufacturer's data.

Pressurized cooling systems

Pressurized cooling systems are used because they allow the engine to operate at a higher temperature. Figure 13.16 shows the layout and main components of a modern pressurized system.

Pressure cap

The cap contains two valves; one is the **pressure valve** the other is the **vacuum valve** (Fig. 13.17). As the temperature of the water increases it expands and in a sealed system this expansion increases the pressure until it reaches the relief pressure of the cap. As the system cools down it contracts and opens the vacuum valve, drawing in air. If no vacuum valve was fitted the depression in the system, caused by the contracting effect of the water as it cools, could cause the rubber hoses in the system to collapse. Most pressure caps operate at $28-100$ kN/m. The pressure is usually stamped on the cap, indicating the maximum relief pressure for the system to which it is fitted. Figure 13.17(c) shows a pressure cap in the open position.

Safety note

Never remove the cap when the coolant temperature is above $100°C$ as this will allow the water to boil violently; the resulting jet of steam and water from the open filler can cause very serious scalds. The system should be allowed to cool down and the cap removed slowly. It is designed so that the spring disc remains seated on the top of the filler neck until after the seal has lifted. This allows the pressure to escape through the vent pipe before it can escape from the main opening.

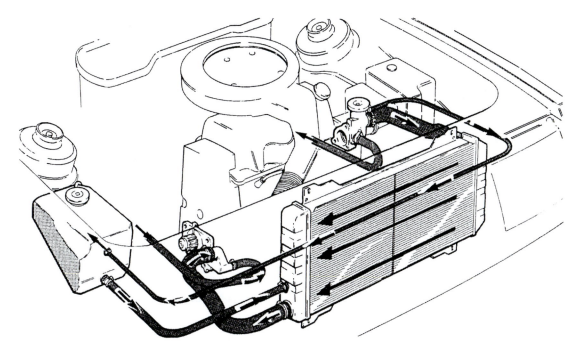

Fig. 13.16 'Degas' system with coolant at normal operating pressure and temperature

The temperature at which a liquid boils rises as the pressure acting on it rises. This is shown in Fig. 13.18. The cooling system's pressure is maintained by the use of a pressure cap fitted to the top of the radiator or expansion bottle. This closes the system off from the atmosphere, creating a sealed system.

The advantages of using a pressurized system are:

- Elimination of coolant loss by surging of the coolant during heavy braking.
- Prevention of boiling during long hill climbs, particularly in regions much above sea level.
- Raising of the working temperature, improving engine efficiency.
- Allowing the use of a smaller radiator to dissipate the same amount of heat as a larger one operating at a lower temperature.

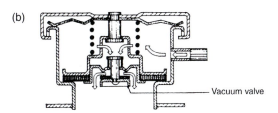

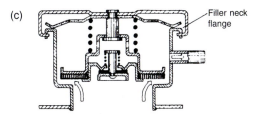

Fig. 13.17 Open-type pressure cap used in the semi-sealed system. (a) Pressure-relief valve open. (b) Vacuum-relief valve open. (c) Cap-removal precaution action

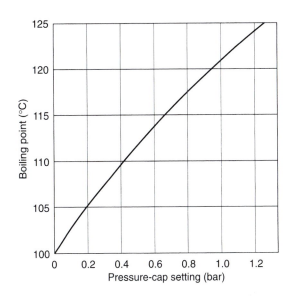

Fig. 13.18 Variation of boiling point of water with pressure

Semi-sealed system

The advantage of the modern sealed cooling system is that it reduces the need for frequent inspection of the coolant level and the risk of weakening the antifreeze solution by topping up. Although the pressure cap provided a semi-sealed system, the fully sealed system has in effect a means of recovering the coolant that is lost when the engine is at its operating temperature and the pressure cap is lifted. It consists of an **expansion tank** that is mounted independently from the radiator and vented to the atmosphere. A flexible hose connects the overflow pipe of the radiator to a dip tube in the overflow tank. As the coolant heats up the pressure valve in the cap opens and excess coolant passes to the expansion tank. When the system cools down the coolant is drawn back into the radiator again through the vacuum valve in the cap. This is shown in Fig. 13.19.

Fully sealed system

In the fully sealed system, the pressure cap is fitted to the expansion tank and a simple cap is used on the radiator. The operation of the system is much the same as in the semi-sealed system. The reservoir is vented to the atmosphere and should be approximately two-thirds full. If the coolant falls below the minimum level indicated on the tank, air may be drawn into the radiator.

> **Learning tasks**
>
> 1. What is the correct method for testing the pressure and vacuum valves in the cap? What would be the effect of fitting a cap with: (a) A stronger spring? (b) A weaker spring?
> 2. How would you test for water/oil contamination and which component would you suspect to be faulty?

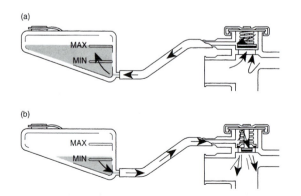

Fig. 13.19 Operating function of the overflow bottle. (a) Water heating up. (b) Water cooling down

Fans and their operations

Fans and temperature control

An important aspect of the cooling capabilities of the radiator is the volume of air, in unit time, which can be caused to flow through the matrix. Hence, the purpose of the cooling fan is to maintain adequate air flow through the matrix, at low and engine idling speeds. The speed of the fan ranges from slightly less to rather more than that of the engine. Excessive fan speeds are avoided because of noise and the power required to drive it.

Electrically driven fans

With this type of fan arrangement an electric motor is used together with a fan and a temperature-sensitive control unit. The advantages of this arrangement are:

- The fan only operates when the engine reaches its predetermined temperature.
- The engine will be more efficient as the fan is not being driven all the time.
- The radiator and fan can now be fitted in any convenient position, ideal for transversely mounted engines.
- The fan assembly can be mounted in front of or behind the radiator.
- Engine temperature is more closely controlled as the temperature sensor will automatically switch the fan on and off within very close limits as required.

A typical layout is shown in Fig. 13.20.

To improve the efficiency of the fan, a cowl is often fitted. Its function is to prevent heated air from reversing its flow past the fan and recirculating through the matrix, which could lead to engine overheating. The fan can be formed from some of the following materials:

- One-piece steel pressing
- Aluminium casting
- Plastic moulding with a metal insert.

An uneven number of irregularly spaced blades (Fig. 13.21) are often employed to minimize fan noise. In all applications the fan assembly must be accurately balanced and the blades correctly aligned to avoid vibration.

Flexible fan blades

Used to reduce frictional power loss from the fan, these are made from fibreglass, metal, and moulded plastic such as polypropylene. The plastic fans are lighter, easier to balance, look better, are more efficient aerodynamically, have a reduced noise level, a reduction in vibration, and offer less risk of serious injury. They

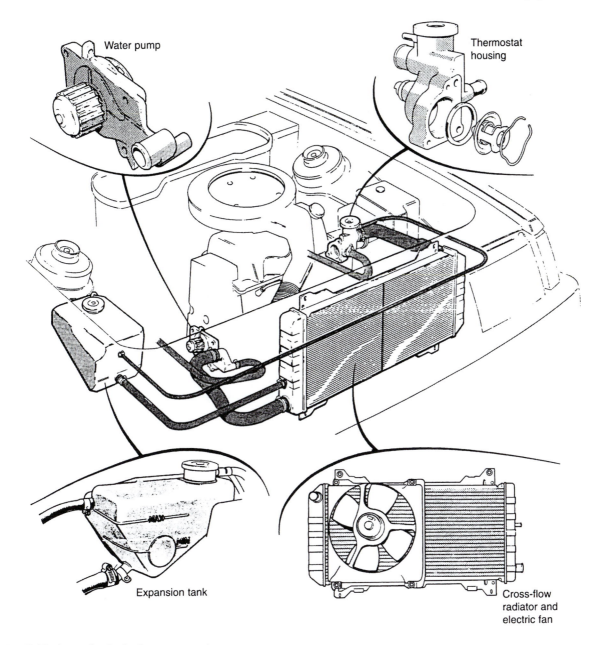

Water pump

Thermostat housing

Expansion tank

Cross-flow radiator and electric fan

Fig. 13.20 Layout for the Ford transverse engine

are also cheaper to manufacture. For these reasons the plastic fan is the most common of this type fitted to light vehicles.

A number of vehicles have a cover surrounding the fan (called a **shroud**) fitted to make sure the fan pulls air through the *entire* radiator. Figures 13.20 and 13.22 show arrangements commonly used. When no shroud is used the fan will only pull air through the radiator directly in front of the blades. There is very little air moving through the corners of the matrix.

With flexible fan blades, as the speed of the engine increases, the blades flatten and therefore move less air; this give a reduction in the power lost to friction. An example of a flexible fan with a shroud is that used in the Mini; the air is forced through the radiator, not drawn through as in most vehicles.

Cooling fan with swept tips

The flexible tips of the fan tend to straighten out as speed increases and this has the effect of reducing the amount of air drawn through the fan.

Cooling fan surrounded by cowl or shroud attached to the radiator

The shroud is a close fit around the radiator and fan to prevent the air from just circulating around the fan

Fig. 13.21 Unevenly spaced fan blades

and not passing through the radiator matrix and cooling the water.

Directly driven fans

Directly driven fans are not normally used in light vehicles because they have the following disadvantages:

- Rising noise level
- Increase in power used
- Tendency to over-cooling at higher speeds.

Some engine manufacturers still use directly driven fans mainly because of the type of engine fitted or because space is limited. The opposed piston engine often has the fan blades bolted directly on the end of the crankshaft, usually with some means of limiting the maximum speed of the fan. LGVs (large goods vehicle) may also use the same method of driving the fan.

Fan drives

The drive for the fan is normally located on the water pump shaft. Its purpose is to draw sufficient air through the matrix for cooling purposes in heavy traffic on a hot day and to ensure adequate cooling at engine idling speeds.

At moderate speeds — say, driving down the motorway — the engine does not require the same amount of assistance from the fan, as the natural flow of air passing through the radiator will often be sufficient to cool the water. Power would be lost unnecessarily to the fan under these conditions. A number of different methods of controlling fan operation and speed are becoming more popular.

Automatically controlled fans are generally classified into four types:

- Free-wheeling
- Variable speed
- Torque limiting
- Variable pitch.

Free-wheeling

The free-wheeling type is mounted on the coolant pump shaft and may form part of an **electromagnetic clutch** with the drive pulley. Connected into the electrical supply circuit to the clutch assembly is a thermostatically controlled switch. When the fan is no longer required, the supply to the electromagnet is interrupted, so that it disengages and allows the fan to free-wheel on the pump shaft.

Variable speed

The variable-speed type generally has a **viscous coupling**, which permits motion or **slip** to occur between the driving and the driven members. The driving member consists of a disc mounted on the

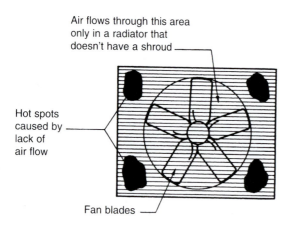

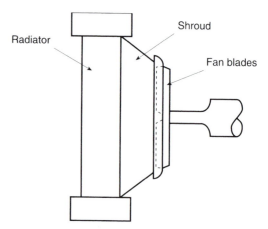

Fig. 13.22 Fan shrouds

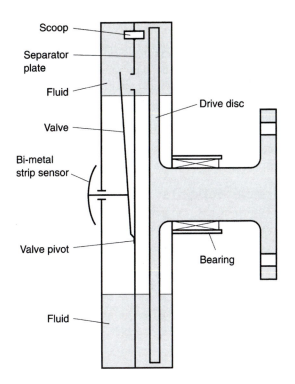

Scoop
Separator plate
Fluid
Valve
Bi-metal strip sensor
Valve pivot
Drive disc
Bearing
Fluid

Fig. 13.23 Simplified diagram showing internal details of a viscous fan drive

pump shaft. Attached to the fan is a sealed cylindrical casing or chamber, freely mounted on the pump shaft. The driving disc revolves within the coupling chamber, which is partially filled with a highly stable **silicone fluid** (this means it retains a nearly constant viscosity with varying operating temperatures). The disc initially revolves in the stationary mass of fluid and sets up a **drag**, so that the fluid begins to circulate in the chamber. As a result of centrifugal force it moves outwards to fill the gaps between the drive faces of the coupling members. Since these faces are very close to one another, drive is transmitted between them by the viscous drag of the silicone fluid. This drag is maintained with increasing fan speed, until the resisting torque imposed upon the fan by the air flow rises to an extent that causes shear breakdown of the fluid film, and the coupling slips. Fan speed is thus controlled by the slipping action of the coupling, which is predetermined by the viscosity of the fluid and the degree to which the coupling is filled. Where no provision is made for the quantity of oil, the coupling is termed a torque-limiting type. The layout of a viscous fan drive is shown in Fig. 13.23.

Torque limiting

The torque-limiting type has the disadvantage of operating within a fixed speed range, regardless of cooling system demands. Thus, air circulation may be inadequate when car speed is low and engine speed is high.

To overcome these difficulties, a temperature-sensitive fan control is used in some installations.

Variable pitch

In the variable-pitch fan the volume of air displaced is controlled by twisting its blades. The term pitch is considered as the distance the blade can twist, if it were turning one revolution in a solid substance (that is with no slip taking place). When the least amount of cooling is required, the blades are automatically adjusted to a low pitch, thereby reducing airflow to a minimum. The variable-pitch fan may be actuated by one of the following methods:

- Centrifugal
- Torque limiting
- Thermostatic means.

An alternative to the engine-driven fan is a separate electrically driven fan, which can be controlled automatically by the thermostat that responds to changes in engine coolant temperature. Although fan installations of this type absorb no engine power, they do impose an electrical load upon the car battery. It has a fixed higher operating speed and it may be noisier than a conventional driven fan, but it has the advantages of greater convenience of mounting position.

> ### Learning tasks
>
> 1. Give reasons for the use of an electrically driven fan over other methods of operation.
> 2. Draw up in logical sequence the way an electric fan should be tested, apart from running the engine up to its normal operating temperature.
> 3. How should a viscous coupling be tested/repaired?

Radiator blinds and shutters

A very simple arrangement of controlling the air flow through the radiator is shown in Fig. 13.24. A spring-loaded **roller blind** is carried at the lower end of a rectangular channel-section frame attached to the front of the radiator. A simple cable control raises the blind to close off as much of the radiator as may be necessary to maintain the correct running temperature of the engine. This is normally operated by the driver from inside the cab.

Another method of controlling the air flow is by the use of a **shutter**. This operates in a similar way to a venetian window blind, each shutter rotating on a separate shaft. Mounted in front of the radiator it can be manually or thermostatically operated.

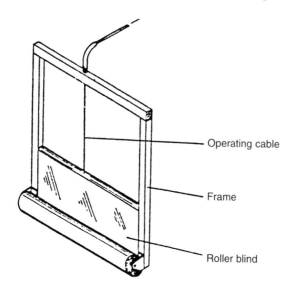

Fig. 13.24 A simple radiator blind

Fan belts

Most fan belts are of the V-type in construction and use the friction produced between the sides of the belt and pulley to provide the drive. The **V-belt** has a larger area of contact and therefore provides a more positive drive between the belt and the pulley. Another popular type is the flat **serpentine** belt; about 35 mm wide, it has several grooves on one side. A larger area is provided by the grooves to give a more positive grip. This type is also used to transmit a drive around a smaller pulley than a conventional V-belt.

Corrosion in the cooling system

This can be very damaging to the engine. It can be caused in several different ways.

Direct attack

This means the water in the coolant is mixed with oxygen from the air. This process produces rust particles that can damage water pump seals and cause increased leakage.

Electromechanical attack

This is a result of using different metals in the construction of the engine. In the presence of the coolant, different metals may set up an electrical current in the coolant. If this occurs, one metal may deteriorate and deposit itself on the other metal. For example, a core plug may deteriorate to the point of causing a leak.

Cavitation

Cavitation is defined as high shock pressure developed by collapsing vapour bubbles in the coolant. These bubbles are produced by the rapid spinning of the water pump impeller. The shock waves cause small pin holes to form in nearby metal surfaces such as the pump impeller or the walls of the wet-type cylinder liner, which could reach all the way through into the cylinder.

Mineral deposits

Calcium and silicate deposits are produced when a hard water is used in the cooling system. Both deposits restrict the conduction of heat out of the cooling system. As can be seen from Fig. 13.25, the deposits cover the internal passages, causing uneven heat transfer.

Antifreeze

Function

When an **ethylene glycol** antifreeze solution is added to the coolant many corrosion problems are overcome. Chemicals are added to the antifreeze to reduce corrosion. It is not therefore necessary to add a corrosion inhibitor to the coolant when using an antifreeze solution. In some cases, mixing different corrosion inhibitors or antifreezes produces unwanted sludges within the cooling system.

Fortunately, the freezing point of water can be lowered quite considerably by the simple addition of certain liquids such as ethylene glycol. The recommended mixture varies with each manufacturer, but is usually 33–50% of antifreeze in the cooling water.

Checking the level

The proportion of ethylene-glycol-based antifreeze present in a cooling system can be determined by checking the **specific gravity** of the coolant and by reference to its temperature. The percentage of antifreeze is measured by the use of a hydrometer and a thermometer.

The graph in Fig. 13.26 shows what happens to water when an antifreeze solution is added. The water changes from a liquid to a mush before becoming solid ice. This ability to form mush before becoming ice gives some warning of freezing and consequently of the danger of damage to the engine.

Methanol-based antifreeze is also used but has the disadvantage of losing its antifreeze effect due to evaporation. It is also inflammable. Both types are toxic and if spilt on the paint work of the vehicle would damage it.

Figure 13.27 shows how to check the antifreeze content using the 'Bluecol' hydrometer. The coolant is drawn into the hydrometer to a level between the two

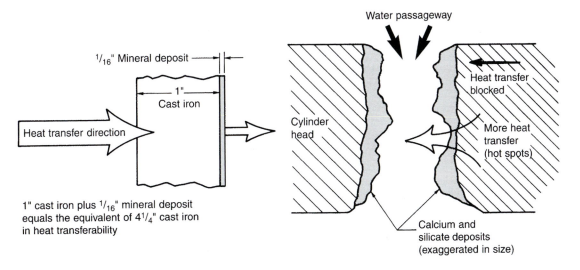

Fig. 13.25 Build-up of deposits in the cooling system

lines. Note the letter on the float at the water line and the temperature of the coolant on the thermometer. Using the slide rule, line up the two readings of temperature and letter. The true percentage content of antifreeze can be identified at the 'read-off' point.

Faults and their possible causes are shown in Table 13.1.

Engine temperature gauge

The engine temperature gauge permits the driver to observe engine operating temperature. In the temperature indicator circuit shown in Fig. 13.28, the main features are the thermal type gauge, the negative temperature coefficient sensor (thermistor), the voltage stabilizer (regulator), and the interconnecting circuit.

The sensor is normally situated in the cylinder head of the engine and the sensing element is surrounded by coolant. In a typical cooling system sensor, the resistance of the sensing element varies from approximately

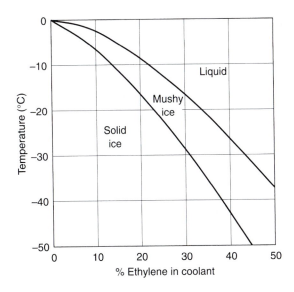

Fig. 13.26 Variation of freezing range with different strengths of antifreeze

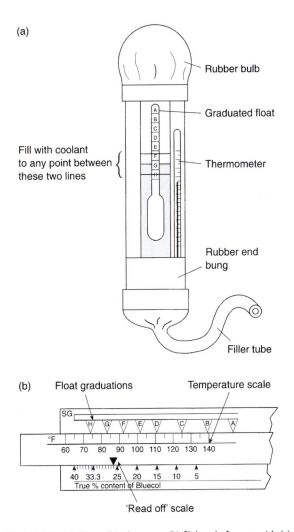

Fig. 13.27 (a) 'Bluecol' hydrometer. (b) Slide rule for use with (a)

Table 13.1 Liquid cooling systems fault chart

Fault	Possible cause
External leakage	1. Loose hose clips or split rubber hose 2. Damaged radiator (cracked joints or corroded core) 3. Water pump leaking (bearing worn) 4. Corroded core plugs 5. Damaged gaskets 6. Interior heater, hoses, valves 7. Temperature sensor connection leaking
Internal leakage	1. Defective cylinder head gasket 2. Cylinder head not correctly tightened 3. Cracked water jacket internal wall 4. Defective cylinder liner seals
Water loss	1. Boiling 2. Leaks – internal and external 3. Restriction in radiator 4. Airways in radiator matrix blocked
Dirty coolant (corrosion)	1. Excessive impurity in coolant water 2. Infrequent draining and flushing of system (where required) 3. Incorrect antifreeze mixtures 4. Lack of inhibitor
Overheating	1. Loose, broken, worn or incorrect fan-belt tension 2. Defective thermostat 3. Water pump impeller loose on shaft 4. Restricted circulation – through radiator, hoses, etc. 5. Radiator airways choked 6. Incorrect ignition timing 7. Incorrect valve timing 8. Tight engine 9. Low oil level 10. Insufficient coolant in system
Over-cooling	1. Defective thermostat 2. Temperature gauge incorrect 3. Electric fan operating continuously

230 ohms at 50°C to approximately 20 ohms at 110°C. This variation of resistance causes the current flowing through the bimetallic element of the gauge to vary. The variation of current causes the temperature of the bimetallic element to change. The change of temperature causes the bimetallic element to bend and so causes the gauge pointer to move to indicate coolant temperature. Because the operation of the gauge is dependent on a steady voltage being applied, the circuit includes a vibrating contact voltage stabilizer, which also relies on a bimetallic strip for its operation.

Learning tasks

1. Explain in your own words what is meant by semiconductor.
2. How should the temperature sensor/transmitter be checked for correct operation? Name any special equipment that might be required.
3. Remove a temperature sensor and test for electrical resistance, both when cold and when hot. Note the readings and check your results with the manufacturer's data. Make recommendations on serviceability.
4. Where in the cooling system would a temperature sensor be located? Why would it be fitted in this position?

Engine core plugs

These are fitted for the following reasons:

• They may blank off the holes left by the jacket cores during casting or machining.

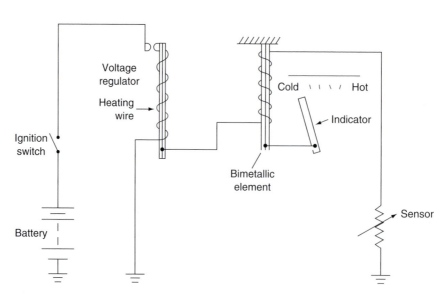

Fig. 13.28 A thermal-type gauge and voltage regulator

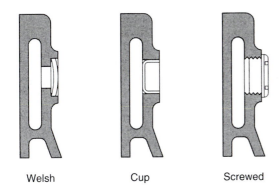

Welsh   Cup   Screwed

Fig. 13.29 Types of core plugs fitted to the engine block

- They may be removed for cleaning out corrosive deposits from the jacket.

The plugs may be of the **welsh plug**, drawn steel **cup** or, less commonly, the **screwed plug** type. The first two are expanded and pressed into core holes that have been machined to size. All three are shown in Fig. 13.29.

Servicing and maintaining the cooling system

In addition to regular checks that should be made by the driver, the following items should be checked at regular intervals as specified in the service manual:

1. Check for evidence of coolant leaks. Examine all hoses and clips, radiator, engine core plugs (where visible), cylinder head gasket, water pump and gasket, thermostat housing, radiator filler cap, and overflow tank.
2. Check all hoses, including interior heater, for deterioration or cracks by flexing manually and by visual examination.
3. Check level of coolant against the indicator line in the overflow tank. The strength of the antifreeze solution should be checked by hydrometer and the specific gravity compared with the figures that are recommended for the conditions in which the vehicle normally operates.
4. Changing and flushing out the coolant. The corrosion inhibitors that are contained in the antifreeze liquid lose their effectiveness after a time and it is recommended that the system should be drained and flushed before refilling with new antifreeze. It is recommended that the system is 'back flushed' by removing the bottom hose and applying a low-pressure water supply that will pass water through the system, in the reverse direction, until it becomes clean.
5. Check that the temperature gauge is operating correctly.
6. Check that the water pump drive belt is in good condition and correctly tensioned.

Safety note

Antifreeze contains toxic substances that are dangerous if taken internally; they can also be absorbed by prolonged skin contact. The following precautions should be taken:

- Antifreeze must never be taken internally. If antifreeze is swallowed accidentally, medical advice must be sought immediately.
- Protective measures such as the use of gloves should be taken to avoid skin contact with antifreeze. In the event of accidental spillage on to the skin, the antifreeze should be washed off immediately. If clothing is splashed with antifreeze it should be removed and washed thoroughly before being used again.
- For regular and frequent handling of antifreeze, protective clothing (plastic or rubber gloves, boots, and impervious overalls) must be used to minimize the risk of skin contact.

Fault diagnosis and repair

Possible faults

- Overheating
- Undercooling
- Frost damage.

Probable causes of overheating

- Low coolant level
- Leaks
- Combustion gas in the coolant
- Defective pressure cap
- Poor circulation of coolant
- Defective thermostat
- Defective water pump
- Defective radiator fan — electric motor, thermal switch, defective fan belt
- Collapsed hose
- Blocked air passages through the radiator grill.

Diagnostic checks

Low coolant level and leaks

Coolant leaks are the most likely cause of a low coolant level. In the first instance, carry out a visual check for obvious signs such as coolant on the floor under the vehicle, followed by a thorough examination of hoses, hose clips, gasket joints, the radiator, internal heater, and the water pump. If these checks are not conclusive, further checks such as the system pressure check can be applied.

A cooling system pressure check

Testing the pressure cap

The cooling system pressure cap can be tested by means of the hand-held apparatus shown in Fig. 13.30. The apparatus consists of a small hand-operated pump, a pressure gauge, and some adaptors that permit the pressure cap to be attached to it. When securely attached, the pressure is raised to its maximum and then recorded for comparison with the engine manufacturer's data, or the figure that is normally stamped on the pressure cap.

Pressure testing the entire system

The pressure cap test instrument can be used to place the cooling system under pressure, as shown in Fig. 13.31.

The equipment for this test contains the hoses and adaptors that permit the instrument to be securely attached to the cooling system in place of the pressure cap. The pressure is then applied by means of the pump until it reaches the recommended test pressure which, in this example is 1.4 bar (20 lbf/in^2). This pressure may be different for other systems and it should be checked against the engine manual before the test proceeds. Once the system has been brought to the recommended pressure it is held there for approximately

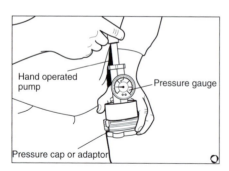

Fig. 13.30 A pressure cap tester (Ford)

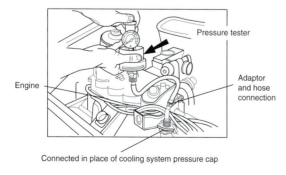

Fig. 13.31 The cooling system pressure test (Ford)

10 seconds. This should be followed by visual examination of all possible leakage points.

Combustion gas leaking into coolant

Combustion gas may enter the cooling system by way of a defective cylinder head gasket, a leaking wet liner seal, a cracked cylinder head or block. If combustion gas is leaking into the cooling system it will give rise to overheating and loss of coolant. The presence of combustion gas in the coolant can be detected by several methods, which include the following:

- The cylinder leakage test
- Chemical analysis.

The cylinder leakage test

The cylinder leakage test (see Fig. 13.32) is performed by applying compressed air to the cylinders through an adaptor that fits into the spark plug or injector hole. The equipment for this test contains a pressure gauge and a leakage meter that shows leakage as a percentage. When all preliminary work has been completed, the adaptor and gauges are attached to the engine and the compressed air pressure is raised to approximately 7 bar (100 lbf/in^2). With the cooling system pressure cap removed, any leak from the cylinders into the cooling system should be evident.

This test should only be conducted in accordance with the manufacturer's instructions.

Chemical analysis of coolant to check for combustion leaks

The upper chamber of the apparatus is filled with the test fluid up to the level that is indicated on the side of the vessel. A sample of steam from the engine cooling system is drawn into the upper sampling chamber and it is allowed to percolate through the test fluid. The test fluid is coloured blue and it changes to yellow if combustion gas (CO_2) is detected in the coolant.

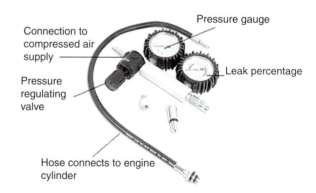

Fig. 13.32 Cylinder leakage test kit (Sealey garage equipment)

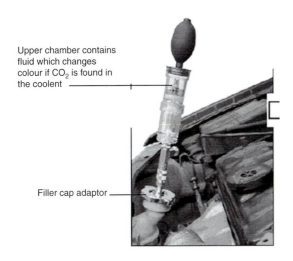

Upper chamber contains fluid which changes colour if CO_2 is found in the coolent

Filler cap adaptor

Fig. 13.33 Combustion gas leak detector (Sealey)

The apparatus shown in Fig. 13.33 is supplied by Sealey Power Products and a full description together with instructions for use can be seen on their website at sealey.co.uk.

Combustion gas leaks into the engine coolant may also be accompanied by coolant entering the combustion chamber; if this is the case it will probably cause the engine to misfire and emit white smoke from the exhaust system.

Poor circulation of coolant

Some probable causes of poor coolant circulation are:

- Defective thermostat
- Defective water pump
- Collapsed hose
- Defective radiator fan − electric motor, thermal switch, defective fan belt
- Blocked coolant passages in radiator and engine.

Defective thermostat

If the thermostat is stuck in the closed, or partly closed, position it will seriously impede coolant flow and will quickly cause the engine to overheat. At the opposite extreme, a thermostat that is stuck in the open position may cause delay in the engine reaching its normal working temperature, with consequent effects on cold running performance of the engine. The operation of the thermostat can be checked with simple equipment as shown in Fig. 13.15.

The water is gradually heated while the temperature is observed; the opening of the valve can be observed as the valve begins to move away from its frame. The valve will begin to open at approximately 80°C, or at a figure that will be found in the appropriate service manual.

Defective water pump

Should the water pump (Fig. 13.34) bearings become worn, the shaft will drop out of alignment and this will damage the seal; this is the most likely cause of coolant leakage. Any leakage from the pump should be visible in the region where the pump shaft protrudes from the casing. When the pump drive belt is removed it should be possible to check for bearing wear by applying a rocking action to the drive pulley. The drive belt and pulley should also be checked to make sure that they are not worn and when the belt is refitted its tension should be set correctly to ensure that it does not impose excessive stress on the bearings.

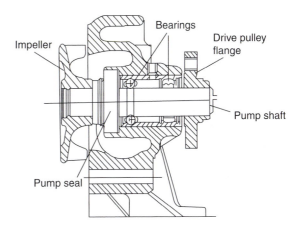

Bearings

Drive pulley flange

Impeller

Pump shaft

Pump seal

Fig. 13.34 A typical coolant pump

Defective fan

The belt drive to the cooling fan may slip because it, or the pulley, is worn. The belt should not be too deep in the V-groove so that it touches the base of the groove. This may be caused by wear of the belt and it can be checked visually − the V-groove can be checked by placing an unworn belt in it to see if it is resting in the correct position. The belt itself should be tensioned enough to prevent slip, but not too tight so that it places stress on the pump bearings.

Electric fan

Electric cooling fans were originally controlled by a thermal switch, mounted in the radiator header tank, that reacted to changes in engine temperature. Because they only operate when the temperature requires them to, they absorb less engine power than the belt-driven alternative. The electric cooling fan is designed to come on at about 95°C and turn off at about 90°C.

In computer-managed engine controls, the electronic control module (ECM) fan circuit (Fig. 13.35) may be designed so that the fan may continue running after the engine is switched off, to provide after cooling. In the system shown in Fig. 13.36 the fan can be operated at two different speeds: a low speed where the fan motor

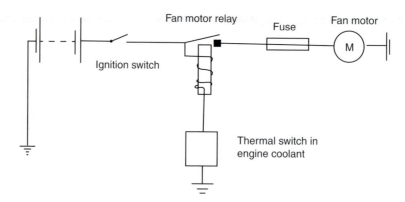

Fig. 13.35 Electric fan circuit

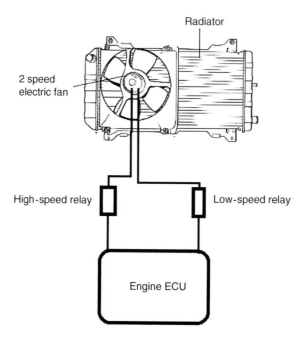

Fig. 13.36 ECM-controlled electric fan

absorbs 200 W and a higher speed when the power is 300 W. In the event of a cooling system malfunction the ECM will output a standard fault code, which will indicate a circuit failure — full circuit diagnosis could then proceed with the aid of a circuit diagram and the necessary meters.

Safety note

It is important to be aware that the electric fan may operate even though the engine is switched off and the necessary precautions should be taken to avoid contact with the fan when working under the bonnet.

Blocked radiator

A reasonable assessment of the condition of the radiator core can be gained by passing a volume of water through it. A simple apparatus of the form shown in Fig. 13.37

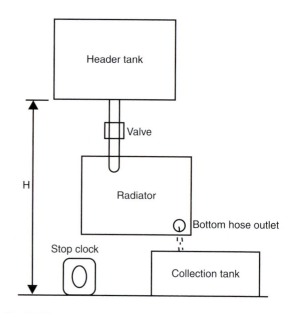

Fig. 13.37 Radiator flow test

used to be recommended by a vehicle manufacturer. If the flow rate is known it can be checked by using a stop clock to record the time taken for a given volume to pass through the radiator — in the absence of such information the best that one could achieve is to observe the rate at which water leaves the radiator, when treated to a similar process. If the water appears not to flow freely it is reasonable to assume that the radiator tubes are partly blocked.

Blocked air passages through the radiator grill

If the areas around the fins and tubes of the radiator are obstructed the air flow across it will be reduced and this will result in a loss of heat transfer properties. A visual examination will normally reveal any obstructions to air flow. One manufacturer suggests using a low-pressure compressed air supply to gently blow material out of the radiator matrix.

Heating and ventilation

The most common method of providing a comfortable atmosphere in the car is through the heating and ventilation system. In some countries it could mean that a full air-conditioning unit is required where refrigeration cooling is fitted to the vehicle. Fresh-air ventilation comes under two headings, **direct** and **indirect**, the heat source being the hot water from the engine. But it may also be gained from a number of other sources such as the exhaust system, as in the air-cooled engine, or in a separate heater where fuel is burnt to heat a chamber over which air is passed, or electrically, in which an element is heated and air passed over the element. Water type systems are normally fitted to the bulkhead or behind the fascia panels and are the most common arrangement in light vehicles.

Direct ventilation

This can be achieved by simply opening one or more of the windows, but this could cause draughts, noise, and difficulties in sealing against rain when open. A number of different methods have been tried to overcome these problems, such as the swivelling quarter light window, but this created problems with water leaking in when the window was shut.

Indirect ventilation

This is routed through a **plenum chamber** located at the base of the windscreen. The internal pressure in the chamber is higher than that of the surrounding atmosphere. It is important therefore that the position of the plenum chamber and its entrance is chosen to coincide with a high-pressure zone of air flow over the car

body, and also where it is free from engine fumes. A heating and ventilation system is shown in Fig. 13.38.

The air flow through the interior of the car can be derived simply from the ram effect of air passing over the car and spilling into the plenum chamber or, at low speeds, it may be boosted by the use of an electrically driven fan connected into the intake of the plenum chamber. Directional control for the air flow is adjustable by deflectors fitted to the outlets on the fascia. Provision for the extraction of the stale air is provided for in grills incorporated in the rear quarter panels of the body, their siting coinciding with **neutral pressure zones** in the air flow over the car.

Learning tasks

1. The volume of air delivered by the interior heater is controlled by the speed of the fan. Identify two methods that are used to control the temperature of the air delivered by the heater.
2. How would you check for the correct operation of the three-speed switch which operated the fan at only two speeds?
3. What symptoms would indicate that the cooling system needs bleeding?
4. How would you identify a suspected problem with a noisy heater fan? Suggest methods you would use to rectify the fault?

Interior heating and air-conditioning

The heater matrix is a small radiator through which engine coolant is circulated (Fig. 13.39). A system of flaps that are operated manually direct the air into the regions of the interior. The temperature control operates the flap that determines the amount of air that passes through the heater matrix.

Air-conditioning
Air-conditioning fluid

The fluid used in air-conditioning systems is called a refrigerant. The refrigerants used in modern systems are known as **F-gas**. F-gases are synthetic gases known as hydrofluorocarbons (HFCs) and they are similar to those used in domestic refrigerators. Because F-gases are dangerous to handle if not treated correctly, and are harmful to the environment, there are strict rules about working with them. As far as the vehicle repair industry is concerned, from 4 July 2010, all automotive technicians working with refrigerants will need an F-gas certificate to continue doing their job.

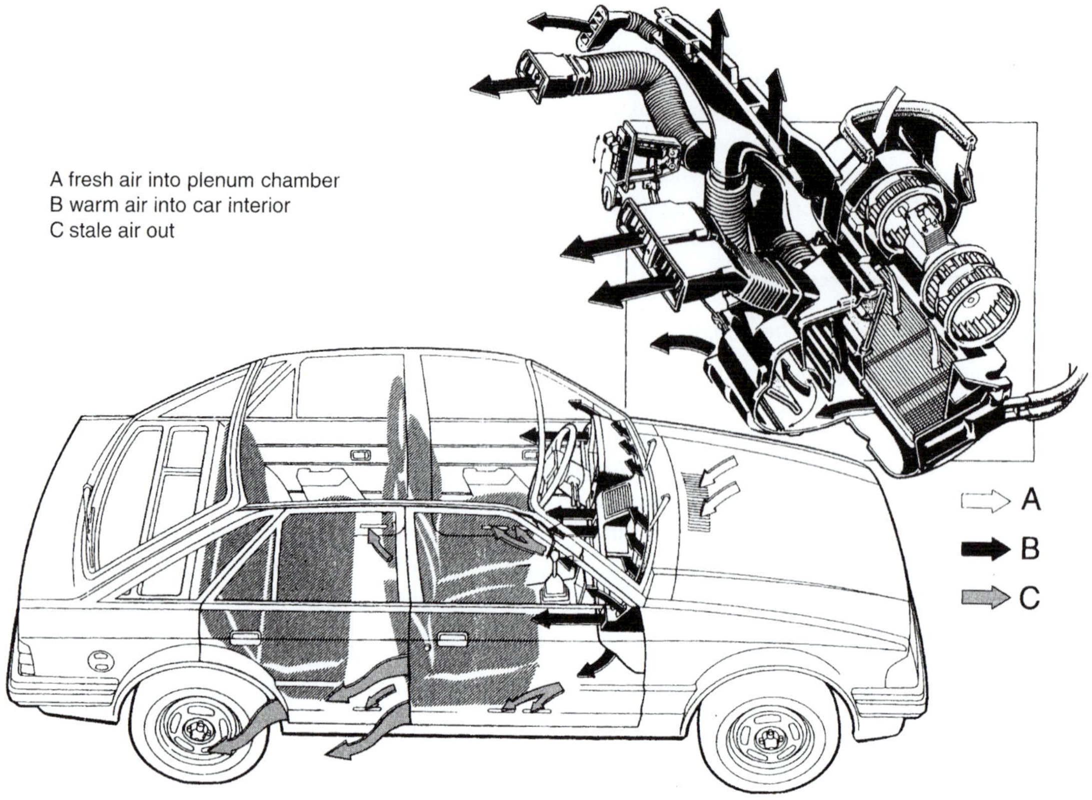

A fresh air into plenum chamber
B warm air into car interior
C stale air out

⇨ A
➡ B
⇨ C

Fig. 13.38   Heating and ventilation system

In common with other substances, F-gas can exist in three states:

- Solid
- Liquid
- Gas (vapour).

The state of a substance depends on its temperature and pressure. In air-conditioning systems the F-gas exists in two of these states – liquid and vapour. In the cycle of events in the system, the F-gas changes from liquid to vapour and back again continuously and, in the process, takes in and gives out heat. The description that follows is intended to provide an insight into the operation that will provide a platform of knowledge for those who wish to obtain the required certification.

A simple air-conditioning system

Cooling down the interior of the vehicle normally requires the use of an extra machine-driven cooling system that will take heat from the interior and transfer it to the atmosphere surrounding the exterior of the vehicle. It is the air-conditioning system that performs this function. See Fig. 13.40.

The liquid (refrigerant) that is used to carry heat away from the vehicle interior and transfer it to the outside is circulated around the closed system by means of a compressor that is driven by the engine of the vehicle.

Inside the system the refrigerant constantly changes state between liquid and vapour as it circulates.

The reducing valve is an important agent in the operation of the system. The 'throttling' process that takes place at the reducing valve causes the refrigerant to vaporize and its presssure and temperature to fall. After leaving the reducing valve, the refrigerant passes into a heat exchanger called the evaporator, where it collects heat from the vehicle interior and thus cools the interior in the process. The heat that refrigerant collects causes it to vaporize still further and it returns to the compressor where its pressure and temperature are raised.

From the compressor, the refrigerant passes into another heat exchanger where it gives up heat to the atmosphere. This heat exchanger is known as a condenser because the loss of heat from the refrigerant causes it to become wet. After the condenser, the refrigerant passes through the accumulator, which serves to separate liquid from vapour. The refrigerant next returns to the reducing valve and evaporator, thus completing the cycle.

Computer control of air-conditioning

Because the compressor takes a considerable amount of power from the engine it is necessary for the air-conditioning computer to be aware of the operational state of the engine. For example, the idling speed of the engine will be affected if the air-conditioning compressor is operating and the engine ECM will normally cause an

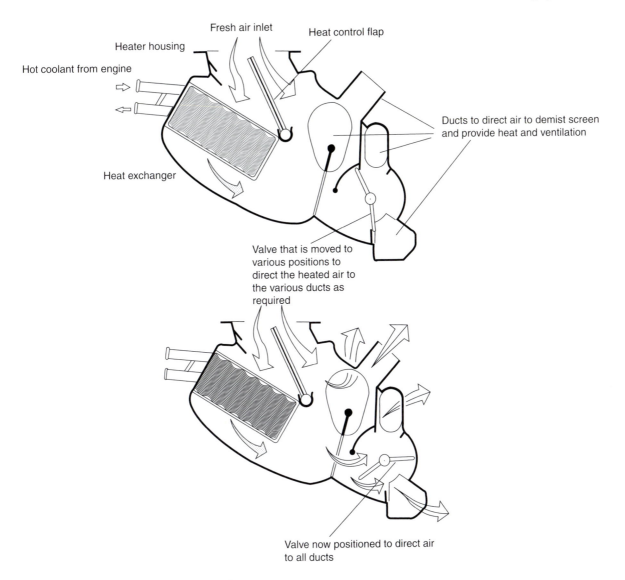

Fresh air inlet

Heat control flap

Heater housing

Hot coolant from engine

Ducts to direct air to demist screen and provide heat and ventilation

Heat exchanger

Valve that is moved to various positions to direct the heated air to the various ducts as required

Valve now positioned to direct air to all ducts

Fig. 13.39 The heater matrix and air flow control

increase in idle speed to prevent the engine from stalling. To allow the air-conditioning compressor to be taken in and out of operation, it is driven through an electromagnetic clutch, which is shown in Fig. 13.41.

This clutch permits the compressor to be taken out of operation at a speed just above idling speed and, in order to protect the compressor, it is also disconnected at high engine speed. In some cases where rapid acceleration is called for, temporary disengagement of the compressor may also occur.

In addition to engine operating considerations, the temperatures of the interior of the vehicle must constantly be compared with the required setting and the exterior temperature, and this is achieved by temperature sensors that are similar to those used for engine coolant temperature sensing. The following list gives an indication of the functions that are controlled by the ECU:

1. Calculation of required outlet air temperature
2. Temperature control
3. Blower control
4. Air inlet control
5. Air outlet control
6. Compressor control
7. Electric fans control
8. Rear defogger control
9. Self-diagnosis.

Dealing with air-conditioning refrigerant

Refrigerants that are used in air-conditioning systems can be harmful to persons who come into contact with them and they are also considered to be harmful to the environment. For these reasons the servicing of air-conditioning systems requires the use of specialized equipment and technicians must be trained for the specific application that they are working on. Most garage equipment manufacturers market air-conditioning service equipment; the Bosch Tronic R134 kit is an example. Equipment suppliers and vehicle

6. Liquid-cooled
7. Corrosive
8. Antifreeze
9. Centrifugal pump
10. Heat exchanger
11. Core plug
12. Pressure
13. Vacuum
14. Shroud
15. Matrix.

Practical assignment — cooling systems

Introduction

At the end of this assignment you will be able to:

- recognize water cooling and heater systems
- test the system for leaks
- test the system for anti-freeze content
- remove and test the thermostat
- test the pressure and vacuum valves
- check the operation of the temperature sensor and electric fan
- remove and check the drive belts for serviceability
- refit and set drive-belt tension
- remove the radiator and test the flow rate
- bleed the cooling and heater system
- remove the water pump and state the type of impeller fitted
- flush the cooling and heater system.

Tools and equipment

- A water-cooled engine
- Suitable drain trays
- Selection of tools and spanners
- Cooling system pressure tester
- Equipment for testing thermostat
- Hydrometer for testing antifreeze content
- Relevant workshop manual or data book.

Objective

- To check the correct operation of each of the components in the cooling system
- To prevent loss of water and overheating, which could cause damage to the engine
- To prevent freezing of the water during very cold weather.

Activity

1. Before starting the engine, remove the radiator pressure cap and pressure test the cooling system.
2. Test the pressure and vacuum valves that are situated in the cap. Note any leaks that occur in tests 1 and 2 on the report sheet.
3. Remove the thermostat and check for correct operation using the equipment provided.
4. Remove the drive belts and check for cracks, splits, wear, and general serviceability.
5. Remove the radiator, reverse flush it, and where necessary complete a flow test. Remove any dirt, leaves, etc. from between fins of the matrix.
6. Remove the water pump, check the bearings for play and the seal for signs of leakage.
7. Refit the thermostat, water pump (using new gaskets), radiator, and drive belts, setting the correct torque on the bolts and drive-belt tension.
8. Flush the heater system, refit all the hoses, and refill with water–antifreeze mixture, bleeding the system as necessary.
9. Check the operation of the temperature sensor by running the engine up to its normal operating temperature.
10. Remove the fan. If it is an electric fan check for any play or tightness in the bearings, undue noise, or loose mountings. Check for damage to the fan blades. Where necessary lubricate the bearings and refit the fan.
11. Complete the report sheet on the cooling system, identify any faults, and report on serviceability.
12. Answer the following questions:
    (a) State the types of cooling systems used on motor vehicles and give two advantages for each.
    (b) State the purpose of:
        (i) the pressure valve
        (ii) the vacuum valve
        (iii) the thermostat
        (iv) the temperature sensor
        (v) the water pump.
    (c) What is the purpose of pressurizing the cooling system?
    (d) Give two advantages of fitting an electric fan compared with belt-driven fans.
    (e) Describe a possible cause and the corrective action to be taken for the following faults:
        (i) Squealing noise from the front of the engine
        (ii) External leak from the bottom of the radiator
        (iii) Internal leak in the engine
        (iv) Overheating with no loss of water
        (v) Heater does not get warm enough
        (vi) Coolant is very dirty.

Checklist

Visual checks on Vehicle

- Type of radiator fitted
- Type of fan fitted
- Type of cooling system.

Practical tests on the cooling system

- Antifreeze content
- Pressure test of cap
- Pressure test of system
- Fan-belt tension and condition
- Operation of heater
- Operation of fan
- Operation of thermostat
- Radiator flow test
- Manufacturer's technical data.

Comments on serviceability

Student's signature

Supervisor's signature

Practical assignment – liquid cooling systems

Objective

To carry out a number of tests on the cooling system to ascertain its serviceability:

- Thermostat setting
- System pressure testing
- Radiator flow test.

Tools and equipment

- Running engine/vehicle
- Assorted hand tools
- Thermostat testing equipment
- Pressure tester
- Header tank, drain tank, and fittings
- Stopwatch.

Safety aspects

- Cables, hands, hair, and loose clothing must be kept clear of fans and other rotating parts. Stationary engines require guards to be fitted to the fan drive.
- Keep hands clear of hot parts of the engine.

- If a vehicle is used, the gear lever must be in the neutral position, the handbrake applied, and the wheels chocked before operating the starter motor.
- Arrangements must be made for the exhaust gases to pass directly out of any enclosed space.
- Remove radiator cap *slowly* and cover with a cloth if the engine is hot.

Safety questions

1. Why must the exhaust gases not be discharged into the garage?
2. What are the dangers from unguarded fan blades and fan belts?
3. What is likely to happen if the radiator cap is removed quickly when the engine is hot?

Activity

1. (a) Disconnect hoses and housing and remove the thermostat from the engine.
   (b) Place the thermostat in the tester and heat the water to the opening temperature of the thermostat, watch for commencement of opening and fully open point, note the temperatures, and record the results.

Note

It may be necessary to first subject the thermostat to boiling water so that the fully open position can be assessed.

   (c) The recommended opening temperatures should be obtained from the data book.
   (d) Replace the thermostat in the correct position, refit the housing using a new gasket as required, fit hoses and tighten hose clips, run engine and check for water leaks.
2. (a) Remove radiator cap (observe safety instructions).
   (b) Fit pressure testing equipment in place of the radiator cap, operate pump to pressurize system to recommended pressure.
   (c) Carry out visual inspection of components in the cooling system and joints for coolant leakage.

Note

Look for a steady reading of the pressure tester gauge; if the pressure falls steadily and there is no coolant leakage, then an internal leakage of the coolant can be suspected. (Check that the tester unit is fully sealed on the radiator neck.)

(d) To identify internal leakage there are products on the market that can be mixed with the coolant, the system is closed and the engine is run until it reaches its normal operating temperature. After the system cools the radiator cap is removed and the colour of the coolant is inspected. The chemical changes colour when it comes into contact with oxygen. Therefore, if the colour of the coolant changes it can be assumed that air is entering the system, most probably through the cylinder head gasket.

(e) The radiator cap can be checked for correct operation by fitting the correct adaptor to the pump and fitting the cap in place on the tester. The pressure is then raised by operating the pump and recorded when the stage is reached for the seal and the spring in the cap to lift and so relieve the pressure.

3. To flow test the radiator:
   (a) Remove the radiator and fit it to the test rig.
   (b) Fill the header tank of the rig with a known quantity of water.
   (c) Open the tap to discharge the water through the radiator and measure the draining time with a stopwatch. Compare this with the manufacturing data (20 litres takes approximately 20 seconds for a car radiator with a water head of 0.7 metres).
   (d) The mineral deposits in the coolant tend to block the water ways, so an approximation of the flow rate can be ascertained by comparing the mass of the radiator under test with the mass of a new radiator of the same type. The deposits are heavy so a 25% increase in weight will give an indication of several mineral deposits.

Questions

1. What are the effects of pressurizing the cooling system?
2. Explain the function of the two valves in the pressure cap.
3. List five reasons for an engine becoming overheated.

Self-assessment questions

1. A thermostat is fitted to cooling systems:
   (a) to provide a variable current in the temperature gauge circuit
   (b) to control circulation of coolant to help the system to warm up quickly and to maintain a constant operating temperature
   (c) to act as a control to switch the cooling fan on and off
   (d) to operate the compressor on an air-conditioning system.

2. The lubrication system of a certain engine contains 0.8 kg of oil that has a specific heat capacity of 1.7 kJ/kg$^{\circ}$C. During the warm-up period the temperature of the oil rises by 30°C. The amount of energy transferred to the oil during this warm-up period is:
   (a) 32.5 J
   (b) 325 MJ
   (c) 40.8 kJ
   (d) 408 J.

3. Inside a pressurized cooling system:
   (a) the coolant boiling temperature is above 100°C
   (b) the increased pressure slows down the circulation of coolant
   (c) the boiling point of the coolant is lowered
   (d) the convection currents cease to operate.

4. Ethylene-glycol-based antifreeze when added to the coolant causes:
   (a) less evaporation of the coolant
   (b) the freezing temperature of the coolant to be raised
   (c) the freezing temperature of the coolant to be lowered
   (d) the boiling point temperature of the coolant to be lowered.

5. The purpose of the water pump in a cooling system is to:
   (a) pressurize the cooling system
   (b) circulate the coolant
   (c) eliminate the need for a radiator
   (d) reduce heat loss.

6. After combustion the temperature inside the engine cylinder is:
   (a) approximately 200°C
   (b) approximately 12 000°C
   (c) approximately 1400–2000°C
   (d) 250°C.

7. Cooling systems rely on heat transfer by:
   (a) pressure differentiation
   (b) electrical conductivity
   (c) Archimedes' principle
   (d) conduction, convection, and radiation.

8. Radiator surfaces are often finished in matt black because:
   (a) this surface is effective against corrosion
   (b) this surface is most effective in radiating heat
   (c) this type of finish reflects most heat
   (d) this surface finish prevents heat escaping into the atmosphere.

9. What feature of the electric cooling fan system assists the engine to reach its normal working temperature more quickly than the belt-driven type of fan?

10. If an electric fan operates at 300 W in a 12-volt system, the current drawn in the pump circuit is:

(a) 25 amperes
(b) 2.5 amperes
(c) 0.25 amperes
(d) 250 amperes.

11. What minimum fuse rating would be suitable for the electric fan circuit?

14
Internal combustion engines

Topics covered in this chapter

- Engine power
- Brake power
- Horse power — PS and DIN power rating
- Indicated power
- Mean effective pressure
- Morse test
- Characteristic curves of engine performance
- Torque and engine speed
- Specific fuel consumption
- Brake torque and brake mean effective pressure (bmep)
- Thermal efficiency.

Engine power

Brake power

The engine power that actually reaches the output shaft or flywheel of an engine is known as the brake power. It is the power that is measured by a dynamometer and a dynamometer is also known as a brake, hence the term brake power. The simplest form of dynamometer is shown in Fig. 14.1. Here a rope is wound around the circumference of the engine flywheel; one end of the rope is supported by a spring balance and the other end has weights attached to it. The load that the engine is working against is varied by increasing, or decreasing, the amount of weight. The effective load (force) that the engine is working against $= W - S$ newtons.

Rotating the flywheel inside the rope, against the force $(W - S)$ newtons, is equivalent to moving the force $(W - S)$ around the circumference of the flywheel. The distance travelled by the force during one revolution of the flywheel is equal to the circumference of the flywheel. If the radius of the flywheel is R metres, the circumference $= 2\pi R$ metres. From this, it may be seen that:

$$\text{Work done per revolution of the flywheel}$$
$$= \text{force} \times \text{distance}$$
$$= (W - S) \times 2\pi R \text{ joules}$$

$(W - S) \times R$ also $= T$, the torque that the engine is exerting. By substituting T for $(W - S) \times R$ the work done per revolution becomes $= T \times 2\pi$ joules

$$\text{Brake power} = \text{work done per second}$$
$$= (T \times 2\pi) \times \text{number of revolutions per second}$$

This is normally written as brake power (bp):

bp $= 2\pi TN$, where $N =$ number of revolutions per second.

When the torque is in newton metres (Nm), the formula bp $= 2\pi TN$ gives the power in watts. As engine power is given in thousands of watts, i.e. kilowatts, the formula for brake power normally appears as:

$$\text{bp} = 2\pi TN/1000 \text{ kW}.$$

Dynamometers for high-speed engines

Hydraulic dynamometers of the type manufactured by the Heenan–Froude company are commonly used for engine testing purposes. The energy from the engine is

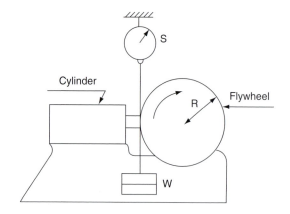

Fig. 14.1 Simple dynamometer for low-speed engines

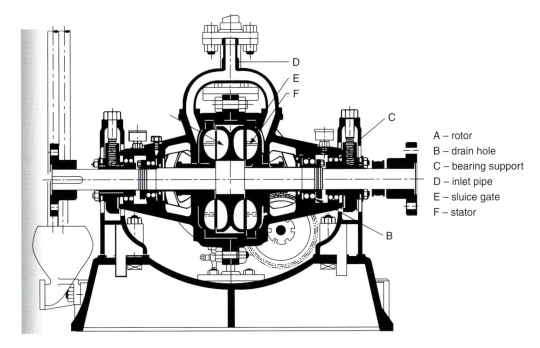

Fig. 14.2 Hydraulic Heenan—Froude dynamometer

converted into heat by the action of the dynamometer rotors and the action is somewhat similar to the action of a torque converter. Figure 14.2 shows a cross-section of the dynamometer; the engine under test is connected to the shaft that drives rotor A. Rotor A acts as a pump that directs pressurized water into the stator F. The stator is connected to a torque arm that registers the torque acting on it by means of weights and a spring balance. The power absorption capacity of the dynamometer is controlled by means of a form of sluice gate E that is interposed between the stator F and rotor A. These sluice gates are controlled manually. The energy absorbed by the dynamometer is converted into heat in the water, which is supplied under pressure through the inlet pipe D.

Electrical dynamometers are also used for highspeed engine testing; these convert the engine power into electricity which is then dissipated through heat or electro-chemical action.

A – rotor
B – drain hole
C – bearing support
D – inlet pipe
E – sluice gate
F – stator

Example 14.1

A certain engine develops a torque of 120 Nm while running at a speed of 3000 rev/min. Calculate the brake power. Take $\pi = 3.142$.

Solution

$$bp = 2\pi TN/1000.$$

From the question, $T = 120$ Nm; $N = 3000/60 = 50$ rev/s

Substituting these values in the formula gives
bp = $2 \times 3.142 \times 120 \times 50/1000 = 37.7$ kW

Horsepower

One horsepower = 746 watts. The conversion from kilowatts to horsepower is: Power in kilowatts ÷ 0.746.

Taking the above example of 37.7 kW, the horsepower equivalent is 37.7 kW ÷ 0.746 = 50.5 bhp.

PS – the DIN

The abbreviation PS derives from *pferdestrake,* which is German for the pulling power of a horse; 1 PS is slightly less than 1 imperial horsepower.

Indicated power

Indicated power (ip) is the power that is developed inside the engine cylinders. It is determined by measuring the pressures inside the cylinders while the engine is on test, on a dynamometer. The device that is used to measure the pressure is called an indicator, from which the term indicated power is derived. The indicator diagram for a four-stroke engine is shown in Fig. 14.3.

Because the pressure varies greatly throughout one cycle of operation of the engine, the pressure that is used to calculate indicated power is the mean effective pressure.

Mean effective pressure

The mean effective pressure is that pressure which, if acting on its own throughout one complete power stroke, would produce the same power as is produced

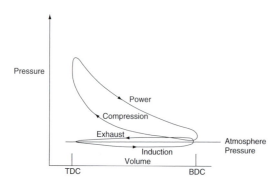

Fig. 14.3   Indicator diagram for a four-stroke engine

by the various pressures that occur during one operating cycle of the engine. This information is usually shown in an indicator diagram of the type shown in Fig. 14.3.

In Fig. 14.4(a), the loop formed by the pressure trace for the exhaust and induction strokes is known as the pumping loop. This represents power taken away and the effective area is determined by subtracting A_2 from A_1. This calculation then gives the indicated mean effective pressure. The resulting area ($A_1 - A_2$) is then divided by the base length of the diagram. The resulting mean height is then multiplied by a constant that gives the indicated mean effective pressure (imep).

Example 14.2

An indicator diagram taken from a single-cylinder four-stroke engine has an effective area of 600 mm². If the

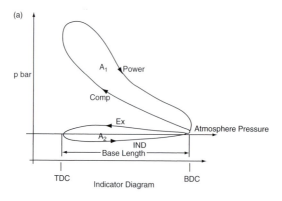

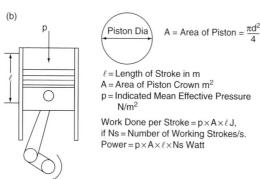

Fig. 14.4   Indicated power calculation

base length of the indicator diagram is 60 mm and the constant is 80 kPa/mm calculate the indicated mean effective pressure.

Solution

Indicated mean effective pressure

$$= \frac{\text{effective area of indicator diagram}}{\text{base length of the diagram}} \times \text{constant}$$

$$\text{imep} = \frac{600}{60} \times 80 \text{ kPa/mm}$$

$$\text{imep} = 800 \text{ kPa} = 8 \text{ bar.}$$

Calculation of indicated power

Indicated power is calculated from the formula

$$\text{Indicated power (ip)} = \frac{PlaN}{1000} \text{ kW,}$$

where P = mean effective pressure in N/m², l = length of engine stroke in m, a = cross-sectional area of cylinder bore in m², N = number of working strokes per second.

Example 14.3

In a test, a certain single-cylinder four-stroke engine develops a mean effective pressure of 5 bar at a speed of 3000 rev/min. The length of the engine stroke is 0.12 m and the cross-sectional area of the cylinder bore is 0.008 m². Calculate the indicated power of the engine in kW.

Solution

The engine is single-cylinder four-stroke, so there is one working stroke for every two revolutions. N, the number of working strokes per second = 3000 ÷ 60/2 = 25; P = 5 bar = 500 000 N/m²; l = 0.12 m; a = 0.008 m². Substituting these values in the formula gives

$$\text{ip} = 500\,000 \times 0.12 \times 0.008 \times 25/1000 \text{ kW}$$

$$= 12 \text{ kW.}$$

Number of working (power) strokes

As mentioned, each cylinder of a four-stroke engine produces one power stroke for every two revolutions of the crankshaft. In a multi-cylinder engine, each cylinder will produce one power stroke in every two revolutions of the crankshaft. A formula to determine the number of power strokes per minute for a multi-cylinder, four-stroke engine is:

number of power strokes per minute

$$= \frac{\text{number of cylinders}}{2} \times \text{rev/min.}$$

Example 14.4

A four-cylinder, four-stroke engine develops an indicated mean effective pressure of 8 bar at 2800 rev/min. The cross-sectional area of the cylinder bore is 0.01 m^2 and the length of the stroke is 120 mm. Calculate the indicated power of the engine in kW.

Solution

The required formula is:

$$ip = \frac{PlaN}{1000} \text{ kW}.$$

From the question, N, which is the number of power strokes per second, is

$$\frac{\text{number of cylinders}}{2} \times \frac{\text{rev/min}}{60},$$

which gives $N = 4/2 \times 2800/60 = 93.3$ per second.
Length of stroke in metres $= 120/1000 = 0.12$ m.
$P = 8$ bar $= 800\,000$ N/m^2.
And the area of the piston $a = 0.01$ m^2.
Substituting these values in the formula gives:

$$ip = 800\,000 \times 0.12 \times 0.01 \times 93.3 \div 1000 \text{ kW}$$
$$= 89.6 \text{ kW}.$$

Because engine power is still frequently quoted in horsepower, it is useful to know that 1 horsepower = 746 watts = 0.746 kW.

This engine's indicated power therefore = 89.6/0.746 = 120.1 hp.

Cylinder pressure vs. crank angle

The behaviour of the gas in an engine cylinder is an area of study that enables engineers to make detailed assessments of the effects of changes in engine design and types of fuel, etc. Devices such as pressure transducers and oscilloscopes permit cylinder gas behaviour to be examined under a range of engine operating conditions. The basic elements of an indicator that permit this type of study are shown in Fig. 14.5(a). The sparking plug is drilled as shown, and the hole thus made is connected to a small cylindrical container in which is housed a piezo-electric transducer. Gas pressure in the cylinder is brought to bear on this piezo transducer through the drilling in the sparking plug and the small pipe. Gas pressure on the transducer produces a small electrical charge that is conducted to an amplifier, which is calibrated to convert an electrical reading into an input to the oscilloscope that represents gas pressure. An additional transducer is attached to the engine crankshaft. The electrical output from the crank transducer is amplified as necessary and fed to the X input of the oscilloscope.

Figure 14.5(b) shows the type of oscilloscope display that is produced by a high-speed indicator. The display shows part of the compression and power strokes of a diesel engine. Injection commences

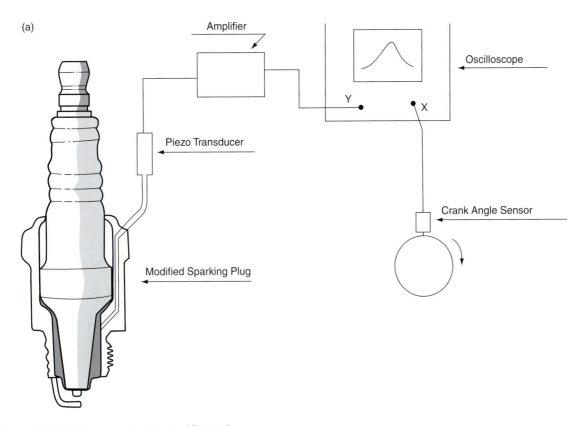

Fig. 14.5 (a) A high-speed engine indicator (*Continued*)

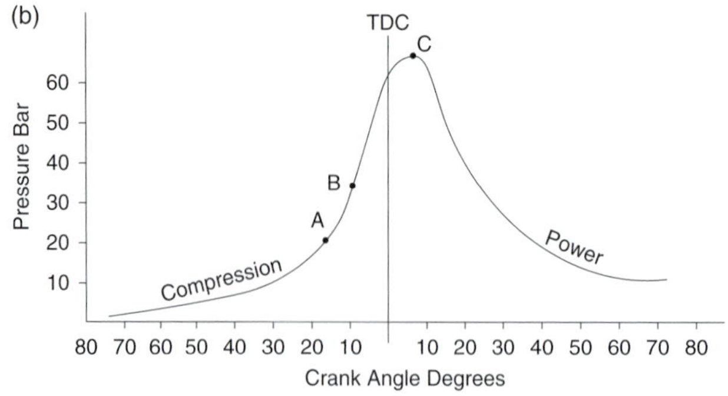

Fig. 14.5 *(Continued).* (b) Cylinder pressure vs. crank angle

at point A and combustion starts at point B; point C represents the end of effective combustion. The periods between each of these points are the three phases of combustion.

Mechanical efficiency of an engine

The mechanical efficiency of an engine is defined as brake power/indicated power.

Example 14.5

A certain engine develops a brake power of 120 kW at a speed of 3000 rev/min. At this speed, the indicated power is 140 kW. Calculate the mechanical efficiency of the engine at this speed.

Solution

$$\text{Mechanical efficiency} = \frac{\text{brake power}}{\text{indicated power}} \times 100\%$$
$$= \frac{120}{140} \times 100$$
$$= 85.7\%$$

Morse test

Frictional losses in the engine bearings, the valve train, and the piston and piston rings are the main causes of the power loss that makes the brake power of an engine smaller than the indicated power. The Morse test is an engine test that gives an approximate value for the frictional losses and which also provides an approximate value for the indicated power of a multi-cylinder engine.

The Morse test is conducted at constant engine speed on a dynamometer. The first phase of the test records the brake power of the engine when all cylinders are firing. Subsequently, one cylinder is prevented from firing and the dynamometer load is adjusted to bring the engine up to the same speed as it was when all cylinders were firing, the brake power then being recorded. The difference between brake power with all cylinders working and that obtained when one cylinder is cut out is the indicated power of the cylinder that is not working. This procedure is repeated for each of the cylinders and the indicated power for the whole engine is the sum of the

power of the individual cylinders. A typical Morse test calculation is shown in the following example.

Example 14.6

The results shown in Table 14.1 were obtained when a Morse test was conducted on a four-cylinder petrol engine.

Table 14.1 Morse test data for a four-cylinder petrol engine

Cylinder cut-out	None	No. 1	No. 2	No. 3	No. 4
Brake power, bp (kW)	60	41	40	43	42
Indicated power, ip (kW)		19	20	17	18

Solution

The total indicated power = sum of each ip of the individual cylinders

$$\text{Total ip} = 19 + 20 + 17 + 18 = 74 \text{ kW}$$

$$\text{Mechanical efficiency} = \frac{\text{bp}}{\text{ip}} \times 100\% = (60/74) \times 100 = 81\%.$$

Characteristic curves of engine performance

Figure 14.6 shows a graph of brake power against engine speed for a petrol engine. The points to note are:

- The graph does not start at zero because the engine needs to run at a minimum idle speed in order to keep running.
- The power increases almost as a straight line up to approximately 2500 rev/min; after that point the power increases more slowly and the curve begins to rise less sharply.
- Maximum power is reached at a point that is near the maximum engine speed.

The main reason why the power does not increase in direct proportion to the engine speed is because the amount of air that can be drawn into the cylinder is

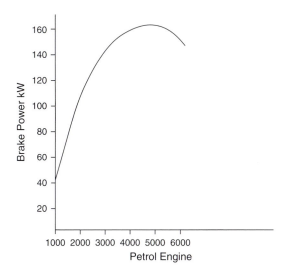

Fig. 14.6 Brake power vs. engine speed

limited by factors such as valve lift and port design. The amount of air actually drawn into the cylinders compared with the amount that could theoretically be drawn into them is known as the volumetric efficiency. Above a certain speed, volumetric efficiency drops and this, coupled with frictional and pumping losses, affects the amount of power produced. Figure 14.7 shows a graph of volumetric efficiency plotted against engine speed.

Volumetric efficiency and air flow

Volumetric efficiency

$$= \frac{\text{volume of charge admitted at STP}}{\text{swept volume}}$$

$$= \frac{\text{actual air flow}}{\text{theoretical air flow}}.$$

Where STP = standard temperature pressure, where standard temperature is $0\,^\circ$C and pressure is 1.01325 bar.

Example 14.7

A four-cylinder four-stroke petrol engine with a bore diameter of 100 mm and a stroke of 110 mm has a volumetric efficiency of 74% at an engine speed of 4000 rev/min. Determine the actual volume of air at STP that flows into the engine in 1 minute.

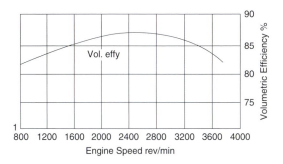

Fig. 14.7 Volumetric efficiency − diesel engine

Solution

$$\text{Volumetric efficiency} = \frac{\text{actual air flow}}{\text{theoretical air flow}}.$$

Theoretical air flow = swept volume of one

$$\text{cylinder} \times \text{number of cylinders} \times \frac{\text{rev/min}}{2}.$$

$$\text{Swept volume of one cylinder} = \frac{\pi d^2 l}{4}$$

$$= 0.7854 \times 0.1 \times 0.1 \times 0.11 \text{ m}^3.$$

$$= 8.64 \times 10^{-4} \text{m}^3.$$

$$\therefore \text{Theoretical volume of air} = 8.64 \times 10^{-4} \text{m}^3 \times 4 \times 4000/2$$

$$= 6.91 \text{ m}^3/\text{min}.$$

Where d = diameter of cylinder bore, in metres, L = length of stroke, in metres.

Actual air flow = theoretical air flow × volumetric efficiency.

$$\text{Actual air flow} = 0.74 \times 6.91 = 5.11 \text{ m}^3/\text{min}.$$

Torque vs. engine speed

Torque is directly related to brake mean effective pressure (bmep). The graph in Fig. 14.8 shows that torque varies across the speed range of an engine. Torque (T) may be determined by direct measurement at a dynamometer or by calculation; for example,

$$T = \frac{\text{brake power}}{2\pi n},$$

where the brake power is determined by dynamometer tests and n = engine speed in rev/min.

Specific fuel consumption vs. engine speed

Brake specific fuel consumption (bsfc) is a measure of the effectiveness of an engine's ability to convert the chemical energy in the fuel into useful work. Brake specific fuel consumption is calculated as shown, for example,

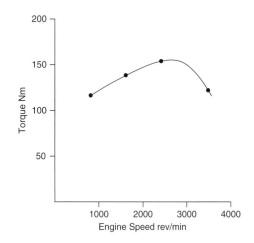

Fig. 14.8 Torque vs. engine speed

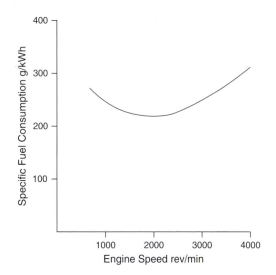

Fig. 14.9 Specific fuel consumption vs. engine speed

$$bsfc = \frac{\text{mass of fuel consumed per hour}}{\text{brake power}}.$$

Figure 14.9 shows how bsfc varies with engine speed.

Example 14.8

The results shown in Table 14.2 were obtained during a fuel consumption test on an engine.

Table 14.2 Results of fuel consumption test on an engine

Brake power (kW)	10	15	20	25	30	35
Total fuel consumed per hour (kg)	4	5.21	6.74	8.4	10.59	12.6

Determine the specific fuel consumption (sfc) in kg/kWh in each case and plot a graph of specific fuel consumption on a base of brake power.

Solution

Specific fuel consumption:

$$sfc = \frac{\text{mass of fuel used per hour}}{\text{brake power}}.$$

Results are shown in Table 14.3.

Table 14.3 Specific fuel consumption calculations based on Table 14.2 (Example 14.8)

Brake power (kW)	10	15	20	25	30	35
Specific fuel consumption (kg/kWh)	0.4	0.347	0.337	0.336	0.353	0.36

Brake power, torque, and sfc compared

In order to assess the importance of these curves, it is normal practice to plot brake power, torque, and specific fuel consumption on a common base of engine speed. The vertical scales are made appropriate to each variable. Figure 14.10 shows such a graph.

The points to note are:

- the maximum power of 52 kW occurs at 4000 rev/min;
- the minimum specific fuel consumption of 237 g/kWh occurs at 2000 rev/min;
- the maximum torque of 147 Nm occurs at 2100 rev/min.

This data highlights details that affect vehicle behaviour; e.g. when the specific fuel consumption is at its lowest value the engine is consuming the least possible amount of fuel for the given conditions − this is considered to be the most economical operating speed of the engine. The maximum torque also occurs at approximately the same speed, because torque converts to tractive effort; the maximum tractive effort is obtained at the same engine speed as is found for the maximum torque speed.

The maximum power is reached at a high engine speed.

The ratio of maximum power speed to maximum torque speed is known as the engine speed ratio.

The most useful part of the power curve of the engine lies between the maximum power speed and the maximum torque speed. The difference between maximum power speed and maximum torque speed is known as the engine operating range.

In this case the engine operating speed range = 4000 rev/min − 2100 rev/min = 1900 rev/min.

The actual shapes of the graphs of these engine characteristics are determined by engine design. The graphs for a heavy transport vehicle engine may be expected to be quite different from those for a saloon car.

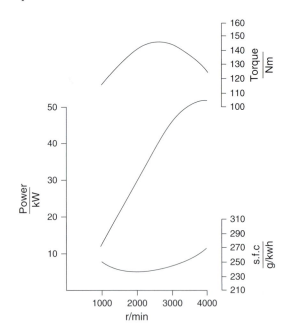

Fig. 14.10 Characteristic curves of diesel engine variables

Brake mean effective pressure

The brake mean effective pressure (bmep) may be obtained from the brake power curve of the engine as follows:

$$bmep = \text{brake power in kW} \times 1000 \div lan \text{ Nm}$$

In this equation, l = length of engine stroke in metres, a = cross-sectional area of the cylinder bore in square metres, and n = the number of working strokes per second.

When bmep is plotted against engine speed, the curve produced is the same shape as the torque curve because torque is related to bmep. Engine performance data such as specific fuel consumption, and its relationship to bmep, at a given engine speed, may be shown in graphical form as in Fig. 14.11. Here the engine is run at constant speed, on a dynamometer, and the air–fuel ratio is varied.

The main point to note here is that maximum bmep is developed when the mixture is rich. The minimum fuel consumption occurs when the air–fuel ratio is slightly weaker than the chemically correct air–fuel ratio of 14.7:1 for petrol.

Thermal efficiency

The thermal efficiency of an engine is a term that is used to express the effectiveness of an engine's ability to convert heat energy into useful work. Thermal efficiency is the ratio of energy output of the engine to energy supplied to the engine in the fuel:

Brake thermal efficiency

$$= \frac{\text{brake energy/s}}{\text{mass of fuel/s} \times \text{calorific value of the fuel}}.$$

Example 14.9

During a 10-minute dynamometer test on a petrol engine, the engine develops a brake power of 45 kW and uses 3 kg of petrol. The petrol has a calorific value of 43 MJ/kg. Calculate the brake thermal efficiency.

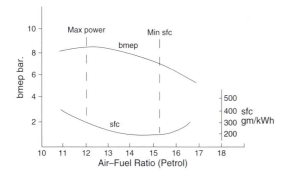

Fig. 14.11 Brake mean effective pressure vs. sfc at constant engine speed

Solution

Fuel used per second 3/600 = 0.005 kg/s.
 1 kW = 1 kJ/s. Brake energy per second = 45 kJ/s.

Brake thermal efficiency

$$= 45 \times 10^3 \div 0.005 \times 43 \times 10^6$$
$$= 45\,000 \div 21\,5000$$
$$= 0.209 \text{ or } 20.9\%.$$

Indicated thermal efficiency

Indicated thermal efficiency

$$= \frac{\text{indicated energy/s}}{\text{mass of fuel/s} \times \text{calorific value}}.$$

Example 14.10

During a dynamometer test, a certain four-cylinder, four-stroke diesel engine develops an indicated mean effective pressure of 8.5 bar at a speed of 2000 rev/min. The engine has a bore of 93 mm and a stroke of 91 mm. The test runs for 5 minutes during which time 0.8 kg of fuel are used. The calorific value of the fuel is 43 MJ/g. Calculate the indicated thermal efficiency.

Solution

$$\text{Indicated power} = \frac{PlaN}{1000} \text{ kW}$$

N = number of working strokes/second
 = number of cylinders/2 × revs/s
 = 4/2 × 2000/60 = 66.7.

Length of engine stroke $(i) = 0.091$ m.

Area of bore $(a) = \pi/4 \times 0.093 \times 0.093 = 0.0068$ m^2.

imep $= 8.5 \times 10^5$ N/m^2

Indicated power

$$= 8.5 \times 10^5 \times 0.091 \times 0.0068 \times 66.7/1000 \text{ kW}$$
$$= 35.1 \text{ kW}.$$

Mass of fuel/s $= 0.8/300 = 0.00027$ kg.

Indicated thermal efficiency

$$= 35.1 \times 1000 \text{ J/s}/0.0027 \times 43 \times 10^6$$
$$= 0.302 \text{ or } 30.2\%.$$

Brake thermal efficiency: petrol vs. diesel

For reasons that are explained in Chapter 15, which deals with engine cycles, the brake thermal efficiency of a diesel engine is normally higher than it is for a comparably sized petrol engine. The higher thermal efficiency of the diesel engine results in good fuel economy. The figures shown in Table 14.4 refer to two vehicles of the same model, one fitted with a diesel engine and the other with a petrol engine of the same size.

Table 14.4   Fuel economy in petrol and diesel engines

Fuel economy	Petrol engine	Diesel engine
Urban cycle miles/US gallon	15	25
Highway cycle	21	34
Composite	18	29

Heat energy balance

In Example 14.9 the engine has a brake thermal efficiency of 20.9%. This means that 20.9% of the energy supplied to the engine in the fuel is converted into useful work. In order to determine what happens to the remainder of the energy supplied in the fuel, a heat energy balance test is conducted. The energy balance is conducted under controlled conditions in a laboratory. The test apparatus is equipped with a dynamometer and with meters for recording air and fuel supply, cooling water flow, and exhaust gas temperature.

Energy supplied to the engine
= mass of fuel used × calorific value.
Energy to the cooling water = mass of water
× specific heat capacity × temperature increase.
Energy to useful work
= brake power as measured by dynamometer.
Specific heat capacity of water
= 4.18 kJ/kg$^\circ$C.
Energy to the exhaust
= mass of exhaust gas × specific heat capacity
× temperature increase.
Specific heat capacity of the exhaust gas
= 1.012 kJ/kg$^\circ$C approximately.
Mass of fuel used as recorded on a fuel flow
meter = m_f kg.
Calorific value is obtained from the fuel
specification.
Mass of water flowing through the engine
cooling system = m_w kg.
Temperature rise of cooling water
= $T_{w_{out}} - T_{w_{in}}$ $^\circ$C.
Temperature rise of exhaust gas is recorded by a thermocouple sensor that is placed in the exhaust stream, close to the engine = $Tex_{out} - Tex_{in}$ $^\circ$C.
Mass of exhaust gas: this is determined by adding the mass of air supplied to the mass of fuel supplied. The mass of air supplied is determined by the air-flow meter through which all air for combustion passes.
Mass of exhaust gas = mass of air + mass of fuel
= m_{ex}.

Example 14.11

The data shown in Table 14.5 was obtained during a 1-hour laboratory test on an engine.

Construct a heat balance for this engine. Take the specific heat capacity of the exhaust = 1.01 kJ/kg$^\circ$C; specific heat capacity of the water = 4.2 kJ/kg$^\circ$C.

Table 14.5   Data from engine test

Brake power	60 kW
Mass of fuel used	18 kg/h
Air used (air–fuel ratio = 16:1 by mass)	288 kg/h
Temperature of air at inlet	20°C
Temperature of exhaust gas	820°C
Mass of water through cooling system	750 kg/h
Cooling water temperature in	15°C
Cooling water temperature out	85°C
Calorific value of fuel	43 MJ/kg

Solution

Heat energy supplied in the fuel = 18 × 43 = 774 MJ/h.
Heat energy to the exhaust
= (18 + 288) × 1.01 × 10^3 × (820 − 20)
= 247.25 MJ/h.
Heat energy to cooling water
= 750 × 4.2 × 10^3 × (85 − 15) = 252 MJ/h.
Heat energy equivalent of brake power
= 60 × 1000 × 3600 = 216 MJ/h.

Where 3600 is the number of seconds in one hour. Results shown in Table 14.6.

Table 14.6   Tabulated results of heat balance calculations

Heat supplied in the fuel	774 MJ/h	100%
Heat to brake power	216 MJ/h	28%
Heat to exhaust	247.25 MJ/h	32%
Heat to cooling water	252 MJ/h	32.6%
Heat to radiation etc. (by difference)	58.75 MJ/h	7.4%

Comment

This balance sheet provides a clear picture of the way in which the energy supplied to the engine is used. It is clear that approximately two-thirds of the energy is taken up by the exhaust and cooling system. In turbocharged engines, some of the exhaust energy is used to drive the turbine and compressor and this permits the brake power output of the engine to rise.

Effect of altitude on engine performance

The density of air decreases as the air pressure decreases and this causes a loss of power in naturally aspirated engines when they are operating at high altitude. It is estimated that this power loss is approximately 3% for every 300 m (1000 ft) of altitude. Turbocharged engines are affected to a lesser extent by this effect.

Summary of main formulae

Volumetric efficiency

$$= \frac{\text{volume of charge admitted at STP}}{\text{swept volume}}.$$

$$\text{Brake power (bp)} = \frac{2\pi T n}{60 \times 1000},$$

where T is torque in Nm, n = rev/min.

$$\text{Indicated power (ip)} = \frac{Plan}{60 \times 1000},$$

where P = imep in N/m^2, l = length of stroke in m, a = area of bore in m^2, n = number of working strokes per min; for a four-stroke engine,

$$n = \frac{\text{number of cylinders} \times \text{rev/min}}{2}.$$

$$\text{Mechanical efficiency} = \frac{\text{bp}}{\text{ip}} \times 100$$

Brake thermal efficiency

$$= \frac{\text{energy to brake power}}{\text{energy supplied in the fuel}}$$

$$= \frac{\text{bp} \times 1000 \times 3600}{\text{mass of fuel/h} \times \text{calorific value}}.$$

Brake specific fuel consumption

$$= \frac{\text{mass of fuel used per hour}}{\text{bp}}.$$

Self-assessment questions

1. A four-cylinder, four-stroke engine with a bore diameter of 80 mm and a stroke length of 75 mm develops an indicated mean effective pressure of 6.9 bar at an engine speed of 3600 rev/min. Calculate the indicated power at this engine speed.

2. The results shown in Table 14.7 were obtained when a Morse test was conducted on a six-cylinder engine operating at a constant speed of 3000 rev/min. Calculate the indicated power of the engine and the mechanical efficiency at this speed.

Table 14.7 Results of a Morse test conducted on a single-cylinder engine (Example 14.3)

Cylinder cut-out	None	No. 1	No. 2	No. 3	No. 4
bp (kW)	120	85	87	88	86

3. (a) Calculate the brake power of a six-cylinder engine that produces a bmep of 7 bar at a speed of 2000 rev/min. The bore and strokes are equal at 120 mm. (b) If the mechanical efficiency at this speed is 80%, calculate the indicated power.

4. A diesel engine develops a brake power of 120 kW at a speed of 3000 rev/min. At this speed the specific fuel consumption is 0.32 kg/kWh. Calculate the mass of fuel that the engine will use in 1 hour under these conditions.

5. As a result of a change in design, the bmep of a four-cylinder, four-stroke engine rises from 7.5 bar to 8 bar at an engine speed of 2800 rev/min. The bore and stroke are 100 mm and 120 mm respectively. Calculate the percentage increase in brake power at this speed that arises from the design changes.

6. A six-cylinder, four-stroke petrol engine develops an indicated power of 110 kW at a speed of 4200 rev/min, the imep at this speed being 9 bar. Calculate the bore and stroke given that the engine is 'square', i.e. the bore = stroke.

7. An engine develops a torque of 220 Nm during an engine trial at a steady speed of 3600 rev/min. The trial lasts for 20 minutes during which time 8.3 kg of fuel are consumed. The calorific value of the fuel is 43 MJ/kg. Calculate the brake thermal efficiency.

8. The data in Table 14.8 shows the results of a Morse test conducted on a four-cylinder engine. Calculate the indicated power and the mechanical efficiency of the engine.

Table 14.8 Results of a Morse test conducted on a four-cylinder engine (Example 14.8)

Cylinder cut-out	None	No. 1	No. 2	No. 3	No. 4
bp (kW)	25	17	16	18	16

9. A four-cylinder, four-stroke petrol engine with a bore diameter of 100 mm and a stroke of 110 mm has a volumetric efficiency of 70% at an engine speed of 4200 rev/min. Determine the actual volume of air at STP that flows into the engine in 1 minute.

10. The data in Table 14.9 shows the results obtained during a dynamometer test on a four-cylinder diesel engine. Calculate the brake thermal efficiency for each value of brake power and plot a graph of (a) specific fuel consumption on a base of bp and (b) brake thermal efficiency on the same base. Calorific value of the fuel = 44 MJ/kg.

Table 14.9 Results obtained during a dynamometer test on a four-cylinder diesel engine (Example 14.10)

bp (kW)	10	20	40	60	80
Specific fuel cons. (kg/kWh)	0.32	0.29	0.27	0.28	0.30

15
The compression ignition engine – diesel engine

Topics covered in this chapter

The four-stroke compression ignition engine (CIE)
The Diesel cycle and the dual combustion cycle
The two-stroke CIE
Diesel engine construction
Direct injection engine
Indirect injection engine
Turbulence
Induction stroke turbulence
Combustion in a compression ignition engine
Three phases of combustion
Diesel fuel and products of combustion
Flash point
Pour point
Cloud point
Products of combustion
Emissions limits
Emissions control on CIE
Particulate trap and selective catalyst reduction
Exhaust gas recirculation (EGR)

Rudolf Diesel (1858–1913) is generally accepted as the person who first developed an internal combustion engine that worked by injecting fuel into compressed air that was hot enough to ignite the fuel. His patent used these words: 'Compressing in a cylinder pure air to such an extent that the temperature thereby produced is far higher than the burning or igniting point of the fuel.' The original diesel engines of the type shown in Fig. 15.1(a) were large and heavy, and they operated at slow speed; they were used for stationary engines and ships but were not considered suitable for use in road vehicles.

Around the 1920s, developments in fuel injection equipment and other technologies led to the development of engines of the type shown in Fig. 15.1(b) that were suitable for use in road vehicles – these engines work on a cycle of operations that is known as the dual combustion cycle. For this reason some authorities

suggest that the vehicle engines that work on the principle of compression ignition should be called compression ignition engines (CIEs). In modern literature the terms are used interchangeably.

The four-stroke compression ignition engine

The four strokes (Fig. 15.2) are:

1. The inlet valve is opened and pure air is drawn into the cylinder as the piston moves down. This is the induction stroke.
2. Both of the valves are closed and the piston moves up the cylinder, compressing the air so that the temperature rises above the ignition point of the fuel. The high pressure and temperature is achieved by a high compression ratio of approximately 20:1. As the piston approaches top dead centre (TDC), fuel is injected so that ignition has started by the time that the piston starts on the next stroke.
3. Both of the valves are closed and the piston is forced down the cylinder by the expanding gas and fuel injection continues for a short period. The fuel is shut off after a few degrees of crank rotation and the high pressure of the gas forces the piston down the cylinder on the power stroke.
4. The exhaust valve is opened and the piston rises in the cylinder, expelling the spent gas through the exhaust port – at the end of this stroke the engine is ready to start the next cycle.

The Diesel cycle and the dual combustion cycle

The difference between the two theoretical cycles can be seen in pressure–volume diagrams, which are graphs that are used to show how the gas pressure and the cylinder volume are related as the piston moves along the cylinder.

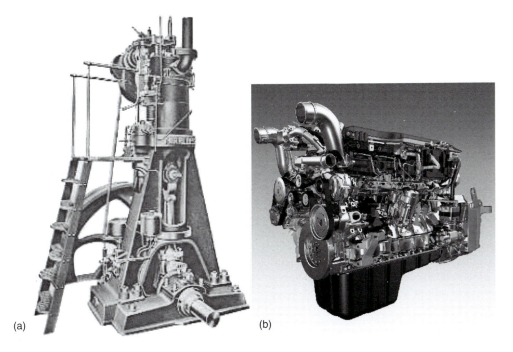

Fig. 15.1 (a) Diesel engine, 1897 (Mirlees). (b) Diesel engine, 2005 (MAN)

The Diesel cycle

In the Diesel cycle the fuel is injected into the hot air in the cylinder and the pressure remains constant until fuel injection ceases. This is represented by the line from 1 to 2 in the pressure–volume diagram shown in Fig. 15.3(a). At point 2 the gas in the cylinder continues to expand, pushing the piston along the remainder of the power stroke. Because the combustion takes place while the pressure remains constant, the Diesel cycle is also known as the constant-pressure cycle. This cycle is close to the sequence of events that takes place in very large diesel engines that operate at slow speeds of a few hundred revolutions per minute.

The dual combustion cycle

In the dual combustion cycle (Fig. 15.3(b)) the fuel injection starts at point 1; this causes the pressure to rise rapidly, as shown by the vertical line that ends at point 2. In this first stage the combustion has taken place while the volume in the cylinder remains constant. Between points 2 and 3 further injection takes place, pushing the piston along the cylinder, while the pressure remains constant and the gas expands, pushing the piston down the cylinder. At point 3 combustion ceases and the gas in the cylinder continues to expand, doing work on the piston until it reaches the end of the stroke.

Inlet valve   Injector

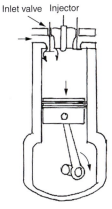
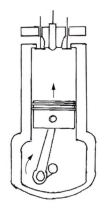
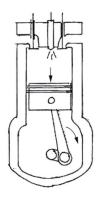
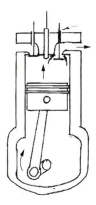

First stroke – Induction. Inlet valve open – pure air drawn into cylinder as piston moves down the stroke.

Second stroke – Compression. Inlet and exhaust valves closed. Piston moves up the stroke. Air is compressed until it is very hot. Fuel sprayed into cylinder and ignites.

Third stroke – Power. Both valves closed. Further fuel injected. The burning fuel raises the pressure in the cylinder and forces the piston down the cylinder on the power stroke.

Fourth stroke – Exhaust. Exhaust valve opens. Piston raises in the cylinder and expels the spent gas. Exhaust valve closes at end of stroke and inlet valve opens ready to start the next cycle.

Fig. 15.2 The four-stroke compression ignition cycle

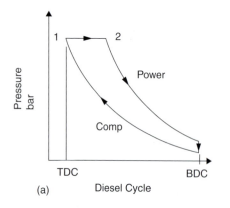

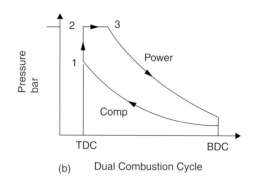

Fig. 15.3 Diesel and dual combustion cycles

This theoretical cycle is considered to be close to the cycle of events that takes place in the compression ignition engines used in motor vehicles. Indicator diagrams for the Diesel cycle and the dual combustion cycle are shown in Fig. 15.3.

The two-stroke CIE

The two-stroke CIE shown in Fig. 15.4 employs end-to-end scavenging, where there is a compressor that feeds compressed air to a ring of ports around the circumference of the cylinder and a pair of exhaust valves that are operated by a camshaft. The air ports are controlled by the piston. When the piston is at bottom dead centre the air ports are uncovered and air under pressure enters the cylinder. At this stage the cylinder is occupied by the exhausted gas from the previous power stroke − the entering fresh air is denser and at higher pressure than the exhaust and this, coupled with the upward motion of the piston, pushes the exhaust gas out through the open exhaust valves. As the piston rises further it covers the air ports; at this stage the exhaust valves close and the

compression process begins. Towards the end of the compression stage, fuel is injected and combustion takes place to drive the piston down the cylinder on the power stage of the cycle. At the end of the compression stage the air ports are uncovered, the exhaust valves are opened, and the process is completed in two strokes of the piston. The compressor is similar to a supercharger and it is normally gear driven from the crankshaft.

Diesel engine construction

Much of the diesel engine mechanism and structural details are similar to those found in petrol engines. The principal differences are concerned with the way in which combustion takes place and the stronger components that are required to cope with the high pressures that are needed to produce the temperature required for combustion. Diesel engines can conveniently be divided into two types:

1. Direct injection engines
2. Indirect injection engines.

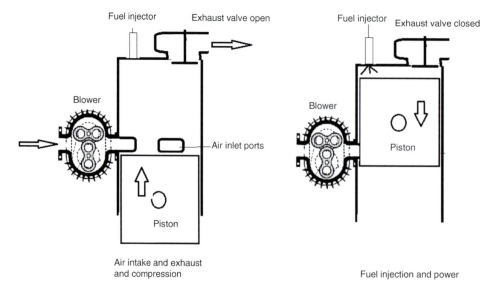

Fig. 15.4 Principle of the two-stroke CIE

Direct injection engines

In the direct injection engine shown in Fig. 15.5 the fuel is sprayed directly into the cylinder. The circular space in the piston crown forms part of the combustion chamber; it is designed to produce turbulence when air is forced in towards the end of the compression stroke. This type of turbulence is known as 'squish' turbulence because it is produced by the 'squashing' of air as the piston forces the air down into the piston cavity. The injector is normally of the multi-hole type and operates at a pressure of approximately 180 bar — the compression ratio of the direct injection engine is of the order of 16:1, which is somewhat lower than that used in indirect injection engines.

Indirect injection engines

In the indirect engine (Fig. 15.6) the fuel is sprayed into a small pre-combustion chamber that is placed in the cylinder head above the piston. As the piston approaches TDC, air is forced into the pre-combustion chamber, which is designed to produce the swirling action necessary for good combustion.

The combustion that starts in the pre-chamber rapidly heats the air and the burning fuel, and air is forced into

the space at the top of the piston, where it mixes with the main body of air to complete the combustion process. The injection pressure in the indirect injection engine is of the order of 120 bar — compression ratios between 22:1 and 28:1 are normal in these engines.

Turbulence

Turbulence is required to ensure that all droplets of fuel are surrounded by sufficient air to provide the oxygen that is required for efficient combustion. There are basically two methods of creating turbulence:

1. Turbulence created on the induction stroke
2. Turbulence created on the compression stroke.

Induction stroke turbulence

Figure 15.7(a) shows the arrangement of the induction ports on a modern diesel engine. The tangential port is designed to set up a rotary motion in the air as it passes through the port into the cylinder on the induction stroke, as indicated in Fig. 15.7(b).

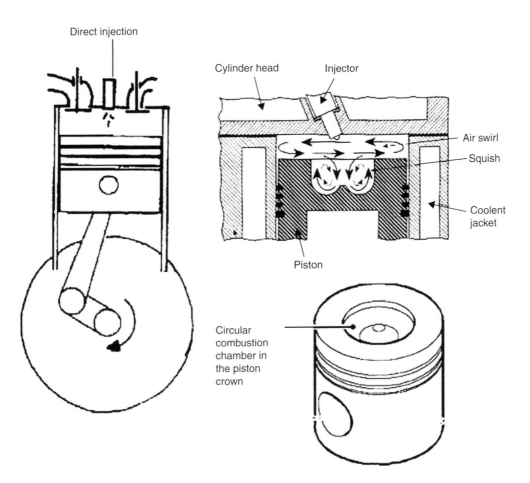

Fig. 15.5 A direct injection engine

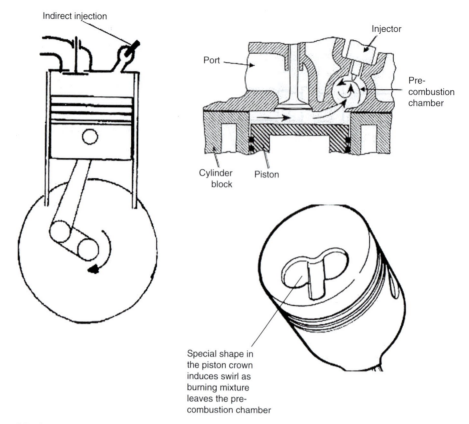

Fig. 15.6   An indirect injection system

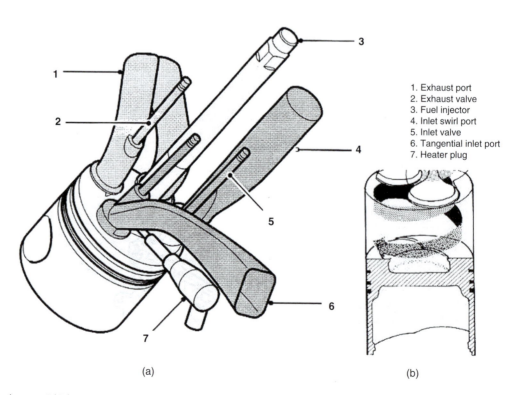

1. Exhaust port
2. Exhaust valve
3. Fuel injector
4. Inlet swirl port
5. Inlet valve
6. Tangential inlet port
7. Heater plug

(a)   (b)

Fig. 15.7   A tangential inlet port

Combustion in a compression ignition engine

The power output of a compression ignition engine is determined by the amount of fuel that is injected — for low power output such as engine idling a very small amount of fuel is injected, while for high power output a large amount of fuel is required. The amount of fuel injected is determined by the length of time for which the fuel is injected and this is controlled by the design of the fuel injection system, which is covered in a later section. For the time being I wish to concentrate on the process of combustion.

Three phases of combustion

Sir Harry Ricardo, the founder of the Ricardo research laboratories at Shoreham in Sussex, first put forward the idea that combustion in a compression ignition engine takes place in three separate phases. The graph in Fig. 15.8 shows how pressure, temperature, and heat release from combustion changes from the point at which injection of fuel starts. Images (a)–(d) are photographs of the cylinder contents from the start of injection at about

25° before TDC, through to full combustion, which continues some way down the power stroke. The three phases of combustion shown in Fig. 15.8 are:

1. The first phase, from (a) to (b). This is known as the delay period. In this period the fuel is sprayed into the dense high-pressure and high-temperature air, and a small period of time elapses during which the tiny particles of fuel are being evaporated. The resultant fuel vapour must then be brought into contact with oxygen so that combustion can start. The length of the delay period depends on several factors, such as:
   - The ignition quality of the fuel (cetane rating).
   - The relative velocity between the fuel and the air in the cylinder (turbulence).
   - The fineness of the atomization of the fuel.
   - The air–fuel ratio.
   - The temperature and pressure of the air in the cylinder.
   - The presence of residual exhaust gas from the previous cycle.
2. The second phase, from (b) to (c). This is the period when combustion spreads rapidly through the combustion space, leading to a rapid rise in pressure. The rate at which pressure rises in this phase governs

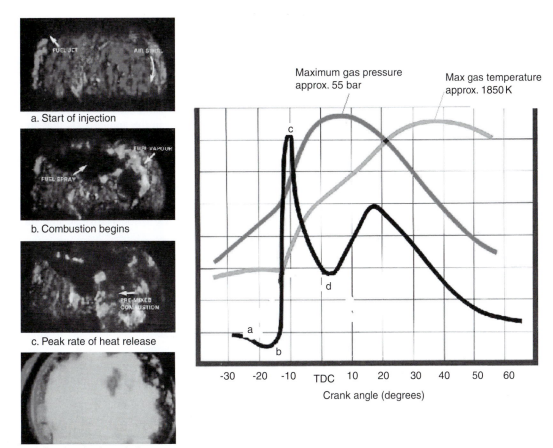

a. Start of injection

b. Combustion begins

c. Peak rate of heat release

d. Combustion by turbulent diffusion

Fig. 15.8 Compression ignition engine combustion (Lucas CAV)

the extent of combustion knock, which is a feature of compression ignition engines and is known as 'diesel knock'.

3. The third phase, from (c) to (d), is the period when combustion is fully operational and the flame spreads to encompass all of the fuel. In this phase the pressure continues to rise at a more gradual rate until injection ceases a few degrees later. After this the expansive working effect of the gas drives the piston down the remainder of the power stroke.

Diesel fuel and products of combustion

Diesel fuel

Diesel fuel has a calorific value of approximately 45 MJ/kg and a specific gravity of about 0.8 g/cm^3. The ignition quality of diesel fuel is denoted by the **cetane** number; a figure of 50 indicates good ignition properties. Among other properties of diesel fuel that affect normal operation are flash point, pour point, and cloud point or cold filter plugging point.

Flash point

The flash point of a fuel is the lowest temperature at which sufficient vapour is given off to cause temporary burning when a flame is introduced near the surface. A figure of $125°F$ ($52°C$) minimum is quoted in some specifications.

Pour point

The pour point of a fuel is the temperature at which the fuel begins to thicken and congeal and can no longer be poured from a container; a pour point of $-18°C$ is considered suitable for some conditions.

Cloud point

The cloud point, which is sometimes known as the cold filter plugging point (CFPP), is the temperature at which the fuel begins to have a cloudy appearance and will no longer flow freely through a filtering medium. The cloud point is normally a few $°C$ higher than the pour point.

Note

These figures for diesel fuel are approximate and are presented here as a guide only. Readers who require more detailed information are advised to contact their fuel supplier.

Products of combustion

Exhaust gases are the products of combustion and under ideal circumstances they would comprise carbon dioxide, steam (water), and nitrogen. However, owing to the large range of operating conditions that engines experience, exhaust gas contains several other gases and substances, such as:

- CO – carbon monoxide due to excess fuel and incomplete combustion.
- NO_x – oxides of nitrogen arising from extremely high combustion temperature.
- HC – hydrocarbons arising largely from incomplete combustion.
- PM – particulate material. The bulk of PM is soot, which is incompletely burnt carbon. Other particulates arise from lubricating oil on cylinder walls and metallic substances from engine wear.
- SO_2 – sulphur dioxide. Some diesel fuels contain small amounts of sulphur, which combines with oxygen during combustion to form SO_2. This in turn can combine with water to form sulphurous acid.
- CO_2 – carbon dioxide is not treated as a harmful emission but it is considered to be a major contributor to the greenhouse effect and efforts are constantly being made to reduce the amount that is produced. In the UK the quantity of CO_2 that a vehicle produces in a standard test appears in the specification, and vehicle taxation (road tax) is less for small CO_2 emitters than it is for large ones.

Emissions limits

In the UK the emissions limits are set by the European Union. The limits are the subject of constant review – those shown in Table 15.1 are for the standards known as Euro 4. The figures apply to vehicles as they leave the manufacturer; once in service the standards set by the UK Department for Transport apply and it is their figures that are used in the annual tests that are known as the MOT.

At the time of writing the test is conducted by passing the exhaust gas through an approved apparatus such as the Hartridge smoke meter, which measures the opacity of the exhaust gas. Figure 15.9 shows how the opacity of the exaust gas relates to the Hartridge and the Bosch scales.

Emissions control on the CIE

The air–fuel ratio in compression ignition engines varies from very weak (probably 50:1) to slightly rich (12:1), and combustion temperatures are high. The three-way catalyst used with petrol engines is not suitable for use with compression ignition engines because

Table 15.1 Emissions limits (g/km)

	CO	HC	HC + NO_x	NO_x	PM
Petrol	1.0	0.10	–	0.08	–
Diesel	0.50	–	0.30	0.25	0.025

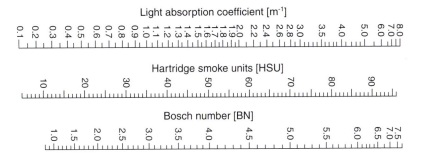

Fig. 15.9 Smoke meter scales

it requires the air–fuel ratio to be held near to 15:1 and alternative methods of dealing with harmful emissions are used. The two systems that are used on heavy vehicles are:

1. Selective catalyst reduction (SCR) and a particulate matter (PM) filter.
2. Exhaust gas recirculation (EGR).

Particulate trap and selective catalyst reduction

The three-way catalyst used on petrol engines requires an air–fuel ratio of about 15:1. Diesel engines operate on mixture strengths that may be as low as 40:1, which means that an alternative system is used to reduce NO_x. On light vehicles there is a tendency to rely on exhaust gas recirculation to limit NO_x and an oxidation catalyst to control HC; in addition, a particulate filter may be used to deal with soot and other particulates. This system is shown in outline in Fig. 15.10.

There is some debate in the heavy vehicle field about the most suitable system for exhaust gas after treatment,

and some large vehicles are equipped with an alternative system of the type shown in Fig. 15.11.

The exhaust gas is first passed through the oxidation catalyst and particulate filter. The high concentration of oxygen in the fuel–air mixture and the relatively high temperature allow the oxidation catalyst to convert HC and CO into CO_2 and H_2O. The gas then enters the particulate filter, where the soot and other materials are filtered out. Any PM that is deposited in the filter can be removed later by active regeneration, which is combustion with oxygen at approximately $600\,^{\circ}C$; this is achieved by a temporary increase in the amount of fuel injected. The regeneration process is performed by the engine management system at intervals dictated by operating conditions. After passing through the oxidation catalyst and particulate filter, a solution of pure water and urea is injected into the exhaust stream, where it reacts with the catalyst to reduce the NO_x to water vapour and nitrogen. In Europe the solution of pure water and urea is called AdBlue and is carried in a small tank that is about 20% of the capacity of the main fuel tank, and it is normally placed next to it, as shown in Fig. 15.12.

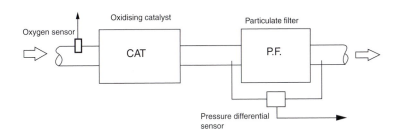

Fig. 15.10 Light diesel exhaust engine emission control

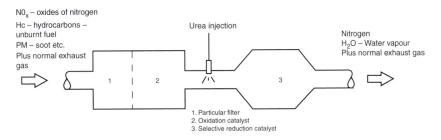

Fig. 15.11 A heavy diesel engine selective catalyst reduction system

Fig. 15.12 The AdBlue tank

Exhaust gas recirculation (EGR)

Oxides of nitrogen (NO_x) are formed when combustion temperatures are high, as they are in compression ignition engines. Exhaust gas contains considerable amounts of CO_2 and H_2O and small amounts added to the incoming air charge reduce the combustion temperature and the production of NO_x. An electrically operated EGR valve of the type shown in Fig. 15.3 that operates under the control of the engine management computer is placed between the exhaust and air intake systems. The engine computer is programmed to recirculate exhaust gas when operating conditions are suitable – in most cases a quantity of exhaust gas equivalent to about 15% of the air intake is recirculated

when the engine is running between idling speed and full load.

Self-assessment questions

1. Why are diesel engines sometimes referred to as compression ignition engines?
2. How is ignition of the fuel achieved in a diesel engine?
3. How does the compression ratio of a diesel compare with that of a petrol engine?
4. How does exhaust gas recirculation help to reduce NO_x emissions?
5. What is meant by the term 'particulate matter'?
6. How does the air–fuel ratio of a diesel engine vary across the engine speed and power range?
7. Give an approximate value of the air–fuel ratio for a diesel engine at idling speed.
8. What effect on cold weather starting will low compression pressure have on a diesel engine?
9. Why doesn't a three-way catalyst work on diesel exhaust?
10. Describe the procedure for conducting the MOT exhaust gas test on a light vehicle equipped with a turbocharged engine.
11. What is the approximate maximum temperature reached in a diesel engine?

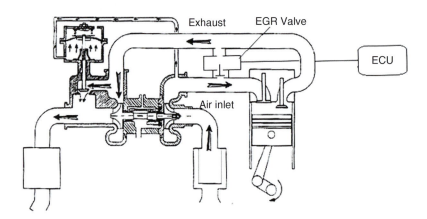

Fig. 15.13 Exhaust gas recirculation

16
Diesel fuel systems

Topics covered in this chapter

Common rail electronically controlled systems
Commonly used injector types
Diesel fuel system maintenance
Fuel filters
Cold starting aids
Rotary and in-line injection pumps

Diesel fuel systems

The purpose of the injection equipment in a diesel fuel system is to supply quantities of fuel oil into the combustion chamber in the form of a very fine spray at precisely timed intervals. To achieve this, the following components are usually employed:

- A fuel tank or tanks
- A fuel feed pump

- A fuel filter or filters
- High- and low-pressure fuel supply lines
- An injector pump
- Injectors
- A timing device
- A governor.

Systems and components vary in design and performance; however, layouts of the components that might be found in a typical system are shown in Figs 16.1 and 16.2. The purposes of the fuel tank and low-pressure pipes are similar to those of the petrol engine.

Lift pump

One of two types is used. The first is the diaphragm type, similar in operation to the petrol fuel system except that it commonly has a **double diaphragm** fitted that is resistant to fuel oil. These are used in low-pressure systems and deliver fuel at a pressure of approximately 34.5 kN/m^2. The other type used is the **plunger-operated**

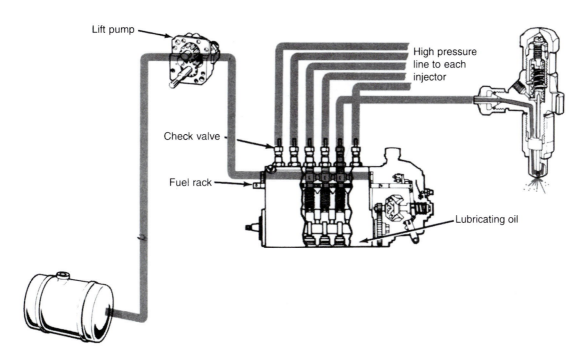

Fig. 16.1 In-line fuel injection pump system

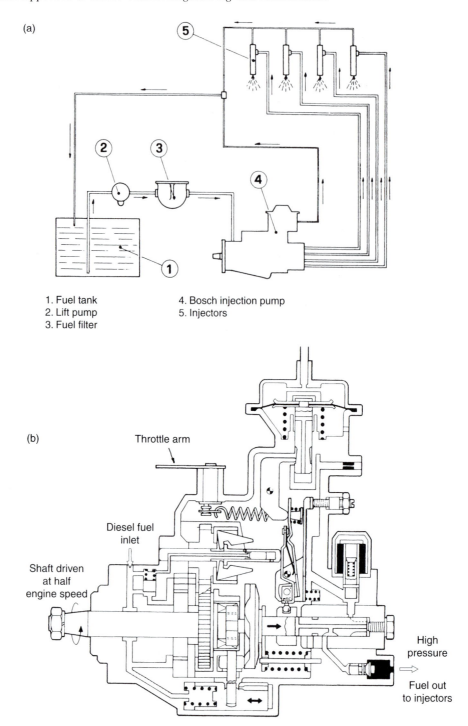

(a)

1. Fuel tank
2. Lift pump
3. Fuel filter

4. Bosch injection pump
5. Injectors

(b)

Throttle arm

Diesel fuel inlet

Shaft driven at half engine speed

High pressure

Fuel out to injectors

Fig. 16.2   (a) The basic components of a diesel fuel system are a fuel tank, lift pump with priming lever, fuel filter, injection pump, injector pipes, and injectors. Additonal pipes connected to the injection pump and injectors return excess fuel to the fuel tank. (b) Distributor-type fuel injection pump

pump, where higher delivery pressures are required. The plunger is backed up by a diaphragm to prevent fuel leakage. These deliver at approximately 104 kN/m^2.

Fuel filters

Working clearances in the injector pump are very small, approximately 0.0001 mm (0.00004 inch); therefore, the

efficiency and life of the equipment depends almost entirely on the cleanliness of the fuel. The fuel filter therefore performs a very important function, that of removing particles of dirt and water from the fuel before they get to the injector pump. After much research, it was found that specially impregnated paper was the best filtering material, removing particles down to a few microns in size. The element consists of the specially treated paper wound around a central core in a spiral form, enclosed in a thin metal canister giving maximum filtration within minimum overall dimension.

Most filters are of the **agglomerator** type (Fig. 16.3), i.e. as the fuel passes through the element, water, which is always present, is squeezed out of the fuel and agglomerates (joins together) into larger droplets, which then settle to the base of the filter by sedimentation. Choking of the filters is caused not only by the solid matter held back by the element, but also by the sludge and wax in the fuel that, under very cold conditions, form a coating on the surfaces of the fuel filter element, thus reducing the rate of fuel flow. Where this becomes too much of a problem, a simple sedimenter-type filter, i.e. one that does not incorporate an element, is fitted between the fuel tank and the lift pump. The object of a simple sedimenter is to separate the larger particles of dirt, wax, and water from the fuel; they are then periodically drained off. Provision is made for the venting of the filter of air (this is commonly called **bleeding** the system, where all the air in the fuel system is removed). If air enters the high-pressure fuel lines then the engine will not run.

Learning tasks

1. Remove and refit the fuel filter on a diesel engine vehicle, bleed the system, and run the engine.
2. Remove and refit the lift pump of a diesel system, test the operation of the pressure/vacuum valves, record results, and check with the manufacturer's data. Give reasons for any differences and make recommendations to the customer on serviceability.
3. Draw up a simple checklist for tracing leaks on a diesel fuel system.
4. Make a list of personal safety/hygiene precautions that a mechanic should consider when working on a diesel fuel system.
5. What are the symptoms of a blocked fuel filter?
6. What are the service intervals for the diesel fuel system and what would these involve?

Injector pump

The function of the injector pump is to:

- Deliver the correct amount of fuel
- At the correct time
- At sufficiently high pressure to enable the injector to break up the fuel into very fine droplets to ensure complete combustion (i.e. complete burning of all the fuel injected).

Multi-element or DPA (distributor pump application) injector pumps are the ones most commonly used.

Multi-element injector pump

This consists of a casing containing the same number of pumping elements as there are cylinders in the engine. Each element consists of a **plunger and barrel** machined to very fine tolerances, and specially lapped together to form a mated pair. A **helix** (similar to a spiral)

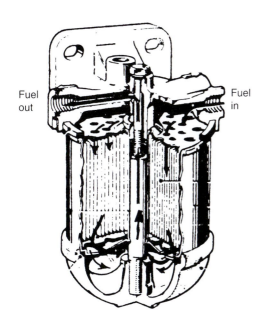

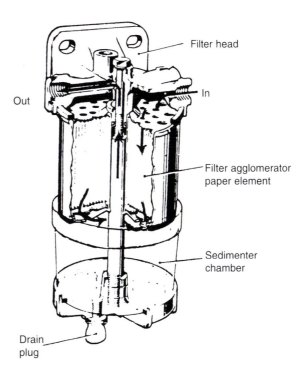

Fig. 16.3 Bowless-type filter agglomerator sedimenter showing agglomerator flow through element

is formed on the outside of the plunger, which communicates with the plunger crown either by a drilling in the centre of the plunger or a slot machined in the side. There are two ports in the barrel, both of which connect to a common fuel gallery feeding all the elements. The plunger is operated by a cam and follower tappet, and returned by a spring. This system is shown in Fig. 16.4.

When the plunger is at bottom dead centre (BDC), fuel enters through the barrel ports filling the chamber above, and also the machined portion forming the helix of the plunger. As the plunger rises, it will reach a point when both ports are effectively cut off (this is known as the **spill cut-off** point and is the theoretical start of injection). Further upward movement of the plunger forces the fuel through the **delivery valve**, injector pipes, and injectors into the combustion chamber. With the plunger stroke being constant, any variation in the amount of fuel being delivered is adjusted by rotating

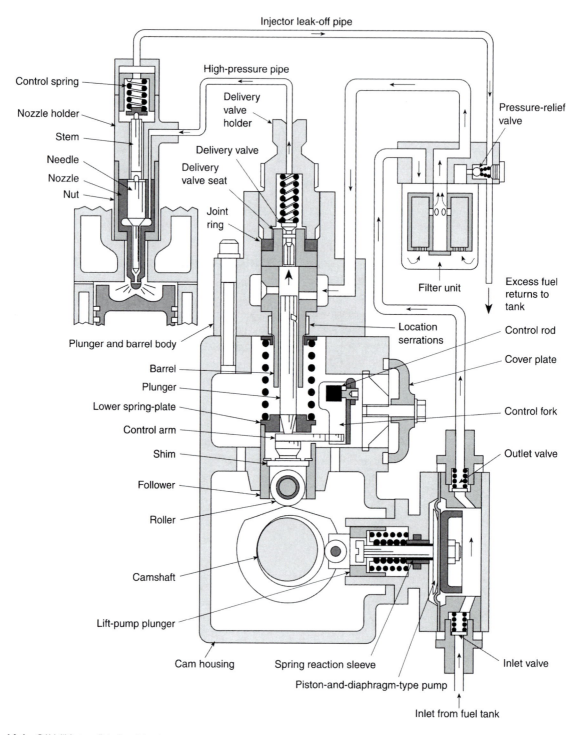

Fig. 16.4 CAV ('Minimec') in-line injection pump

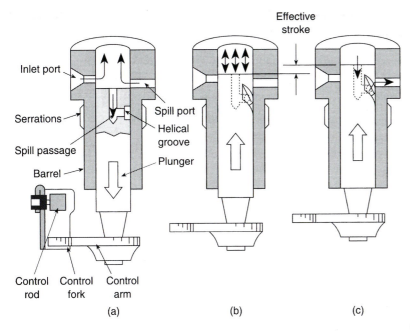

Fig. 16.5 Plunger and barrel. (a) Pumping element filling. (b) Injection. (c) Spill: no fuel out

the plunger, causing the helix to uncover the spill port sooner or later depending on rotation. Immediately the helix uncovers the spill port, the fuel at high pressure above the plunger spills back to the common gallery, the delivery valve resumes its seat, and injection stops without any fuel dribbling from the injector. Each injector is rotated simultaneously by a rack or **control rod**. No fuel or engine stop position is obtained by rotating the plunger so that the helix is always in alignment with the ports in the barrel; in this position, pressure cannot build up and hence no fuel will be delivered. Figure 16.5 illustrates this.

Delivery valve

The purpose of a delivery valve (Fig. 16.6) fitted above each element is to:

- prevent fuel being drawn out of the injector pipe on the downward stroke of the plunger;
- ensure a rapid collapse of the pressure when injection ceases, thus preventing fuel from dribbling from the injector;
- maintain a residual pressure in the injector pipes.

The valve and guide of the delivery valve are machined to similar tolerances as the pumping elements. Approximately two-thirds of the valve is machined to form longitudinal grooves. Above the grooves is the **unloading collar**; immediately above the collar is the valve seat. When the pump is on the delivery stroke, fuel pressure rises and the delivery valve moves up until the fuel can escape through the longitudinal grooves. Immediately the plunger releases the fuel pressure in the barrel, the delivery valve starts to resume its seat under the influence of the spring and the difference in

the pressure above and below the valve. As the unloading collar enters the guides dividing the element from the delivery pipe, further downwards movement increases the volume above the valve (by an amount equal to the volume between the unloading collar and the valve seat). The effect of this increase in volume is to suddenly reduce the pressure in the injection pipe so that the nozzle valve 'snaps' closed on to its seat, thus instantaneously terminating injection without dribble.

Excess fuel device

Multi-element injection pumps are normally fitted with an excess fuel device which, when operated, allows the pumping elements to deliver fuel in excess of normal maximum. This ensures that the delivery pipes from

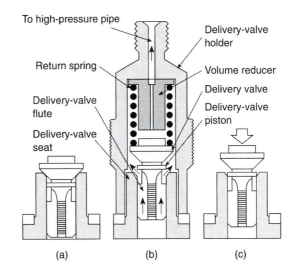

Fig. 16.6 Delivery-valve action

the injection pump to the injectors are quickly primed if the engine has not been run for some time or if assistance is needed for easy starting in cold conditions. When the device is operated (with the engine stationary), the rack or control rod of the fuel injection pump moves to the excess position under the pressure of the **governor spring**. On operating the starter, excess fuel is delivered to the engine. As soon as the engine starts, governor action moves the control rod towards the minimum fuel position, making the device inoperative.

Cam shapes

To prevent the possibility of reverse running, it is normal practice to fit a camshaft with profiles designed so that the plunger is at TDC for approximately two-thirds of a revolution. In the event of a backfire, the engine will not run. Certain CAV pumps are fitted with reversible-type camshafts. In order to prevent reverse running, a spring-loaded coupling, similar to the **pawl-type free-wheel**, is fitted between the pump and the engine. Figure 16.7 shows a cam profile.

Phasing and calibration

Phasing is a term used when adjustment is made to ensure injection occurs at the correct time, i.e. on four-cylinder engines each element injects at 90°

intervals while on a six-cylinder engine each element injects at 60° intervals. This adjustment is carried out by raising or lowering the plunger so the spill cut-off point is reached at the correct time. Simms pumps have spacers in the tappet blocks (Fig. 16.6). CAV pumps have normal tappet adjustment. Phasing should not be confused with spill timing (this is when the injection pump is timed to the engine).

Calibration refers to the amount of fuel that is injected. Correct calibration ensures that the same amount of fuel is injected by each element at a given control rod setting. It is effected by rotating the plunger independently of the control rod. Both phasing and calibration can only be carried out on proper equipment and using data sheets to obtain speed and fuel delivery settings for any given injection pump. When settings are adjusted correctly the maximum fuel stop screw is sealed and must not be adjusted under any circumstances.

Lubrication

The delivery valves and pumping elements are lubricated by the fuel oil, a small quantity of which leaks past the plunger and barrel into the cambox. The camshaft, bearings, tappets, etc. are lubricated by engine oil contained in the cambox. To prevent a build-up of oil

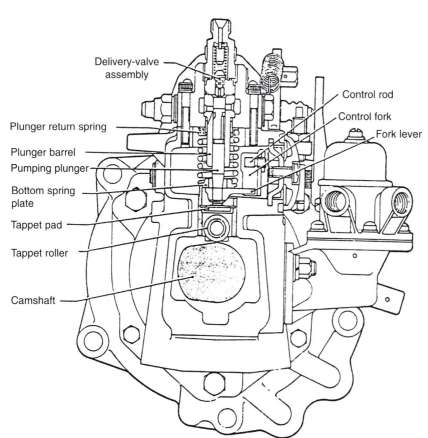

Fig. 16.7 Sectioned view showing cam profile

due to the fuel leaking past the elements, the level plug incorporates a leak-off pipe.

Spill timing the multi-element pump to the engine

After mounting the injection pump to the engine and checking that the alignment and drive coupling clearance are correct, it is necessary to adjust the pump to ensure that number 1 cylinder is on compression. Refer to the workshop manual for the correct static timing (e.g. it may read 28° BTDC). The correct method would be as follows:

1. Set the engine to 28° BTDC (before top dead centre) with number 1 cylinder on compression stroke.
2. Remove the delivery valve from number 1 cylinder pump element.
3. Replace the delivery valve body and fit the spill pipe.
4. Loosen the pump coupling and fully retard the pump.
5. Ensure that the stop control is in the run position.
6. Operate the lift pump; fuel will now flow from the spill pipe.
7. Whilst maintaining pressure on the lift pump, slowly advance the injection pump, when a reduction in the flow of fuel from the spill pipe will be noticed as the plunger approaches the spill cut-off point. Continue advancement until approximately one drop every 10–15 seconds issues from the spill pipe.
8. Tighten the coupling bolts, remove the spill pipe, and refit the delivery valve. The pump is now correctly timed in relation to the engine.

Spill timing is shown in Fig. 16.8.

Bleeding the fuel system

Bleeding the fuel system refers to removing all the air from the pipes, lift pump, filters, injection pump, and injectors. This operation must be carried out if the fuel system is allowed to run out of fuel, any part of the system is disconnected, or the filter elements are changed. Assuming the system has been disconnected for some reason, the correct method is as follows:

1. Disconnect the pressure side of the lift pump, operate the lift pump until fuel, free from air bubbles, flows from the outlet. Reconnect the fuel line.
2. Slacken off the bleed screw of the fuel filter and operate the lift pump until all the air is expelled from the filter. Re-tighten the bleed screw.
3. Open the bleed screw (sometimes called the vent screw) on the injection pump and operate the lift pump again. When fuel free from air bubbles comes out, re-tighten the screw.
4. It may also be necessary to bleed the high-pressure pipes to the injectors by slackening the union at the injector and operating the starter until small amounts of fuel can be seen to be coming from the union. Re-tighten the unions and operate the starter to run the engine. Small amounts of air in the fuel system, though not necessarily enough to prevent the engine starting, may cause loss of power and erratic running. It is therefore necessary to carry out this operation methodically and with care.

Learning tasks

1. Explain in your own words the difference between phasing, calibration, and spill timing. When would each term be used in the course of servicing the diesel fuel system?
2. Remove a multi-element injection pump from a diesel engine, turn both the pump and engine over, re-time the pump to the engine, bleed the fuel system, and run the engine.

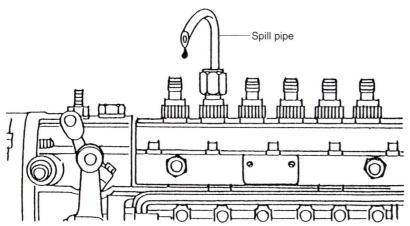

Fig. 16.8 Spill timing number 1 pumping element to number 1 cylinder

The DPA or rotary fuel injection pump

The DPA fuel injection pump (Fig. 16.9) serves the same purpose as the multi-element type and offers the following advantages:

- It is smaller, more compact, and can be fitted in any position, not just horizontal.
- It is an oil-tight unit, lubricated throughout by fuel oil.
- Only one pumping element is used, regardless of the number of cylinders to be supplied.
- No ball or roller bearings are required and no highly stressed springs are used.
- No phasing is required; calibration once set is equal for all cylinders.
- An automatic advance device can be fitted.

In this pump the fuel at lift pump pressure passes through a nylon filter, situated below the inlet union, to the **transfer pump**. Fuel pressure is increased by the transfer pump, depending on the speed of rotation of the pump, and controlled by the **regulating valve**. The regulating valve maintains a relationship between pump speed and transfer pressure, which at low revolutions is between 0.8 and 1.4 kg/cm^2 (11–20 lb/inch^2), increasing to between 4.2 and 7.0 kg/cm^2 (60–100 lb/inch^2) at high revolutions. From the transfer pump, fuel flows through a gallery to the **metering valve**. The metering valve, which is controlled by the **governor**, meters the fuel passing to the rotor, depending on engine requirements. The fuel is now at metering pressure, this being lower than transfer pressure. As the rotor rotates, the inlet ports come into alignment and fuel enters the rotor, displacing the plungers of the pumping elements outwards until the ports move out of alignment. Further rotation brings the outlet ports of the rotor into alignment with one of the outlet ports, which are spaced equally around the hydraulic head. At the same time, contact between the plunger rollers and the cam ring lobes forces the pumping elements inwards. Fuel pressure between the plungers increases

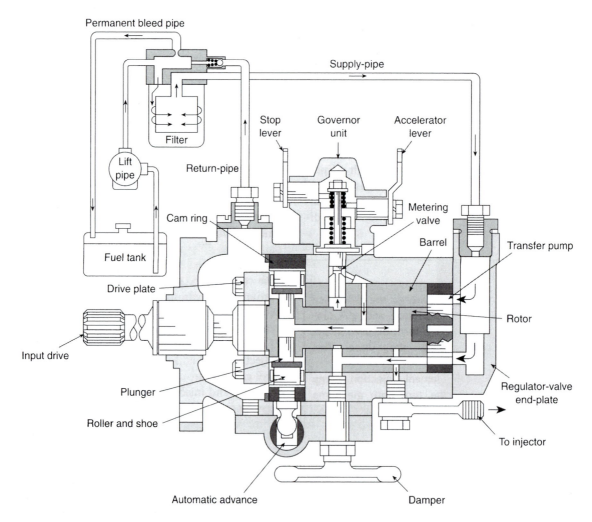

Fig. 16.9 CAV ('DPA') distributor-type injection pump

to injection level and fuel is forced along the control gallery, through the outlet port to the injector pipe and injector. As the next charge port in the rotor aligns with the metering valve port, the cycle begins again. Figure 16.10 illustrates this.

The inside of the cam ring has as many equally spaced lobes as there are cylinders in the engine. Each lobe consists of **two peaks**, the recess between them being known as the retraction curve. As the pumping element rollers strike the first peak, injection takes place. On reaching the retraction curve, a sudden drop in pressure occurs and injection stops without fuel dribbling from the injector. Further movement of the rotor brings the rollers into contact with the second peak, which maintains residual line pressure until the outlet port moves out of alignment. The cam ring rotates within the pump housing, varying the commencement of injection. Movement is controlled by the advance/retard device.

It should be remembered that this type of pump is lubricated by the fuel oil flowing through the pump and if it runs out of fuel at any time, due to a new pump being fitted, any parts of the fuel system being disconnected or the filter elements having been changed, then to prevent damage to the pump occurring the following method should be used to bleed the system of air:

1. Slacken the vent screw on the fuel filter. This may be a banjo union of the 'leak-off' pipe. Operate the lift pump until fuel free from air flows from the filter. Tighten the vent screw.

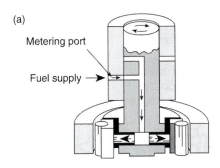

(a)

Metering port

Fuel supply

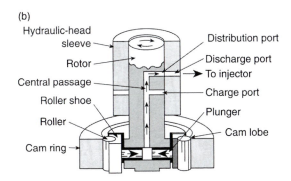

(b)

Hydraulic-head sleeve

Rotor

Central passage

Roller shoe

Roller

Cam ring

Distribution port

Discharge port

To injector

Charge port

Plunger

Cam lobe

Fig. 16.10 Cycle of operation. (a) Charge phase. (b) Injection phase

2. Slacken the main feed pipe union nut at the pump end and operate the lift pump until fuel free from air flows from the union. Tighten the union.
3. Slacken the vent screw of the governor control housing and the vent screw in the hydraulic head. Operate the lift pump until fuel free from air flows from the vent screws. Tighten the hydraulic head vent screw and governor control housing vent screw.
4. Crank the engine by hand one revolution and repeat the operations listed in (3).
5. Slacken all high-pressure injection pipes at the injector end and turn the engine over with the starter until fuel free from air flows from unions. Tighten unions and start engine.

Note

Always ensure that the stop control is in the start position and the throttle is wide open when bleeding the pump.

Figure 16.11 shows one type of injector pump. This is fitted with a mechanical means of governing the maximum and minimum revs of the engine. The automatic advance operates when the engine is stationary, enabling injection to take place earlier than normal when starting the engine. This gives easier starting and reduces the amount of smoke being passed through to the atmosphere.

Injectors

The injector can be considered as an **automatic valve** that performs a number of tasks. It may vary in design but all conform to the following requirements:

- It ensures that injection occurs at the correct pressure.
- It breaks up the fuel into very fine droplets in the form of a spray that is of the correct pattern to give thorough mixing of the fuel with the air.
- It stops injecting immediately the injection pump pressure drops.

The **nozzle** is the main functional part of the injector. It consists of a **needle valve** and **nozzle body** machined to fine tolerances and lapped together to form a mated pair. Extreme care must be exercised when handling this component.

Lubrication is achieved by allowing a controlled amount of fuel to leak past the needle valve. Provision is made for this **back-leakage** to return either to the filter or fuel tank; it also circulates heated fuel to help overcome the problem of the fuel freezing in very cold weather.

Adequate cooling of the injector is most important and is catered for by careful design and positioning in the cylinder head. To prevent overheating, it is essential that the correct injector is fitted and that any joint

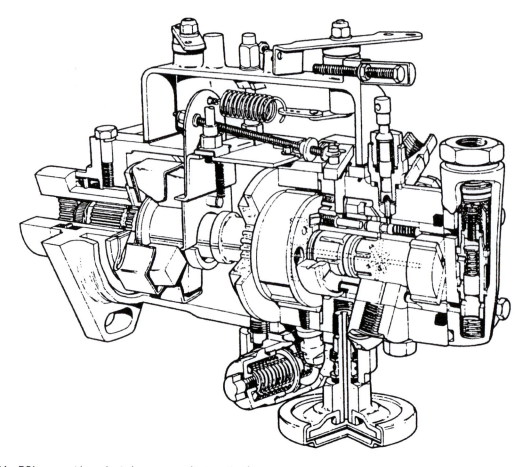

Fig. 16.11   DPA pump with mechanical governor and automatic advance

washers are replaced each time the injector is removed (make sure the old joint washer is removed before fitting a new one). No joint washer is required on those engines fitted with copper sleeves that form part of the injector housing. In addition to the normal copper sealing washer, a corrugated-type steel washer is sometimes located between the nozzle and the heat shield. It is essential that the outside edge faces *away* from the nozzle, otherwise serious overheating of the injector will occur. To prevent the washer turning over whilst fitting it should be fed down a long-bladed instrument such as a screwdriver or a suitable length of welding rod.

Method of operation

The needle is held on its seat by spring pressure acting through the spindle. Fuel, when delivered at high pressure from the injector pump, acts on the shoulder at the lower end of the needle. When the fuel pressure exceeds the spring tension, the needle lifts off its seat and fuel is forced through the hole(s) in a finely atomized spray. The spring returns the needle valve back on to its seat at the end of each injection. The spring tension is adjustable by releasing the lock nut and screwing the spring cap nut in or out as required. This determines at

what pressure injection commences. The pressures at which the injector operates are very high, typically between 125 and 175 atmospheres. It should be noted that the spring tension only determines at what pressure injection starts. Pressure may momentarily increase up to a maximum of approximately 420 atmospheres (to start the needle moving), depending on engine speed and load. At these high pressures it is very important that the testing is done with the proper equipment and in the correct way. If the hand is placed in front of the injector whilst it is being tested then it is possible for the fuel to be injected directly into the bloodstream, with serious consequences and possible death. Use of safety equipment such as gloves and goggles is essential to reduce any possible risks of injury to a minimum.

Types of injector

These are classified by the type of nozzle fitted and fall into four main groups:

- Single-hole
- Multi-hole
- Pintle
- Pintaux.

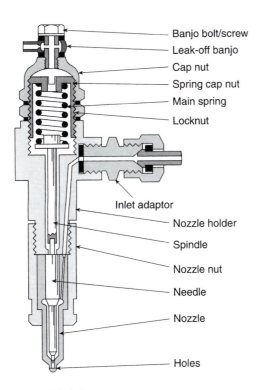

Fig. 16.12 Multi-hole injector

Single- and multi-hole injectors

The **single-hole** nozzle has one hole drilled centrally in its body, which is closed by the needle valve. The hole can be of any diameter from 0.2 mm (0.008 in.) upwards. This type is now rarely used in motor vehicle engines. **Multi-hole** injectors (Fig. 16.12) have a varying number of holes drilled in the bulbous end of the nozzle beneath the needle valve seating. The actual number, size, and position depends on the requirements of the engine concerned. There are usually three or four. This is the type that is fitted to the **direct injection** engine, which, due to the larger combustion chamber, requires the fuel to be injected in a number of sprays at high pressure to ensure even distribution and good penetration of fuel into the rapidly moving air stream. They are often of the long stem type to give good cooling of the injector.

Pintle

This nozzle is designed for use with **indirect injection** combustion chambers. The needle valve stem is extended to form a pintle, which protrudes beyond the mouth of the nozzle body. By modifying the size and shape of this pintle, the spray angle can be altered from parallel to a 60° angle or more. A modified pintle nozzle, known as the **delay** type, gives a reduced rate of injection at the beginning of delivery. This gives quieter running at idling speed on certain engines.

Pintaux nozzles

A development of the pintle-type nozzle, these have an auxiliary hole to assist starting in cold conditions. At cranking speeds, the pressure rise is slow and the needle valve is not lifted high enough for the pintle to clear the main discharge port. The fuel passing the seat is sprayed from the auxiliary hole towards the hottest part of the combustion chamber (which is within the area of the heater plug). At normal running speeds, the rapid pressure rise lifts the pintle clear of the main discharge port, allowing the fuel to form the appropriate spray pattern. Approximately 10% of the fuel continues to pass through the

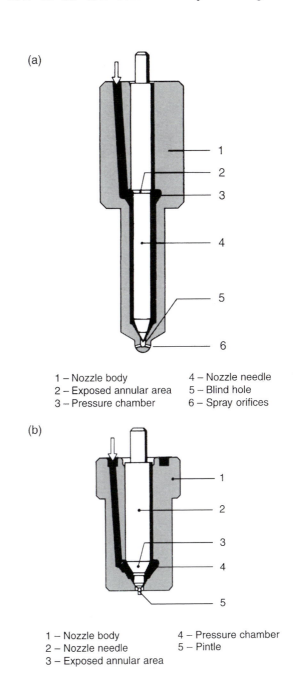

(a)
1 – Nozzle body
2 – Exposed annular area
3 – Pressure chamber
4 – Nozzle needle
5 – Blind hole
6 – Spray orifices

(b)
1 – Nozzle body
2 – Nozzle needle
3 – Exposed annular area
4 – Pressure chamber
5 – Pintle

Fig. 16.13 (a) Hole-type nozzles. (b) Pintle-type nozzles

auxiliary hole at normal running speeds to keep it free from carbon.

The efficiency of the injector deteriorates with prolonged use, making it necessary to service the nozzles at periodic intervals. The frequency of maintenance depends on factors such as operating conditions, engine condition, cleanliness of fuel, etc.

Examples of nozzles are shown in Fig. 16.13.

Cold starting aids

Some form of cold starting aid is usually fitted to compression ignition engines to assist starting in cold conditions. There are numerous types, the two most common being:

- Heater plugs
- Thermo-start unit.

Heater plugs

These are located in the cylinder head, the heating element of the plug being located just inside the combustion chamber. When an electrical current is supplied to the plug, the element heats up, thus heating the air trapped in the chamber. Typical heater plugs are shown in Fig. 16.14.

The 'pencil'-type heater plug is similar in design to the 'coil' type, except that the heating coil is contained in a tube which when heated glows red. The plugs are wired in **parallel** and operate at 12 V with a loading of approximately 60 W. This gives a maximum element

temperature of 950–1050°C. When the heater plug has been in service for some time, the small air gap between the element sheath and the cylinder head becomes filled with carbon. This reduces the efficiency, making starting more difficult in very cold weather. There is also a risk of the element burning out through overheating. This can be avoided by removing the heater plugs and cleaning out the plug hole from time to time. Before refitting the plug remove any particles of carbon that may have lodged in the conical seating in the cylinder head. A faulty plug may be located by removing each feed wire in turn and fitting a test lamp or ammeter in the circuit. If the plug is in working order, the lamp will light or the ammeter will show a reading of approximately 5 amps when the heater circuit is in operation. A quick test can be carried out when the engine is cold. Operate the heater circuit and after approximately 30 seconds each plug should feel warm to the touch.

In the double-coil type the plugs are connected in **series**, each one operating at 2 V. A resistance unit is wired into the circuit to reduce battery voltage to plug requirements. Apart from keeping the exterior of the plug and electrical connections tight, no other servicing or maintenance is required.

Thermo-start unit

The unit, shown in Fig. 16.15, is screwed into the inlet manifold below the butterfly valve. Fuel is supplied to the unit from a small reservoir fed from the injector leak-off pipe. The thermo-start comprises a valve

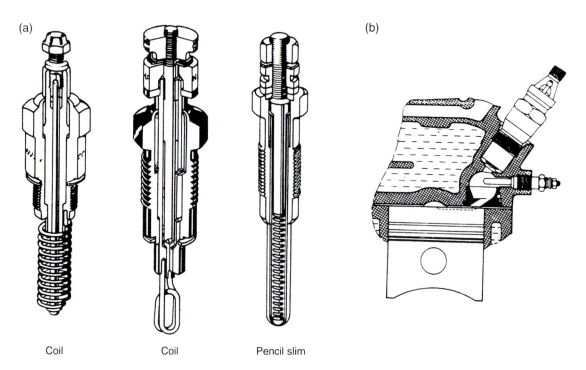

(a)   (b)

Coil   Coil   Pencil slim

Fig. 16.14 (a) Three types of heater plugs used on indirect injection diesel engines. (b) Sectional view showing the position of the plug in the cylinder head

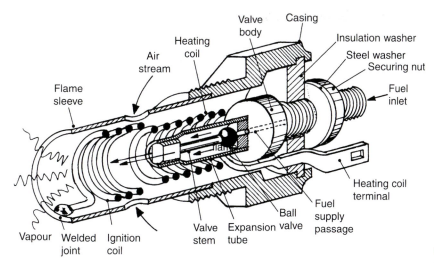

Fig. 16.15 Manifold heat-operated thermostat (CAV) (cold starting aid)

surrounded by a heater coil, an extension of which forms the igniter. The valve body houses a spindle, which holds a ball valve in position against a seat, preventing fuel entering the device. When an electric current is supplied to the unit, the valve body is heated by the coil and expands. This releases the ball valve from its seat, allowing fuel to enter the manifold, where it is vaporized by the heat. When the engine is cranked, air is drawn into the manifold and the vapour is ignited by the coil extension, thus heating the air being drawn into the engine. On switching off the current to the unit, the valve body contracts and the spindle returns the valve to its seat, cutting off the fuel supply. The reservoir must be positioned 10–25 cm above the thermo-start unit to provide a positive fuel supply. This unit gives very little trouble in service provided the preheat time before operating the starter does not exceed that recommended (approximately 15 seconds).

Electrical connections

The heater plugs are connected so that when the ignition is switched on the warning lamp on the dashboard is illuminated; this warns the driver that the heater plugs are operating (warming up). The starter should not be operated until the warning lamp goes out, which happens automatically after approximately 10–20 seconds. The heater plugs will now be at their correct temperature for cold starting. This is shown in Fig. 16.16.

Common rail diesel fuel system

Another recent development in computer-controlled diesel systems is the common rail system shown in Fig. 16.17.

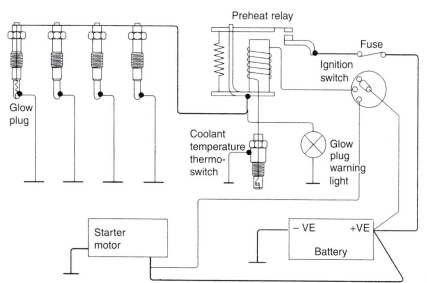

Fig. 16.16 Basic cold start glow plug preheated electrical circuit

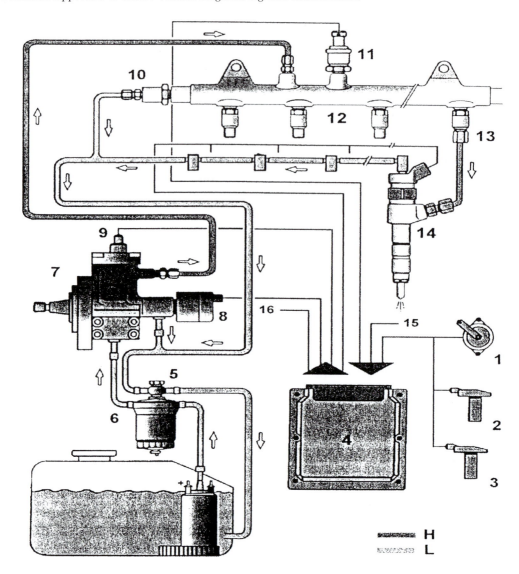

1. Accelerator
2. Engine speed (crank)
3. Engine speed (cam)
4. Engine control module
5. Overflow valve
6. Fuel filter
7. High pressure pump
8. Pressure regulating valve
9. Plunger shut-off
10. Pressure limiting valve
11. Rail pressure sensor
12. Common rail
13. Flow limiter
14. Injector
15. Sensor inputs
16. Actuator outputs

Fig. 16.17 The Rover 75 common rail diesel fuel system

In this common rail system, the fuel in the common rail (gallery) is maintained at high pressure (up to 1500 bar). A solenoid-operated control valve that is incorporated into the head of each injector is operated by the ECM. The point of opening and closing the injector control valve is determined by the ROM program and the sensor inputs. The injection timing is thus controlled by the injector control valve and the ECM, and the quantity of fuel injected is determined by the length of time the injector remains open; this is also determined by the ECM.

The high injection pressures used and electronic control of the amount of fuel that is injected means that combustion efficiency is good, thus leading to improved fuel economy and reduced emissions.

Details of the common rail fuel system

The system can be viewed as two parts: the low-pressure part that contains a primary pump in the fuel tank and a secondary pump that is located near the high-pressure pump. In addition, the low-pressure side contains the fuel filter and the necessary pipe work. These items can be identified in Fig. 16.18. The high-pressure side contains the high-pressure pump, the fuel gallery (rail), the injectors, some sensors for ECM control, and a pressure control valve.

The high-pressure pump

The high-pressure pump input shaft is driven from the crankshaft. The pumping action is produced by an eccentric cam that rotates inside an actuating collar — this collar has three flat faces; as shown at 2 in Fig. 16.19, the rotation of the eccentric cam inside this collar causes the three pistons to reciprocate in their chambers to produce the high pressure. This form of pump action absorbs a good deal less engine power than the conventional type of high-pressure fuel injection pump. The secondary pump operates at slightly higher pressure than the primary pump in the tank and it helps to ensure that there is an adequate supply of high-pressure fuel when the engine speed is low.

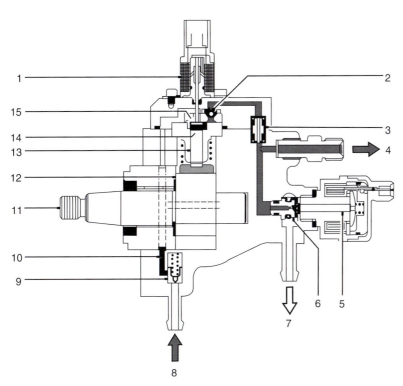

1. Cut out valve
2. High-pressure outlet valve
3. Seal
4. High-pressure fuel to common rail
5&6. Pressure control valve
7. Fuel return to tank
8. Fuel supply from low pressure pump
9. Safety valve
10. Low pressure duct
11. Pump drive shaft
12. Cam driven by eccentric shaft
13&14. Pump piston and chamber
15. Suction chamber

Fig. 16.18 High-pressure pump for common rail fuel system

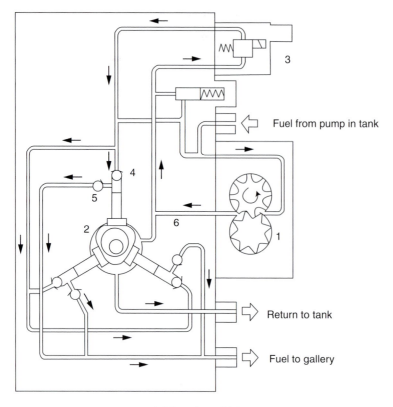

1. Low-pressure pump
2. High-pressure pump
3. Pressure regulating valve
4. Inlet valve to high-pressure pump
5. Outlet valve from high-pressure pump
6. Lubrication circuit

Fig. 16.19   High-pressure pump

The pressure-regulating valve

The pressure-regulating valve is operated by a solenoid which, when commanded to do so by the ECM, will open the valve to release fuel back to the low-pressure side — in this way the pressure in the fuel rail gallery can be adjusted to suit demand.

The fuel rail and pressure-relief valve

The fuel rail (Fig. 16.20) is a tube that has sufficient capacity to ensure an adequate supply of high-pressure fuel for all operating conditions. The pressure-relief valve ensures that a maximum safe pressure is not exceeded and, if required, it will open and release high-pressure fuel back to the low-pressure side. The pressure sensor produces a signal that is used in the ECM to operate the pressure-regulating valve. There are separate connections to each injector and these also operate under the control of the ECM.

Sensors

The power output of a diesel engine is controlled by the amount of fuel that is injected. In the common rail system the amount of fuel to be injected for any set of conditions is determined by the program in the ECM. In order for the ECM to function it must receive

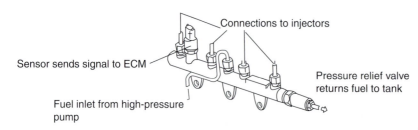

Fig. 16.20   The fuel rail

information from the sensors. The sensors shown in this example are:

- The throttle position sensor
- The engine speed sensor
- The fuel pressure sensor
- The camshaft sensor that determines the injection point.

In response to signals received from these and other sensors, the ECM determines the type of electrical pulse to send to the injectors.

In the common rail system the injection normally takes place in two stages, one called pilot injection just before TDC and the other after a short interval when the combustion has started. This improves cold starting performance and reduces emissions.

Common rail diesel system injectors

Figure 16.21 shows two views of an electronically controlled injector that is used in typical Common Rail diesel fuel system. In diagram A, the control valve is closed and no fuel is being injected while, in diagram B the control valve is open and fuel is being injected into the engine cylinder.

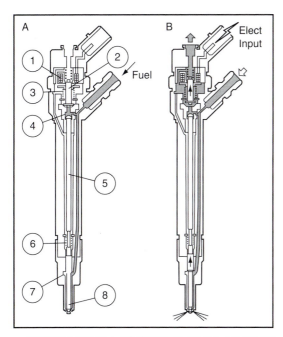

1	Coil	5	Push rod
2	Valve (upper chamber)	6	Spring (lower chamber)
3	Spring (upper chamber)	7	Lower chamber
4	Upper chamber	8	Fuel needle

Fig. 16.21 A common rail fuel injector

Operation of injector in the closed position (diagram A)

The solenoid (1) is not energised and the control valve (2) is held on its seat. Fuel from the common rail fills the upper chamber (4) and the lower chamber (7). The pressures in the upper and lower chambers are equal. The injector needle valve (8) is held firmly on its seat by control spring (6) and no injection occurs.

Operation of the injector in the open position (diagram B)

An electrical pulse from the ECM energises the solenoid (1), lifting the control valve (2) and releasing pressure in the upper chamber (4). The imbalance of pressure between the upper and lower chambers causes the injector needle valve (8) to be lifted against the spring (6) and atomised fuel is injected into the engine cylinder.

Safety precautions

'Frequent and prolonged contact with used engine oil may cause dermatitis and other skin disorders, including skin cancer, so avoid unnecessary contact. Adopt safe systems of work and wear protective clothing, which should be cleaned or replaced regularly. Maintain high standards of personal hygiene and cleanliness.' This advice from the Health and Safety Executive **also applies to diesel fuel**.

It is possible to obtain special oils for testing diesel pumps and injectors; these are designed to minimize health risks and they should always be used when testing diesel equipment.

Self-assessment questions

1. The electronically controlled injectors on a certain common rail diesel injection system operate at 80 volts with a current of 20 amperes. The power of these injectors is:
   - (a) 16 kW
   - (b) 1600 W
   - (c) 4 W
   - (d) 0.25 kW.
2. In indirect injection diesel engines:
   - (a) the injectors spray fuel into the induction manifold
   - (b) the injectors spray fuel into a pre-combustion chamber
   - (c) the fuel is injected into a cavity in the piston crown
   - (d) the injection pressure is 10 bar.
3. Diesel injectors are lubricated:
   - (a) by a separate supply of engine oil
   - (b) by means of a grease nipple mounted on the injector nozzle

(c) by leak back of diesel fuel

(d) by coating the needle valve with PTFE.

4. In a diesel fuel system equipped with an in-line injection pump, dribble from the injectors is prevented by:

   (a) a strong injector spring

   (b) reducing the pressure at the lift pump

   (c) a cylindrical collar on the delivery valve

   (d) a sharp cut-off profile on the injection pump camshaft.

5. Checking the phasing of an in-line fuel injection pump is carried out to:

   (a) ensure that fuel injection starts at the correct number of degrees after top dead centre

   (b) ensure that each element commences delivery at the correct angular interval

   (c) ensure that the engine cannot run backwards

   (d) ensure that the lift pump is working.

6. The low-pressure, diaphragm-type lift pump operates at:

   (a) 345 bar

   (b) 34.5 kN/m^2

   (c) 104 kN/m^2

   (d) 0.104 bar.

7. Spill timing is a process that:

   (a) permits the operator to determine the point at which an element of the injector pump starts to inject so that the pump can be timed to the engine

   (b) determines the amount of time that fuel flows through the spill port

   (c) sets the amount of fuel that each pump element delivers

   (d) sets the time constant of an electronic control unit.

8. After fitting a new fuel filter element, the fuel system should be:

   (a) completely drained down

   (b) bled to remove any air that may be trapped

   (c) bled by slackening the connections at the injectors

   (d) put back into commission by starting the engine.

9. The excess fuel device on an injection pump:

   (a) is used to increase pulling power on steep hills

   (b) comes into operation when the turbocharger is operating

   (c) is a cold start aid

   (d) increases the engine's thermal efficiency.

10. Diesel fuel injectors operate at:

   (a) approximately 180 bar

   (b) 7.5 lbf/in^2

   (c) 0.05 Mpa

   (d) 18 kN/m^2.

11. Using Table 16.1 and the list of possible causes, write down the possible causes of blue–white smoke in the exhaust gas.

Table 16.1 Diesel engine faults

Fault	Number reference of possible cause
Engine will not start	1, 3–7, 9–11, 13–15, 21, 22
Engine misfires — lacks power	7–11, 14–19, 21, 22, 33, 35, 36
High fuel consumption	8, 10, 11, 14, 15, 19, 20, 22
Black smoke from exhaust	8, 10, 11, 14, 15, 35, 36
Blue–white smoke from exhaust	10, 12–15, 18, 20–23, 28, 34
Excessive diesel knock	10, 11
Engine runs erratically	7–11, 15, 16, 18
Excessive engine oil pressure	26, 37
Low cylinder compression pressures	14, 18–22
Excessive engine temperature	10, 11, 18, 27–31, 35, 36
Engine starts then stops	4, 7, 9, 16, 17
Excessive crankcase pressure	18, 20, 22, 32

Key:

1. Battery discharged
2. Poor electrical connections
3. Faulty starter motor or solenoid
4. Stop/start solenoid fault
5. No fuel supply
6. Faulty lift pump
7. Dirty fuel filter element
8. Blocked air cleaner
9. Air in fuel system
10. Faulty injection pump/incorrect timing
11. Faulty injectors
12. Faulty cold start advance unit
13. Faulty glow plugs
14. Incorrect valve timing
15. Poor cylinder compressions
16. Fuel tank vent restricted
17. Throttle cable incorrectly adjusted
18. Cylinder head gasket leaking
19. Tappet clearances incorrect
20. Worn cylinder bores
21. Valves not seating/broken spring
22. Piston rings worn
23. Valve guides/stem oil seal worn
24. Main and/or big-end bearings worn
25. Low oil level
26. Faulty oil pump/pressure-relief valve
27. Piston height incorrect
28. Faulty/incorrect thermostat
29. Alternator drive belt incorrect adjustment
30. Radiator blocked
31. Coolant level low
32. Faulty brake exhauster or vacuum pipe leaking
33. Turbocharger fault
34. Induction leak (turbocharged engine)
35. Faulty wastegate operation
36. Boost sensing pipe fault
37. Faulty gauge
38. Incorrect grade of lubricating oil

12. Make a list of the possible causes of black smoke in the exhaust gas.

13. Figure 16.22 shows an injector tester that can be used to perform the following checks:
- Leak back
- Braking pressure

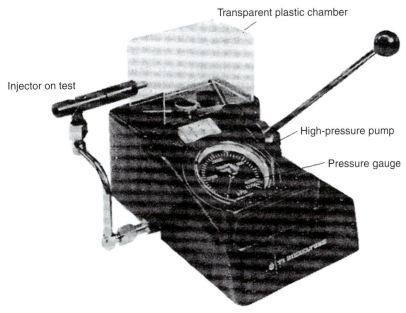

Fig. 16.22 A diesel injector tester

- Spray pattern
- Injector nozzle valve seat condition.

By reference to a workshop manual, and by discussion with colleagues, write a few lines to describe how each of the above tests are performed.

14. A certain amount of fuel is allowed to pass between the nozzle valve and the nozzle body. What is the purpose of this leak back and where does the fuel go when it leaks back?

15. Among the possible causes for poor starting are listed worn cylinder bores, and valves not seating/broken spring. A cylinder compression test may help to confirm that these are the cause of a fault. Describe the procedure for conducting a compression test on a diesel engine.

16. With the aid of sketches describe the type of combustion chamber that is used with each of the injector nozzles shown in Fig. 16.23.

17. Each pumping element of an in-line diesel injection pump must deliver the same amount of fuel as all other pumping elements on the pump. The process of checking and setting the pump so that this happens is called **calibration**. The pump is mounted on a special test rig that powers the pump, and the oil used for testing is forced through injectors and then into transparent measuring cylinders. The machine is capable of rotating the pump a set number of times as required. The number of rotations of the pump camshaft determines the number of pumping strokes of each element. A particular test may require the test oil output to be tested over 100 pumping strokes. Table 16.2 shows some figures for in-line fuel pump calibration checks. Complete

Table 16.2 to show the difference and the percentage.

18. The angle between pumping actions of the elements on an in-line diesel injection pump must be the same for each element. The process of ensuring that this happens is called **phasing**. On a four-cylinder in-line pump for a four-cylinder engine, each element will pump once on each revolution of the pump camshaft. The phase angle on this pump will be $360/4 = 90°$. What will be the phase angle for a six-cylinder in-line diesel injection pump?

19. Figure 16.24 shows a single element of an in-line diesel fuel injection pump. The phase angle is adjusted by the addition or removal of shims that

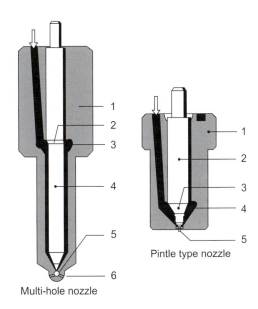

Fig. 16.23 Diesel injector nozzles

Table 16.2   Fuel pump calibration

Speed of pump camshaft rotation (rev/min)	Permitted oil delivery per 100 strokes of pumping elements (cm³)	Permitted difference between maximum and minimum amount of oil delivery (cm³)	Difference as a % of the minimum oil delivery
300	8.8–10.4	1.6	$\frac{1.6}{8.8} \times 100 = 18.2\%$
600	10.6–11.4		
900	11.4–12.8		

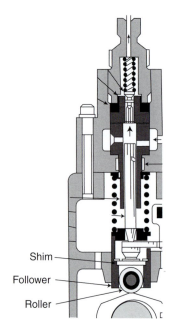

Fig. 16.24   Shims used to adjust phase angle

are placed between the pump plunger and the cam follower. Decreasing the gap increases the phase angle and increasing it reduces the phase angle. In the type of pump shown, shims are available in sizes that allow the phase angle to be altered in steps of 1°. Table 16.3 shows the phase angle test results for an in-line fuel injection pump for a four-cylinder engine that has a firing order of 1–3–4–2. Complete Table 16.3 by filling in the error column and the column that states the action to be taken.

20. Diesel fuel injectors operate at very high pressure in order to atomize the fuel and force the fuel into the high-pressure air in the combustion chamber. The pressure is regulated by the spring at the top of the injector and the operating pressure may be altered by the screw adjustment. A special rig that incorporates a pump and pressure gauge is used to test injectors. Table 16.4 shows some test results for a set of six injectors. Complete Table 16.4 by filling in the error and percentage error columns. Give some thought to the effect on engine performance that these uneven injector settings may have.

21. Table 16.5 shows the four pressure readings that were taken during a compression test. Write down the readings in the spaces provided:
    Cylinder 1 pressure =
    Cylinder 2 pressure =

Table 16.3   Phase angles for in-line pump

Pump element for engine cylinder number	Required phase angle (°)	Actual phase angle (°)	Error + or −	Plunger gap to be increased or decreased?
1	0	0	0	No action
3	90	89	−1	
4	180	182		
2	270	269		

Table 16.4   Diesel injector test results

Injector number	Standard pressure (bar)	Test pressure (bar)	Error (bar)	Percentage error
1	150	148	−2	$\frac{2}{150} \times 100 = 1.33\%$
2	150	160		
3	150	138		
4	150	151		
5	150	142		
6	150	149		

Table 16.5 Compression pressures

Cylinder number	Pressure (lbf/in²)
1	460
2	395
3	380
4	495

Cylinder 3 pressure =
Cylinder 4 pressure =
In order for the engine to function properly the pressure should not be more than 15% below 490 lbf/in².

(a) Calculate each of the above pressures as a percentage of 490 lbf/in²:

Percentage reading cylinder 1 $\frac{cyl1}{490} \times \frac{100}{1} = \%$

Percentage reading cylinder 2 $\frac{cyl2}{490} \times \frac{100}{1} = \%$

Percentage reading cylinder 3 $\frac{cyl3}{490} \times \frac{100}{1} = \%$

Percentage reading cylinder 4 $\frac{cyl4}{490} \times \frac{100}{1} = \%$.

(b) Which of the cylinder pressures is below the recommended pressures?

(c) As this is a diesel engine, what effect is this likely to have on the cold starting and general performance of the engine?

(d) Using the conversion factor of 14.45 lbf/in² = 1 bar, convert the cylinder pressures from lbf/in² to bar correct to two decimal places.

22. Effect of altitude on engine power. Atmospheric pressure and air density decrease with height above sea level. Table 16.6 shows some approximate figures for loss of power of a non-turbocharged engine. If the engine has a power output of 200 kW

Table 16.6 Effect of altitude on engine power

Altitude (m)	300	600	1000	1300	1700	3000
Loss of power (%)	3	6	9	12	15	25

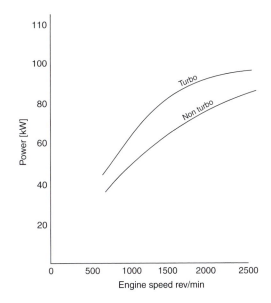

Fig. 16.25 Power curves

at sea level, calculate its power at an altitude of 1000 m.

23. The graph in Fig. 16.25 shows the power curves for two versions of a 6-litre diesel engine. One engine is fitted with a turbocharger and the other one is not. From the graph determine the increase in power output that the turbocharger produces at an engine speed of 2000 rev/min.

17
Engine fault diagnosis

Topics covered in this chapter

Detailed examination of some examples of engine failures
Systematic approaches to fault diagnosis

Engine faults

The in-depth consideration of two typical automotive failures in this chapter is designed to demonstrate principles of automotive fault tracing and rectification.

In order for an internal combustion engine to work it must:

1. Be in good mechanical condition
2. Have a fuel and air system
3. Have a means of igniting the fuel — the ignition system
4. Have a means of disposing of exhaust products — the exhaust system.

This chapter deals with the first of these; other chapters deal with the points 2–4.

Mechanical condition

Engine components

In normal use, all of the mechanical components shown in Fig. 17.1 wear and in time this causes them to cease to function as intended. Careful routine maintenance keeps wear to a minimum, but eventually the wear of components will cause a lowering of performance and repairs will be needed to bring the engine or other unit back to its original performance level. **Faults** may also occur through misuse, failure to maintain, and defects in manufacture or materials.

Whilst accepting that almost any mechanical component may fail, it is not reasonable to expect to cover every possible contingency. However, by examining a number of examples it is possible to see that there is

a procedure that is applicable to all fault diagnosis of automotive systems and that this procedure comprises six steps:

1. Collect evidence
2. Analyse the evidence
3. Locate the fault
4. Find the cause of the fault
5. Rectify the fault
6. Thoroughly test the system.

I propose to show how this six-step procedure applies to fault diagnosis by examining a number of examples.

Example 17.1. Cylinder head gasket failure (Fig. 17.2)

Step 1. Collect the evidence

What are the symptoms? They could be one or more of the following:

- Engine malfunction indicator light (MIL)
- Loss of power
- Difficulty in starting the engine
- Blow back into the intake system
- Defective emissions control.

How did the technician find out about the symptoms?

- Report from the driver
- Detected during other work on the vehicle.

Step 2. Analyse the evidence

Assuming that the driver has reported the problem it will be necessary to discuss the details with her/him. Some drivers will deal with this better than others and a technician should always be aware that this an area of information about faults that can be of considerable value. An assessment of the different types of customer that an automotive technician may expect to encounter is given in the Introduction.

A check sheet on which to make notes is a useful aid that can help when deciding how to deal with the defective vehicle.

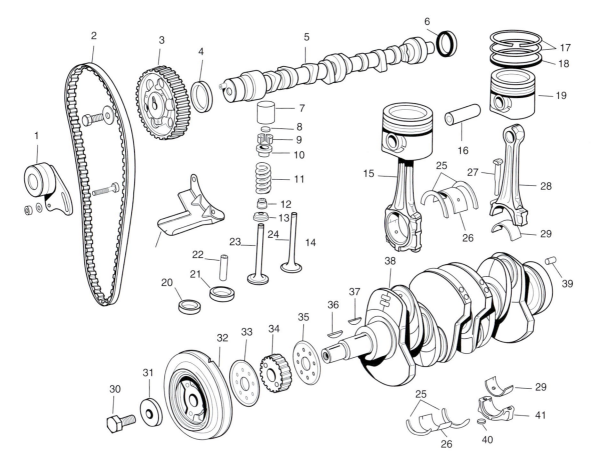

Fig. 17.1 Major engine components

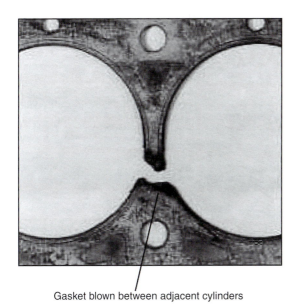

Gasket blown between adjacent cylinders

Fig. 17.2 Cylinder head gasket failure

Once the customer has been attended to, the vehicle can be examined. In this particular case the blown gasket will probably cause the engine to idle badly and the loss of compression in the adjacent cylinders that are affected by the blown gasket will probably

affect the combustion; this will cause the check engine lamp to illuminate. Other symptoms will be loss of power and misfire that may be detected in the exhaust note. These are symptoms that most technicians will notice and it may not be necessary to road test the vehicle.

Step 3. Locate the fault

The aural assessment that was made in assessing the engine behaviour at idle speed should direct attention towards the engine and preparation to conduct further examination should take place. Part of this preparation will be to decide which tests and equipment to use. As the engine problem is an apparent loss of compression it will be necessary to conduct appropriate tests. These tests may include:

- Preliminary visual inspection of vehicle and engine
- On-board diagnostics (OBD), PO code
- Cylinder balance
- Cylinder compression.

Preliminary visual inspection

A preliminary examination to check that there are no obvious contributory factors such as disconnected or

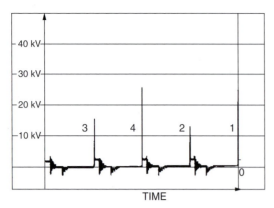

Cylinder balance oscilloscope pattern.

Fig. 17.3 Cylinder balance test

damaged cables and other components is a useful first step. It also helps to ascertain general vehicle condition, which may be a factor to consider when the vehicle is returned to the owner after repairs have been completed.

On-board diagnostics

If the engine check lamp is illuminated some fault codes will be available. These codes are called diagnostic trouble codes (DTCs) and they are downloaded to a code reader through the diagnostic port that is normally located in a readily accessible place near the instrument panel. Details of fault code readers and scanners are given in the section that deals with tools and equipment. At the moment it is sufficient to consider what the DTCs may tell us.

As we are probably dealing with an emissions-related fault, the DTC will be in the form of an EOBD or OBD2 code similar to these examples:

- PO 302 — cylinder 2 misfire detected
- PO 303 — cylinder 3 misfire detected.

This indicates a problem in cylinders 2 and 3 which, in a four-cylinder in-line engine, are adjacent to each other.

Cylinder balance test

The cylinder balance test makes use of the fact that cylinder pressure and sparking plug HT (high tension) voltage are related; low cylinder pressures produce lower sparking voltages than cylinders that have higher pressure. The oscilloscope pattern in Fig. 17.3 shows that the HT voltage in cylinders 2 and 3 is significantly lower than that in cylinders 1 and 4. This supports the information gathered from the DTCs and is further evidence that the fault is probably very low compression in cylinders 2 and 3. A cylinder compression test will produce further evidence to help locate the fault.

Cylinder compression test

A compression test (Fig. 17.4) is a useful method of checking engine condition. The spark plugs are first removed and the compression tester is inserted into the spark plug hole in turn. Each cylinder is checked and the figures recorded ready for examination. The actual pressure reached varies according to the make and type of engine; the workshop manual should be referred to.

The type of result expected here is shown in Table 17.1. The **workshop manual** shows that the cylinder pressure should be within 10% of 12.5 bar. Cylinder pressure in 2 and 3 is lower than this, which is further evidence that the fault is low compression in cylinders 2 and 3.

This leads to the next step.

Step 4. Find the cause

At this stage it is necessary to consider possible causes of low compression (Table 17.2) and then, in the light of

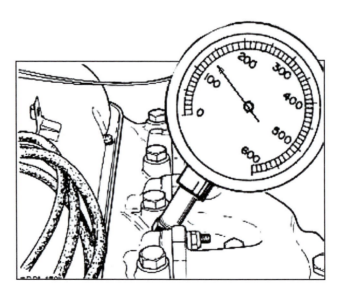

Fig. 17.4 Petrol engine compression test

Table 17.1 Compression pressure

	Cylinder number			
	1	2	3	4
Pressure (bar)	12.2	7.8	7.2	12.6

Table 17.2 Possible causes of low compression in engine cylinders'

1. Damaged or worn pistons and piston rings
2. Damaged or worn valve gear
3. Incorrect valve clearances
4. Damaged cylinder head gasket

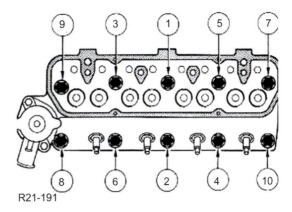

R21-191

Fig. 17.5 Cylinder head bolt-tightening sequence

Table 17.3 Torque settings

Cylinder head bolt tightening	Torque wrench setting (Nm)
Stage 1	10–15
Stage 2	40–50
Stage 3	80–90
Stage 4 (after 10–20 min)	100–110

knowledge and experience, decide which is the most probable.

By considering each item in this list it is possible to arrive at the possible causes that are most likely in this case:

1. Cylinders 1 and 4 are in good condition; it is unlikely that excessive engine wear is the cause.
2. Further inspection could be used – if necessary.
3. Check valve clearances – if necessary.
4. The symptoms are often associated with this fault.

Among the factors to consider here are:

- Are there known problems associated with the particular vehicle?
- Are there any service bulletins from the manufacturer that relate to this problem?
- Has any work been done on the vehicle before – has the engine been altered?

It is not possible to proceed far beyond this point without internal examination of the engine. The point is – how far to go?

Removal of the cylinder head appears to be the answer. In order to gain access to the cylinder head bolts it is necessary to remove the rocker cover; this will permit examination of the valve gear and checking of valve clearances. If the fault is found to be here it may not be necessary to proceed with removal of the cylinder head.

If the valves and valve gear are in satisfactory condition, work to remove the cylinder head would continue, in accordance with the **workshop manual and health and safety** procedures.

When the cylinder head has been removed and placed in a safe position the gasket defect should be evident. The next step, after cleaning and preparing components for inspection, is to look for the causes of gasket failure.

Cylinder head and block examination

It is believed that 80% of head gasket failures can be attributed to incorrect tightening of the cylinder head bolts. The correct procedure follows the pattern of bolt tightening shown in Fig. 17.5. This applies to

a four-cylinder in-line engine but the procedure on other engines will follow a similar pattern that will be shown in the workshop manual.

It is good practice to follow this sequence and to tighten the bolts in stages as follows. The actual torque settings will vary according to the maker's specification; the manual I am referring to here gives the figures in Table 17.3.

If these steps are not followed the result may be a warped cylinder head and this could be the cause of gasket failure.

Checking for flatness of the cylinder head

The cylinder head must be thoroughly cleaned and placed in a suitably secure position. The tools required are an engineer's straight edge and a set of feeler gauges. The flatness should be checked across the length and width as shown in Fig. 17.6; a guide to acceptable figures is 0.06 mm distortion on the length and 0.02 mm on the width. The manufacturer's recommended figures should always be used for reference purposes.

Step 5. Rectifying the problem

Having located the cause of the fault, the next step is deciding what work is required to restore the vehicle to working order. If, as an example, the cylinder head is found to be distorted, the following factors should be taken into account:

- Extent of distortion
- Can distortion be removed by machining?
- What does the engine manufacturer recommend?

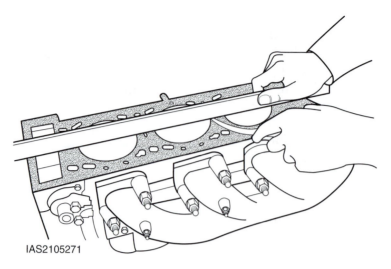

IAS2105271

Fig. 17.6 Checking flatness of cylinder head

- Will removing metal from the head surface raise the compression ratio?
- If machining is permissible, what thickness of metal can be removed?
- How old is the vehicle — would a new cylinder head be an option?

Table 17.4 lists other possible causes of gasket failure.

The practical work required in automotive repair involves many skills that are acquired during practical training — they are important elements of successful fault tracing and rectification, and they are dealt with in apprenticeships and other forms of on- and off-the-job training.

Step 6. Thoroughly test the system

The final step is to thoroughly test the system. In the case of major engine work this would include a comprehensive road test and engine analysis to ensure that emissions are within legal limits. The engine would be examined to ensure that there are no leaks of oil or engine coolant. When all is found to be satisfactory the vehicle should be checked to make sure that the interior and exterior are clean and ready to be returned to the customer.

Example 17.2. Bearing failure

According to the bearing manufacturer's literature, the type of bearing failure shown in Fig. 17.7 is often

caused by lubrication problems that lead to metal-to-metal contact, high friction, and wear. This suggests that lubrication system failures are the general area to examine.

The examination in Example 17.1 shows that the six-step approach is a valuable procedure that lends itself to the treatment of other faults and their diagnosis; this is the approach that is used for this and other examples in this book.

Step 1. Collect evidence

This is probably best done in stages:

1. Consult the customer.
2. Perform aural and visual checks.
3. Use a stethoscope to attempt to refine the evidence.
4. Establish the conditions under which the knock is most evident.
5. Check to find out if the particular engine is prone to the type of knock identified.

Step 2. Analyse the evidence

In order to deal with this step, I shall assume that the fault displays the following symptoms:

1. The oil warning lamp is slow to go out when the engine is first started and the knocking noise is loud.

Table 17.4 Other possible causes of gasket failure

Possible cause	Remedy
Wet cylinder liner unseated. The liner is not protruding enough to allow the gasket to seat properly	Find cause and carry out repairs in accordance with the workshop manual
Combustion problems. Detonation and pre-ignition cause high temperatures and pressure in the combustion chamber. Excessively high temperature can burn the gasket. Extremely high local pressure offsets the bolt clamping force	Check ignition setting. Ensure that correct grade of fuel has been used. Check to see if engine has been altered to raise compression ratio
Material trapped between cylinder head or block and the gasket	Make sure this does not happen on reassembly

Fig. 17.7 Big-end bearing failures

2. After the engine has run for a second or two the warning lamp goes out and the knocking noise subsides a little.
3. Sometimes, when the engine is hot, the oil warning lamp comes on again.

These symptoms suggest that oil pressure is low and the causes of low oil pressure should be considered in deciding how to locate the fault.

Some possible causes of low oil pressure are:

- Pressure-relief valve defective
- Oil pump defective
- Oil strainer in sump blocked
- Low oil level − defective dipstick sensor
- Incorrect grade of oil
- Worn engine components.

There are other possible causes, but this list will suffice for current purposes.

Step 3. Locate the fault

There are many possible causes but for the purposes of this book I shall assume that the oil pump is at fault, because it is a known problem on the engine under investigation. A good deal of successful fault diagnosis is based on similar expert knowledge of particular vehicles and systems.

Step 4. Find the cause

After cleaning, the oil pump can be opened up for inspection. Figure 17.8 shows an example of the type of checks that can be made on the mechanical condition of a rotary-type oil pump. The clearance between the

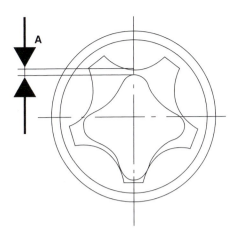

Fig. 17.8 Oil pump checks

pump components is critical in the effective operation of oil pumps and similar checks can be made on other types of pump.

- The gap A between the inner rotor and the outer rotor should not exceed 0.15 mm.
- The rotor end float should fall between 0.08 and 0.09 mm.
- The gap between the outer rotor and the pump casing should be between 0.10 and 0.16 mm. These checks would be made with feeler gauges.

Step 5. Rectify the fault

If the oil pump is found to be defective it should be replaced. However, because we are looking at bearing failure there is likely to be damage to other components, particularly the crankshaft. When the necessary preliminary dismantling work has taken place the condition of the crankshaft can be assessed by taking measurements of the form shown in Fig. 17.9. The crankpin shown is measured at A, B, C, and D. Dimensions A and B are taken to determine taper, and dimensions C and D to determine ovality. In this case the dimensions are: A = 42.99 mm, B = 42.75 mm, C = 42.99 mm, D = 42.60 mm. The taper on this crankpin is A − B = 0.24 mm, and the ovality is C − D = 0.39 mm.

These dimensions show a considerable amount of crankshaft wear and probably mean that regrinding

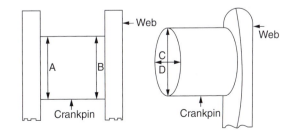

Fig. 17.9 Crankshaft measurement

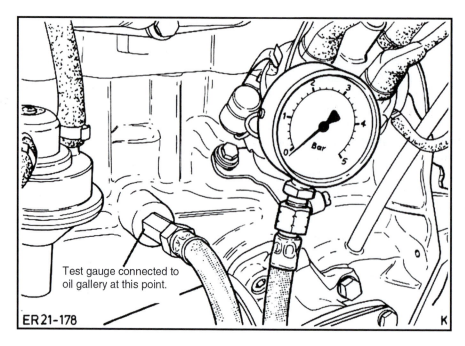

Fig. 17.10   Oil pressure test

or replacement is necessary. The bearings are obviously badly damaged and a decision about repair or replacement of the engine may need to be made.

Step 6. Thorough testing

Testing should follow broadly the steps outlined in Example 17.1, with the addition of a separate oil pressure check. A test gauge is connected to the main oil gallery by temporarily removing the oil warning light sensor and replacing it with the test gauge. With the engine at operating temperature the oil pressure is recorded at idle speed and approximately 2500 rev/min. The figures to hand for the engine shown in Fig. 17.10 are:

- Minimum oil pressure at idle speed is 1 bar
- Minimum oil pressure at 2500 rev/min is 2.8 bar.

Summary

These detailed examinations of automotive faults reveal:

- The need for underpinning knowledge
- Use of tools and equipment
- Sources of information − manuals, bulletins
- Information from owner − vehicle history (see information about customer types in the Introduction)
- The need to work safely and observe the rules and conditions of company safety policy and Health and Safety at Work (HASAW) regulations
- There are six identifiable steps to the fault-finding and rectification process
- The need for measuring skills.

Learning tasks
Reading a micrometer
When taking measurements the force exerted on the thimble must not be excessive and to assist in achieving this, most micrometers are equipped with a ratchet that ensures that a constant force is applied when measurements are being taken.

   When taking a reading, the procedure is as follows. First note the whole number of millimetre divisions on the sleeve. These are called Major divisions. Next observe whether there is a 0.5 mm division visible. These are called Minor divisions. Finally, read the thimble for 0.01 mm divisions. These are called Thimble divisions.

   Taking the metric example shown in Fig. 17.11, the reading is:

- Major divisions = 10 × 1.00 = 10 mm
- Minor divisions = 1 × 0.50 = 0.50 mm
- Thimble divisions = 16 × 0.01 = 0.16 mm.

These are then added together to give a reading = 10.66 mm.

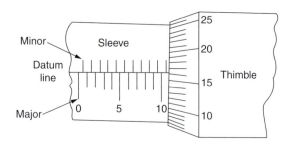

Fig. 17.11   Micrometer readings

Write down the four readings shown in Fig. 17.12.

Crankshaft end float

Figure 17.13 shows the procedure for checking the end float on a crankshaft. End thrust on a crankshaft occurs when the clutch is operated and when the vehicle is operating on an incline. In the engine shown, the end thrust is taken by half-washers made from bronze faced with bearing metal. The thrust faces are lubricated by a film of oil and a small gap between the thrust washers and the crankshaft is provided so that the oil film can be maintained; additional gap is also required to allow for thermal expansion.

The following figures relate to the engine shown in Fig. 17.13:

- **Crankshaft.** Permissible end float 0.09–0.30 mm
- **Thrust half-washer thickness:**
  - **Standard size** 2.301–2.351 mm
  - **Oversize size** 2.491–2.541 mm.

If the end float is found to be greater than the permitted figure, new thrust washers must be fitted to restore the end float to the correct figure.

Self-assessment questions

1. Table 17.5 shows a set of compression figures for a four-cylinder engine. The percentage difference in readings should not be greater than 15% of the maximum pressure recorded. Calculate the percentage difference between the pressure in cylinder number 3 and the other three cylinders, and enter the values into the table. Make a list of the factors that may be causing the low pressure in cylinder number 4.

2. A four-cylinder in-line four-stroke engine has a firing order of 1–3–4–2. What stroke is number 2 cylinder on when number 3 cylinder is on power stroke?

3. Describe the difference between the terms detonation and pre-ignition in the context of engine performance.

4. Figure 17.14 shows a dial test indicator (DTI) being used to check wet cylinder liner protrusion above the cylinder block face. Why is this measurement important?

5. Sometimes people contemplate raising the compression ratio of an engine to increase the power output. Make a list of some of the factors that need to be taken into account when considering such a major alteration to an engine.

6. Table 17.6 shows some torque wrench settings that are given in lbf-ft. Complete the table by converting these settings to Nm. Use the conversion factor 1 lbf-ft = 1.36 Nm.

7. A customer contacts the garage because the check engine lamp in their car is illuminated. When checked the scan tool shows fault 'PO301 – Cylinder misfire cylinder number 1'. Make a list of the factors that may be causing the misfire.

8. Pressure gauges that are used for compression tests require calibration from time to time to ensure that they are working correctly. How should this calibration take place in the context of garage work?

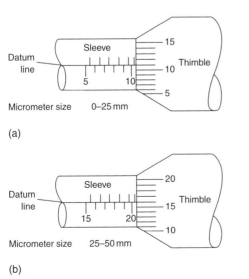

(a)

Micrometer size   0–25 mm

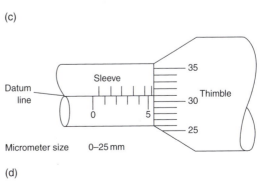

(c)

Micrometer size   75–100 mm

(b)

Micrometer size   25–50 mm

(d)

Micrometer size   0–25 mm

Fig. 17.12   External micrometer readings

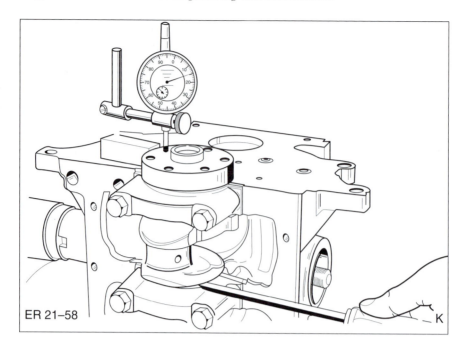

Fig. 17.13 Checking crankshaft end float

Table 17.5 Compression pressure

	Cylinder number			
	1	2	3	4
Pressure (bar)	11.2	11.6	12	9
Percentage difference				

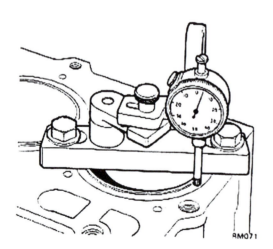

Fig. 17.14 Dial test indicator

9. Refer to Fig. 17.1. How many revolutions does camshaft gear 3 make when the crankshaft gear 34 makes four revolutions?

10. A crankpin has a diameter of 58.170 mm and the big-end bearing fitted to it has an internal diameter of 58.178 mm. The clearance between the crankpin and the big-end bearing is:
    (a) 1.0013 mm

Table 17.6 Torque wrench settings

Component	Torque wrench setting (lbf-ft)	Torque wrench setting (Nm)
Main bearing cap studs	88–102	
Inlet manifold bolts	16–170	

    (b) 0.008 mm
    (c) 0.08 mm
    (d) 0.48 mm.

11. (a) The recommended grade of oil for use in a certain engine is SAE 20W/50. What does this grading reference mean in terms of viscosity at certain temperatures?
    (b) What effect on engine operation is the use of an incorrect grade of oil likely to have?

12. Look at the Health and Safety Executive website and read the section on handling and disposal of used engine oil in the publication HSG 1761. Describe the rules in your area for disposal of waste oil.

13. Make a list of the tools and equipment that are mentioned in this chapter.

Table 17.7 Torque wrench settings

Component	Torque wrench setting (lbf-ft)	Torque wrench setting (Nm)
Main bearing cap	65–75	
Connecting rod bolts	26–33	
Oil pump set screws	16–20	
Exhaust manifold bolts	14–17	
Spark plugs	15–20	
Oil pressure switch	13–15	

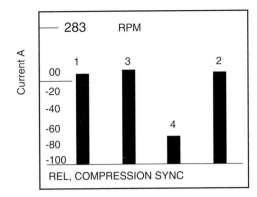

Fig. 17.15 Cylinder compression test based on starter motor current

14. Table 17.7 contains some torque wrench settings in lbf-ft. Complete the table by converting these values to Nm using the conversion factor given in question 6.

15. What measures should be taken to ensure that a pressure gauge used for diagnostic tests remains accurate throughout its working life?

16. In an overhaul of an engine of the type shown in Fig. 17.13 the crankshaft end float is found to be 0.42 mm. On being measured the thrust washers are found to be 2.296 mm thick. It is decided to replace these thrust washers with oversize ones that are each 2.498 mm thick. What will be the amount of crankshaft end float after fitting the new washers?

17. Figure 17.15 shows an oscilloscope bar chart that is based on starter motor current at an engine speed of 283 rev/min as the starter motor is operated. Describe how starter motor current is a reliable measure of variations in engine compression pressure. How do you think that the engine is prevented from starting during this test?

18
Transmission

Topics covered in this chapter

The clutch

Most clutches used in the modern motor car are called **friction clutches**. This means that they rely on the friction created between two surfaces to transmit the drive from the engine to the gearbox. The clutch fulfils a number of different tasks. The three main ones are:

- It connects/disconnects the drive between the engine and the gearbox.
- It enables the drive to be taken up gradually and smoothly.
- It provides the vehicle with a temporary neutral.

The three main component parts of a clutch are:

- The **driven plate**, sometimes referred to as the **clutch**, **centre**, or **friction plate**.
- The **pressure plate**, which comes complete with the clutch cover, springs or diaphragm to provide the force to press the surfaces together.
- The **release bearing**, which provides the bearing surface which, when the driver operates the clutch pedal, disconnects the drive between the engine and the gearbox.

Through the centre of the pressure plate the **input shaft** of the gearbox (sometimes called the **spigot**, **first motion** or **clutch shaft**) is splined to the middle of the driven plate. On the conventional vehicle layout it is located on a bearing (called the **spigot bearing**) in the flywheel.

Multi-spring clutch

The driving members of the multi-spring clutch consist of a **flywheel** and pressure plate (both made of cast iron) with the driven plate trapped between. The pressure plate rotates with the flywheel, by means of projections on it locating with slots in the clutch cover, which is bolted to the flywheel. A series of springs located between the cover and the pressure plate force the plate towards the flywheel, trapping the clutch plate between the two driving surfaces. The **primary** (or input) shaft of the gearbox is splined to the hub of the clutch disc and transmits the drive (called **torque**, which means turning force) to the gearbox. The drive is disconnected by withdrawing the pressure plate and this is achieved by the operation of the release levers. Figure 18.1 shows a multi-spring clutch.

Diaphragm-spring clutch

The diaphragm-spring clutch (shown in Fig. 18.2) is similar in many ways to the coil-spring type. The **spring pressure** is provided by the **diaphragm**, which also acts as the lever to move the pressure plate. On depressing the clutch pedal the release lever is forced towards the flywheel, thus pulling the pressure plate away from

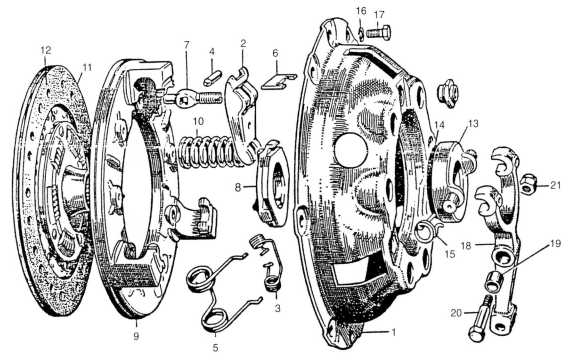

16 17

12 11 7 4 2 6 14 13 21

10 8 18 19

9 5 3 15 20 1

1 Cover – clutch	8 Plate – bearing thrust	15 Retainer
2 Lever – release	9 Plate – pressure	16 Washer – spring – cover screw
3 Retainer – lever	10 Spring – pressure plate	17 Screw – cover by flywheel
4 Pin – lever	11 Plate assembly – driven	18 Lever – withdrawal
5 Spring – anti-rattle	12 Lining	19 Bushes
6 Strut	13 Ring assembly – thrust	20 Bolt for lever
7 Eyebolt with nut	14 Ring – carbon	21 Nut for bolt

Fig. 18.1 The multi-coil spring clutch

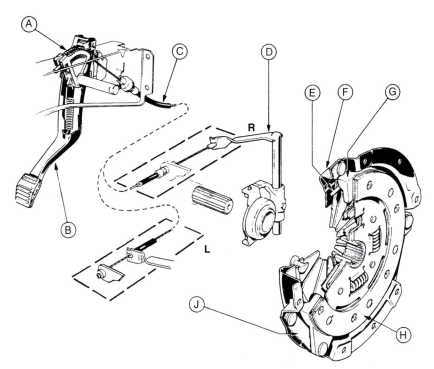

A – Automatic adjuster
B – Clutch pedal
C – Clutch cable
D – Release shaft with fork and thrust bearing
E – Fulcrum ring
F – Clutch cover
G – Pressure plate
H – Clutch disc
J – Spring steel straps

Fig. 18.2 Diaphragm-spring clutch components (L = left-hand drive, R = right-hand drive)

the flywheel. This type has a number of advantages over the coil-spring type:

- It is much simpler and lighter in construction with fewer parts.
- It has a lighter hold-down pressure and is therefore easier to operate, thus reducing driver fatigue.
- The clutch assembly is smaller and more compact.
- It provides almost constant pressure throughout the life of the driven plate.
- Unlike the coil spring, the diaphragm spring is not affected by centrifugal force at high engine speeds. It is also easier to balance. One type of diaphragm spring, the strap drive, gives almost frictionless movement of the pressure plate inside the clutch cover.

Friction plate construction

The important features incorporated in the design of a driven plate can best be seen by considering the disadvantages of using a plane steel plate with a lining riveted to each side. A number of serious problems would very soon be encountered:

- Buckling of the plate can occur due to the heat that is generated when the drive is taken up.
- The drive may not be disconnected completely (called **clutch drag**), caused by the plate rubbing against the flywheel when the clutch should be disconnected.
- Very small movement of the clutch pedal. The clutch is of the 'in or out' type, with very little control over these two points. This may cause a sudden take-up of the drive (called **clutch judder**).

To overcome these problems the plate is normally **slotted** or 'set' in such a way as to produce a flexing action. Whilst disengaging, the driven plate will tend to jump away from the flywheel and pressure plate to give a clean break. Whilst in this position the linings will be held apart and air will flow between them to take away the heat. During engagement, spring pressure is spread over a greater range of pedal movement as the linings are squeezed together. This gives easier control and smoother engagement. In most cases the hub is mounted independently in the centre of the clutch plate, allowing it to twist independently from the plate. To transmit the drive to the plate either springs or rubber, bonded to the hub and plate, are used. This flexible drive absorbs the torsional (twisting) shocks due to the engine vibrations on clutch take-up, which could otherwise cause transmission noise or rattle. These features can be seen in the clutch drive (H) in Fig. 18.2.

Friction plate materials

Coefficient of friction

This is the relationship between two surfaces rubbing together. On a clutch both static and sliding friction

are necessary. Sliding friction occurs when the drive is being taken up and static friction when the clutch is fully engaged.

Wear properties

This relates to the ability of a material to withstand wear. The surfaces of the flywheel and pressure plate should not become scored or damaged due to friction as the surfaces slide over each other. If they do become damaged then any number of clutch faults could become apparent; under these circumstances they would be replaced. A typical coefficient of friction for a motor car would be approximately 0.3.

Linings

These should have a high and stable coefficient of friction against the flywheel and pressure plate surfaces over a wide a range of temperatures and speeds. They should also have good wear resistance and not score or cause thermal damage to any of the surfaces with which they are in contact (e.g. flywheel or pressure plate).

Woven linings

These are formed from a roll of loosely woven asbestos containing brass or zinc wires, which is soaked in resin and other ingredients. The resin is then polymerized (which hardens the resin mixture). This has the following advantages:

- It is very flexible compared to other types
- It is easier to drill
- It can be cut and bent to shape.

They also have a high coefficient of friction when used in conjunction with cast iron surfaces. They are not suitable for heavy-duty use where the temperature is much above $200°C$ (this gives a loss of friction and a high wear rate).

Moulded linings

These are more popular now. In this type the various components are mixed together and cured in dies at high temperatures and pressures to form a very hard material. These can withstand high temperatures and pressures when in use. Different types of linings can be made having widely different coefficients of friction and wear properties because of the much wider range of **polymers** that can be used in their manufacture. They require more careful handling during drilling and riveting as the material is more brittle than the woven type. They must also be formed to the required radius as they cannot be cut or bent. Heavy-duty clutches often use **ceramic materials** as these can withstand higher operating temperatures and pressures. Since the end of November 1999, asbestos has been banned by law in the manufacture of new clutch linings (refer to the section on **Health and Safety** in the Introduction).

Modern clutch lining materials

The use of asbestos in vehicle brake, clutch, and gasket components has been prohibited since 1999, with the exception that pre-1973 vehicles could continue to be fitted with asbestos-containing brake shoes until the end of 2004. It is possible that some old vehicles could contain asbestos and sensible precautions should be observed when working on such vehicles. But remember that all brake and clutch dust is potentially harmful, so:

- Never blow dust out of brake drums or clutch housings with an air line
- Use properly designed brake-cleaning equipment that prevents dust escaping
- Use clean, wet rags to clean out drums or housings and dispose of used rags in a plastic waste bag (HSE).

Operating mechanisms

The clutch release bearing is fitted to the clutch release fork. When the clutch pedal is pressed down by the driver, pressure is exerted against the diaphragm spring fingers by the release bearing forcing them towards the flywheel; this action disengages the drive from the engine.

The pedal linkage may be either **mechanical** or **hydraulic**. It must be flexibly mounted as the engine (which is mounted on rubbers) may move independently of the body/chassis.

Hydraulic linkage

In this arrangement, shown in Fig. 18.3, oil is displaced by the movement of the piston in the **master cylinder** to the **slave cylinder** to operate the release fork. The reservoir on the master cylinder tops up the system as wear in the clutch takes place.

Another form of hydraulic clutch release mechanism is shown in Fig. 18.4. This system dispenses with levers at the gearbox end because the annular slave cylinder applies direct axial pressure to the release bearing.

Mechanical linkage

This operates on the principle of levers and rods or cable. The movement ratio and force ratio can be arranged to give a large force acting on the clutch, with a small force acting on the pedal. Figures 18.2 and 18.5 show this system.

Adjustment

Adjustment of the hydraulic operating mechanism is automatic. As wear of the linings takes place, the piston in the slave cylinder takes up that amount of wear.

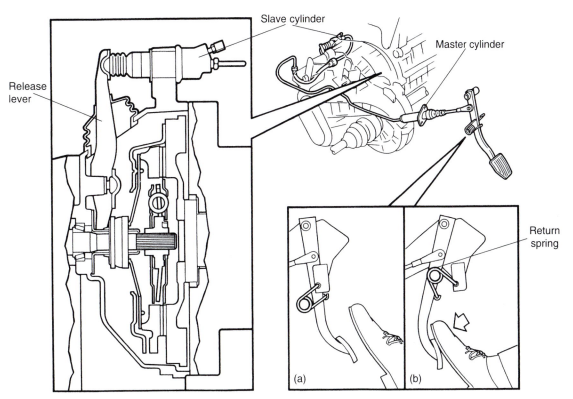

Fig. 18.3 The hydraulic clutch release mechanism. (a) Position of spring in the off (rest) position. (b) Position of spring with pedal depressed

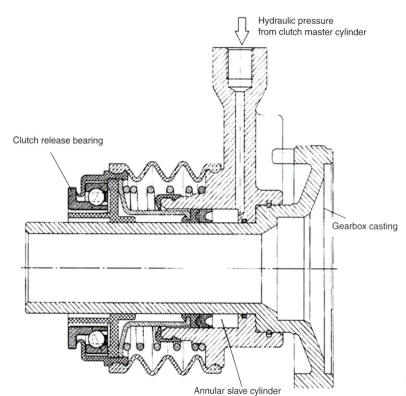

Fig. 18.4 Direct-acting hydraulic clutch release mechanism

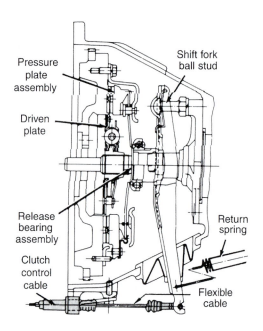

Fig. 18.5 Cable-operated clutch arrangement

The mechanical system is usually a flexible cable and may have an automatic self-adjusting mechanism as in Fig. 18.2 or else it may need to be checked and adjusted at regular service intervals.

Clutch faults

These are the most common faults found on the clutch.

Clutch slip

This occurs when the clutch fails to transmit the power delivered by the engine. It may be caused by one of the following faults:

- Insufficient 'free' movement on the clutch pedal or at the withdrawal (release) lever.
- Oil or grease on the friction linings.
- Friction linings worn out.
- Scored faces on the flywheel or pressure plate.

Clutch drag

This produces some difficulty in engaging and changing gear. It may be caused by one of the following faults:

- Insufficiently effective pedal travel (excessive 'free' movement on the pedal or slave cylinder push rod).
- Hydraulic system defective (due usually to the loss of most, if not all, of the hydraulic oil).
- Release lever broken, incorrectly adjusted or seized.
- Release bearing defective.
- Clutch plate seized on the gearbox input shaft splines.
- Clutch shaft spigot bearing (used where the gearbox input shaft is located in the flywheel) is seized in the flywheel.
- Clutch linings are cracked or buckled.
- Oil or grease on the linings.

Vibration or judder

This occurs when the drive is gradually taken up. It can be caused by one of the following faults:

- Cracked or distorted pressure plate.
- Loose or badly worn linings.
- Loose or protruding rivets.
- Misalignment of the gearbox with the engine, caused by distortion of the bell housing or the dowels not being located correctly.
- Engine or gearbox mountings defective.
- Excessive backlash (play) in the universal joints (UJs) of the drive/prop shafts (bolts may be loose on the UJ flange).
- Flywheel bolts loose on the crankshaft flange.
- Defective release mechanism.

Power transmitted by a clutch

The amount of power that can be transmitted by a clutch is dependent on the torque that it can transmit, which is given by the following formula:

Torque transmitted by a friction clutch,

$$T = \frac{n \times \mu \times W(R_1 + R_2)}{2} \text{ Nm.}$$

where n = number of friction faces, one for each side of the friction plate; μ = coefficient of friction surfaces; W = total spring force pressing the surfaces together; R_1 = outer radius of friction lining and R_2 = inner radius of lining.

Example 18.1

A particular single plate clutch has linings with R_1 = 165 mm and R_2 = 115 mm. The spring clamping force $W = 3.25$ kN and the coefficient of friction of the clutch linings $\mu = 0.35$. Calculate the torque T that this clutch can transmit.

$$T = \frac{n \times \mu \times W(R_1 + R_2)}{2} \text{ Nm.}$$

Using the data from the question and remembering that a single plate clutch has number of friction surfaces $n = 2$:

$$T = \frac{2 \times 0.35 \times 3250 \times (0.165 + 0.115)}{2} = 318.5 \text{ Nm.}$$

Multiple-plate clutches

In some applications, such as large vehicles, a twin-plate clutch (Fig. 18.6) may be used to provide the extra torque capacity that is required. When two friction plates are used the number of friction surfaces is increased to four and this doubles the torque capacity for a given clutch diameter.

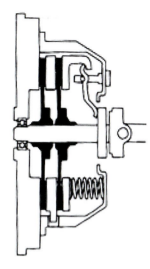

Fig. 18.6 Two-plate (twin-plate) clutch

Example 18.2

A particular twin-plate clutch has linings with R_1 = 165 mm and R_2 = 115 mm. The spring clamping force $W = 3.25$ kN and the coefficient of friction of the clutch linings $\mu = 0.35$. Calculate the torque T that this clutch can transmit.

$$T = \frac{n \times \mu \times W(R_1 + R_2)}{2} \text{ Nm.}$$

Using the data from the question and remembering that a two-plate clutch has number of friction surfaces $n = 4$:

$$T = \frac{4 \times 0.35 \times 3250 \times (0.165 + 0.115)}{2} = 637 \text{ Nm.}$$

Matching the engine to the road-going needs of the vehicle

When a vehicle first starts to move it requires extra pushing force to overcome inertia; once moving the pushing force can be reduced. Extra pushing force is needed when it becomes necessary to make the vehicle go faster or to climb a hill. The pushing force is called the tractive effort (see Fig. 18.7) and it is applied at the point of contact between the driving wheels and the road surface. When the tractive effort (F) is multiplied by the radius (R) of the driving wheel, the driving torque is obtained (see Fig. 18.8).

Unfortunately, the torque directly available from the engine is not always strong enough for acceleration and hill climbing, and some means must be employed to multiply the engine torque so that, by the time it reaches the driving wheels, it is strong enough. One

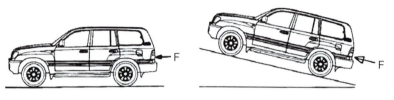

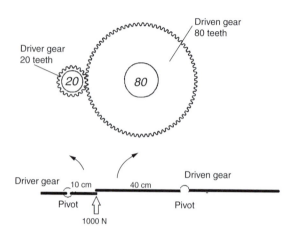

Fig. 18.7 Tractive effort

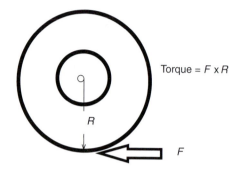

Torque = *F* x *R*

R

F

Fig. 18.8 Torque at the driving wheels of the vehicle

Fig. 18.10 Torque multiplier and gear ratio

function of the gearbox and transmission system is to provide a range of torque multipliers that enable the engine to meet all the demands placed on it — these torque multipliers are the gear ratios.

Leverage and torque

Leverage of the type shown in Fig. 18.9 permits a man to exert a relatively small force at one end of the lever to move a heavy object such as a boulder at the other end.

Gear ratio

Imagine the pair of gears shown in Fig. 18.10 to be replaced by a pair of levers that are hinged at their

Fig. 18.9 Principle of the lever

centres. The driver gear acts like a simple lever 10 cm long and the large driven gear a lever 40 cm long. In this example the small driver gear exerts a force of 1000 N on the large gear, which means that it is applying a torque of 1000 N × 0.10 m = 100 Nm to the large gear. The same force of 1000 N is applied to the large gear, so the torque on the large gear = 1000 N × 0.40 m = 400 Nm. The torque of the large gear is **four times** as great as the torque on the small gear.

Now consider the movement that takes place when the driver gear makes one complete revolution or a distance of 20 teeth. Because its gear teeth are in mesh with the large gear the driven gear will rotate by the same amount of 20 teeth, but 20 teeth of movement on the driven gear amounts to 20/80 = one-quarter of a revolution. In terms of revolutions the driven gear rotates one-quarter of the revolutions of the driver.

The gear ratio = number of teeth on driven gear ÷ number of teeth on driver gear. In this case, the gear ratio = 80/20 = 4:1.

In most cases in motor vehicle practice the gear ratio acts as a torque multiplier and a speed reducer. The exception is an overdrive gear.

A basic gearbox

In addition to providing a range of gear ratios to meet different driving conditions, the gearbox must provide a permanent neutral and a reverse gear. The basic gearbox shown in Fig. 18.11 meets these requirements.

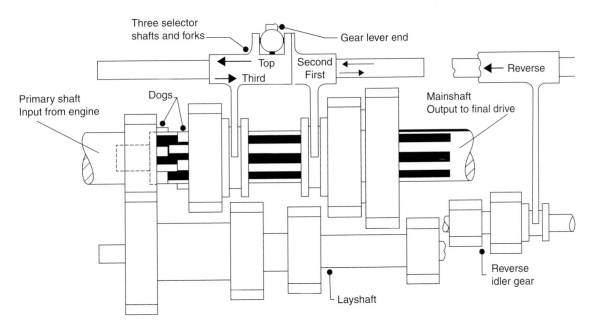

Fig. 18.11 A basic sliding mesh gearbox

The sliding mesh gearbox shown in Fig. 18.11 has three shafts:

- The primary shaft, which transmits the drive from the engine and clutch to the gearbox — it is also called the first motion shaft and it carries the gear that drives the lay shaft gears.
- The layshaft — the gears on the layshaft are connected together and they are all driven at the same speed, by the first motion shaft gear.
- The mainshaft — this is splined so that the gears can slide along it. When a gear ratio is required the required gear on the mainshaft is moved along until it is engaged with the corresponding gear on the layshaft.

The gears are moved along the shaft by means of the selector mechanism, which is operated by the gear lever.

The gearbox shown in Fig. 18.11 has four forward speeds. Figure 18.12 shows how the different ratios are obtained. When the dog clutch is engaged with the first motion shaft the drive is direct to the mainshaft — this provides the fourth gear, which is also called top gear. The next gear ratio is third gear; the mainshaft gear is moved along the shaft until it engages with the third gear on the layshaft — at the same time, the dog clutch disengages from top gear so that only one gear is engaged. When third gear is disengaged and returned to its neutral position, second gear can be engaged. When first gear is required, second gear is moved out of mesh as the first gear is moved into mesh with the corresponding gear on the layshaft — first gear is also called bottom gear.

Considerable skill is required to drive a vehicle equipped with a sliding mesh gearbox because it is difficult to slide the gears into mesh without causing them to crash into each other — that is why they are sometimes referred to as **crash gearboxes**.

The constant mesh and synchromesh gearbox

The constant mesh gearbox (Fig. 18.13) overcomes the difficulty of engaging sliding gears by using a dog clutch of the type shown in Fig. 18.14 to connect the gears to the mainshaft as they are required. The basic constant mesh gearbox is similar in layout to the sliding mesh one, the main difference being that the gears on the mainshaft are free to rotate on it. In between the mainshaft gears are the dog clutches that are splined to it — when a particular gear ratio is required, the appropriate dog clutch is moved along the mainshaft to connect the gear to it.

Synchromesh

Synchromesh mechanisms of the type shown in Fig. 18.15 perform the same function as the dog clutch in the constant mesh gearbox. The main difference lies in the small cone clutch that is designed to match the speed of the mainshaft to that of the mainshaft gear that is being selected — when the speeds are the same (synchronized) the gear can be connected to the shaft without any grating sounds.

The synchromesh hub is splined to the mainshaft and it moves axially along it when moved by the gear selector mechanism. The springs and balls that are placed around the circumference of the hub provide

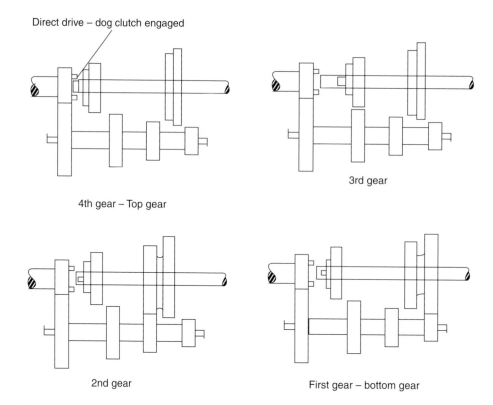

Fig. 18.12   The four gear ratios of the sliding mesh gearbox

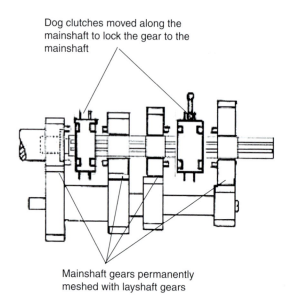

Fig. 18.13   The constant mesh principle

the load that controls the movement of the sleeve across the hub. The inner edge of the selector ring contains the dog teeth that engage with the corresponding ones on the gear that is being selected. Initial movement of the synchromesh unit brings the synchronizing cones into contact and this action synchronizes the speeds of the two sets of dog teeth that are being engaged. Further movement of the selector mechanism causes the sleeve to override the spring and ball to achieve the fully engaged position − the unit remains in this position

until another gear is selected. When this type of synchromesh is used carefully it provides a reliable method of achieving smooth gear changing; however, if hurried gear changes are attempted a certain amount of clashing of the dog teeth may result. The baulk ring type of synchromesh is a device that is designed to overcome this difficulty and thus provide for faster gear changes, where they are required.

The baulk-ring synchromesh

The principal differences between the baulk-ring type of synchromesh (Fig. 18.16) and the constant-load one are the baulking rings and the shifting plates. The baulk rings and spring-loaded shifting plates are designed to prevent the dog teeth from engaging until the two members that are being connected are rotating at the same speed. Unlike the constant-load synchromesh, the synchronizing cones (baulk rings) are separate from the hub and they can rotate a few degrees relative to it. It is this feature that should ensure synchronization at whatever speed of gear change is required.

Reverse gear

In addition to providing a range of gears for forward motion of the vehicle it is necessary for the gearbox to provide a means of reversing, and this is achieved by means of an additional gear called an idler that is interposed between the mainshaft gear and the layshaft, as shown in Fig. 18.17.

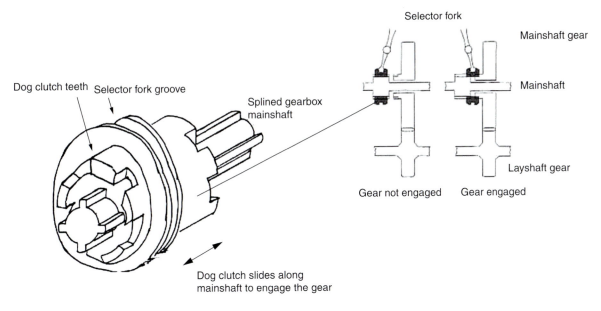

Fig. 18.14   A simple dog clutch for constant mesh gears

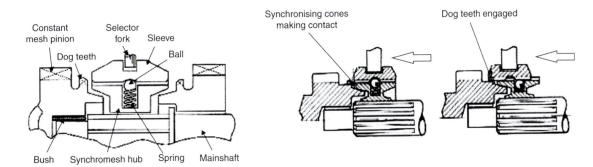

Fig. 18.15   A constant-load synchromesh mechanism

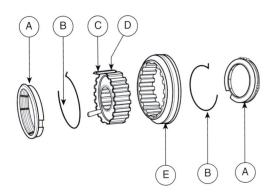

A. Baulk rings – blocker rings – synchronising rings
B. Shifting plates retaining springs
C. Shifting plates – also called blocker bars
D. Synchroniser hub – splined to mainshaft
E. Selector ring

Fig. 18.16   A baulk-ring synchromesh mechanism

Selector mechanisms

The selector forks that are attached to the selector shafts fit into the groove on the synchromesh. When the gear lever is operated, the selector shaft and fork move the synchromesh device into the required position.

Gear selector mechanisms are designed to prevent more than one gear at a time being engaged and to prevent reverse gear being accidentally engaged while the vehicle is moving forward. In order to engage another gear when one is already in use, the selector mechanism returns to the neutral position and reverse gear selection normally requires the gear lever to be lifted, or depressed, when reverse gear is being selected.

The interlocking device shown in Fig. 18.18 has two balls that rest in grooves in the two outer selector shafts – the interlocking pin rests in a hole that is drilled through the centre shaft. In the position where no gear is engaged, any of the shafts can move. When a gear is

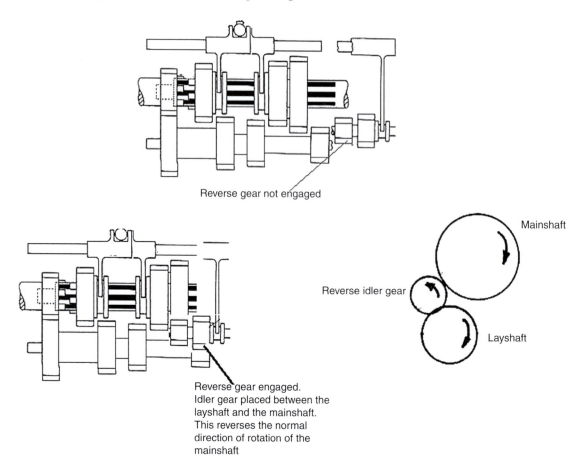

Fig. 18.17   A reverse gear arrangement

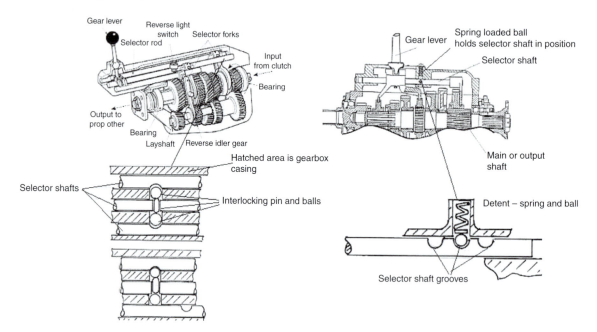

Fig. 18.18   A selector mechanism

selected the ball on that shaft pushes on the interlock pin, which in turn pushes the other ball into the groove on its shaft. The centre shaft and the upper one are then locked until the other shaft returns to the central position.

The detent spring and ball

When a gear is selected the selector shaft pushes the ball out of the groove against the spring. The additional pressure that is exerted on the ball and shaft serves to prevent the gear from sliding out of engagement due to vibration or other effect. Should the ball and groove wear, or the spring break or become weak, the retaining function may fail and cause the vehicle to 'slip out of gear'.

Gear ratio in a gearbox

In Fig. 18.19 the fourth gear is direct drive, giving a ratio of 1:1. In the other gears the drive is from the primary gear to the layshaft gear and from there to the respective gear on the mainshaft. Because several gears are involved in each gearbox ratio the gear ratio is calculated as shown in the example.

Example of gear ratio calculation

Gearbox ratio = number of teeth on each **driven** gear multiplied together ÷ number of teeth on each **driver** gear multiplied together.

In the gearbox shown in Fig. 18.20 the driver gears have 20 and 25 teeth respectively, and the driven ones have 40 and 35 teeth. In this case,

$$\text{Gear ratio} = \frac{40 \times 35}{20 \times 25} = 2.8 : 1.$$

Overdrive

Under certain driving conditions such as light load cruising speed it is advantageous to let the engine run a little slower whilst maintaining a fairly high speed. This can be achieved by making the gearbox output shaft rotate faster than the input shaft. The device that achieves this is called an overdrive and it can be fitted to the outside of the gearbox or, as is the case in most modern gearboxes, it can be incorporated as a fifth gear in the conventional four-speed gearbox.

The epicyclic type of overdrive is operated by hydraulic pressure under the control of a solenoid, which is operated by a switch near the control column. This type of overdrive normally only operates in the higher gears at speeds above 30 mph — engagement at lower speeds is normally prevented by an inhibitor switch. The input shaft to the overdrive operates an oil pump that generates the oil pressure required to move the cone clutch to the overdrive, or direct drive position.

In the internal type of overdrive the fifth gear and its synchromesh are housed at the end of the mainshaft, as shown in Fig. 18.21.

Transmitting the power to the driving wheels

The output from the gearbox is eventually transferred to the final drive and differential gear before passing to the road wheels. There are three basic layouts for this aspect of the transmission system:

1. Front-wheel drive
2. Rear-wheel drive
3. Four-wheel drive.

Front-wheel drive

Front-wheel drive vehicles normally have the engine, gearbox, and final drive mounted transversely as a unit at the front of the vehicle. The output from the gearbox connects directly to the final drive pinion, which means that there is no need for a propeller shaft. The drive to

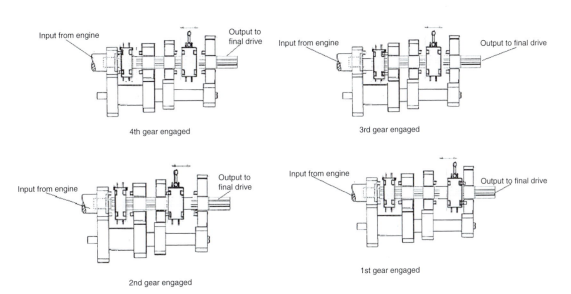

4th gear engaged

3rd gear engaged

2nd gear engaged

1st gear engaged

Fig. 18.19 Gear ratios in a synchromesh gear box

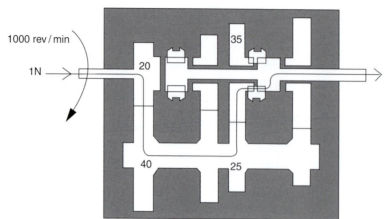

Fig. 18.20   Second gear ratio

the wheels is carried by the drive shafts, which incorporate constant-velocity joints that allow the wheels to be steered while they are being driven. A typical example of a front-wheel drive arrangement is shown in Fig. 18.22.

Rear-wheel drive

The power from the gearbox is transferred to the rear axle by means of the propeller shaft using a layout similar to the Hotchkiss drive shown in Fig. 18.23. In the example shown, the axle is attached to the leaf springs, which hold the axle in place and provide the springing necessary for suspension. When in motion the axle is subject to movements that necessitate the use of a sliding joint and two universal joints. The sliding joint is required to allow for a slight increase in the distance between the gearbox and the axle that occurs as the axle rises and falls when the vehicle moves across bumps in the road surface. The second universal joint at the axle end of the propeller shaft allows for the partial rotation of the axle casing that occurs under acceleration and braking. This motion causes the springs to flex and this also affects the distance between the gearbox and the axle. In addition, the second

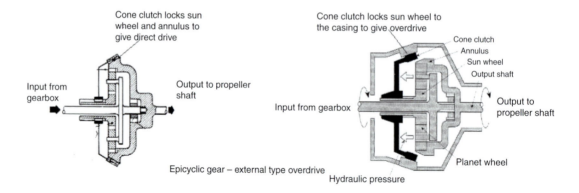

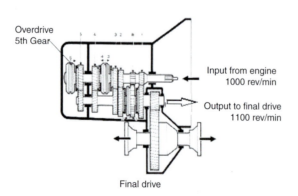

Fig. 18.21   Two types of overdrive

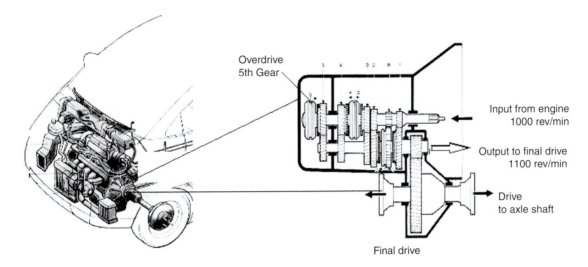

Fig. 18.22   A front-wheel drive system

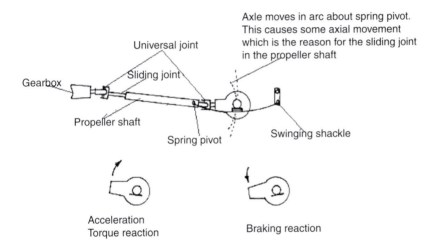

Fig. 18.23   The Hotchkiss drive

universal joint provides the necessary constant-velocity motion of the propeller shaft.

Rear-wheel drive with coil spring suspension

Coil springs are not able to transmit driving and braking forces from the axle to the vehicle frame. To overcome this difficulty the axle is attached to the vehicle frame by means of the type of suspension arms shown in Fig. 18.24; this arrangement allows vertical movement and is capable of dealing with driving and braking forces and torque reaction.

Four-wheel drive

Originally, four-wheel drive systems were limited to vehicles, such as those used by the military, that

were expected to be able to operate in many different types of terrain. The main advantages of four-wheel drive over the two-wheel drive system are:

- Better traction on normal road surfaces, particularly in wet or icy conditions
- Better balanced axle load distribution
- Better hill-climbing performance
- Better for cross-country work in general.

These are some of the advantages that make the four-wheel drive system attractive to drivers and many normal road-going light vehicles now incorporate four-wheel drive.

Some disadvantages of four-wheel drive are:

- Additional cost
- Higher fuel consumption.

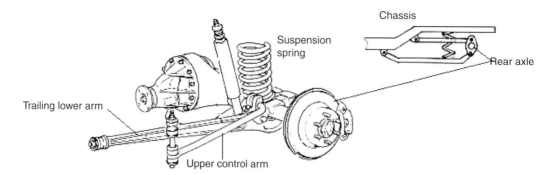

Fig. 18.24 Control linkage for coil spring suspension on rear drive axle

The two basic types of four-wheel drive are:

1. Occasional four-wheel drive, where the driver selects it as required.
2. Permanent four-wheel drive, where the four wheels are permanently driven whenever the vehicle is in motion.

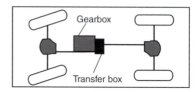

Fig. 18.25 The transfer box in a four-wheel drive system

Operation of four-wheel-drive systems

When a vehicle turns a corner the front and rear axles do not follow exactly the same path, which means that the front wheels will complete more revolutions than the rear ones; this results in a condition known as transmission 'wind-up'. If this condition continues it can cause damage to the transmission system. The problem can be overcome by fitting a differential gear between the front and rear axles; however, this brings with it another problem, which is the loss of traction when slip occurs at one axle — this can be overcome by fitting a differential, which is known as a third differential. Unfortunately, a vehicle fitted with a third differential suffers from the problem that is caused when either the front or rear axle loses traction, because the action of the differential causes all drive to be lost. This leads to the need for a differential lock, which prevents the differential from operating under such conditions.

These arrangements are normally incorporated into four-wheel drive systems that are used in some modern light vehicles.

Occasional four-wheel drive

The occasional four-wheel-drive system incorporates an additional gearbox of the type shown in Fig. 18.26, this gearbox is called a transfer gearbox, or just transfer box. The transfer box takes the drive from the main gearbox and it is normally mounted on or adjacent to it. The transfer box may be simple device that permits two- or four-wheel drive to be selected, or it may be slightly more complicated to incorporate the choice of high or low speed in four-wheel drive.

The transfer box

The transfer box shown in Fig. 18.25 is used in a general-purpose vehicle that is often used in off-road conditions. There are two gear ratios; 2.488:1 and a 1:1 ratio for normal road driving. These ratios are engaged and disengaged by the selector mechanism that is situated between the two gears. Four-wheel drive is obtained by connecting the front and rear drives through a coupling device, which is similar to the dog clutch used in constant mesh gearboxes.

Permanent four-wheel drive

Permanent four-wheel drive systems can be divided into two categories, one that distributes the driving effort equally between front and rear axles, and the other which divides the driving effort with approximately 40% going to the front axle and 60% to the rear. The transfer box shown in Fig. 18.27 divides the drive equally between the axles; the third differential is the device that achieves this — a differential lock is incorporated to allow traction to be maintained should one axle encounter slippery conditions.

A differential lock

In its simplest form the differential lock is a dog clutch that connects one axle shaft to the gear that carries the differential assembly, thus preventing relative motion between the two (see Fig. 18.28). The sun wheel on the other axle shaft then rotates at the same speed as the one that is now fixed to the driving member.

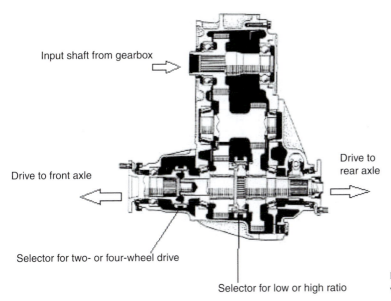

Input shaft from gearbox

Drive to front axle

Drive to rear axle

Selector for two- or four-wheel drive

Selector for low or high ratio

Fig. 18.26 A transfer gearbox for occasional four-wheel drive

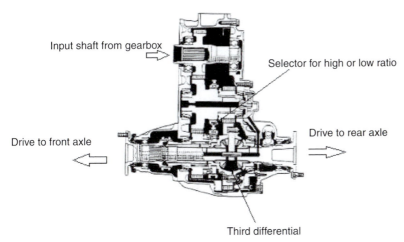

Input shaft from gearbox

Selector for high or low ratio

Drive to front axle

Drive to rear axle

Third differential

Fig. 18.27 A permanent four-wheel drive transfer box with third differential

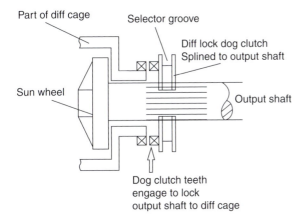

Part of diff cage

Selector groove

Diff lock dog clutch
Splined to output shaft

Sun wheel

Output shaft

Dog clutch teeth
engage to lock
output shaft to diff cage

Fig. 18.28 Principle of a differential lock

Permanent four-wheel drive via a viscous coupling

As the demand for four-wheel drive vehicles for private motorists increased, several different systems were developed that made use of two devices, one to apportion the drive between the axles, the other to overcome the problem of the axles running at different speeds. An epicyclic type of differential may be used to apportion the torque between the axles and a viscous coupling (see Fig. 18.29) may be used to deal with differences in axle speeds.

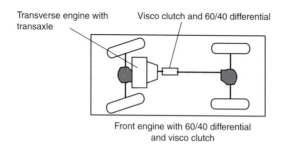

Fig. 18.29   Four-wheel drive systems with viscous coupling (visco clutch)

The viscous coupling (visco clutch) and epicyclic differential

The visco clutch contains a number of discs that are contained in a sealed cylindrical chamber and are keyed to it. In the example shown in Figs 18.30 and 18.31, the chamber acts as the annulus of the differential and the drive to the rear axle. The other set of clutch discs is keyed to the sun wheel, which is also the drive to the front axle. Should the speed between front and rear axles be different, the clutch plates will begin to move relative to each other and this will cause the viscous fluid to heat up. As the fluid becomes hotter the pressure in the chamber rises; this makes the fluid harder to shear. The discs can no longer move relative to each other and are effectively locked together. In effect, the visco clutch makes a progressively acting differential lock.

In the epicyclic differential the planet carrier is the input drive from the gearbox. The drive to the front wheels is from the sun wheel, and the rear-wheel drive is taken from the annulus. The force exerted by the planet wheel is the same on the sun and the annulus – as the annulus radius is greater than the sun radius, the torque on the annulus is greater than that on the sun. In this way the third differential is able to transmit more torque to the rear axle than to the front axle.

- Certain vehicles should not be tested on a roller brake tester, e.g. vehicles with more than one driving axle permanently engaged.
- Limited-slip differential.

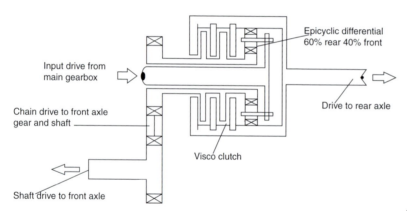

3rd differential and visco clutch

Fig. 18.30   A simplified type of permanent four-wheel drive

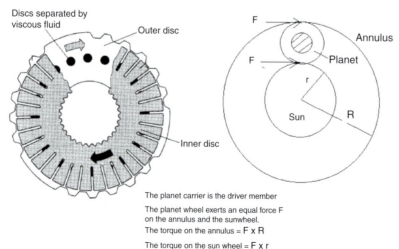

The planet carrier is the driver member
The planet wheel exerts an equal force F on the annulus and the sunwheel.
The torque on the annulus = F x R
The torque on the sun wheel = F x r

Fig. 18.31   Third differential and visco clutch

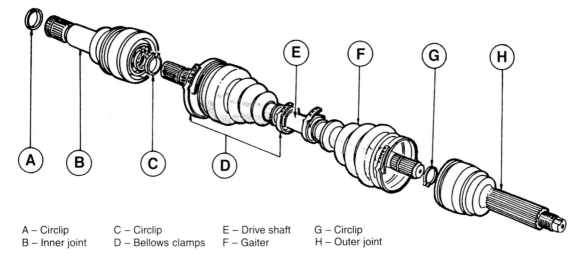

A – Circlip   C – Circlip   E – Drive shaft   G – Circlip
B – Inner joint D – Bellows clamps F – Gaiter   H – Outer joint

Fig. 18.32 Drive shaft assembly and CV joints as used on front-engine four-wheel drive

- Belt-driven transmission. Brakes for which the servo operates only when the vehicle is moving.
- These vehicles should be tested using a properly calibrated and maintained decelerometer or a plate brake tester designated as acceptable for the statutory tests.

Propeller shafts and drive shafts

In the front engine four-wheel drive, drive shafts are used instead of a prop shaft. These are normally fitted with constant-velocity (CV) joints to enable the drive to pass through a greater angle without the problems associated with universal joints. Figure 18.32 illustrates this.

Constant-velocity universal joints

Variation in speed

When a single Hooke-type joint is used to transmit a drive through an angle, you will find that the output shaft does not rotate at a constant speed. During the first $90°$ of its motion, the shaft will travel faster, and on the second $90°$ slower. Correction of this speed variation is normally done by a second coupling, which must be set so that when the front coupling increases the speed the rear coupling decreases the speed (Fig. 18.33). When the two Hooke-type couplings are aligned correctly, the yoke at each end of the propeller shaft is fitted in the same plane. When fitted to a vehicle, the two or more joints cannot always be operated such that the two angles remain equal. To allow for angle variation and shaft wind-up, especially on larger

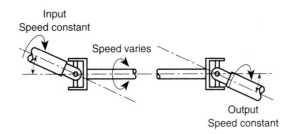

Fig. 18.33 Driving through an angle. Both angles are equal, giving no variation between input and output

vehicles such as LGVs, the joints are sometimes set slightly out of phase.

Apart from the Hooke joint, a number of other types are used. The doughnut and the Layrub coupling are similar in that they are made from steel and rubber bonded together. So long as the angle of drive is not too large, these will give satisfactory service with the added advantage of not requiring any lubrication.

Constant-velocity (CV) joints

A single Hooke-type joint driving through a comparatively large angle would give severe vibration due to the inertia effect produced by the speed variation. This variation increases as the drive angle enlarges — for example, the percentage variation in speed is about nine times greater at 30 mph than at 10 mph. Any drive shaft subjected to speed changes during one revolution will cause vibration, but provided the drive angle is small and the shaft is light in weight, then drive-line elasticity will normally overcome the problem. Large drive angles, as used in front-wheel drive (FWD) vehicles, need special compact joints that maintain a constant speed when driving through an angle — these joints are called CV joints.

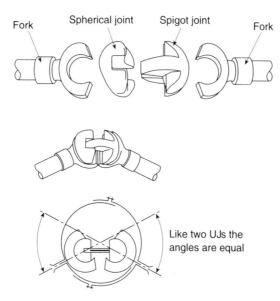

Fig. 18.34 The Tracta joint

Tracta CV joints

The need for CV joints was discovered in 1926 when the first FWD vehicle was made in France by Fenallie and Gregoire – the Tracta (tractionavant) car. When the second car was made, the drive-line incorporated CV joints, and the type used on that car is now known as a Tracta joint. This type is now made by Girling; details are shown in Fig. 18.34.

Reference to the figure shows that the operating principle is similar to two Hooke-type joints: the angles are always kept constant and the yokes are set in the same place.

The joint is capable of transmitting a drive through a maximum angle of about $40°$ and its strong construction makes it suitable for agricultural and military vehicles, but the friction of the sliding surfaces makes it rather inefficient. The joint is also too large to be accommodated in a small FWD vehicle with wheel diameters of less than approximately 15 inches (38 cm).

Rzeppa-type CV joint

This joint was patented by A. H. Rzeppa (pronounced Zeppa) in America in 1935. The development of this type is in common use today; it is called a Birfield joint and is made by Hardy Spicer. Figure 18.35 shows the construction of a Birfield joint.

Constant velocity is achieved if the device (**steel balls** in this case) connecting the driving shaft to the driven shaft rotates in a plane that bisects the angle of drive: the Birfield joint achieves this condition. Drive from the inner to outer race is by means of longitudinal, elliptical grooves that hold a series of steel balls (normally six) that are held in the bisecting plane by a cage. The balls are made to take up their correct positions by offsetting the centres of the radii for inner and outer grooves.

A Birfield joint has a maximum angle of about $45°$, but this angle is far too large for continuous operation because of the heat generated by the balls. Lubrication is by grease – the appropriate quantity is packed in the joint 'for life' and a synthetic rubber boot seals the unit.

Plunge joints

These universal joints allow a shaft to alter its length due to the up and down movement of the suspension. A Birfield joint has been developed to give a plunge action (in and out) as required. It will be seen that the grooves holding the balls in place are straight instead of curved;

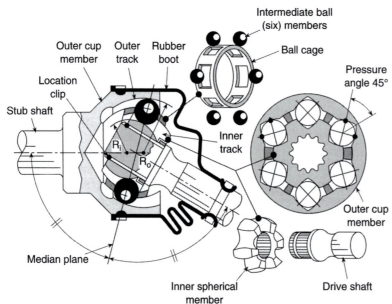

Fig. 18.35 Birfield Rzeppa-type constant-velocity joint

this allows the shaft length to vary. Positioning of the balls in the bisecting plane is performed by the cage. Since about 20° is about the maximum drive angle, this type of joint is positioned at the engine end of the drive shaft used for FWD vehicles.

Weiss CV joint

This type was patented in America in 1923 by Weiss and later developed by Bendix. It is now known as the Bendix−Weiss and produced in this country by Dunlop. The two forks have grooves cut in their sides to form tracks for the steel balls; there are four tracks so four balls are used to transmit rotary motion. A fifth ball is placed in the centre and locates the forks, resisting any inward force placed on the drive shaft. The driving balls work in compression, so two balls take the forward drive and the other two operate when reverse drive is engaged. The complete joint is contained in a housing filled with grease. The maximum angle of drive is approximately 35°. Constant velocity is achieved in a manner similar to the Rzeppa joint − the balls always take up a position in a plane that bisects the angle of drive.

Learning tasks

1. How would the driver notice if there was severe wear in the UJ of the prop shaft?
2. Draw up a schedule for removing and refitting the CV joint on a front-engine FWD vehicle. Make a special note of any safety precautions that should be observed.
3. What checks would be made on the drive-line of a front-engine rear-wheel drive (RWD) or FWD vehicle when undertaking an MOT?
4. What should you look out for when reassembling a divided-type prop shaft. What would be the result if these were ignored?
5. If the vehicle, on being driven round a corner, made a regular clicking noise, what component in the drive-line would you suspect as being worn and why? How would you put the fault right?

Axles and axle casings

Axles

Axles may be divided into two types, the **live axle** and the **dead axle**. The difference is that a dead axle only supports the vehicle and its load, whereas a live axle not only supports the vehicle and its load, but also contains the drive.

Axle casings

Three main types of casings are in use:

- **Banjo.** Normally built up from steel pressings and welded together. The crown wheel assembly is mounted in a malleable housing, which is bolted to the axle. This is the most common type fitted to light vehicles and is shown in Fig. 18.36.
- **Split casing.** These are formed in two halves and bolted together to contain the final drive and differential (see Fig. 18.37). They are used more in heavy vehicle applications as they are more rigid in construction and can withstand heavy loads.
- **Carrier.** This is more rigid than the banjo casing. The final drive assembly is mounted directly into the axle, and the axle tubes are pressed into the central housing and welded into place. A cover is fitted to the rear of the housing to allow for access and repair. It is used in off-road 4 × 4 and LGVs, and is shown in Fig. 18.38.

Final drive

The purpose of the final drive is two-fold: first to transmit the drive through an angle of 90°, and second to provide a permanent gear reduction and therefore a torque increase, usually about 4:1 in most light vehicles. Several types of bevel gearing have been used in the drive between the pinion and the crown wheel. These are the straight bevel, the spiral bevel, the hypoid, and the worm and wheel.

Straight bevel

The straight bevel (Fig. 18.39(a)) is not designed for continuous heavy-duty high-speed use as in the normal crown wheel and pinion. This is because only one tooth is in mesh at any one time. The gearing is noisy and suffers from a high rate of wear, but the design is the basis from which the final drives are formed.

Spiral bevel

Shown in Fig. 18.39(b), this has more than one tooth in the mesh at a time. It is quieter, smoother, can operate at much higher loadings, and is used where the shaft operates centre-line as the crown wheel.

Hypoid bevel

This is similar to the spiral bevel in action but the shaft operates at a lower level than the crown wheel. This has the advantage of the prop shaft being lower and not intruding into the floor space of the passenger compartment. It is stronger and is the quietest in operation. Extreme pressure (EP) oil must be used as a lubricant

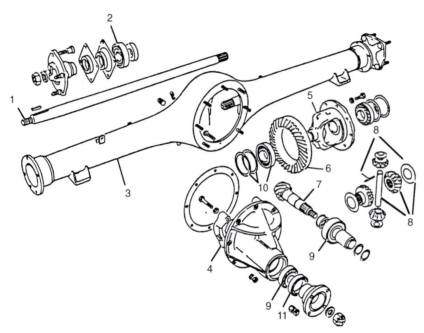

1 – Half-shaft
2 – Half-shaft bearing
3 – Axle casing
4 – Final drive casing
5 – Differential housing
6 – Crown wheel
7 – Pinion
8 – Final drive assembly
9 – Pinion bearings
10 – Differential housing bearings
11 – Pinion oil seal

Fig. 18.36   Banjo axle casing

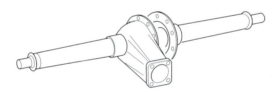

Fig. 18.37   Split axle casing

because of the frictional forces and high loadings generated between the gear teeth (a disadvantage of this arrangement). The materials used in their manufacture are a nickel–chrome alloy, which is carburized after machining and case-hardened to give long life. Figure 18.39(c) shows a hypoid bevel.

Worm and wheel

The worm in this arrangement, shown in Fig. 18.39(d), is driven by the prop shaft and looks like a very coarse screw thread. The wheel has a toothed outer edge with which the screw thread of the worm meshes. There are a number of advantages with the worm and wheel, the two main ones being: first it provides a very large gear reduction and therefore a large torque increase in one gear set, and second it gives a high prop shaft; this is especially useful for off-road vehicles and LGVs.

Servicing and adjustments

Excessive wear will result on the gear teeth if the backlash between the crown wheel and pinion is set incorrectly.

To adjust the backlash, shims must be added or removed to obtain the correct readings. The crown wheel teeth shown in Fig.18.40 give the different markings, the centre diagram showing where the marks should occur. To obtain the markings on the gear teeth a little engineers' blue is placed on the pinion teeth and the pinion rotated until the crown wheel has completed one revolution.

A dial test indicator (DTI) should be used to check the crown wheel and pinion backlash (this is the amount of play between the gear teeth). It should be positioned as shown in Fig. 18.41.

The task of setting up the crown wheel and pinion is quite skilled. This is why the exchange unit comes complete with the differential and all the settings are done by the manufacturer. All the mechanic has to do when changing the final drive is fit the unit to the axle.

Differential

The purpose of the differential is to allow the wheels to rotate at different speeds, whilst still transmitting an equal turning force (torque) to both wheels. The half-shafts are splined to the sun gear, the planets

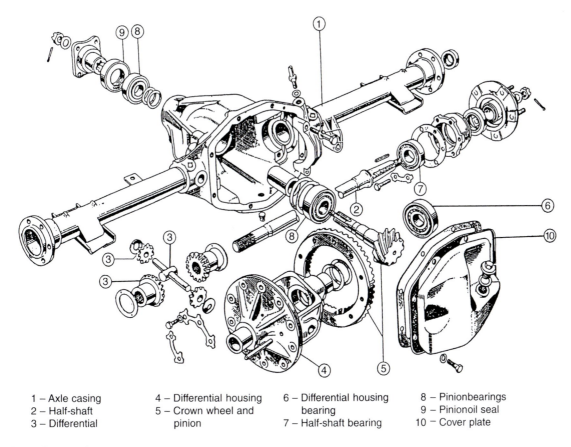

1 – Axle casing	4 – Differential housing	6 – Differential housing	8 – Pinionbearings
2 – Half-shaft	5 – Crown wheel and	bearing	9 – Pinionoil seal
3 – Differential	pinion	7 – Half-shaft bearing	10 – Cover plate

Fig. 18.38 Carrier axle casing

transmitting the motion from one sun gear to the other when the vehicle is turning a corner. This is shown in Fig. 18.42.

When both drive shafts are travelling at the same speed, the planet gears orbit (rotate) with the sun gears but do not rotate on their shafts. The whole unit acts as a solid drive (see Fig. 18.43).

If one shaft is stopped, the planet gears turn on their shafts, orbiting round the stationary sun gear and driving the other sun gear but twice as fast.

In the final drive the differential is in a housing (sometimes called a cage) to which the crown wheel is bolted. When the car is travelling in a straight line the planet gears orbit, but do not rotate on their shafts, and the unit drives both half-shafts at the same speed as the crown wheel and with the same turning force.

When turning a corner, the sun gear on the inner half-shaft turns more slowly than the crown wheel; the outer half-shaft, driven by the other sun gear, turns correspondingly faster. The crown wheel turns at the average of the half-shaft speeds.

A sectioned view of the complete final drive and differential arrangement is shown in Fig. 18.44. Both the crown wheel and pinion are held in the housing by taper roller bearings. The cross pin acts as the drive for and the shaft on which the planet gears rotate.

Rear hub bearing and axle shaft arrangements

Three layouts are now commonly used and are subjected to various forces acting on the axle shaft — for example, **twisting** when accelerating or braking, **bending** due to cornering and some of the load, and **shear** due to the vertical load imposed on the axle shaft.

Semi-floating axle

This hub arrangement, shown in Fig. 18.45(a), is used on many small light vehicles. The road wheel is attached to the **half-shaft** rather than the hub and the bearing is fitted between the half-shaft and the **axle casing**. Therefore, if a break should occur in the half-shaft inside the axle casing, the wheel will tilt and very often become detached due to the lack of support at the inner end of the half-shaft. As can be seen, the shearing point is positioned between the shaft and the axle casing, and the shaft is subject to shearing, bending, and twisting forces.

Three-quarter floating axle

Used on cars and light vans, the main difference is the position of the bearing, now shown on the outside of

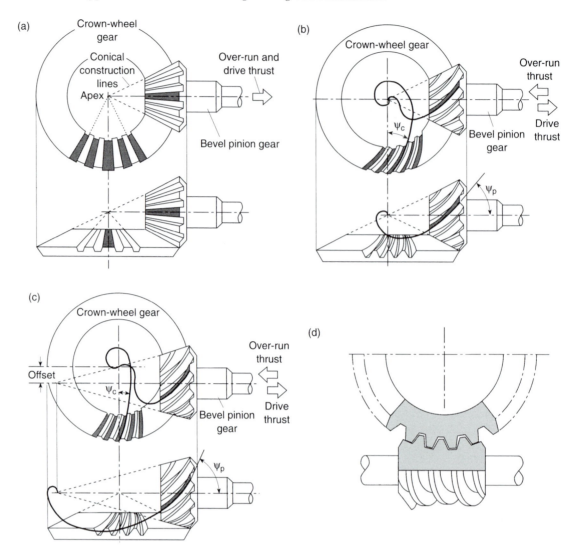

Fig. 18.39 (a) Geometry of straight-tooth bevel-gear crown wheel and pinion. (b) Geometry of spiral-tooth bevel-gear crown wheel and pinion. (c) Geometry of hypoid-tooth bevel-gear crown wheel and pinion. (d) Hourglass worm and worm—wheel final drive

the axle between the hub and the outside of the axle casing. The casing therefore takes most of the weight of the vehicle and its load. The shaft is still subject to twisting and bending forces. This type is shown in Fig. 18.45(b).

Fully-floating axle

In this arrangement (Fig. 18.45(c)) the bearings (normally taper roller) are fitted between the hub and the outside of the axle casing. In this way the only force to which the axle shaft is subject is a twisting action.

Lubrication

To prevent oil from leaking between the shaft and housing, lip-type oil seals are pressed into the housing.

In some cases an oil slinger washer is located just inside and next to the bearing to prevent flooding of the seal. Pressure build-up due to temperature changes during operation is prevented by a breather in the top of the axle casing.

The action of the gear teeth meshing together tends to break down the oil film on the gear teeth (due to the very high pressures and forces created between the gear teeth). The oil used in the axle therefore must prevent a metal-to-metal contact from taking place. An EP additive in the oil reacts with the metal surfaces at high temperatures to produce a low-friction film or coating on the teeth. This prevents scuffing and rapid wear from taking place; in effect, the additive becomes the oil. It is essential that no other type of oil is used as this could cause rapid wear and early failure of the final drive to take place.

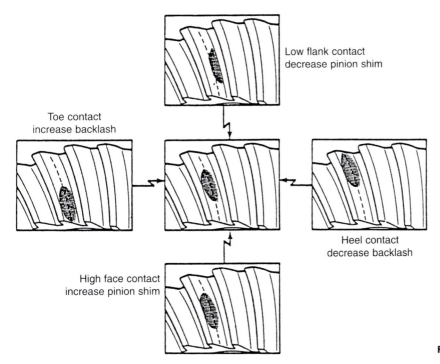

Low flank contact
decrease pinion shim

Toe contact
increase backlash

Heel contact
decrease backlash

High face contact
increase pinion shim

Fig. 18.40   Crown wheel teeth workings

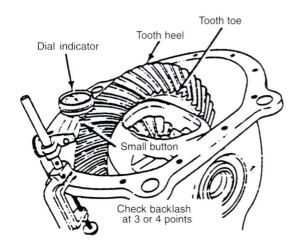

Dial indicator

Tooth heel

Tooth toe

Small button

Check backlash
at 3 or 4 points

Fig. 18.41   Position of the DTI

Types of bearings used in the transmission system

Plain bearings

A plain bearing can be described as a round hole lined with bearing metal or fitted with a bearing bush. These metals reduce friction in the bearing and also enable easy replacement as wear takes place. Typical examples of where these are used are in the engine (big-end and main bearings), in the gearbox (main shaft and lay shaft bearings), and in the flywheel (spigot shaft bearing).

Ball and roller bearings

Ball and roller bearings reduce friction by replacing sliding friction with rolling friction. There are several types of ball and roller bearings in use.

Single-row deep groove radial or journal bearing

In this bearing (Fig. 18.46(a)) the outer 'race' is a ring of hardened steel with a groove formed on its inner face. The inner race is similar to the outer race except it has a groove on its outer face. Hardened steel balls fit between the two rings. The balls are prevented from touching each other by the cage. Although designed for radial loads, journal bearings will take a limited amount of end-thrust, and they are sometimes used as a combined journal and thrust bearing.

Double-row journal bearing

This bearing is similar to the single-row journal bearing and is shown in Fig. 18.46(b). There are two grooves formed in each race, together with a set of balls and cages. These are able to take larger loads and are often used where space is limited. A typical example would be in the wheel bearing or hub bearing. Some axle shafts are fitted with this type.

Double-row self-aligning journal bearing

This bearing (Fig. 18.46(c)) has a double row of grooves and balls. The groove of the outer race is ground to form

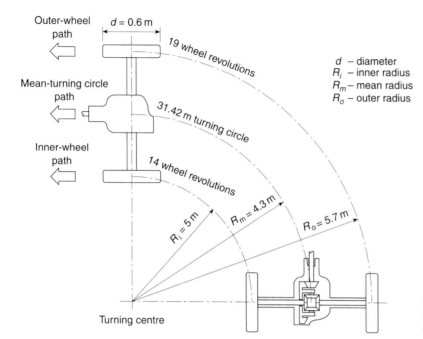

Outer-wheel path

Mean-turning circle path

Inner-wheel path

$d = 0.6$ m

19 wheel revolutions

31.42 m turning circle

14 wheel revolutions

$R_i = 5$ m $R_m = 4.3$ m $R_o = 5.7$ m

Turning centre

d – diameter
R_i – inner radius
R_m – mean radius
R_o – outer radius

Fig. 18.42   Illustration to show the distance travelled by each wheel. The differential allows for this difference

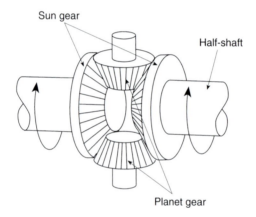

Sun gear

Half-shaft

Planet gear

Fig. 18.43   Operation of the differential when the vehicle is travelling in a straight line

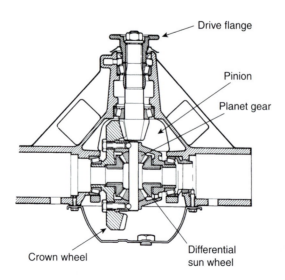

Drive flange

Pinion

Planet gear

Crown wheel

Differential sun wheel

Fig. 18.44   Final drive gear. Crown wheel and pinion, and differential

part of a sphere whose centre is on an axis with the shaft. This allows the shaft to run slightly out of alignment, which could be caused by shaft deflection. This is commonly used as the centre bearing in the two/three-piece prop shaft.

Combined radial and one-way thrust bearing

This type (Fig. 18.46(d)) is designed to take end-thrust in one direction only, as well as the radial load. They are normally mounted in pairs and are commonly used in the rear axles and clutch withdrawal thrust bearings. It is very important that these are fitted correctly otherwise the bearing will fall to pieces. They are also used as wheel bearings fitted back to back to enable them to take the side-thrust imposed on the wheel.

Parallel-roller journal bearing

This bearing (Fig. 18.46(e)) consists of cylindrical rollers having a length equal to the diameter of the roller. These can withstand a heavier radial load than the corresponding ball bearing, but cannot withstand any side-thrust. They are used in large gearboxes and in some cases have replaced the plain bearing.

Taper roller bearing

The rollers and race of this bearing (Fig. 18.46(f)) are conical (tapered) in shape with a common apex (centre) crossing at a single point on the centre line of the shaft. This ensures that the rollers do not slip. The rollers are

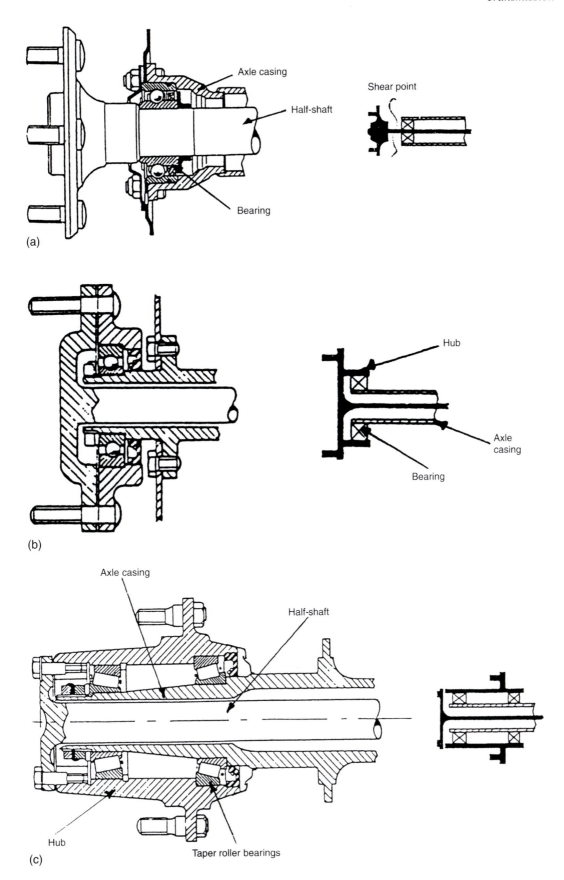

Fig. 18.45 (a) Semi-floating axle. (b) Three-quarter floating axle. (c) Fully-floating axle

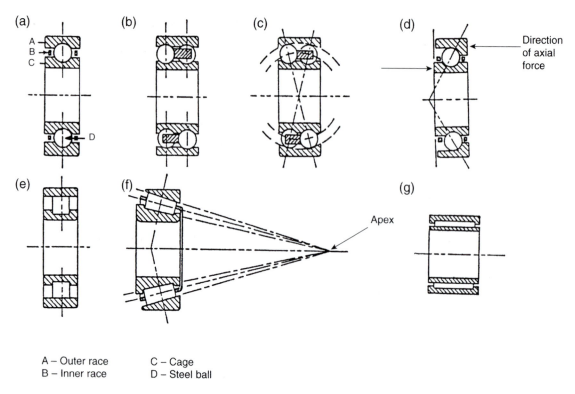

A – Outer race   C – Cage
B – Inner race   D – Steel ball

Fig. 18.46   (a) Single-row deep groove radial or journal bearing. (b) Double-row radial or journal bearing. (c) Double-row self-aligning journal bearing. (d) Radial and one-way thrust bearing. (e) Parallel-roller journal bearing. (f) Taper roller bearing. (g) Needle roller bearing

kept in position by flanges on the inner race. This type of bearing can withstand larger side-thrusts than radial loads and they are normally used in pairs facing each other. They are most often used in the final drive, rear- and front-wheel hub assemblies. Some form of adjustment is normally used to reduce movement to a minimum or to provide some pre-load to the bearing.

Needle roller bearing

In this bearing (Fig. 18.46(g)) the rollers have a length of at least three roller diameters, packed in cages between inner and outer sleeves. The bearings can be found in universal joints and gearboxes, especially where space is limited.

Learning tasks

1. Draw up a simple service schedule for checking the axle and final drive for correct operation. Include such things as type, grade and amount of oil used, and time/mileage base for changing/ checking oil level.

2. How should the differential be checked for abnormal wear? What would be the main signs of early failure?
3. If a steady whining noise was heard as the vehicle was driven at a constant speed along a level road, what would you suspect as being the fault and what would your recommendations be to put it right?
4. What main safety precautions would you observe when removing a live axle from a vehicle?
5. Why is it that most vehicles now use the hypoid type of gearing on the final drive?
6. Draw up the procedure for replacing the half-shaft on a live axle; assume the vehicle has been brought into the workshop, and the half-shaft has sheared between the final drive and the hub bearing.
7. Investigate the differences between a front-engine FWD and a front-engine RWD final drive and differential. Itemize the differences in servicing and adjustments that may be necessary when removing, stripping, rebuilding, and refitting.
8. An axle requires a new crown and pinion fitting. Remove and dismantle an assembly using the procedure illustrated in Figs 18.38 and 18.39 to give the correct markings on the gear teeth.

Practical assignment — removing a clutch

When completing the task of removing and refitting the clutch, complete the following worksheet.

Questions

1. Why are the splines of the gearbox input shaft left clean and dry and not lubricated with engine oil?
2. State the type of clutch spring fitted to the vehicle.
3. Make a simple sketch of the clutch release mechanism. Label your sketch with the main components.
4. Inspect a diaphragm spring clutch assembly when attached and removed from the flywheel. Sketch the shape (a side view will do) of the diaphragm spring in the normal run position and in the removed position.
5. State the method used to adjust the clutch and the manufacturer's recommended clearance.
6. Record the condition of the following components:
   (a) the driven plate
   (b) the pressure plate assembly
   (c) the release bearing and mechanism
   (d) the oil seals of both the engine and the gearbox.
7. Make your recommendations to the customer to give adequate operation of the clutch for say the next 50 000 miles.

Practical assignment — removing and refitting a gearbox

Complete this when undertaking the removal and refitting of the gearbox.

Questions

1. Dismantle a front-engine FWD gearbox. Identify wear on moving components such as bearings, shafts, gear teeth, thrust washers, and oil seals. List the equipment used to measure this wear.
2. Identify the type of synchromesh unit fitted to the gearbox. From what material is it made?
3. Make a simple sketch of the method used to prevent the engagement of two gears at the same time. When does this unit operate?
4. Sketch and label the position of the selector forks, rails, and location devices. Describe how the gear, once it is engaged, is held in position.

5. Identify first, second, third, fourth, and where necessary fifth gear positions. State the gear ratios for each gear. The formula for calculating gear ratios is:

$$\frac{\text{driven gear}}{\text{driver gear}} \times \frac{\text{driven gear}}{\text{driver gear}}.$$

6. Reassemble the gearbox and state any special precautions to be observed during reassembly and safety points to look out for.

Practical assignment — axles and final drives

Remove and dismantle the final drive and differential unit from a front-engine RWD vehicle.

Questions

1. The type of final drive is:
   (a) straight bevel
   (b) spiral bevel
   (c) hypoid
   (d) worm and wheel.
2. The bearings used to support the crown wheel are:
   (a) ball bearing taking axial thrust only
   (b) taper roller bearings taking axial thrust only
   (c) ball bearings taking axial and radial thrust
   (d) taper roller bearings taking axial and radial thrust.
3. Why should the pinion be pre-loaded?
4. The type of bearing arrangement supporting the half shafts is:
   (a) semi-floating
   (b) three-quarter floating
   (c) fully floating
   (d) a combination of (a), (b) and (c).
5. The axle half-shaft is splined to the:
   (a) sun gear
   (b) planet gear
   (c) crown wheel
   (d) pinion gear.
6. The type of axle is:
   (a) split axle
   (b) live axle
   (c) dead axle
   (d) trans-axle.
7. Describe the procedure for reassembling the final drive and differential. State the following information:
   (a) the pre-load on the pinion bearing
   (b) the tolerance for backlash
   (c) the torque setting for the pinion nut
   (d) the final drive ratio
   (e) the type of oil seal used on the pinion

(f) the type of oil seal used on the axle half-shaft
(g) the type and grade of oil used in the axle
(h) the service interval when the oil should be changed.

Self-assessment questions

1. A gearbox ratio of 4:1 means that:
   (a) the power at the gearbox output shaft is four times the engine power
   (b) the power at the gearbox output shaft is a quarter of the engine power
   (c) the torque at the gearbox output shaft is four times the engine torque
   (d) the speed of the gearbox output shaft is four times the speed of the engine crankshaft.

2. A differential gear in the final drive:
   (a) allows the wheel on the outside of a bend to rotate slower than the one on the inside of the bend
   (b) prevents skidding during acceleration
   (c) permit's the drive to be transmitted through an angle of 90°
   (d) allows the wheel on the outside of the bend to rotate faster than the one on the inside of the bend.

3. In a synchromesh gearbox:
   (a) the gears are engaged by sliding them into mesh as required
   (b) the gears are engaged by double declutching and blipping the throttle
   (c) the gears are engaged by the operation of the synchromesh units
   (d) the gears are engaged by the interlock device on the selector shaft.

4. In the Hotchkiss type drive:
   (a) the distance between the driven axle and the gearbox remains constant
   (b) a sliding joint is fitted in the propeller shaft to allow for variations in the distance between the gearbox and the driven axle
   (c) only one Hooke-type coupling is required
   (d) a torque tube is always required.

5. The technical details of a certain vehicle state that a road speed of 30 mph is achieved when the engine speed is 2000 rev/min and the vehicle is in top gear. On test, this vehicle achieves 25 mph in top gear when the tachometer shows an engine speed of 2300 rev/min. This is an indication that:
   (a) the clutch is slipping
   (b) the engine is producing too much power
   (c) the driver is using too much throttle
   (d) the alternator charging rate is too high and it is causing false speedometer readings.

6. In a hypoid bevel final drive:
   (a) the centre line of the pinion is offset from the centre line of the crown wheel
   (b) the drive is transmitted through an angle of 100°
   (c) a differential gear is not required
   (d) the pressure between the gear teeth is very low.

7. In a semi-floating live axle:
   (a) most of the load is taken by the axle casing
   (b) shearing and bending forces are carried by the axle casing
   (c) most of the load is taken by the axle shaft
   (d) a shear pin is fitted at the differential end of the half-shaft.

8. The stalling speed of a torque converter is:
   (a) the speed at which the idle control valve is set to operate
   (b) the speed at which torque multiplication is at a maximum
   (c) the highest speed that the engine can reach while the vehicle is in gear and prevented from moving
   (d) lowest speed at which the vehicle can be driven.

9. When moving away from a standing start, a torque converter increases engine torque by a factor of 2:1. When the gearbox ratio is 2.5:1 and engine torque is 100 Nm:
   (a) the torque at the gearbox output shaft will be 50 Nm
   (b) the torque at the gearbox output shaft will be 250 kW
   (c) the torque at the gearbox output shaft will be 500 Nm
   (d) the torque at the gearbox output shaft will be 4.5 times the engine torque.

10. In a fully floating drive axle:
    (a) shearing, bending, and torsional forces are all taken by the axle shafts
    (b) a broken axle shaft may cause the wheel to fall off
    (c) shearing and bending forces are taken by the axle casing, and torsional forces only are taken by the axle shafts
    (d) the axle has to be dismantled in order to replace a half-shaft.

19
Electronically controlled automatic transmission

Topics covered in this chapter

Fluid couplings
The fluid flywheel
Torque converter
Epicyclic gear trains and ratios
Band brake and clutches
A selection of examples of gear ratios
Sensors and actuators used in automatic gearbox operation
The hydraulic system and its pumps and valves
Automatic gearbox maintenance – performance checks and maintenance

The details contained in this chapter are intended to provide a broad overview of the subject that should provide adequate background knowledge for the specialist training courses that are provided by manufacturers' training schools.

The majority of automatic transmission systems of the type shown in Fig. 19.1 employ a fluid coupling to make the connection between the engine and the gearbox, and an epicyclic gearbox where the gears are changed without them being taken in and out of mesh.

Fluid couplings

The fluid flywheel

The fluid coupling is called a fluid flywheel because it is normally attached to the engine flywheel and forms part of it. The fluid flywheel is a sealed container that is filled with the oil that provides the driving force that operates the device when the engine is running. As the impeller rotates, the centrifugal force on the oil in the curved segments causes the oil to move to the outer edge of the impeller. Because of the speed of rotary motion of the impeller, the oil leaves at an angle; as the oil leaves the impeller it strikes the curved segments in the turbine and, when the force of the oil is sufficient, it will cause the turbine to rotate. When a gear is engaged the torque (force) on the turbine will gradually increase until it is strong enough to make the vehicle move. The fluid flywheel is, in effect, an automatic clutch (see Fig. 19.2).

As the rotary speed of the turbine rises, the pressure on the oil in the segments begins to match that of the

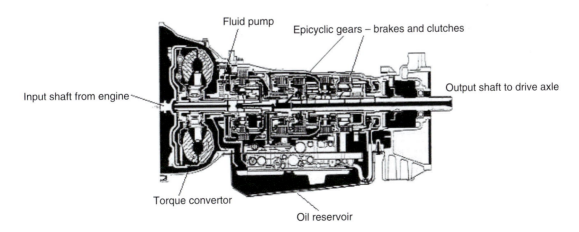

Fig. 19.1 A typical automatic transmission

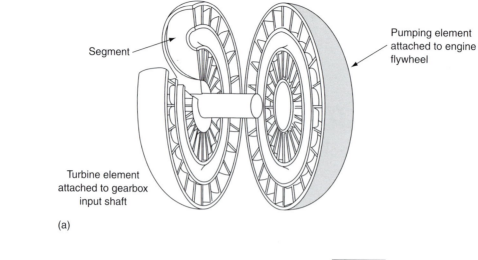

Segment

Pumping element attached to engine flywheel

Turbine element attached to gearbox input shaft

(a)

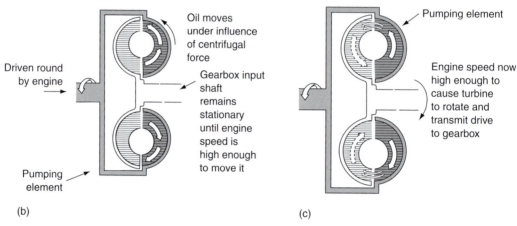

Driven round by engine

Oil moves under influence of centrifugal force

Gearbox input shaft remains stationary until engine speed is high enough to move it

Pumping element

(b)

Pumping element

Engine speed now high enough to cause turbine to rotate and transmit drive to gearbox

(c)

Fig. 19.2   The principle of the fluid flywheel

impeller and flow from one member to the other virtually ceases – at this point the coupling begins to slip. This occurs when the turbine speed is about 95% of the impeller speed. The speed at which this occurs is called the stalling speed and engine power is converted into heat by the slipping action – in most cases this is not a problem because the small amount of slip that occurs in normal operation means that the turbine speed is always slightly less than that of the impeller.

The torque converter

The main difference between the fluid flywheel and the torque converter is the third element that is called the stator. The stator consists of a set of curved blades on a circular hub, and it is placed between the pumping element and the turbine. As the fluid leaves near the centre of the turbine rotor, it is directed across the curved blades of the stator. The change of direction of the mass of fluid as it passes across the stator blades generates a force; this force attempts to rotate the stator on its hub, but it is prevented from doing so by the one-way clutch. The force that would have rotated the stator

is now directed towards the input side of the pumping member and is added to the force that is generated by the pump – it is this increased force that now acts on the turbine to give an increase in torque, hence the term 'torque converter' (see Fig. 19.3).

The amount of torque multiplication that is produced is approximately 2:1 at its maximum, which occurs when the difference in speed between the impeller and the turbine is great. It gradually decreases up to the point when the two speeds nearly match; at this point the torque converter begins to act as a fluid flywheel with no torque multiplication – the fluid leaving the inner edge of the turbine blades now acts on the back of the stator blades so that the free wheel operates to allow the stator to rotate with the other two members of the converter.

Slip in a fluid coupling

In order for a fluid coupling to function there must be a difference in rotational speed between the impeller and the turbine; as the speed of the turbine begins to approach that of the impeller the pumping action of

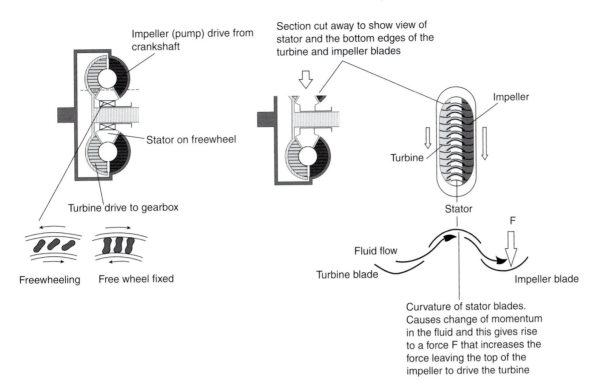

Fig. 19.3 Action of the torque converter

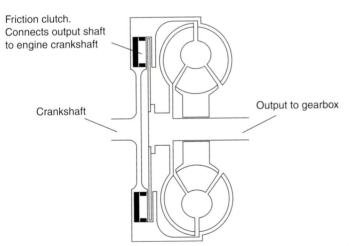

Fig. 19.4 The converter lock-up clutch

the impeller becomes weak and the turbine will start to slow down. This results in an unavoidable amount of slip, which results in a difference in speed between the two members of 4—5%. In some applications a friction clutch of the type shown in Fig. 19.4 is added to the torque converter to overcome the problem of slip.

Automatic gearbox

Automatic gearboxes are those in which the various gear ratios are selected by the control system on the vehicle. When the conditions demand it the control system operates a set of valves that determine the ratio to be selected. Most automatic gearboxes make use of epicyclic gear trains, because gear ratios can be changed without sliding gears into mesh — this is achieved by a system of hydraulically operated clutches and brakes that lock and unlock the various components of the epicyclic gear trains.

Epicyclic gear ratios

The simple epicyclic gear train can be used to produce several different gear ratios depending on which member is fixed, which is the driver, and which is the driven member. Some examples of the different ratios that can be obtained in an epicyclic gear set are shown in Fig. 19.5.

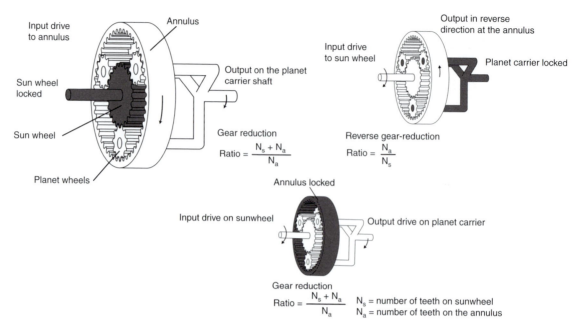

Input drive
to annulus

Annulus

Output on the planet
carrier shaft

Sun wheel
locked

Sun wheel

Planet wheels

Gear reduction

$$\text{Ratio} = \frac{N_s + N_a}{N_a}$$

Output in reverse
direction at the annulus

Input drive
to sun wheel

Planet carrier locked

Reverse gear-reduction

$$\text{Ratio} = \frac{N_a}{N_s}$$

Annulus locked

Input drive on sunwheel

Output drive on planet carrier

Gear reduction

$$\text{Ratio} = \frac{N_s + N_a}{N_a}$$

N_s = number of teeth on sunwheel
N_a = number of teeth on the annulus

Fig. 19.5 Epicyclic gear ratios

A gear set that produces the gear ratios

The following descriptions relate to an automatic gearbox that is used in some KIA vehicles. The gear system is a combination set that comprises an annulus, a large sun gear, a small sun gear, long and short planet pinions, and their carrier (see Fig. 19.6). The forward and reverse ratios are obtained by locking and unlocking the elements of the gear set by means of the friction brakes, disc clutches, and one-way free wheels.

The friction elements that lock and unlock elements of the gear set

The friction elements are the band brakes and multi-disc clutches that are hydraulically operated under the control of the ECU. Simplified forms of these are shown in Fig. 19.7. The positions of the friction elements and one-way clutches in the gearbox are shown in Fig. 19.8.

The gear ratios

In keeping with my intention to provide a general over-view of automatic transmission systems I have decided to describe two forward ratios and reverse, rather than the four forward ratios and reverse that this gear system provides.

When studying the diagrams in Figs 19.9–19.11 it should be understood that the dotted lines represent hollow shafts that allow one shaft to rotate inside the other.

The hydraulic system

The hydraulic system relies on the supply of oil from the transmission oil pump that is driven from the torque converter. The pump is driven at engine speed and it draws the oil from the oil pan through a fine mesh filter that prevents foreign matter from entering the hydraulic circuit. Because the transmission oil pump only operates when the engine is running, it is not possible to tow start the vehicle.

Oil pressure in the system

The hydraulic pressure is conveyed to the various parts of the transmission through internal oilways where it is directed, by a number of electrically operated shift valves and pressure regulators, to the clutches and brakes. Figure 19.12 shows the general principle of a typical automatic transmission hydraulic circuit.

In the system shown in Fig. 19.12, the line pressure that operates the clutches and band brakes is controlled by the pressure-regulating valve, which is designed to ensure that the correct pressure to operate the clutches and brakes is available to suit all conditions — for example, higher torque and pressure are required at the brake bands and clutches under acceleration than are required under cruising conditions. The throttle valve is connected to the accelerator linkage and it outputs a pressure that is dependent on the position of the accelerator pedal. The throttle pressure is applied to the main pressure-regulating valve through the throttle modulator valve, which serves to eliminate sudden pressure rises that may adversely affect the line

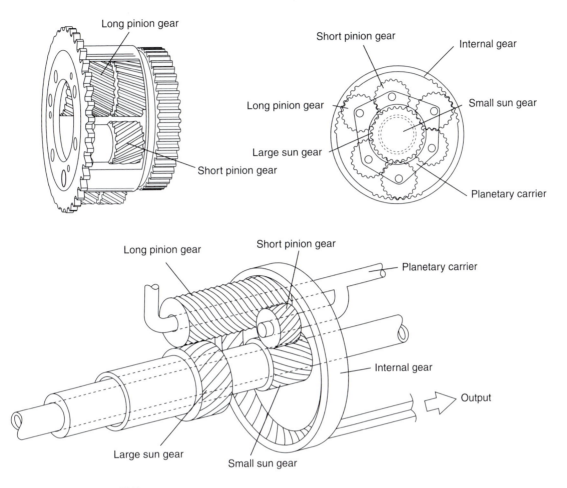

Fig. 19.6 An epicyclic gear set (KIA)

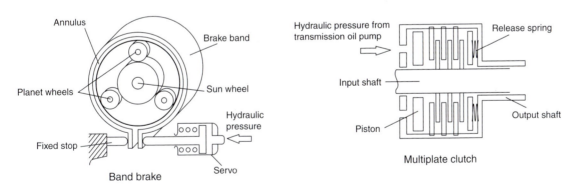

Fig. 19.7 Principle used for locking elements of epicyclic gear set

pressure. At the right-hand end of the pressure-regulating valve is the valve that controls the oil pressure in the torque converter, which is normally lower, at approximately 1.5 bar, than the line pressure that operates the gear change system.

The shift valves

The shift valves (Fig. 19.13) are controlled by the solenoid valve that operates under the control of the ECU.

The sensors send information to the ECU, which then compares them with the operating program to decide on the output that is required. The following description relates to the procedure for changing gear from first to second.

In first gear the solenoid valve is switched off. Line pressure (12) from the regulating valve pushes the shift valve to the left, against the pressure from the throttle valve. This action uncovers the port (9) and releases the second gear brake.

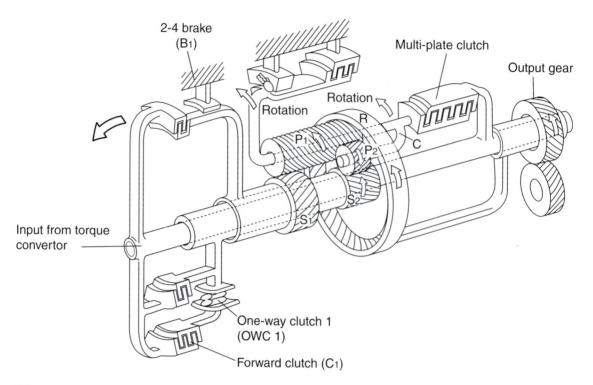

Fig. 19.8 Multi-plate clutches and brakes in an automatic gearbox (KIA)

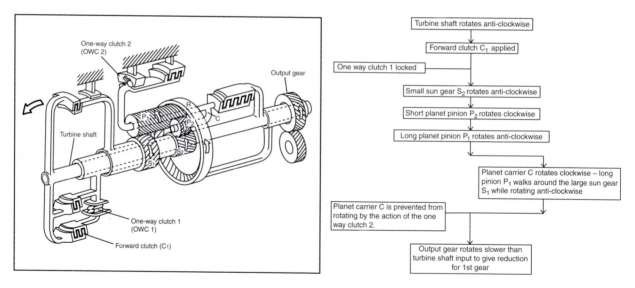

Fig. 19.9 Obtaining first gear (KIA)

When a shift from first to second is initiated, the solenoid valve is opened by the ECU. This allows the line pressure to be diverted from the right-hand end of the shift valve and allows the shift valve to move to the right. Line pressure is now applied to the second gear brake band while the first gear brake is released. Similar actions apply to the other shift valves that select higher gears and reverse.

The sensors

The operation of the shift valves is determined by the ECU in response to signals received from the sensors and driver's controls. The KIA system that I have outlined here has the following list of sensors for the automatic transmission system:

- Throttle position — creates a voltage that is sent to the ECU to represent the position of the accelerator pedal.
- Idle switch — the signal generated here tells the ECU when the throttle is fully closed.
- Pulse generator on the gearbox — sends a signal that represents road speed to the ECU.
- Vehicle speed sensor — uses the speedometer reading to act as a check on the pulse generator signal.

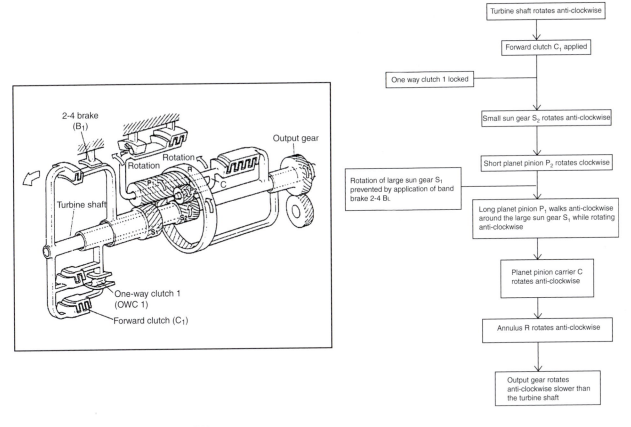

Fig. 19.10 Power flow in second gear (KIA)

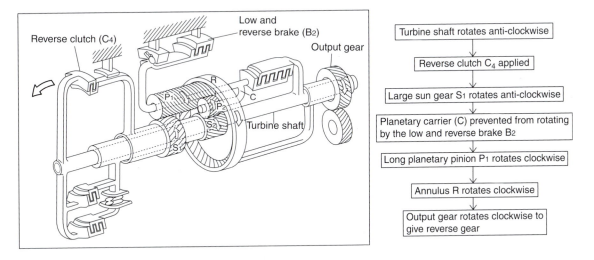

Fig. 19.11 Power flow in reverse gear

- Inhibitor switch – sends a signal to the ECU. Prevents engine start unless selector is in the park position.
- Stop light switch – electrical signal tells the ECU when the brakes are applied.
- Coolant temperature sensor – electrical signal tells the ECU the state of the engine.

In order for the transmission to function, all of these sensors must be operating correctly.

Maintenance and fault tracing

The operation of automatic transmission systems relies heavily on the hydraulic system and the fluid that it uses. Automatic transmission fluids (ATFs) are oils that contain several chemical compounds that help the oil to meet the severe demands that are placed on it. Examples of these conditions are:

- Transmission of engine torque

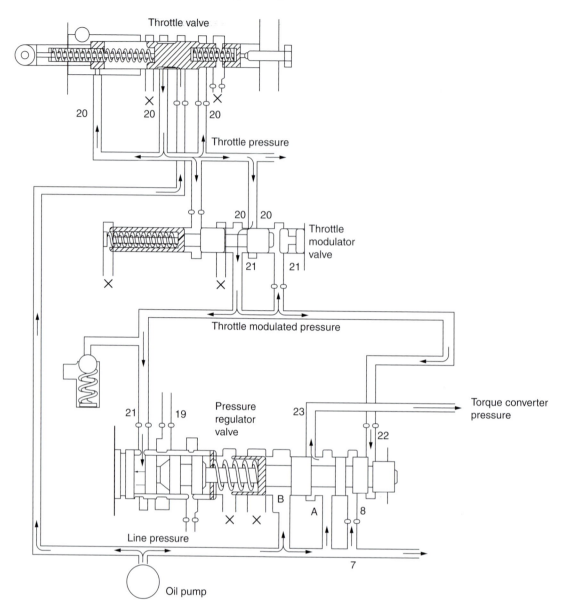

Fig. 19.12   Automatic transmission hydraulic circuit (KIA)

- Operation of the valves, clutches, and brakes
- Lubrication of the mechanical parts
- Cooling action − removal of heat away from the areas of friction
- Keeping working parts clean.

Chemical additives are used for the following purposes:

- **Viscosity index improvers.** The transmission operates across a wide range of temperatures, from below freezing to as high as 200°C. At low temperature the ATF may not flow easily and this can cause problems with the clutches and brakes, which may lead to delayed gear changes. On the other hand, at high temperature, the oil may be too thin to prevent metal-to-metal contact, leading to friction and wear. The viscosity index improvers are added to ensure

that the fluid is not too viscous at low temperature and still sufficiently viscous at high temperature.
- **Oxidation inhibitors.** When oil is heated in the presence of oxygen a chemical reaction occurs that leads to the formation of varnish, sludge and acids, which, in time, will seriously affect the ability of the ATF to perform its functions.
- **Anti-foaming agent.** When in operation, the fluid in the torque converter is churned and sheared and these actions generate large amounts of heat. During the churning action air becomes trapped in the ATF and foam is produced. Foam in the fluid causes problems at the transmission pump and in the operation of the brake bands and clutches, as well as accelerating oxidation of the fluid. The anti-foaming agent helps to limit the production of foam and quickly reduces the life span of any foam that is formed.

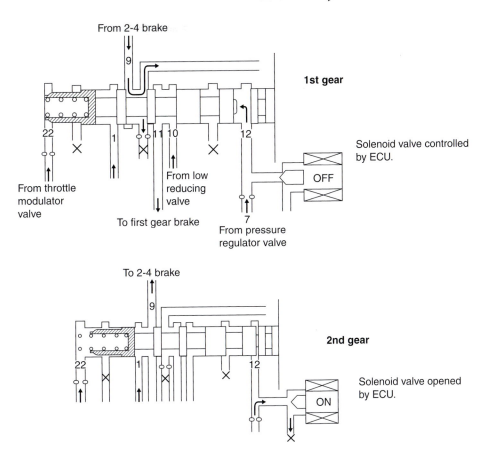

From 2-4 brake

9

1st gear

22

1

11 10

12

Solenoid valve controlled
by ECU.

OFF

From throttle
modulator
valve

From low
reducing
valve

To first gear brake

7

From pressure
regulator valve

To 2-4 brake

9

2nd gear

22

1

12

Solenoid valve opened
by ECU.

ON

Fig. 19.13 Solenoid shift valves

- **Corrosion inhibitors.** Corrosion inhibitors enable the ATF to cling to metal surfaces to protect them against rust and etching, and to prevent moisture from accumulating in the system.

Fluid level is checked at recommended service intervals and while it is being checked it is possible to make an assessment of the condition of the gearbox. The details that I have to hand specify that the fluid level should be checked after the transmission has been warmed up to a temperature of approximately 80°C.

When checking the fluid level, the state of the fluid on the dipstick can be examined for signs of problems. If the fluid is dark brown to black in colour it may indicate that the fluid has not been changed at the recommended mileage. Further examination may also reveal the presence of sediment and metallic particles, suggesting serious clutch and brake band wear, and possible damage to metallic components.

Some automatic transmission systems incorporate a liquid-to-liquid oil cooler; if the ATF from such a system is a milky colour it indicates that coolant is leaking into the fluid.

Should the fluid on the dipstick show signs of small bubbles they are a sign of excessive aeration, which may be caused by the fluid level being too high. This should be remedied because excessive aeration causes oxidation and build-up of varnish.

Stall test

The condition of the torque converter and some aspects of gearbox operation can be checked by a stall test. When the transmission is in gear and the vehicle is prevented from moving, the engine speed can be increased until it reaches a speed of rotation beyond which no further increase in engine speed will occur. At this point the converter is said to be stalled and the speed at which it occurs is called the stalling speed. At the stalling speed all of the engine power is being converted into heat — **in order to prevent damage from overheating the duration of a stall test must not last for more than 3–5 seconds.**

A guide to the procedure for conducting a stall test is:

- Make sure that all safety precautions are observed — no one is standing in front or immediately behind the vehicle.
- Connect exhaust extraction equipment.
- Check that the engine and transmission are at normal working temperature.
- Connect a tachometer to the engine — if there is not one on the instrument panel.
- Chock the wheels and firmly apply the handbrake.
- Run the engine and select the gear position that is recommended in the workshop manual. Fully depress

the throttle and note the maximum engine speed that is reached — bearing in mind that the test is restricted to 3–5 seconds.

- Let the engine idle — move the selector to P and switch off the engine.
- Note the engine speed and compare it with the manufacturer's data.

Pressure checks

Most automatic transmissions are fitted with a sealing plug which, when removed, allows a pressure gauge to be attached so that the pressure generated by the transmission oil pump can be checked. The procedure for performing an oil pressure check varies according to the vehicle and type of transmission system, and the specific manufacturer's procedures must be followed. The following data that relates to a Ford vehicle is given as a guide:

- Measure oil pressure at idle speed.
- With the pressure gauge fitted and the system warmed up to operating temperature and after all preliminary safety precautions have been taken, the engine should be started and allowed to idle.
- With hand and foot brake firmly applied, the gear selector should be moved through all positions while the oil pressure is observed.
- In this particular case, the prescribed oil pressures are:
  - Selector position P-N-D-l, pressure 3.9–5.1 bar
  - Selector position R, pressure 6.2–8.7 bar.

Road test

A road test permits the performance of the transmission to be checked under normal driving conditions. The gear changes take place at speeds that are determined by the program in the ECU. The speed at which gear changes take place can be recorded by an assistant and later compared with those prescribed by the manufacturer. The quality of the gear shift can also be checked to see if they are:

- Normal
- Harsh
- Sluggish as indicated by slippage.

ECU fault codes

The ECU constantly monitors all inputs and outputs, and if a fault is detected a fault code is generated. These codes can then be read out to diagnostic equipment and the necessary steps taken to confirm the cause and effect a remedy. In common with other ECU-controlled systems, the fault codes often point to a problem with a particular component. Further checks are then

required in order to determine the precise cause so that a decision as to whether to repair or replace can be made. Examples of sensor checks are given in the section of the book that is devoted to the subject of electronically controlled systems.

Self-assessment questions

1. Which part of a torque converter is responsible for torque multiplication?
2. What is the purpose of the direct drive clutch that is fitted to some fluid couplings?
3. What is the reason for fitting a heat exchanger to an automatic gearbox?
4. Which parts of an automatic gearbox are used to bring about gear changes?
5. Examine a workshop manual for vehicles that you work on and describe the procedure for conducting a stall test on an automatic transmission system.
6. With the aid of an instruction manual, describe the equipment required and the procedure for checking the throttle sensor on an ECU-controlled automatic transmission system.
7. What is the reason for restricting the length of time that a stall test takes?
8. What should be the temperature of an automatic gearbox when the oil level is being checked?
9. Where is the pump situated that generates the hydraulic pressure for an automatic gearbox?
10. Describe the reasons why an anti-foaming agent is added to automatic transmission fluid.
11. What is the reason for the anti-oxidant chemical being added to automatic transmission fluid?
12. A certain torque converter has a maximum 1.8:1 torque multiplying power. When operating at this maximum the engine torque is 150 Nm. The torque at the gearbox input is:
    (a) 83.3 Nm
    (b) 330 Nm
    (c) 270 Nm
    (d) 27 Nm.
13. What percentage of slip occurs in a fluid coupling when it is operating at vehicle cruising speed? What extra device is sometimes fitted to overcome this problem?
14. Why is it not possible to tow-start a vehicle that is fitted with an automatic gearbox?
15. In a certain automatic gearbox the second gear ratio is 1.7:1. The torque converter produces a torque multiplication of 1.9:1. At a certain engine speed the engine torque is 120 Nm. Assuming 100% efficiency under these conditions, the gearbox output torque is:
    (a) 387.6 Nm
    (b) 432 Nm

(c) 33.3 Nm

(d) 37.15 Nm.

16. Which valve regulates the hydraulic pressure in an automatic transmission system?

17. How are the shift valves operated in an ECU-controlled automatic transmission?

18. What is the purpose of the throttle position sensor in an ECU-controlled automatic transmission?

19. How does the ECU know when the vehicle brakes are being operated?

20. What is the purpose of the speed sensor that is fitted in the automatic gearbox?

20
Suspension systems

Topics covered in this chapter

- Unsprung mass (weight)
- Leaf springs
- Coil springs
- Torsion bar
- Pneumatic (air) suspension
- Rubber suspension
- Hydraulic suspension
- Active suspension
- Dampers (shock absorbers)

The suspension of a vehicle is present to prevent the variations in the road surface encountered by the wheels from being transmitted to the vehicle body. There are two main categories of suspension systems: **independent** and **non-independent**. An independent type has each wheel moving up and down without affecting the wheel on the opposite side of the vehicle. On non-independent systems movement of one wheel will affect the wheel on the opposite side of the vehicle. A typical system consists of:

- a wheel and pneumatic tyre;
- a spring and damper unit together with arms and links that attach the wheel or axle to the vehicle.

Many different types are used on light vehicles.

Unsprung mass

The unsprung weight (mass) is the weight of components that are situated between the suspension springs and the road surface. Unsprung weight has an undesirable effect on vehicle handling and ride comfort, and suspension systems are designed to keep it to a minimum.

Non-independent suspension systems

Leaf springs

Leaf springs of the type shown in Fig. 20.1(a) are manufactured from a number of leaves (layers) of steel strip.

The use of multiple leaves ensures that the bending and shearing strength of the spring is uniform along the span of the spring from the spring eye to the point where the spring is attached to the axle. At each end of the main leaf is a circular hole (eye), which is used to attach the spring to the vehicle frame. As the spring flexes in use, the surfaces of the leaves rub against each other and this provides a degree of damping; the clips on the spring serve the purpose of keeping the leaves in contact with each other. The leaf spring is able to transmit driving and braking forces between the road wheels and the chassis.

Single leaf springs of the type shown in Fig. 20.1(b) are used on the rear suspension systems of many light commercial vehicles. Modern design techniques and materials permit the use of these single leaf springs, which reduce the weight and cost of springs.

Learning tasks

1. The modern trend is to use leaf springs containing a single leaf only. Why is this and what are the advantages?
2. The multi-leaf spring performs a number of functions (tasks). One is to positively locate the axle to prevent it moving sideways independently of the body. Name two other functions and give reasons for your answer.
3. What effect does the movement of the suspension have on the drive line and/or steering of the vehicle?
4. Why should a reduced 'unsprung' weight improve the ride and handling characteristics of the vehicle?
5. Remove a leaf spring assembly from a vehicle and inspect the mounting bushes and rubber for deterioration and wear. Reassemble the spring to the vehicle, replacing worn components as necessary.

Coil springs

This type of spring gives a smoother ride than the multi-leaf due to the absence of **interleaf friction**. It can be used on front and rear systems. The helical spring is normally used in conjunction with independent suspension and is now often used with beam-axle rear suspensions. Figure 20.2 illustrates this type.

The **coil spring** and the **torsion bar** suspension are alike and are superior to the leaf spring as regards the energy stored, but unlike the leaf spring extra members to locate it are required, which adds to the basic weight. The advantages of the coil spring are:

- a reduction in unsprung weight;
- energy storage is high;
- it can provide a softer ride;
- it allows for a greater movement of suspension;
- it is more compact.

Its main disadvantage is the location of the spring (the axle requires links and struts to hold it in place). This is shown in Fig. 20.3.

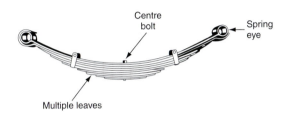

Fig. 20.1(a)   A semi-elliptic leaf spring

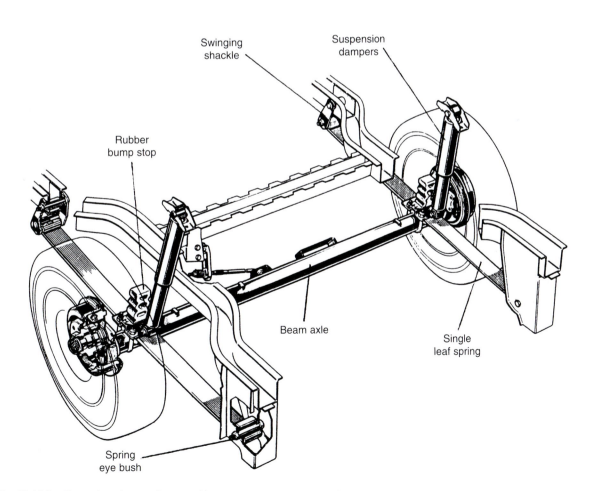

Fig. 20.1(b)   Rear axle and suspension assembly

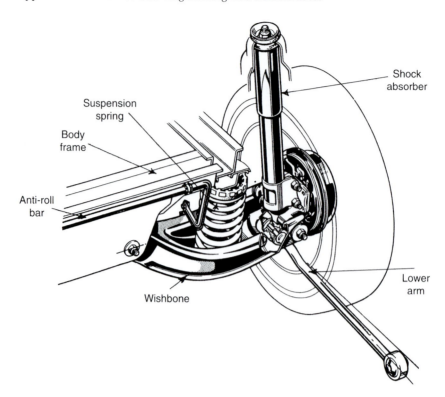

Fig. 20.2   Near-side independent rear suspension

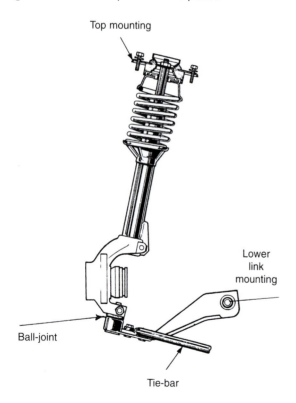

Fig. 20.3   MacPherson strut, offside front suspension

Torsion bar

This is a straight bar that can either be round or square section and fixed at one end to the chassis. The other end is connected by a lever to the axle. At each end of

the bar the section is increased and serrated or splined to connect with the lever or chassis. Adjustment is provided for at the chassis end to give the correct **ride height** for the vehicle. This type of suspension is shown in Fig. 20.4 and is used on the front of vehicles. Figure 20.5 shows the arrangement used on the rear suspension.

Since the coil spring is a form of torsion bar suspension, the rate of both types of spring is governed by the same factors:

- The length of the bar.
- The diameter of the bar. If the length is increased or the diameter decreased, the rate of the spring will decrease, i.e. the spring will become softer.
- The material it is made from.

Learning tasks

1. How is the 'ride height' adjusted on a vehicle fitted with a torsion bar front suspension? Where on the vehicle is the measurement taken?
2. List the main advantages and disadvantages of torsion bar suspension over the leaf and coil spring types?

Independent suspension

Cars and many of the lighter commercial vehicles are now fitted with some form of **independent front suspension** (IFS) and/or **independent rear suspension**

(a)

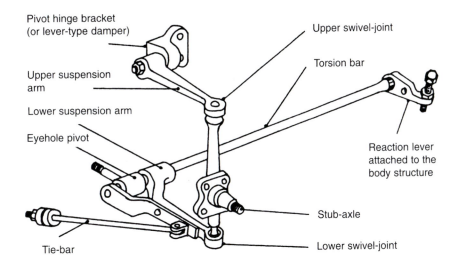

Pivot hinge bracket
(or lever-type damper)

Upper swivel-joint

Torsion bar

Upper suspension
arm

Lower suspension arm

Eyehole pivot

Reaction lever
attached to the
body structure

Stub-axle

Tie-bar

Lower swivel-joint

(b)

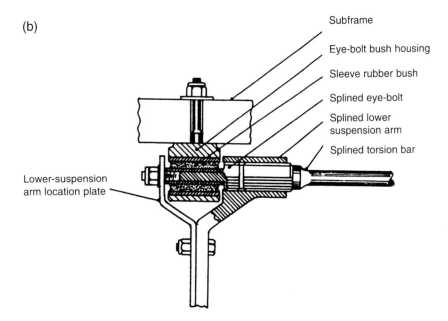

Subframe

Eye-bolt bush housing

Sleeve rubber bush

Splined eye-bolt

Splined lower
suspension arm

Splined torsion bar

Lower-suspension
arm location plate

Fig. 20.4 Torsion-bar double-transverse-arm independent front suspension. (a) Pictorial view. (b) Section view — lower-suspension-arm pivot assembly

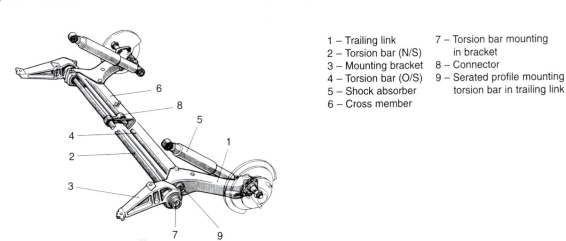

1 – Trailing link
2 – Torsion bar (N/S)
3 – Mounting bracket
4 – Torsion bar (O/S)
5 – Shock absorber
6 – Cross member

7 – Torsion bar mounting
   in bracket
8 – Connector
9 – Serated profile mounting
   torsion bar in trailing link

Fig. 20.5 Transverse torsion bar rear suspension

(IRS). These have the following advantages over the earlier beam axle and leaf spring arrangements:

- Reduced unsprung weight.
- The steering is not affected by the 'gyroscopic' effect of a deflected wheel being transmitted to the other wheel.
- Better steering stability due to the wider spacing of the springs.
- Better road holding as the centre of gravity is lower — due to the engine being mounted nearer the ground. This is because there is no front beam axle in the way.
- More space in the body due to the engine being lower and possibly further forward.
- More comfortable ride due to the use of lower rate springs.

Where a beam axle and parallel leaf spring arrangement is used the springs are subjected to the following forces:

- suspension loads due to vehicle weight;
- driving and braking thrusts due respectively to the forward movement of the chassis and its retardation when braking;
- braking torque reaction — the spring distorting but preventing the rotating of the back plate and axle;
- twisting due to the deflection of one wheel only.

When the springs are strong enough to resist all these forces they are too stiff and heavy to provide a comfortable ride and good road holding. Independent suspension designs must provide for the control or limitation of these same forces, and their action must not interfere with the steering geometry or the operation of the braking system.

As we have already mentioned with the independent suspension, when one wheel moves up or down it has little or no effect on the opposite wheel. In the case of the beam axle suspension, when one wheel rises over a bump the other wheel on the same axle is also affected. Because of this neither wheel is vertical to the road surface and so the road-holding ability of the vehicle is affected. In a truly independent system each wheel is able to move without affecting any of the others. This has a number of obvious advantages:

1. A softer ride giving greater comfort for the passengers.
2. Better road holding especially on rough or uneven road surfaces.
3. The engine can normally be mounted lower in the vehicle.
4. Because of (1), (2), and (3), the vehicle will corner better.
5. It allows for a greater rise and fall of the wheel.

Because of the expense, greater complication and difficulty of mounting, many vehicle manufacturers of smaller cars fit independent suspension to the front and beam axle on the rear, especially where the engine drives the front wheels. This arrangement gives a light,

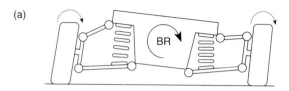

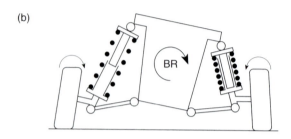

Fig. 20.6 Effects of body roll (a) and irregular road surfaces (b) on suspension geometry

simple layout for the rear suspension using coil springs and links to support the axle to the body.

The effects of uneven road surfaces and body roll on the suspension can be seen in Fig. 20.6. From the illustrations it will be seen that the **MacPherson strut** arrangement gives the most stable body and wheel alignment. This is one reason why the MacPherson strut is the most common type of suspension fitted to light vehicles.

Learning tasks

1. Explain what is meant by 'brake torque reaction'. Illustrate where necessary your answer with a simple sketch.
2. List a typical sequence of checks that should be made to find a reported knocking noise coming from the front offside suspension.
3. What effect would the lowering of the front suspension have on the steering geometry of the vehicle? Give reasons for your answer.

Axle location

As already mentioned the leaf spring locates the axle, but the coil, torsion bar, **rubber,** and **air spring** suspension support the vehicle and its load and remove the unevenness of the road surface. They do not locate the wheel assembly or axle in any way. Extra arms (**radius arms**) or links are fitted to locate the wheels for both fore and aft movement and to resist the turning forces of both driving and braking. A **Panhard rod** gives sideways location between the axle and the body (see Fig. 20.7).

A number of other methods are in use that locate the axle. These include the use of **tie rods, wishbone,** and **semi-trailing links** (Fig. 20.8). When an IRS is fitted (as on some of the more expensive vehicles) then the arrangement for location of the wheel assembly can become more complicated.

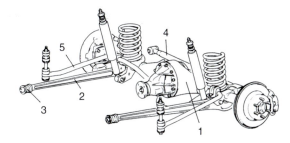

1 – Axle casing
2 – Longitudinal (down the length of the car) control arms
3 – Rubber mountings
4 – Panhard rod
5 – Anti-roll bar

Fig. 20.7 Layout of rigid line axle using coil springs and Panhard rod

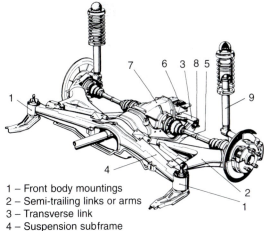

1 – Front body mountings
2 – Semi-trailing links or arms
3 – Transverse link
4 – Suspension subframe
5 – Outriggers
6 – Rear body mountings
7 – Final drive
8 – Rubber mountings
9 – McPhearson strut suspension

Fig. 20.8 Independent rear suspension and final drive using semi-trailing links

The method used to attach the links and arms to the body is by the use of **rubber bushes**. There is always a certain amount of 'give' in these rubber joints and this is termed compliance. This is generally allowed for in the specifications given by the manufacturer when checking the suspension for alignment.

Learning tasks

1. To which types of suspension arrangements is the Panhard rod fitted and why to this particular type?
2. How should the rubber bushes in the suspension be checked for wear?
3. If a driver complained of vehicle instability especially when cornering or driving over rough and uneven road surfaces what would you suspect the fault to be? Give reasons for your answer.
4. Remove and refit a suspension link, inspect the bushes for wear and soundness, the link for signs of accident damage and rust, and the mounting point for signs of wear or rust.

Rubber suspension

Rubber can be used as the suspension medium as well as for mountings and pivots. Its main advantage is that for small wheel movements the ride is fairly soft but it becomes harder as wheel movement increases. It has the advantage of being small, light and compact, and will absorb some of the energy passed to it, unlike a coil spring, which gives out almost as much energy as it receives. The rubber spring is also commonly used on LGVs and trailers. A rubber suspension unit is shown in Fig. 20.9.

Hydrolastic suspension

This is a combination of rubber and fluid (which is under pressure). Each wheel is fitted with a **hydrolastic unit,** which consists of a steel cylinder mounted on the body of the car. A tapered piston, complete with a rubber and

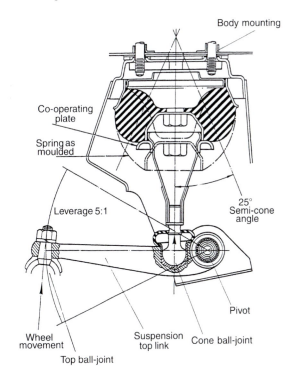

Fig. 20.9 Rubber suspension unit

nylon diaphragm connected to the upper suspension arm, fits in the bottom of the unit. When the wheel moves upwards the piston is also moved up and via fluid action compresses the rubber spring. The units are connected together on the same side by large bore pipes and some of the fluid is displaced down the pipe to the other unit. In this way the tendency for the vehicle to pitch (this is the movement of the body fore and aft, that is front to back) is reduced. A hydrolastic system is shown in Fig. 20.10.

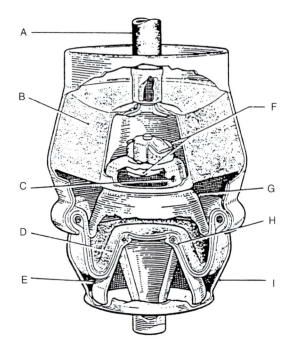

A – Fluid outlet to rear unit   F – Damper valves
B – Rubber spring   G – Dividing member
C – Fluid bleed hole   H – Reinforcing in diaphragm
D – Rubber diaphragm   I – Tapered outer cylinder
E – Tapered piston

Fig. 20.10 Hydrolastic spring displacer unit

Hydrogas suspension

Sometimes called **hydropneumatic**, this system, shown in Fig. 20.11, is a development of the hydrolastic system. The main difference is that the rubber is replaced by a gas (usually nitrogen), hence the term

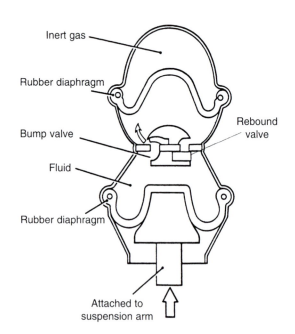

Fig. 20.11 Hydrogas suspension

'pneumatic'. With this type the gas remains constant irrespective of the load carried. Gas pressure will increase as volume is reduced. This means that the suspension stiffens as the load increases. The units are connected together in a similar manner to the hydrolastic suspension and the fluid used is a mixture of water, alcohol, and an anti-corrosive agent.

The ride height in both these systems can be raised or lowered by the use of a hydrolastic suspension pump to give the correct ride height and ground clearance.

Learning tasks

1. What safety precautions must be observed when working on pressurized suspension systems?
2. State the method used for adjusting the 'ride height' on the hydropneumatic suspension system. Identify any specialized equipment and fluid type that must be used. Adjust the ride height to the upper limit specified by the manufacturer in the repair manual.
3. Identify *four* units in the suspension that use rubber bushes to mount the suspension to the body or chassis.
4. Remove a suspension unit from a vehicle with hydrolastic or hydrogas suspension. Identify any worn or defective components. Replace the unit and adjust the ride height as necessary.

Active or 'live' hydropneumatic suspension

This system, shown in Fig. 20.12, allows the driver to adjust the ride height (sometimes inaccurately referred to as ground clearance) of the vehicle. It also maintains this clearance irrespective of the load being carried. First developed by Citroen, it has recently been taken up by a number of other manufacturers. On the Citroen arrangement each **suspension arm** is supported by a **pneumatic spring**.

Connected between the suspension arms at both front and rear are **anti-roll bars.** These are linked to **height correctors** by means of control rods. An engine-driven pump supplies oil under pressure to a **hydraulic accumulator** and this is connected to the height control or **levelling valves**.

As the vehicle is loaded, the downward movement of the body causes the rotation of the anti-roll bar. This moves the slide valve in the height correctors and uncovers the port to supply oil under pressure from the accumulator to the suspension cylinders. When the body reaches the predetermined height (which can be varied by the driver moving a lever inside the vehicle), the valve moves to the 'neutral' position. Removal of the load causes the valve to vent oil from the cylinder back to the reservoir.

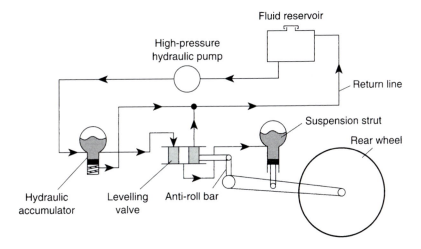

Fig. 20.12 Simplified layout of a hydropneumatic suspension system

A delay device is incorporated to prevent rapid oil flow past the valve when the wheel contacts a bump. This prevents the valve from continuously working and giving unsatisfactory operation.

In some systems a third spring unit is fitted between the two spring units on the front axle and between the two spring units on the rear axle. This gives a variable spring rate and roll stiffness, i.e. the suspension is active. The system is controlled by an ECU (electronic control unit), which senses steering wheel movement, acceleration, speed and body movement, and reacts accordingly via control valves to regulate the flow of oil to and from the suspension units. Under normal driving conditions the ECU operates the solenoid valve, which directs fluid to open the regulator valves. This allows fluid to flow between the two outer spring units and the third spring units via the damper units to give a soft ride. During harder driving the solenoid valve is switched off automatically, relieving the regulator valves which close, preventing fluid flow between the spring units, The third spring unit, being isolated and not in use, gives a firmer ride.

There are a number of benefits of this system:

- It automatically adjusts the spring and damper rate to suit road conditions and driving styles.
- It can provide a soft and comfortable ride under normal driving conditions.
- It will stiffen to give better road holding during hard driving.
- A near-constant ride height can be achieved irrespective of the load on the vehicle.

Learning tasks

1. A number of safety precautions must be observed before dismantling this type of suspension system. What are these and why should they be carried out?

2. What are the main advantages/disadvantages of this type of suspension system over a mechanical system?
3. The oil in the system should be changed at regular intervals. How often is this and what is the correct procedure? State the correct type of fluid that must be used in the system.
4. Remove and refit a hydropneumatic suspension unit. Take care when undoing pipe unions that the pipes are not damaged. Observe all the safety precautions necessary. Make a note of all defects, re-pressurize the system, and check for leaks.

Rear wheel 'suspension steer'

In some systems the rear suspension is arranged to produce a steering effect when cornering. As the suspension is deflected the road wheel **toes-in** due to the arc of movement of the semi-trailing link. This produces **understeer** on cornering (it gives a small degree of same direction rear-steer). Figure 20.13 shows one example that will give:

- toe-in when braking;
- understeer when cornering;
- stability in straight-line running and when changing lanes.

A simpler suspension arrangement is shown in Fig. 20.14 that produces toe-in under similar operating conditions.

Air suspension

Some vehicle manufacturers fit air suspension units in place of the conventional coil springs. A typical example with electronically controlled air suspension is shown in Fig. 20.15. Three switches, positioned near the steering column, control the ride height: the

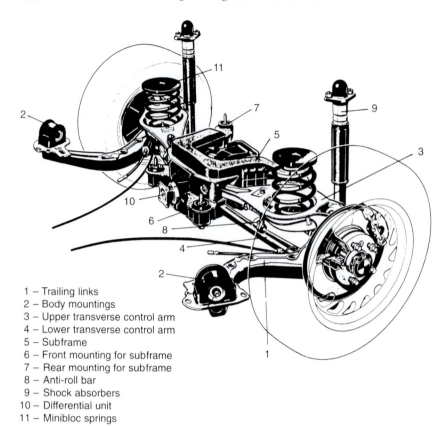

1 – Trailing links
2 – Body mountings
3 – Upper transverse control arm
4 – Lower transverse control arm
5 – Subframe
6 – Front mounting for subframe
7 – Rear mounting for subframe
8 – Anti-roll bar
9 – Shock absorbers
10 – Differential unit
11 – Minibloc springs

Fig. 20.13   Multi-link rear suspension

upper switch raises the vehicle by approximately 40 mm for driving through deep water or over rough ground; the middle switch gives a similar ride height to the coil-sprung model; the lower switch lowers the vehicle approximately 60 mm below the standard setting to enable easier loading and for getting in and out. A number of automatic adjustments such as lowering the vehicle when driving over 50 mph to reduce wind resistance and raising the vehicle if it should become grounded are programmed into the ECU and operate independently of the driver.

Another system uses a digital controller (ECU), which reacts to suspension acceleration and ride height by varying the oil pressure. It will also adjust damper action by reference to body movement.

Some manufacturers make **ride height levellers** (Fig. 20.16), which are shock absorbers that will raise the rear of the vehicle to its normal ride height when loaded. The system is operated through a switch, electric motor, and compressor; the dampers are pumped up via air pipes connecting the units together.

The **suspension damper** has to perform a number of functions. These are:

- absorb road surface variations;
- insulate the noise from the vehicle body;
- give resistance to vehicle body pitch under braking and acceleration;
- offer resistance to roll when cornering.

A passive mechanical system to tackle each of these tasks is impossible as the rate of the damper must be varied to suit differing circumstances and handling characteristics. Shock-absorber manufacturers have made provision for some control over the damping rate by making them adjustable, but once set even these are not suitable for all applications and uses.

With this in mind many of the new suspension systems are **active**; in other words they will operate continuously, adjusting to preset limits the suspension spring rate. One system uses an ECU that receives information on cruising, suspension condition, body roll, and pitch. It then directs air from the compressor to the suspension units to give the correct spring rate and ride height for the load carried.

Learning tasks

1. What is the purpose of the ECU and electronic sensors in an active suspension system? What attention do these units require and at what intervals should checks be carried out?
2. On some models the system is connected to the anti-lock braking system. Explain in simple terms the reason for this and how it operates.
3. List the maintenance requirements for one of the 'active' suspension systems with which you are familiar.

4. Draw up a work schedule for removing a suspension unit from a vehicle. Special note should be made of the safety aspects of the task.

5. Remove an ECU from an active suspension system, check the terminals for corrosion or damage, and clean with appropriate spray cleaner where necessary. Ensure that the mounting surface is clean and free from rust. Refit the ECU taking special care when refitting the terminal connector.

Suspension dampers

The function of the suspension damper is not to increase the resistance to the spring deflecting but to control the **oscillation** of the spring (this is the continuing up and down movement of the spring after going over a bump or hollow in the road surface). In other words it absorbs energy given to the spring, hence the more common name of **shock absorber**. It does this by forcing oil in the damper to do work by passing it through holes in the piston and converting the energy of the moving spring into heat, which is passed to the atmosphere.

Two main categories of dampers are in common use: the **direct-acting** (usually telescopic) and the **lever arm**.

The twin-tube telescopic damper

This type is usually located between the chassis and the axle so that on both bump and rebound oil is forced through the holes in the piston. The reservoir is used to

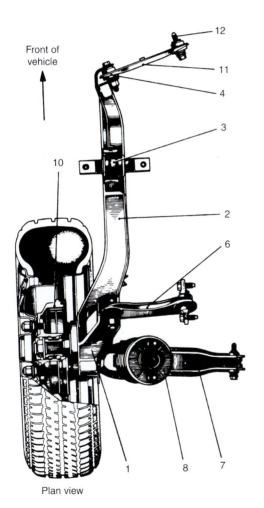

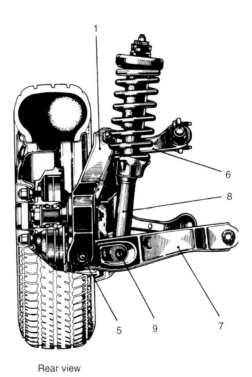

1 – Wheel hub carrier
2 – Trailing arm
3 – Rubber bearing
4 – Trailing arm front mounting
5 – Trailing arm rear mounting
6 – Upper transverse control arm
7 – Rear lower transverse control arm
8 – Shock absorber
9 – Shock absorber lower mounting
10 – Brake drum
11 – Front lower transverse control arm
12 – Inner mounting

Fig. 20.14 Multi-link rear suspension giving toe-in during braking and cornering

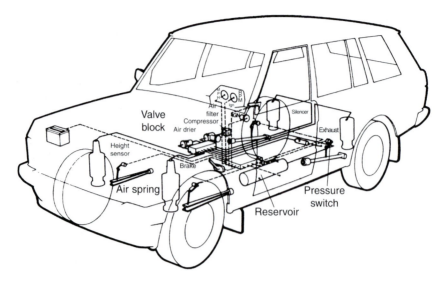

Fig. 20.15   Air suspension

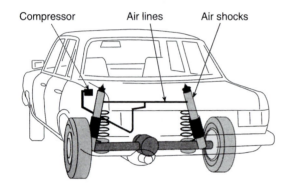

Fig. 20.16   Height levellers in the rear axle

accommodate the excess oil that is displaced as a result of the volume in the upper cylinder being smaller. If oil is lost and air enters the damper it will affect its performance and it will become 'spongy' in operation. This will have an effect on the stability of the vehicle, especially when cornering, travelling over rough ground, uneven surfaces, and on braking or accelerating. When checking for correct operation disconnect one end of the damper and check the amount of resistance to movement. Figure 20.17 shows a telescopic shock absorber.

Gas-pressurized dampers

A **single tube** is used as the cylinder in which the piston operates. It is attached to the car body and suspension by rubber bushes to reduce noise and vibration, and to allow for slight sideways movements as the suspension operates. A chamber in the unit, sealed by a free piston, contains an inert gas that is under pressure when the damper is filled with oil. As the suspension operates the piston moves down the cylinder and oil is forced through the 'bump' valve to the upper chamber; excess oil that cannot be accommodated in the upper chamber because of the rod moves the free piston to compress the

gas. On rebound the oil is made to pass through the 'rebound' valve in the piston; by varying the size of the holes in the valves the resistance of each stroke can be altered to suit different vehicle applications. This arrangement is most commonly fitted to the MacPherson strut suspension systems. The main advantages of the single-tube damper are:

- It can displace a large volume of fluid without noise or fluid aeration.
- It is fairly consistent in service even when operating at large angles to the suspension movement.
- It has good dissipation of heat to the air flow.

Checking damper operation

Two tests are normally used to check whether the dampers are serviceable without removing them from the vehicle. The **bounce test** involves pushing down on each corner of the vehicle and observing the up and down movement of the body as it comes to its rest position. The second method involves the use of some form of **tester**, of which there are several types available. One type uses an eccentric roller arrangement that the vehicle is driven onto to produce the required movement of the suspension. In another the vehicle is driven up a collapsible ramp; when set the ramp is dropped. Both types give a graphical printout of the test that shows the oscillations of the suspension.

Materials used in suspensions

- **Road springs**. These are made from a silicon manganese steel as they must withstand very high stresses and fatigue yet still retain elasticity and strength.
- **Suspension links**. Here a nickel steel is used, which gives elasticity together with toughness.

(a)

(b)

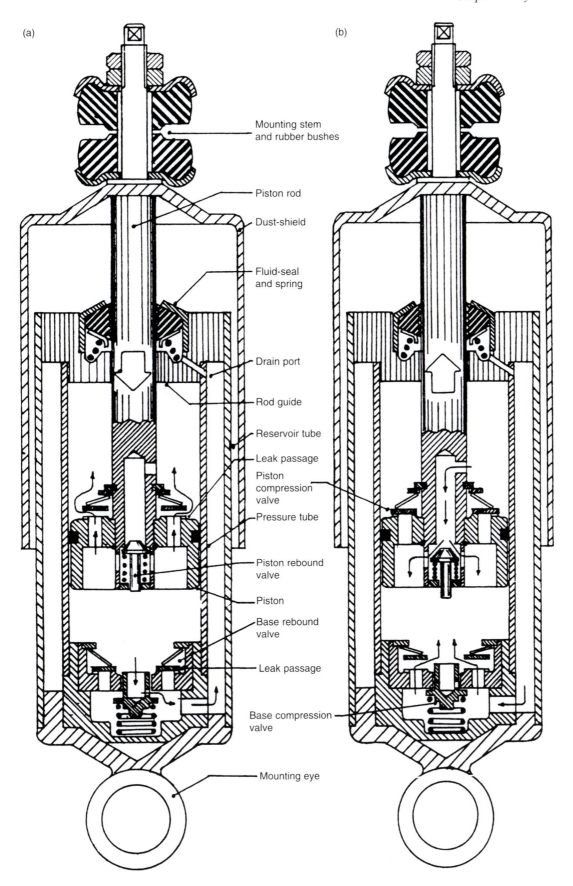

Mounting stem
and rubber bushes

Piston rod

Dust-shield

Fluid-seal
and spring

Drain port

Rod guide

Reservoir tube

Leak passage

Piston
compression
valve

Pressure tube

Piston rebound
valve

Piston

Base rebound
valve

Leak passage

Base compression
valve

Mounting eye

Fig. 20.17 Telescopic shock absorber. (a) Bump or compression stroke. (b) Rebound or extension stroke

- **Bushes**. These are usually made from rubber to reduce vibration, noise, and the need for lubrication.

Maintenance checks

General rules

There are a number of general rules that should be observed when carrying out maintenance checks and adjustments; there are also methods of protecting the system against accidental damage during repair operations on the suspension system. These are listed below:

- Preliminary vehicle checks — see the vehicle report sheet.
- Safety — the proper use of lifting, supporting, and choking equipment.
- Safe use of special tools such as a coil spring compressor, high-pressure lubricants, and high-pressure hydraulic equipment.
- Care when working with and disposing of toxic and corrosive fluids.
- Checking of the suspension thoroughly before road testing.

Checking suspension alignment and geometry

To check the alignment of the suspension a four-wheel aligner is used. The gauges are located on each rear wheel and the light on each front wheel; the light shines onto the gauge down each side of the vehicle. This will show any misalignment between the front and rear wheels, which can then be identified, helping to prevent tyre scrub, steering pulling to one side, or crabbing of the vehicle down the road. The lights also shine across the front of the vehicle to check front wheel alignment. When all four gauges give the correct readings, the vehicle will drive down the road in a straight line.

Learning tasks

1. What are the symptoms of faulty dampers? How will the driver notice this problem? How should they be checked for correct operation?
2. Draw up a work schedule for stripping and rebuilding a MacPherson strut-type suspension. Name any special tools that should be used and identify any safety precautions that should be observed.
3. For each of the following suspension faults: describe the fault (what is wrong?); identify the symptom (how will the driver notice?); state the probable cause (what has caused the fault?); and give the preventative or corrective action that should be taken.

(a) Excessive uneven tyre wear
(b) Excessive component wear
(c) Premature failure of component
(d) Vibration and/or noise from suspension
(e) Uneven braking
(f) Steering pulling to one side
(g) Incorrect trim height
(h) Axle misalignment
(i) Excessive pitch or roll
(j) Vehicle instability over rough road surfaces
(k) Poor handling and ride quality
(l) Noisy suspension.

Practical assignment — suspension system

Introduction

At the end of this assignment you should be able to:

- carry out a visual inspection
- remove and replace a
  - MacPherson strut
  - coil spring
  - shock absorber, checking for correct operation
  - suspension bush
- check alignment of suspension
- make a report together with recommendations.

Tools and equipment

- A vehicle fitted with suitable suspension
- A vehicle lift or jack and axle stands
- Workshop manual
- Selection of tools to include specialist equipment such as a spring compressor.

Objective

- To check operation of damper
- To replace suspension bushes
- To check 'free height' of coil spring
- To investigate oil leaks
- To replace worn/damaged components.

Activity

1. After suitably raising and supporting the vehicle, remove the wheels (*do not* support the vehicle under the suspension to be removed).
2. Observe and note the type of suspension, e.g. coil spring, leaf spring, torsion bar, rubber, etc.

3. Make a simple sketch of the layout.
4. Before cleaning around the mounting points look for signs of:
   (a) insecure or loose components
   (b) places where dirt may be rubbed off by something catching
   (c) bright or rusty streak marks where body or chassis may be cracked
   (d) rust where thickness of material may be reduced to a failure level
   (e) excessively worn components
   (f) accident damage to body or components.
5. Remove suspension and dismantle where necessary.
6. Check operation of damper.
7. Reassemble suspension (replacing any worn/broken components) according to manufacturer's manual.
8. Check for correct assembly and tightness of all mounting bolts before fitting wheel.
9. Complete an inspection report and recommendations for the customer.

Checklist

Vehicle

Removal and refitting

MacPherson strut
Coil spring
Leaf spring
Shock absorber
Rubber suspension bush

Dismantle and reassemble

MacPherson strut
Shock absorber

Checking and adjusting

Castor angle
Camber angle
King pin inclination
Toe-out on turns
Wheel alignment
Axle alignment
Body alignment
Ride height

Investigating and reporting

Rubber
Hydrogas
Hydrolastic
Hydropneumatic
Coil spring
Leaf spring
Air

Student's signature

Supervisor's signature

Self-assessment questions

1. The rear suspension system shown in Fig. 20.13:
   (a) is a trailing arm type
   (b) is a torque tube transaxle type
   (c) has no unsprung weight
   (d) is a dead axle.
2. A principal task of the suspension dampers:
   (a) is to dampen out spring vibrations
   (b) is to allow the use of weaker springs
   (c) is to keep the vehicle upright on corners
   (d) is to eliminate axle tramp under braking.
3. Coil spring suspension systems:
   (a) are lighter than leaf spring suspension systems
   (b) require the use of radius arms to transmit driving and braking forces from the wheels to the vehicle frame
   (c) do not require the use of suspension dampers
   (d) are only suitable for small vehicles.
4. With independent front suspension:
   (a) the engine can be set lower in the frame because there is no axle beam
   (b) vertical movement of a wheel on one side of the vehicle will cause the wheel on the opposite side to move in a similar way
   (c) double-acting dampers are essential
   (d) tyre wear is lower than it is on beam axle systems.
5. In leaf spring-type suspension systems:
   (a) extra tie rods are needed in order to transmit driving and braking forces from the wheels to the vehicle body
   (b) both ends of the spring are rigidly fixed to the chassis
   (c) driving and braking forces are transmitted to the chassis through the spring leaves
   (d) extra strong dampers are required to cope with the greater tendency for this type of suspension to vibrate.
6. Figure 20.7 shows a Panhard rod fitted to a coil spring suspension. The purpose of this rod is to:
   (a) limit vertical movement of the vehicle body relative to the road wheels
   (b) transmit cornering forces between the vehicle body and the axle
   (c) absorb braking torque reaction
   (d) secure the brake caliper to the vehicle frame.

7. In a hydropneumatic suspension unit:
   (a) ride height is fixed at a constant level
   (b) ride height is varied by introducing or removing hydraulic fluid from the chamber below the gas diaphragm
   (c) gas is pumped in and out of the chamber above the gas diaphragm
   (d) the equivalent spring rate is constant.

8. An active suspension system:
   (a) uses an ECU to adjust suspension characteristics to suit driving conditions
   (b) is only suitable for use on racing cars
   (c) has fewer working parts than a non-active-type suspension system
   (d) can only work on torsion bar-type suspension.

21
Braking systems

Topics covered in this chapter

Drum and disc brakes
Hydraulic principles and their application to vehicle braking systems
Master cylinders
Wheel cylinders
Split braking systems
Vacuum servos
Brake testing and braking efficiency
The hand brake (parking brake)
Brake fluid
Brake pipes and hoses
Brake adjustment
Bleeding the brakes
Brake lights
Pressure control valves
Anti-lock braking systems – sensors and actuators

Vehicle braking systems

The purpose of the braking system is to slow down or stop the vehicle and, when the vehicle is stationary, to hold the vehicle in the chosen position. When a vehicle is moving it contains energy of motion (kinetic energy) and the function of the braking system is to convert this kinetic energy into heat energy. It does so through the friction at the brake linings and the brake drum, or the brake pads and the disc.

Some large vehicles are fitted with secondary braking systems that are known as retarders. Examples of retarders are exhaust brakes and electric brakes. In all cases, the factor that ultimately determines how much braking can be applied is the grip of the tyres on the driving surface.

Types of brakes

Two basic types of friction brakes are in common use on vehicles:

1. The drum brake
2. The disc brake.

The drum brake

Figure 21.1 shows a drum brake as used on a large vehicle. This cut-away view shows that the linings on the shoes are pressed into contact with the inside of the drum by the action of the cam. In this case the cam is partially rotated by the action of a compressed air cylinder. The road wheel is attached to the brake drum by means of the wheel studs and nuts.

A brake of this type has a leading shoe and a trailing shoe. The leading shoe is the one whose leading edge comes into contact with the drum first, in the direction of rotation. A leading shoe is more powerful than a trailing shoe and this shows up in the wear pattern, because a leading shoe generally wears more than a trailing shoe owing to the extra work that it does.

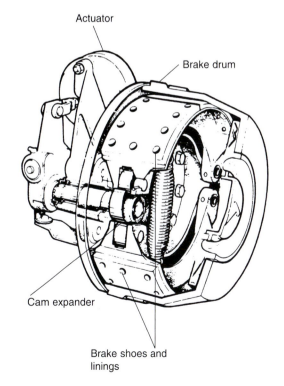

Fig. 21.1   A cam-operated drum brake

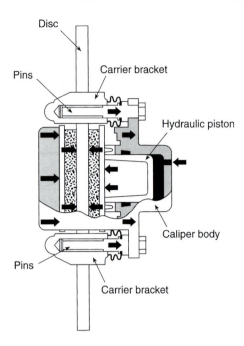

Fig. 21.2 A disc brake

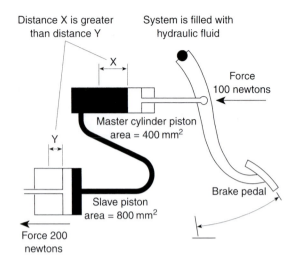

Fig. 21.3 The principle of a hydraulic braking system. The hydraulic pressure is equal at all parts of the system

The disc brake

Figure 21.2 shows the principle of the disc brake. The road wheel is attached to the disc and the slowing down or stopping action is achieved by the clamping action of the brake pads on the disc.

In this brake the disc is gripped by the two friction pads. When hydraulic pressure is applied to the hydraulic cylinder in the caliper body, the pressure acts on the piston and pushes the brake pad into contact with the disc. This creates a reaction force that causes the pins to slide in the carrier bracket and this action pulls the other pad into contact with the disc so that the disc is tightly clamped by both pads.

Hydraulic operation of brakes

The main braking systems on cars and most light commercial vehicles are operated by hydraulic systems. At the heart of a hydraulic braking system is the master cylinder, as this is where the pressure that operates the brakes is generated.

Principle of the hydraulic system

The small-diameter master cylinder is connected to the large-diameter actuating cylinder by a strong metal pipe. The cylinders and the pipe are filled with hydraulic fluid. When a force is applied to the master cylinder piston a pressure is created that is the same at all parts of the interior of the system. Because pressure is the amount of force acting on each square millimetre of surface, the force exerted on the larger piston will be greater than the force

applied to the small piston. In the example shown in Fig. 21.3 the force of 100 newtons on an area of 400 square millimetres of the master cylinder piston creates a pressure of 0.25 newtons per square millimetre. The piston of the actuating cylinder has an area of 800 square millimetres and this gives a force of 200 newtons at this cylinder.

The master cylinder

The part of the hydraulic braking system where the hydraulic operating pressure is generated is the master cylinder. Force is applied to the master cylinder piston by the action of the driver's foot on the brake pedal.

In the example shown in Fig. 21.4 the action is as follows. When force is applied to the push rod the piston moves along the bore of the master cylinder to take up slack. As soon as the lip of the main rubber seal has covered the bypass hole, the fluid in the cylinder, and the system to which it is connected, is pressurized.

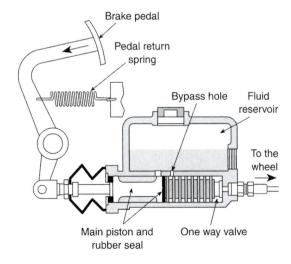

Fig. 21.4 A simple type of hydraulic master cylinder

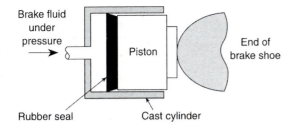

Fig. 21.5 A single-acting hydraulic wheel cylinder

When the force on the brake pedal and the master cylinder push rod is released, the return spring pushes the piston back and the hydraulic operating pressure is removed. The action of the main piston seal ensures that the master cylinder remains filled with fluid.

Wheel cylinders

The hydraulic cylinders that push the drum brake shoes apart, or apply the clamping force in the disc brake, are the wheel cylinders. There are two principal types of wheel cylinders, a single-acting cylinder and a double-acting cylinder.

The single-acting wheel cylinder

In the single acting wheel cylinder shown in Fig. 21.5, the space in the wheel cylinder, behind the piston and seal, is filled with brake fluid. Pressure from the master cylinder is applied to the wheel cylinders through pipes. Increased fluid pressure pushes the piston out and this force is applied to the brake shoe or brake pad.

Double-acting wheel cylinder

Figure 21.6 shows that the double-acting wheel cylinder has two pistons and rubber seals. Hydraulic pressure applied between the pistons pushes them apart. The pistons then act on the brake shoes and moves the linings into contact with the inside of the brake drum.

The handbrake

The handbrake (parking brake) is required to hold the vehicle in any chosen position when the vehicle is stationary. In addition to its function as a parking brake, the handbrake is also used when making hill starts and similar manoeuvres. The handbrake also serves as an emergency brake in the event of failure of the main braking system. Figure 21.7 shows the layout and main features of a handbrake for a car or light van. The vehicle has trailing arm rear suspension and the four swivel sector pivots are needed to guide the cable on these suspension arms. The purpose of the compensator is to ensure that equal braking force is applied to each side of the vehicle. The handbrake

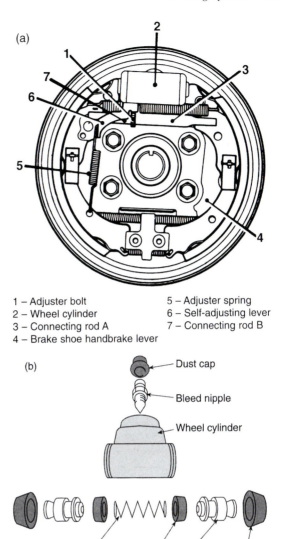

1 – Adjuster bolt
2 – Wheel cylinder
3 – Connecting rod A
4 – Brake shoe handbrake lever
5 – Adjuster spring
6 – Self-adjusting lever
7 – Connecting rod B

Fig. 21.6 A double-acting hydraulic wheel cylinder. (a) Rover 200 series drum brake. (b) A double-acting wheel cylinder

normally operates through the brakes at the rear of the vehicle.

With normal use, the brake linings will wear and it is also possible that the handbrake cable may stretch a little. In order to keep the handbrake working properly it is provided with an adjustment, as shown in Fig. 21.8.

The nut (2) is the adjusting nut and the nut (1) is the lock nut. In order to adjust the handbrake, the rear of the vehicle must be lifted so that the wheels are clear of the ground. The normal safety precautions must be observed and the wheels should be checked for freedom of rotation after the cable has been adjusted.

Split braking systems

The split braking circuit provides for emergencies such as a leak in one area of the braking system. There are

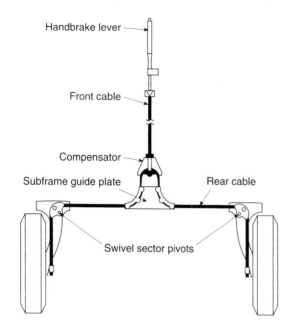

Fig. 21.7 A handbrake system

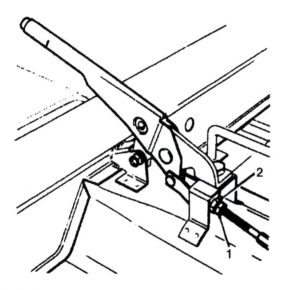

Fig. 21.8 Handbrake cable adjustment

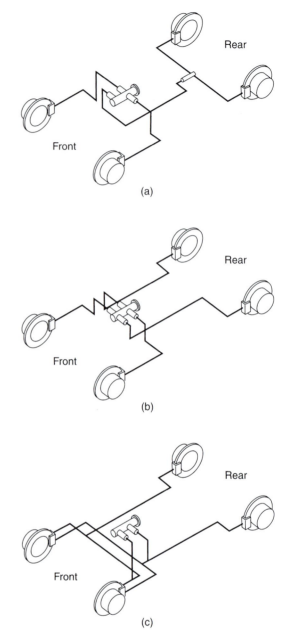

Fig. 21.9 (a) Front/Rear split of braking system. (b) Diagonal split of braking system. (c) Front axle and one rear wheel split

various methods of providing split braking circuits and three of these are shown in Fig. 21.9.

Front/rear split

Figure 21.9(a) shows how the front brakes are operated by one part of the master cylinder and the rear brakes by the other part. This system is used for rear-wheel-drive cars.

Diagonal split

One part of the tandem master cylinder is connected to the offside front and the nearside rear brakes and the other part to the nearside front and the offside rear, as shown in Fig. 21.9(b).

Front axle and one rear wheel split

This is known as the 'L' split. One part of the master cylinder is connected to the front brakes and the offside rear wheel brake and the other part to the front brakes and the nearside rear wheel brake. This arrangement requires four piston calipers for the front disc brakes or two pistons, if they are floating calipers.

The tandem master cylinder

In each split system a tandem master cylinder is used. Figure 21.10 shows a tandem master cylinder.

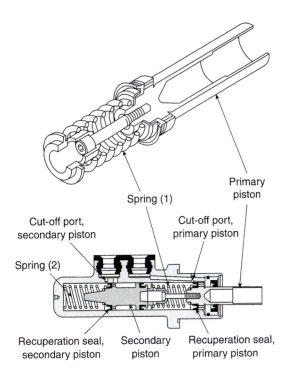

Fig. 21.10 A tandem master cylinder

The master cylinder is designed to ensure that the brakes are applied evenly. There are two pistons, a primary piston and a secondary piston. The spring (1) is part of the primary piston and it is stronger than the spring (2) that is fitted to the secondary piston. Application of force to the piston causes the spring (1) to apply force to the secondary piston so that both pistons move along the cylinder together. The two piston recuperation seals cover their respective cut-off ports at the same time and this ensures simultaneous build-up of pressure in the primary and secondary circuits.

When the brake pedal force is released, the primary and secondary pistons are pushed back by the secondary spring and both cut-off ports are opened at the same time, thus releasing the brakes simultaneously.

In the event of leakage from one part of the master cylinder, the other cylinder remains operative. This is achieved through the design of the stops and other features of the master cylinder.

The brake fluid reservoir

The fluid reservoir is usually mounted directly on to the master cylinder. Figure 21.11 shows a fluid reservoir equipped with a fluid level indicator. The fluid reservoir is normally made from translucent plastic and, provided it is kept clean, it is possible to view the fluid level by under-bonnet inspection. The reservoir is fitted with an internal dividing panel, ensuring that one section remains operative should the other section develop

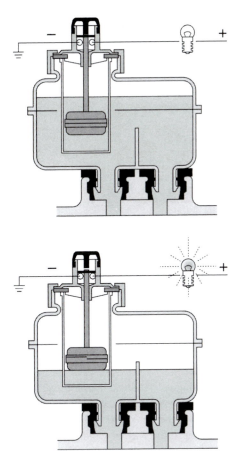

Fig. 21.11 The fluid level indicator

a loss of fluid. The fluid level indicator operates a warning light. Should the fluid level drop below the required level, the switch contacts will close and the warning light on the dash panel will be illuminated.

Brake pipes and hoses

The hydraulic pressure created at the master cylinder is conveyed to the wheel brakes through strong metal pipes, where they can be clipped firmly to the vehicle, and through flexible hoses where there is relative movement between the parts, for example axles and steered wheels and the vehicle frame.

The brake servo

A brake servo of the type shown in Fig. 21.12 is used as a means of increasing the force that the driver applies to the brake pedal. The servo is the device that allows the driver to apply a large braking force by the application of relatively light force from the foot. The amount of increased force that is produced by the servo is dependent on the driver's effort that is applied to the brake pedal. This ensures that braking effort is proportional to the force applied to the brake pedal.

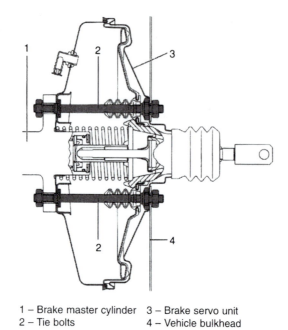

1 – Brake master cylinder   3 – Brake servo unit
2 – Tie bolts   4 – Vehicle bulkhead

Fig. 21.12 A vacuum servo

On petrol-engined vehicles, manifold vacuum is used to provide the boost that the servo generates. On diesel-engined vehicles there is often no appreciable manifold vacuum, owing to the way in which the engine is governed. In these cases, the engine is equipped with a vacuum pump that is known as an exhauster.

The master cylinder (1) is firmly bolted to the servo unit on one side and, on the other side, the servo unit is firmly bolted to the vehicle bulkhead (4). The brake pedal effort is applied to the servo input shaft and this in turn pushes directly on the master cylinder input piston. Figure 21.13 shows the servo in more detail.

At this stage you should concentrate on the flexible diaphragm, the sealed container, the two chambers A and B, and the vacuum connection. The valve body and the control piston are designed so that when the brakes are off, the manifold vacuum will draw out air and create a partial vacuum in chambers A and B on both sides of the diaphragm.

When force is applied to the brake pedal, the control piston closes the vacuum port and effectively shuts off chamber B from chamber A. At this stage, the control piston moves away from the control valve and atmospheric pressure air is admitted into chamber B. The greater pressure in chamber B, compared with the partial vacuum in chamber A, creates a force that adds to the pedal effort applied by the driver. The servo output rod pushes directly on to the master cylinder piston, as shown in Fig. 21.14, so that there is no lost motion and the resulting force applied to the master cylinder is proportional to the effort that the driver applies to the brake pedal.

It is common practice to fit a non-return valve in the flexible pipe, between the manifold and the servo unit,

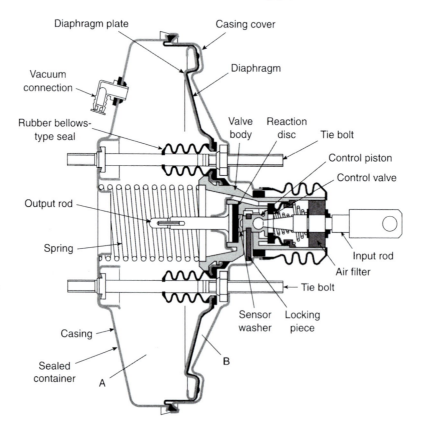

Fig. 21.13 Details of a vacuum servo

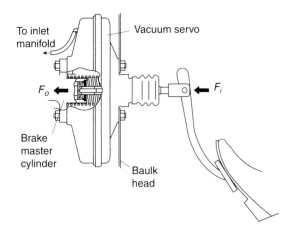

Fig. 21.14 Vacuum servo and master cylinder

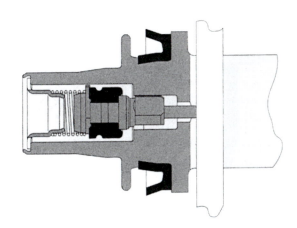

Fig. 21.15 The non-return valve in the vacuum pipe

as shown in Fig. 21.15. This valve serves to retain vacuum in the servo after the engine is stopped and also prevents petrol engine vapours and 'backfire' gases from entering the servo.

Brake adjustment

During normal usage, the brake friction surfaces wear. In disc brakes this means that the actuating pistons push the pads closer to the disc and the light rubbing contact between friction material (pads) and the discs is maintained; disc brakes are thus 'self-adjusting'. There is some displacement of brake fluid from the reservoir to the wheel cylinders and this will be noticed by the lowering of fluid level in the reservoir. When new brake pads are fitted there may be an excess of fluid in the system, if the fluid was topped up when the brakes were in a worn condition.

When the friction linings on brake shoes wear, the gap between the lining and the inside of the brake drum increases. To compensate for this wear, drum brakes are provided with adjusters. Adjusters take two forms: (1) manual adjusters and (2) automatic adjusters.

Manual brake adjuster

Figure 21.16 shows a type of manual brake adjuster that has been in use for many years. The threaded portion is provided with a slotted part that permits it to be rotated by means of a lever. The adjuster is accessed through a hole in the brake drum and the lever, probably a screwdriver, is applied through this hole. The two ends of the adjuster are located on the brake shoes and 'screwing out' the threaded portion pushes the shoes apart until the correct lining to drum clearance is obtained.

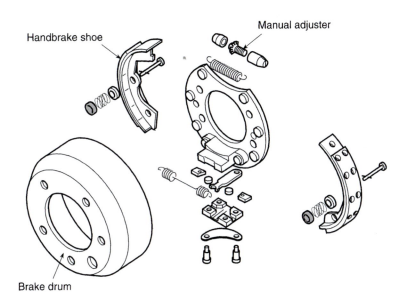

Fig. 21.16 A manual brake adjuster

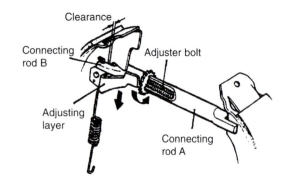

Fig. 21.17   An automatic brake adjuster

Automatic brake adjuster

This adjustment mechanism relies on the movement of the operating mechanism, to operate a ratchet and pawl. The mechanism, together with part of the two brake shoes, is shown in Fig. 21.17.

The ratchet is a small-toothed wheel that is fixed to the adjusting bolt. The pawl is on the end of the adjusting lever and it engages with the ratchet. The connecting rod is in two parts, A and B, and fits between the two brake shoes. The pawl is pulled lightly into contact with the ratchet by the spring. Operation of the brake shoes pushes the shoes apart and this creates a clearance at the end of the connecting rod. When there is sufficient clearance, the ratchet will rotate by one notch, which increases the length of the connecting rod and takes up the excess clearance between the shoes and the drum.

These mechanisms work well when they are properly maintained and service schedules must be properly observed to ensure that the brakes are maintained in good order.

Wear indicators

The purpose of the friction pad wear indicator is to alert the driver to the fact that the pads have worn thin. The warning light on the dash panel is illuminated when the pads have worn by a certain amount. Figure 21.18 shows a pair of brake pads. One of the pads is equipped with a pair of wires whose ends are embedded in the friction material of the pad. When the pad wears down to the level of the ends of the wires, the wires are bridged electrically by the metal of the brake disc. This completes a circuit and illuminates the warning light.

Stop lamp switch

The purpose of the stop lamp switch is to alert following road users to the fact that the driver of the vehicle in front of them is applying the brakes. Figure 21.19 shows an arrangement that is frequently used to operate the

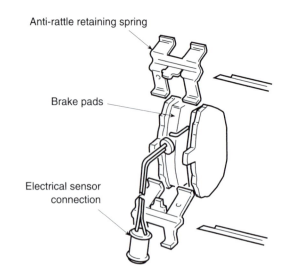

Fig. 21.18   Pad wear indicator light

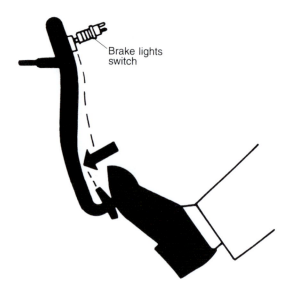

Fig. 21.19   A brake-light switch

brake-light switch. As the foot is applied to the pedal, a spring inside the switch closes the switch contacts to switch on the brake lights.

Brake pressure control valve

A pressure control valve is fitted between the front and rear brakes to prevent the rear wheels from 'locking up' before the front wheels. This arrangement contributes to safer braking in emergency stops. Figure 21.20 shows how such a valve is mounted on a vehicle.

The internal details of the valve are shown in Fig. 21.21.

'Normal' braking fluid under pressure passes through the valve, from port B to port E and the rear brakes. If the deceleration of the vehicle reaches the critical level, the ball (D) will move and seal off the fluid path

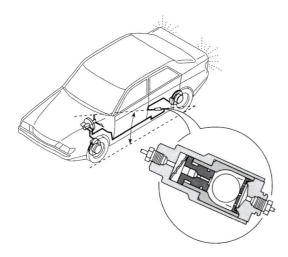

Fig. 21.20 The pressure-limiting valve on the vehicle

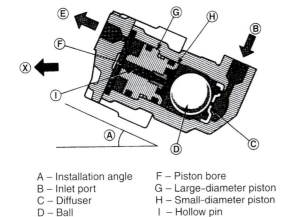

A – Installation angle   F – Piston bore
B – Inlet port   G – Large-diameter piston
C – Diffuser   H – Small-diameter piston
D – Ball   I – Hollow pin
E – Outlet port

Fig. 21.21 Internal detail (cross-sectional view) of inertia-type limiting brake pressure control valve — diagram simplified to show schematic flow of fluid

to the bore of the valve (F). At this point the rear brakes are effectively 'cut off' from the front brakes and the pressure on the rear brakes is held at its original level. Further pressure on the brake pedal increases the front brake pressure without increasing the rear brake pressure. Dependent on the design of the valve, further brake pedal pressure will cause the piston (G—H) to move and increase the pressure in the rear brake line.

When the deceleration falls, the ball will 'roll' back to its seat on the diffuser (C) and the pressure throughout the braking system is stabilized.

Brake fluid

Brake fluid must have a boiling temperature of not less than 190°C, and a freezing temperature not higher than −40°C. Brake fluid is hygroscopic, which means that it absorbs water from the atmosphere. Water in brake fluid affects its boiling and freezing temperatures,

which is one of the reasons why brake fluid needs to be changed at the recommended intervals.

Brake fluid is normally based on vegetable oil and its composition is carefully controlled to ensure that it is compatible with the rubber seals and not corrosive to the metal parts. Some manufacturers use mineral oil as a base for brake fluids and their systems are designed to work with this fluid. It is important always to use only the type of fluid that a vehicle manufacturer recommends for use in their vehicles.

Braking efficiency

The concept of braking efficiency is based on the 'idea' that the maximum retardation (rate of slowing down) that can be obtained from a vehicle braking system is gravitational acceleration, $g = 9.8$ m/s^2. The actual retardation obtained from a vehicle is expressed as a percentage of 'g' and this is the braking efficiency. For example, suppose that a vehicle braking system produces a retardation of 7 m/s^2. The braking efficiency $= (7/9.8) \times 100 = 71\%$. Because of the physics of vehicle dynamics the braking efficiency can be obtained without actually measuring deceleration. This is so because the total weight of the vehicle divided by the sum of the braking forces applied between the tyres and the driving surface produces the same result. The roller brake testers of the type shown in Fig. 21.22 that are used in garages make use of this principle to measure braking efficiency.

The left ($F1$) and right ($F2$) front braking forces are measured and recorded. The vehicle is moved forward so that the rear wheels are on the rollers. The test procedure is repeated and the two rear wheel braking forces ($F3$ and $F4$) are recorded. The four forces are added together and divided by the weight of the vehicle.

For example, in a certain brake test on a vehicle weighing 1700 kg, the four braking forces add up to 1070 kg. This gives a braking efficiency of $(1070/1700) \times 100 = 63\%$. Often the data relating to a particular vehicle is contained on a chart kept near the test bay.

The regulations regarding braking efficiency and the permitted differences in braking, from side to side of the vehicle, together with other data are contained in the tester's manual. Any technician carrying out brake tests must familiarize themselves with the current regulations.

Fig. 21.22 Roller-type brake tester measuring the braking force at each wheel

Anti-lock braking system (ABS)

The term ABS covers a range of electronically controlled systems that are designed to provide optimum braking in difficult conditions. ABS systems are used on many cars, commercial vehicles, and trailers.

The purpose of anti-skid braking systems is to provide safer vehicle handling in difficult conditions. If wheels are skidding it is not possible to steer the vehicle correctly and a tyre that is still rolling, not sliding, on the surface will provide a better braking performance. ABS does not usually operate under normal braking. It comes into play in poor road surface conditions, such as ice, snow, water, etc., or during emergency stops.

Figure 21.23 shows a simplified diagram of an ABS system, which gives an insight into the way that such systems operate.

The master cylinder (1) is operated via the brake pedal. During normal braking, manually developed hydraulic pressure operates the brakes and, should an ABS defect develop, the system reverts to normal pedal-operated braking.

The solenoid-operated shuttle valve (2) contains two valves, A and B. When the wheel sensor (5) signals the ABS computer (ECU) (7) that driving conditions require ABS control, a procedure is initiated which energizes the shuttle valve solenoid. Valve A blocks off the fluid inlet from the master cylinder and valve B opens to release brake line pressure at the wheel cylinder (6) into the reservoir (3) and the pump (4), where it is returned to the master cylinder.

In this simplified diagram, the shuttle valve is enlarged in relation to the other components. In practice, the movement of the shuttle valve is small and movements of the valve occur in fractions of a second.

In practical systems, the solenoids, pump and valves, etc. are incorporated into a single unit, as shown in Fig. 21.24. This unit is known as a modulator.

This brief overview shows that an anti-lock braking system has sensors, an actuator, an ECU, and interconnecting circuits. In order that the whole system functions correctly, each of the separate elements needs to be working correctly.

When deciding whether or not a vehicle wheel is skidding, or on the point of doing so, it is necessary to compare the rotational movement of the wheel and brake disc, or drum, with some part such as the brake back plate that is fixed to the vehicle. This task is performed by the wheel speed sensing system and the procedure for doing this is reasonably similar in all ABS systems, so the wheel speed sensor is a good point at which to delve a little deeper into the operating principles of ABS.

The wheel speed sensor

Figures 21.25 and 21.26 show a typical wheel sensor and reluctor ring installation. The sensor contains a coil and a permanent magnet. The reluctor ring has teeth and when the ring rotates past the sensor pick-up the lines of magnetic force in the sensor coil vary. This variation of magnetic force causes a varying voltage (emf) to be induced in the coil and it is this varying voltage that is used as the basic signal for the wheel sensor. The particular application is for a Toyota but its principle of operation is typical of most ABS wheel speed sensors.

The raw output voltage waveform from the sensor is approximately of the form shown in Fig. 21.27. It will be seen that the voltage and frequency increase as the wheel speed, relative to the brake back plate, increases. This property means that the sensor output is a good representation of the wheel behaviour relative to the back plate and, thus, to provide a signal that indicates whether or not the wheel is about to skid.

In most cases, this raw curved waveform is not used directly in the controlling process and it has first to be shaped to a rectangular waveform, and tidied up before being encoded for control purposes.

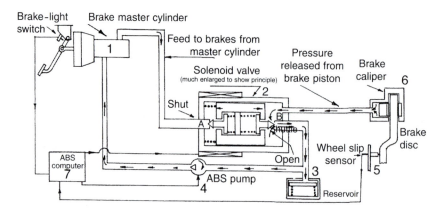

Fig. 21.23   A simplified version of an ABS

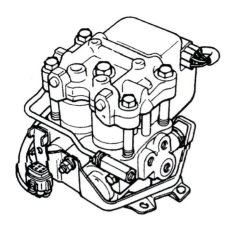

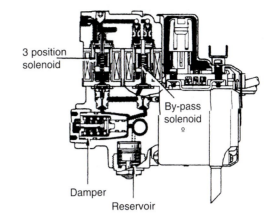

3 position
solenoid

By-pass
solenoid

Damper

Reservoir

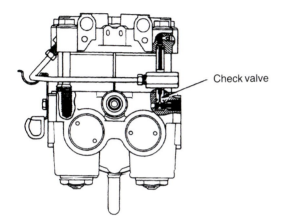

Check valve

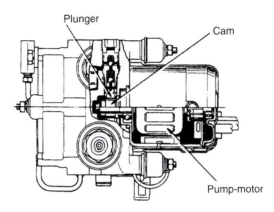

Plunger

Cam

Pump-motor

Fig. 21.24 ABS modulator

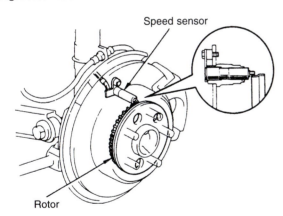

Speed sensor

Rotor

Fig. 21.25 ABS wheel sensor

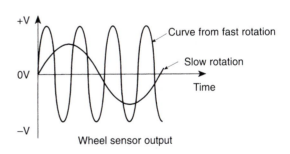

+V

0V

−V

Curve from fast rotation

Slow rotation

Time

Wheel sensor output

Fig. 21.27 The voltage waveform of the ABS sensor output

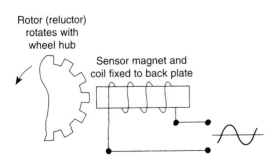

Rotor (reluctor)
rotates with
wheel hub

Sensor magnet and
coil fixed to back plate

Fig. 21.26 The basic principle of the ABS sensor

If the brake is applied and the reluctor (rotor) starts to decelerate rapidly, relative to the sensor pick-up, it is an indication that the wheel rotation is slowing down. If the road surface is dry and the tyre is gripping well, the retardation of the wheel will match that of the vehicle and normal braking will occur. However, if the road surface is slippery, a sudden braking application will cause the reluctor rotor and road wheel to decelerate at a greater rate than the vehicle, indicating that a skid is about to happen. This condition is interpreted by the electronic control unit. Comparisons are made with the signals from the other wheel sensors and the brake line pressure will be released automatically, for sufficient time (a fraction of a second) to prevent the wheel from locking.

In hydraulic brakes on cars the pressure release and re-application is achieved by solenoid valves, a pump and a hydraulic accumulator, and these are normally incorporated into one unit called the modulator. The frequency of 'pulsing' of the brakes is a few times per second, depending on conditions, and the pressure pulsations can normally be felt at the brake pedal.

With air brakes on heavy vehicles, the principle is much the same, except that the pressure is derived from the air braking system and the actuator is called a modulator. The valves that release the brakes during anti-lock operation are solenoid operated on the basis of ECU signals, and the wheel sensors operate on the same principle as those on cars.

As for the strategy that is deployed to determine when to initiate ABS operation, there appears to be some debate. Some systems operate what is known as 'select low', which means that brake release is initiated by the signal from the wheel with the least grip, irrespective of what the grip is at other wheels. An alternative strategy is to use individual wheel control. Whichever strategy is deployed, the aim is to provide better vehicle control in difficult driving conditions and it may be that the stopping distance is greater than it would be with expert manual braking.

The ABS warning light

ABS systems are equipped with a warning light. This lamp is illuminated when the system is not operating. When the vehicle is first started, the ABS warning light is illuminated and the system runs through a self-check procedure. As the vehicle moves away, the ABS warning lamp will remain 'on' until a speed of 3 mph (7 km/h) is reached. If the ABS system is functioning properly, the warning lamp will not 'come on' again until the vehicle stops. The system is constantly monitoring itself when the vehicle is in motion. Should a fault occur, the warning light will again come on. Should this happen, the system reverts to normal braking operation and the vehicle should receive urgent attention to ascertain the cause of the problem.

Bleeding the brakes

During repair work, such as replacing hoses and wheel cylinders, air will probably enter the hydraulic system. This air must be removed before the vehicle is returned for use and the process of removing the air is called 'bleeding the brakes'.

Practical assignment

1. With the aid of sketches, describe the equipment and procedures for bleeding brakes on the types of vehicle that you work on.
2. Braking system check.

This check should be completed in a logical manner and in the correct sequence. Make up a tick sheet to indicate pass/fail for each item. The following list should be used to make up the 'tick sheet'.

Checklist for braking system

Inside the vehicle

1. Check the operation of the brake pedal:
    (i) brake pedal travel
    (ii) pedal feel, e.g. soft or spongy
    (iii) pedal security
    (iv) operation of brake light switch
    (v) security of switch and cables.
2. Check operation of the handbrake:
    (i) distance travelled when applied (no more than 3 to 7 clicks)
    (ii) handbrake resistance (does it feel right?)
    (iii) security of handbrake lever
    (iv) operation of ratchet and release button
    (v) operation of warning lamp switch
    (vi) security of switch and cables.

Under the bonnet

3. Master cylinder:
    (i) check the fluid level in the reservoir
    (ii) check operation of fluid level warning device
    (iii) inspect for security of mounting bolts, brackets, etc.
    (iv) check operation of the servo unit
    (v) look for evidence of hydraulic fluid leaks.
4. Pipes and hoses:
    (i) inspect brake pipes for signs of corrosion
    (ii) check security of pipes (are the retaining clips in position?)
    (iii) look for evidence of leaks and check the security of connections and unions
    (iv) test flexible hoses for signs of ageing and damage
    (v) check the condition of the servo vacuum hose.
5. Front brakes:
    (i) check operation of calipers/brakes
    (ii) measure thickness of brake pads
    (iii) measure thickness of brake discs and check the condition of the disc surfaces
    (iv) movement of caliper on the slide (where fitted)
    (v) check for splits and cracks in flexible hoses

(vi) check for corrosion and damage in metal pipes and check the unions for security and tightness

(vii) check the security and operation of the pad wear indicator light and cables

(viii) check security and condition of anti-rattle devices

(ix) check for evidence of leaks

(x) check for movement and excessive free play in wheel bearings.

6. Rear brakes:

   (i) check condition of brake drums and look for evidence of scoring (ensure that dust removal is done with a vacuum cleaner whilst wearing a mask)

   (ii) measure thickness of brake shoe linings

   (iii) check operation of wheel cylinders (will require an assistant)

   (iv) check to see that all parts are in place and that brakes are correctly assembled

   (v) check for fluid leaks

   (vi) check security and condition of brake pipes and unions.

7. Under the car:

   (i) security of pipes and flexible hoses, check that all retaining clips are properly fitted

   (ii) look for any evidence of fluid leaks

   (iii) check the operation of handbrake cables, compensator, etc.

   (iv) lubrication of cables and clevis pins

   (v) security of handbrake linkages.

Self-assessment questions

1. Brake fluid is said to be hygroscopic because:
   (a) it is like water
   (b) it absorbs water
   (c) it is heavier than water
   (d) it freezes at a higher temperature than water.

2. Brake fluid absorbs water over a period of time. In order to prevent problems that may arise from this it is recommended that:
   (a) brakes are not allowed to overheat
   (b) brake shoes are changed at regular intervals
   (c) brake fluid is replaced at regular intervals
   (d) vehicles are kept under cover in bad weather.

3. A diagonal split braking system requires:
   (a) two master cylinders
   (b) a tandem master cylinder
   (c) two calipers on each front brake
   (d) disc brakes at the front and drum brakes at the rear.

4. In a test to determine braking efficiency a vehicle weighing 1200 kg is placed on a brake testing machine, the brake tester showing the following readings: front right 2120 N; front left 2080 N; rear right 1490 N; rear left 1510 N. The braking efficiency is:
   (a) 60%
   (b) 17%
   (c) 25%
   (d) 75%.

5. On diesel-engined vehicles the vacuum for the brake servo is provided by:
   (a) manifold vacuum
   (b) a compressor
   (c) a vacuum pump known as an exhauster
   (d) a venturi fitted in the engine air intake.

6. Anti-lock braking systems:
   (a) decrease stopping distance in all conditions
   (b) assist the driver to retain control during emergency stops
   (c) are not used on heavy vehicles
   (d) are a form of regenerative braking system.

7. Disc brakes are said to be self-adjusting because:
   (a) they use a ratchet and pawl mechanism
   (b) the pads are kept in light rubbing contact with the disc
   (c) the discs expand when heated
   (d) the brake servo compensates for wear.

22

Steering, wheels, and tyres

Topics covered in this chapter

Ackermann principle
Light vehicle steering layouts
Steering geometry
Castor
Camber kingpin inclination
Centre-point steering
Steering roll axis
Wheel alignment
Steering gearboxes
Rack and pinion
Ball-joints
Front-hub assemblies
Road wheel construction
Tyres
Tyre details, tread wear indicators
Tyre regulations
Wheel balance
Tyre wear patterns as a guide to steering misalignment

The steering mechanism has two main purposes. It must enable the driver to: easily maintain the straight-ahead direction of the vehicle even when bumps are encountered at high speeds; and to change the direction of the vehicle with the minimum amount of effort at the steering wheel. One of the simplest layouts is the **beam axle** arrangement (Fig. 22.1) as used on large commercial vehicles. This is where the hub pivots or swivels on a kingpin (in the case of a car a top and bottom ball-joint) to give the steering action.

As can be seen the two stub axles are connected together by two steering arms and a track rod with ball-joints at each end. The **steering gearbox** converts the rotary movement of the steering wheel into a straight-line movement of the steering linkage; it also makes it easier for the driver to steer by giving a **gear reduction**. The drop arm and drag link connect the steering gearbox to the stub axle. The steering arms, track rod, and ball-joints connect the stub axles together and allow the movement to be transferred from one side of the vehicle to the other, as well as providing for the movement of the linkage as the suspension operates.

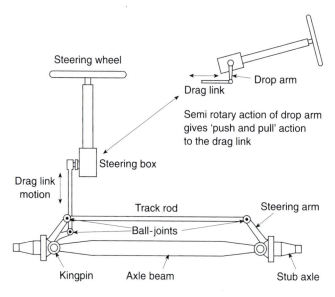

Fig. 22.1 Layout for beam axle steering system

Light vehicle steering layouts

To provide means of turning the front wheels of a vehicle left or right would not be too difficult were it not also necessary to make provision for their movement up and down with the suspension. Most modern cars now have a fully independent front suspension. This creates serious problems when one wheel moves upwards or downwards independent of the opposite wheel. If a single track rod were used the tracking would alter every time the wheels moved, causing the vehicle to wander from the straight-ahead position. This problem has been overcome by the use of two or three part track rods. As can be seen from Fig. 22.2 on a vehicle fitted with a rack and pinion type steering the centre track rod has been replaced by the rack.

Learning tasks

1. Inspect several different types of vehicles and make a simple line diagram of the layout of the steering system, identifying the following components: steering wheel and column, gearbox, linkage and ball-joints, stub axles and swivel joints.
2. Identify how the tracking is adjusted, check the manufacturer's setting for the tracking, and state the tolerance given (the difference between the upper and lower readings). What equipment is used to check the measurements and how should it be set up on the vehicle?

---

3. When undertaking an MOT the steering system should be checked. Look in the MOT tester's manual and list the areas that are subject to testing. What in particular is the tester looking for? Your list should include the following areas:

(a) Inside the vehicle:
- security of steering column mountings
- play in upper steering column bush
- amount of rotary movement in steering wheel before movement of the steered wheels
- amount of lift in steering column
- security of steering wheel to column
- any undue noise or stiffness when operating the steering from lock to lock.

(b) Under the vehicle:
- security of steering box to chassis
- signs of rust in the chassis around steering box mounting
- excessive wear in steering column universal joints (these may be fitted inside the vehicle)
- excessive play in steering box
- amount of play in inner and outer track rod ends
- splits or holes in rubber boots (steering rack arrangement)
- loss of oil from steering box
- play in suspension mountings and pivots that may affect the operation of the steering
- signs of uneven tread pattern wear that may indicate a fault in the steering mechanism.

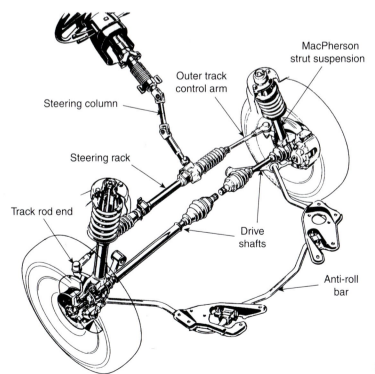

Fig. 22.2 Steering layout for independent front suspension (IFS) system

Steering geometry

The subject of steering geometry is a very complex one in its own right and because of this many manufacturers import the knowledge and skills of specialists. However, the basic principles are relatively simple to understand and apply to most vehicles. The development of the steering and suspension (to which it is very closely linked) is based on past experience; much of the work has been through an evolutionary process, learning from mistakes and modifying systems to suit varying applications. Many of the terms used are applied only to the steering and have no alternative; hopefully most of these are explained in the text.

The geometry of steering may best be understood by looking at Fig. 22.3. A swinging beam mounted on a turntable frame turns the wheels. This keeps all the wheels at right angles to the centre of the turn, which reduces tyre wear especially when turning a corner.

Rudolph Ackermann took out a patent in 1818 in England that is now widely used and is known as the

Ackermann layout. The angles of the front wheels about the turning point depend upon the wheel base (*W*) and the width of the track (*T*). In 1878 Jeantaud showed that the layout should conform to Fig. 22.4. In this arrangement the inner wheel (A) turns through a larger angle than the outer wheel (B) to give true rolling motion. The Ackermann layout does not fully achieve these conditions in all wheel positions; normally it is only accurate when the wheels are straight ahead and in one position on each left and right turn wheel setting. This system gives as near true rolling motion as possible together with simplicity.

Ackermann principle

The Ackermann layout is obtained by arranging for the stub axles to swivel on kingpins or ball-joints to give the steering action of the wheels. The track rod ball-joints are positioned on an imaginary line drawn between the kingpins and the centre line of the vehicle. When the track rod is positioned behind the swivel pins it is made shorter and has the protection of the axle, but the rod must be made stronger as it is in compression when the vehicle is being driven. This is shown in Fig. 22.5(a). When it is positioned in front of the swivel pins (Fig. 22.5(b)), it is made longer and can be out of the way of the engine, and is made thinner as it is in tension when the vehicle is being driven.

Figure 22.6 shows the arrangements for an independent front suspension where the top and bottom ball-joints are placed in a line that forms the kingpin inclination and also allows for the movement of the steered wheels both up and down over the irregularities of the road surface and as the driver turns the wheels to negotiate corners.

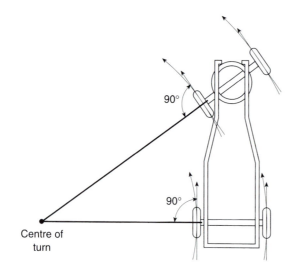

Fig. 22.3   Swing beam steering

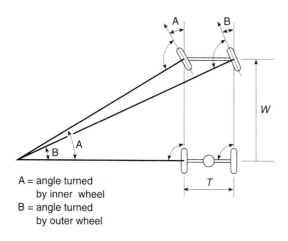

A = angle turned
     by inner wheel
B = angle turned
     by outer wheel

Fig. 22.4   Differing angles of front wheels about centre of turn

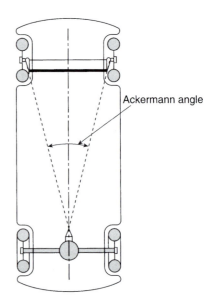

Fig. 22.5(a)   Track rod behind swivel pins still conform to Ackermann principle

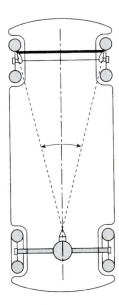

Fig. 22.5(b) Track rod in front of swivel pins still conforms to Ackermann principle

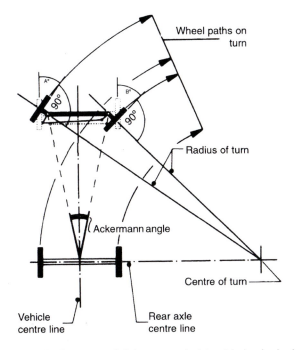

Fig. 22.6 Operation of Ackermann principle; right-hand wheel turns more than left-hand wheel

Centre-point steering

The stub axle arrangement shown in Fig. 22.7 has certain disadvantages due to the **offset** (X). These are:

- There is a large force generated, due to the resistance at the wheel (R) from the road surface trying to turn the steering about the swivel pin (F), especially when the brakes are applied.
- The forces generated produce large bending stresses in the stub axle and steering linkages.

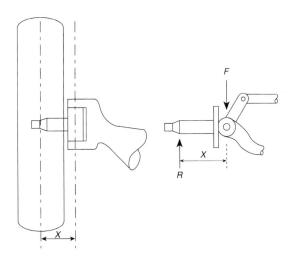

Fig. 22.7 Diagram shows the forces acting on the steering without centre point

- Heavy steering as the steered wheel has to rotate around the kingpin. When this offset is eliminated and the centre line of the wheel and the centre line of the kingpin coincide at the road surface then **centre-point steering** is produced. This condition is achieved by the use of **camber**, swivel axis inclination (often referred to as **kingpin inclination** or **KPI** for short) and **dishing** of the wheel.

Camber

Camber (Fig. 22.8) is the amount the wheel slopes in or out at the top relative to the imaginary vertical line when viewed from the front of the vehicle. This reduces the **bending** and **'splaying out' stresses** on the stub axle and steering linkages. The amount of camber angle is usually quite small as large angles

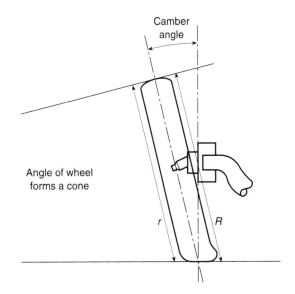

Fig. 22.8 Camber angle

will produce rapid tyre wear on the shoulder of the tyre tread as the inside tread of the tyre (R) travels a greater distance than the outside tread (r), even when travelling in a straight line. A cambered wheel tends to roll in the direction in which it is leaning, which has the effect of producing a side force. Two beneficial effects of this are that it reduces any small sideways forces imposed on the wheel by ridges in the road surface, and that it also produces a small lateral pre-load in the steering linkage. The actual angle varies depending on the suspension system used, but normally it is no more than 2°.

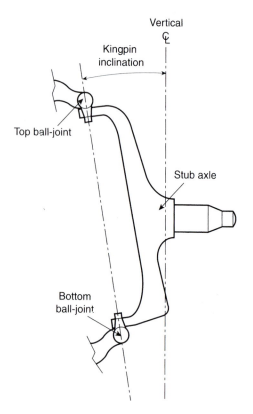

Fig. 22.9 Kingpin inclination (KPI), sometimes referred to as swivel axis

Kingpin or swivel axis inclination (KPI)

When the kingpin is tilted inwards at the top the resulting angle between the vertical line and the kingpin centre line is called the kingpin inclination (Fig. 22.9). Normally this will be between 5° and 15° to produce **positive offset** and to accommodate the brake, wheel bearings, and drive shaft joint. When the kingpin is set outwards at the bottom then **negative offset** is produced.

As the wheel is steered through an angle it will pivot around the kingpin. This will have the effect of lifting the front of the vehicle, helping to give a self-centring action to the steering.

Steering roll radius

If the steering wheel is turned, the front wheels move along an arc around (a) in Fig. 22.10; this is where the extension of the kingpin centre line meets the ground. The point (b) is where the tyre centre line meets the ground; radius (a–b) is the steering roll radius.

Whether the steering roll radius is positive or negative depends on the position of the KPI. It is positive if the extension of the KPI axis meets the ground on the inside of the **tyre contact centre.** It is negative if the extension of the KPI meets the ground outside the tyre contact centre. The advantage of having **negative steering roll radius** is increased directional stability in the case of uneven braking forces on the front wheels, or if a tyre is suddenly deflated.

As can be seen in Fig. 22.11, if the brake force on the right-hand front wheel is greater, then the vehicle tends to slide in an arc around that wheel, which means that the rear of the vehicle veers out to the left. In a vehicle with negative steering roll radius, the force of the car in motion will turn the wheel with the stronger braking force around the lower arm formed by the steering roll radius.

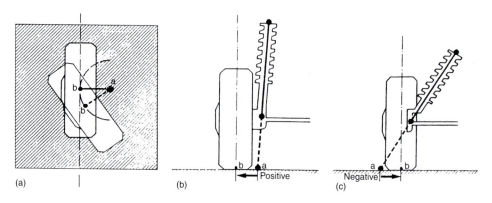

Fig. 22.10 Steering roll radius. (a) No steering roll radius. (b) Positive roll radius. (c) Negative roll radius

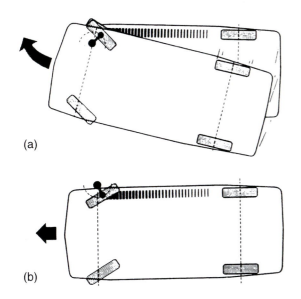

Fig. 22.11 (a) With positive steering roll and uneven braking on one front wheel. (b) With negative roll radius extra braking effort equals direction of steered wheels, giving greater stability

Castor

The action of the castor may best be understood by looking at the castor wheel on a shopping trolley. When the trolley is pushed forwards the castor wheels always follow behind the **pivot points** where they are attached to the trolley. The same action applies to the front wheels of the motor vehicle. When the pivot point is in front of the steered wheel the wheel will always follow behind.

This is called **positive castor** and is used on rear-wheel-drive vehicles; most front-wheel-drive and four-wheel-drive vehicles have **negative castor.**

This angle viewed from the side of the vehicle gives some 'driver feel' to the steering. It enables the steering to **self-centre** and a force must be exerted on the steering wheel to overcome the castor action of the steered wheels. It is produced by tilting the kingpin forwards at the bottom approximately 2–5° so that when the line of the swivel axis of the kingpin is extended it lies in front of the tyre contact point on the road surface; too much castor gives heavy steering and too little causes the vehicle to 'wander'. Castor angle steering geometry is shown in Fig. 22.12.

Tolerances may vary for all the steering angles so it is important to make reference to the manufacturer's specification as even small variations from this can lead to rapid tyre wear and poor handling characteristics. The checking of camber, kingpin inclination, and castor is done by using a set of gauges such as the Dunlop castor, camber, and kingpin inclination gauges together with a set of turn-tables. The vehicle is driven onto the turn-tables and the camber is measured by placing the camber gauge on the side of the wheel. The bubble in the level is adjusted to the central position to give the reading. The castor and KPI gauges are attached to the wheel and set approximately horizontal and the turn-tables are adjusted to zero. Steer the wheel to be checked 20° in (i.e. the right-hand wheel steered to the left and the left-hand wheel steered to the right). Set castor and KPI dials to zero at the pointer and centre the bubbles in the spirit levels by turning the lower knurled

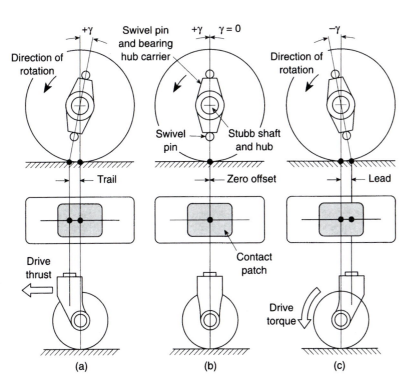

Fig. 22.12 Castor angle steering geometry (a) positive castor. (b) Zero castor. (c) Negative castor

screws. Steer the wheel to be checked 20° out and centre the bubbles in the spirit levels by turning the castor and KPI dials and take the readings off the dials.

Wheel alignment

Adjusting the track

Often referred to as toe-in/toe-out **or** tracking, wheel alignment is a plan view of the wheels and means 'are the wheels correctly aligned with each other?' When the vehicle is travelling in a straight line all the wheels must be parallel, especially the steered wheels.

Arrangement for changing or adjusting the track is made in the track rod or outer track control arms and, when correctly aligned for toe-in, the distance across the front of the steered wheels measures less than the distance across the rear (Fig. 22.13(a)). These measurements are taken using very accurate measuring gauges, of which several types are available (e.g. the **Dunlop optical alignment gauge**, Fig. 22.13(b)). The gauges are set to zero and are positioned on the wheels. The alignment is adjusted to give a reading on the scale, which is then checked against the manufacturer's specifications.

A more accurate method would be to use four **wheel alignment gauges**. This would give the mechanic more precise information as to which wheel needs adjusting, whether the front axle lines up with the rear axle and if the vehicle suspension has been misaligned in an accident.

Toe-out on turns

When turning a corner, the inner wheel turns through a greater angle than the outer wheel. This difference in angles is called 'toe-out on turns' (Fig. 22.14(a)). To check these angles the wheels are placed on turntables and the outer wheel is turned on its axis through 20°. The inner wheel should now read a larger angle (typically approximately 22°). This is because it is rolling around a smaller radius.

When checking steering alignment such as toe-in or toe-out, a number of factors must be attended to, for example:

1. The vehicle must be on a level surface.
2. The vehicle must be loaded to the maker's specification, such as may be the kerbside unladen weight figure.
3. The tyres must be inflated to the correct pressure.
4. The front wheels should be in the straight-ahead position.
5. The vehicle should be rocked from side to side to ease any stresses in the steering system.
6. Prior to checking the track, the vehicle should be rolled forward by about a vehicle's length in order to take up small amounts of slack in the linkage.

If the track requires adjustment (toe-in or toe-out) it is important to ensure that the track rods are maintained at the correct lengths, because this will affect the toe-out on turns. In the example shown in Fig. 22.14(b), the track rods are of equal length. This may not always apply and such details must be checked prior to making any adjustments.

Learning tasks

1. Draw a simple diagram of the steering layout to conform to the Ackermann principle. Show the position of the front and rear wheels, point of turn, approximate angles of the front wheels, and the position of the track rod. Indicate on your diagram where adjustments may be made.
2. Make a list of reasons why the steering mechanism would require adjusting.
3. Draw up a workshop schedule that you could use when checking a steering system for signs of wear and serviceability.
4. What important safety factors would you consider when working under a vehicle on the steering system.
5. Name two steering faults that would cause uneven tyre wear. How would these faults be put right?
6. If a customer came into the workshop complaining that his vehicle was 'pulling to one side' when being driven in a straight line, what would you suspect the fault to be? Describe the tests you would undertake to confirm your diagnosis.
7. Check the tracking on a vehicle using both optical and four-wheel alignment gauges. Adjust the steering to the middle of the limits set by the manufacturer.

Steering gearboxes

The steering gearbox is incorporated into the steering mechanism for two main reasons:

- to change the rotary movement of the steering wheel into the straight-line movement of the **drag link**;
- to provide a gear reduction and therefore a torque increase, thus reducing the effort required by the driver at the steering wheel.

Quite a number of different types of steering gearboxes have been used over the years, the most common ones being:

- worm and roller
- cam and peg
- recirculating ball (half nut)
- rack and pinion.

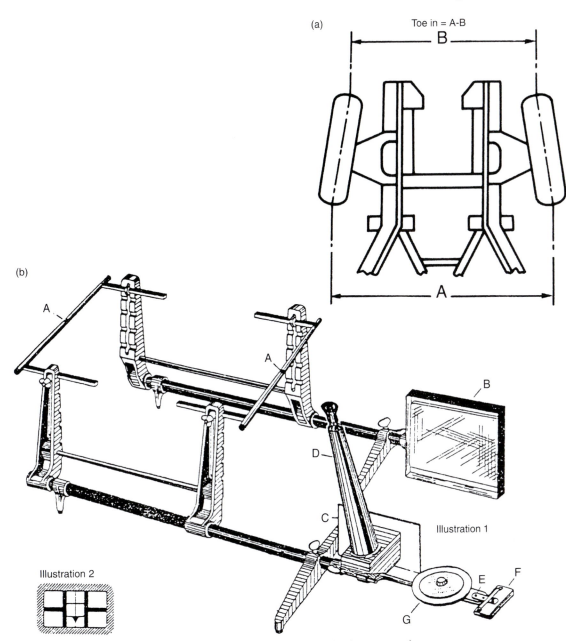

To assemble gauge

Assemble gauge as shown in illustration 1 with the periscope (D) fixed on the left-hand unit and the mirror (B) on the right-hand unit.

The contact bars may be fitted at any of five different height positions to suit the radius of the tyre and wheel assembly being checked. The height of all the bars must be the same and should be selected to bring the bars as near hub centre height as possible.

Each bar may be fitted into the support arms in either of two directions providing a range of width sufficient to cover all tyres on 9" to 24" diameter rims. The contact bars may be both inboard of the support arms, both outboard of the support arms, or one inboard and one outboard according to need.

To check accuracy of gauge

1. Stand the complete gauge on a level, clean floor with contact bars touching as shown in (A) illustration 1.

2. Adjust mirror and periscope until the reflection of target plate (C) is visible through periscope.

3. Sighting through periscope move pointer (E) until the image reflects the hair-line in the centre of the triangle between the vertical lines as in illustration 2.

The pointer should now be at zero on graduated scale (F). If not, slacken the two wing nuts holding the scale, adjust the scale to zero and retighten wing nuts. The gauge is now ready for use.

Fig. 22.13 (a) Toe-in. (b) Dunlop AGO/40 optical alignment gauge

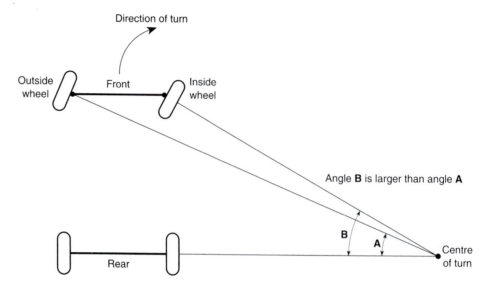

Fig. 22.14(a) Toe-out angle on turns. Inner wheel turns through a greater angle

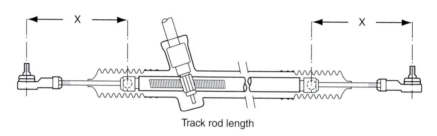

Fig. 22.14(b) Track rod length

The gearbox should have a degree of reversibility to provide some driver 'feel' and at the same time keep the transmission of road shocks through to the steering wheel to a minimum. The arrangement should be positive, i.e. it should have a minimum amount of backlash. Most steering gearboxes are designed with the following adjustments:

- end float of the steering column, usually by shims;
- end float of the rocker shaft, either by shims or adjusting screw and lock nut;
- backlash between the gears — these can be moved closer together, again by the use of shims or adjusting screw and lock nut.

It is essential to keep the 'backlash' to a minimum to provide a positive operation of the steering mechanism and to give directional stability to the path of the vehicle. The components inside the gearbox are lubricated by gear oil and the level plug serves as the topping up and level position for the oil.

Worm and roller

In this arrangement, shown in Fig. 22.15, the **worm** (in the shape of an hourglass) is formed on the inner steering column. Meshing with the worm is a **roller,** which is

attached to the **rocker shaft.** As the steering is operated the roller rotates in an arc about the rocker shaft, giving the minimum amount of backlash together with a large gear reduction. This means that for a large number of turns on the steering wheel there is a very small number of turns on the rocker shaft, with very little free play. Because of the specialized machining geometry and shape, the hourglass worm is really in the form of a cam rather than a gear. This is why the arrangement is sometimes known as a cam-and-roller steering gearbox. It will mainly be found on LGVs as it provides a large gear reduction and can transmit heavy loads.

Cam and peg

In the cam and peg steering box a **tapered** peg is used in place of the roller. This engages with a special **cam** formed on the end of the inner steering column. The peg may be made to rotate on needle roller bearings in the rocker arm to reduce friction as the steering column is rotated and the peg moves up and down the cam.

Recirculating ball

In this arrangement the worm is in the form of a thread machined on the inner steering column. A nut with **steel**

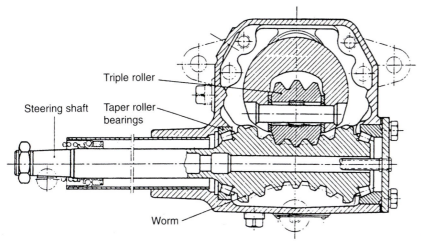

Fig. 22.15 Worm and roller steering gearbox

ball bearings acting as the thread operates inside the nut; as the worm rotates the balls reduce the friction to a minimum. In many cases a half nut is used and a transfer tube returns the balls back to the other side of the nut. A peg on the nut locates in the rocker arm, which transfers a rocking motion to the rocker shaft. In Fig. 22.16 a sector gear is used to transfer the movement to the rocker shaft.

Rack and pinion

This type, shown in Fig. 22.17, is now probably the most common type in use on cars and light vehicles. It has a rack that takes the place of the middle track rod and outer track rods (sometimes called tie rods), which connect to the steering arms at the hub. The pinion is mounted to the steering column by a universal joint as often the steering column is not in line with the input to the steering rack. This gives ease of mounting and operation of the gearbox.

On each end of the rack is a ball-joint to which the track rod is mounted; these are spring loaded to allow for movement together with the minimum of play in the joint. The system is arranged so that in the event of an accident the column, because it is out of alignment (not a solid straight shaft), will tend to bend at the joints. This helps to prevent the steering wheel from hitting the driver's chest and causing serious

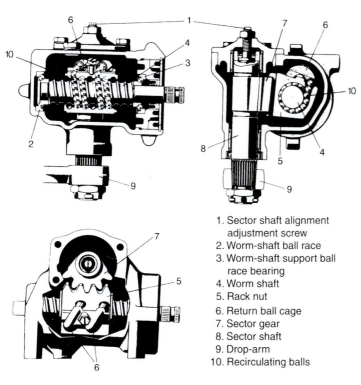

1. Sector shaft alignment adjustment screw
2. Worm-shaft ball race
3. Worm-shaft support ball race bearing
4. Worm shaft
5. Rack nut
6. Return ball cage
7. Sector gear
8. Sector shaft
9. Drop-arm
10. Recirculating balls

Fig. 22.16 Recirculating ball rack and sector steering box

injury. A spring-loaded rubbing pad (called a slipper, Fig. 22.18) presses on the underside of the rack to reduce backlash to a minimum and also to act as a damper absorbing road shocks that are passed back through the steering mechanism from the road surface.

Adjustment of the steering rack

The inner ball-joints of the outer track rod are adjusted to give the correct pre-load and the locking ring of the ball-joint housing is tightened. In some cases it is locked to the rack by locking pins, which must be drilled out to dismantle the joint as shown in Fig. 22.19.

Fig. 22.17 The complete layout of a rack and pinion steering system showing the steering column, with universal joints (UJs)

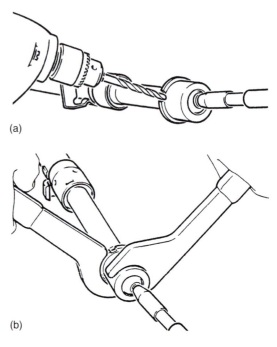

(a)

(b)

Fig. 22.19 (a) Drilling out tie rod inner ball-joint housing locking pins. (b) Unscrewing tie rod inner ball-joint housings

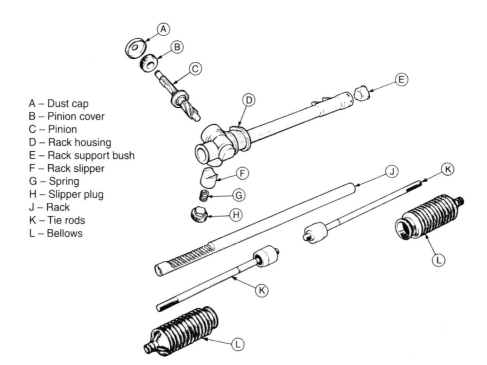

A – Dust cap
B – Pinion cover
C – Pinion
D – Rack housing
E – Rack support bush
F – Rack slipper
G – Spring
H – Slipper plug
J – Rack
K – Tie rods
L – Bellows

Fig. 22.18 Exploded view of steering gear

(a)

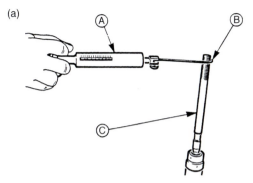

A – Piston pull side or spring balance
B – Wire hook 6mm (0.25") from end of tie rod
C – Tie rod

(b)

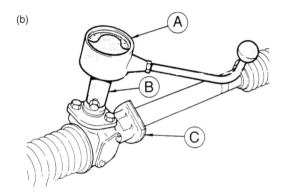

A – Torque gauge, tool number 15-041
B – Adaptor, tool number 13-008
C – Steering gear

Fig. 22.20 (a) Checking tie rod articulation. (b) Checking pinion turning torque

Shims are used to adjust the slipper/rubbing block to give the correct torque on rotating the pinion (Fig. 22.20). Shims are also used to adjust the bearings on the pinion. It is important that the correct data is obtained and used as these may vary from model to model.

Ball-joints

There are a number of requirements that must be fulfilled by steering ball-joints. These are:

- They must be free from excess movement (backlash) to give accurate control of the steering over the service life of the vehicle.
- They must be able to accommodate the angular movement of the suspension and the rotational movement of the steering levers.
- They must have some degree of damping on the steering to give better control of the wobble of the wheels, especially at low speeds.
- Where possible eliminate the need for lubrication at regular intervals – this is achieved through the use of good bearing materials and sealing the bearing with a good-quality grease at manufacture.

Ball-joints that are used on modern cars do not need lubricating as they are sealed for life, although on some medium to heavy vehicles they may require greasing at regular service intervals. Track rod and ball-joint assemblies are shown in Fig. 22.21.

Learning tasks

1. What is the difference between angular movement and rotational movement? Which component in the steering mechanism uses angular movement and which uses rotational movement?
2. Using workshop manuals identify the method for adjusting the types of steering gearboxes identified in this chapter.
3. Dismantle each of the different types of steering gearboxes. Fill in a job sheet for each type. Note any faults or defects and make recommendations for their serviceability.

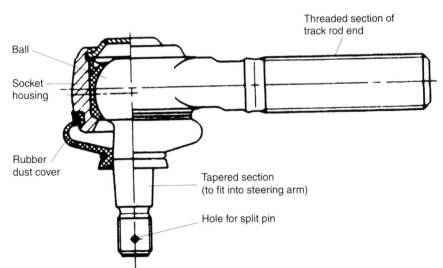

Fig. 22.21 Track rod end ball-joint assembly

Front-hub assemblies

Non-driving hubs

Figure 22.22 shows one type of front-hub assembly used on a front-engine rear-wheel-drive vehicle. As can be seen the hub rotates on a pair of bearings that are either taper roller or angular-contact ball bearings. The inner bearing is usually slightly bigger in diameter than the outer, as this carries a larger part of the load. Various methods of adjustment are used depending on the type of bearing and the load to be carried. Some manufacturers specify a torque setting for the hub nut (usually where a spacer or shims are positioned between the bearings); others provide a castellated nut that has to be tightened to a specified torque and then released before the split pin is located. The hubs are lubricated with a grease that has a high melting point; this is

because the heat from the brakes can be transferred to the hub, melt the grease, and cause it to leak out on to the brakes. A synthetic lip-type seal is fitted on the inside of the hub to prevent the grease escaping. It also prevents water, dust, and dirt from entering into the bearing. The arrangement in Fig. 22.23 is used mainly on larger vehicles, where a beam axle is fitted instead of independent suspension.

Driving hubs

Figure 22.24 shows one arrangement of hub and drive shaft location using thrust-type ball bearings, in this case a single bearing with a double row of ball bearings. When the nut is tensioned/tightened to the correct torque it also adjusts the bearing to the correct clearance.

On a front-wheel-drive vehicle the drive shaft passes through the hub to engage with the flange that drives the wheel. In this way the hub becomes a carrier for the bearings (normally taper roller, Fig. 22.25), which are mounted close together. Where angular-contact bearings are used they are of the double-row type, which have a split inner ring; in this way large-diameter balls can be used to give greater load carrying capacities. They also have the added advantage that they require no adjustment and the torque setting for the drive nut can be high to give good hub-to-shaft security.

Light vehicle wheels

Most cars use wheels with a well base rim of the types shown in Fig. 22.26. The standard type of rim has the form shown in Fig. 22.26(a). The rim has a slight taper and the internal pressure in the tyre forces the bead of

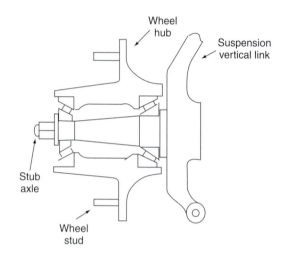

Fig. 22.22   Layout of non-driving hub assembly

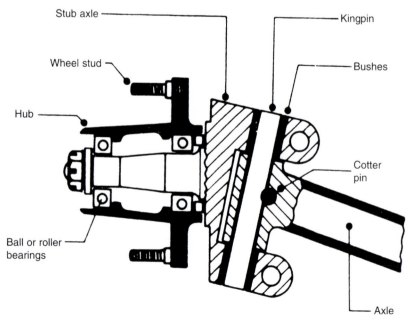

Fig. 22.23   Axle-beam and stub-axle assembly

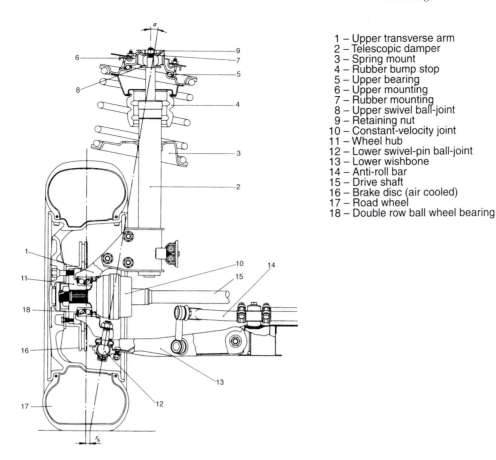

1 – Upper transverse arm
2 – Telescopic damper
3 – Spring mount
4 – Rubber bump stop
5 – Upper bearing
6 – Upper mounting
7 – Rubber mounting
8 – Upper swivel ball-joint
9 – Retaining nut
10 – Constant-velocity joint
11 – Wheel hub
12 – Lower swivel-pin ball-joint
13 – Lower wishbone
14 – Anti-roll bar
15 – Drive shaft
16 – Brake disc (air cooled)
17 – Road wheel
18 – Double row ball wheel bearing

Fig. 22.24 Independent suspension front-wheel-drive stub-shaft and swivel-pin carrier assembly

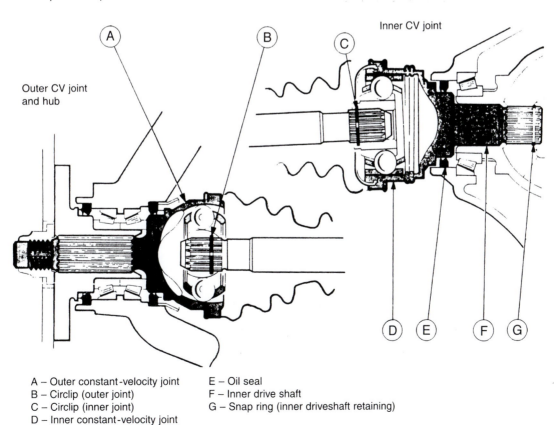

A – Outer constant-velocity joint
B – Circlip (outer joint)
C – Circlip (inner joint)
D – Inner constant-velocity joint
E – Oil seal
F – Inner drive shaft
G – Snap ring (inner driveshaft retaining)

Fig. 22.25 Layout of a driving hub arrangement

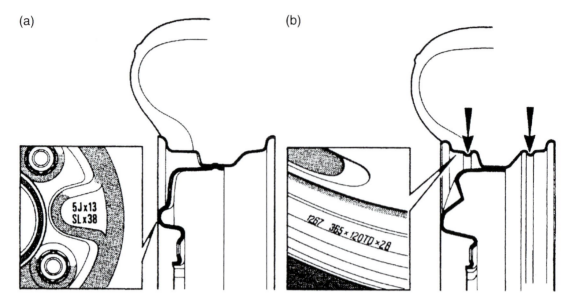

Fig. 22.26 Two types of wheel rim in common use. (a) Standard. (b) Double hump rim

the tyre into tight frictional contact with the wheel. This tight contact is required so as to maintain an airtight seal and also to provide the grip that will transmit driving and braking forces between the tyre and the wheel. The grip is dependent on the correct tyre pressure being maintained and, in the event of a puncture, the tyre bead is likely to break away from the wheel rim. The double hump rim shown in Fig. 22.26(b) is intended to provide a more secure fixing for the tyre in the sense that, should a puncture occur, the bead of the tyre will be held in place by the humps and this should provide a degree of control in the event of a sudden, rapid puncture.

Types of car wheels

Most car wheels are fabricated from **steel pressings** and their appearance is often altered by the addition of an **embellisher.** Figures 22.27 and 22.28 show a range of wheels and wheel trims as used on Ford Escort vehicles. When weight and possibly 'sporty' appearance is a consideration **cast alloy wheels** are used. The metals used are normally **aluminium** alloy, or **magnesium**-based alloy, and the wheels are constructed by casting. An alloy wheel is shown in Fig. 22.27(e).

Figures 22.27(f) and 22.27(g) shows the different types of wheel bolts and balance weights that are required. The steel wheel uses a bolt with a taper under the head. This taper screws into a corresponding taper on the wheel and, when the correct torque is applied, the wheel is secured to the vehicle hub. In order to prevent damage to the alloy wheel its bolts are provided with a tapered washer. The taper on the washer sits in the taper on the wheel and the bolt can be tightened without 'tearing' the alloy of the wheel. Figure 22.27 also shows the two different types of wheel balancing weights that are required. The obvious point to make here is that the steel wheel weight is not suitable for use on the alloy wheel, and vice versa.

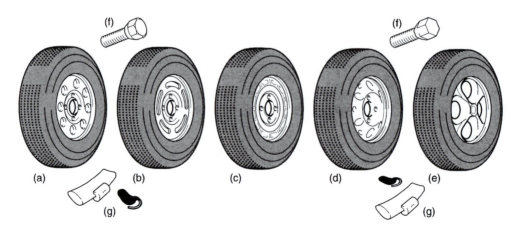

Fig. 22.27 A range of wheels

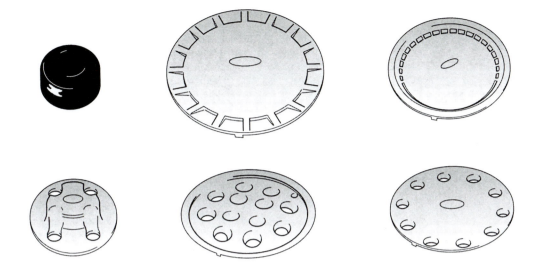

Fig. 22.28   A range of wheel trims (embellishers)

Wire wheels

These are wheels that are built up in a similar way to spoked bicycle wheels. Figure 22.29 shows a typical wire wheel. They are normally to be found on sports cars.

Wheel nut torque

In order to ensure that wheels are properly secure it is advisable to use a torque wrench for the final tightening of the wheel nuts. This is shown in Fig. 22.30.

Tyres

Tyres play an important part in the steering, braking, traction, suspension, and general control of the vehicle. Two types of tyre construction are in use:

- Cross-ply **tyres**
- Radial-ply **tyres.**

The term **ply** refers to the layers of material from which the tyre casing is constructed.

Cross-ply tyre

Figure 22.31 shows part of a cross-ply tyre. The plies are placed one upon the other and each adjoining ply has the bias angle of the cords running in opposite directions. The angle between the cords is approximately 100° and the cords in each ply make an angle of approximately 40° with the tyre bead and wheel rim.

Radial-ply tyre

Figure 22.32 shows part of a radial-ply tyre. The plies are constructed so that the cords of the tyre wall run at right angles to the tyre bead and wheel rim.

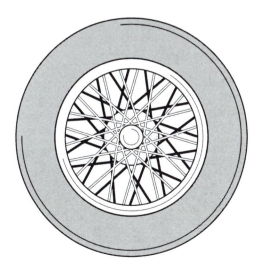

Fig. 22.29   A wire (spoked) wheel

Fig. 22.30   Using a torque wrench for the final tightening

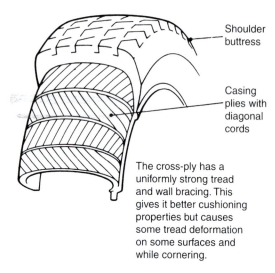

Shoulder
buttress

Casing
plies with
diagonal
cords

The cross-ply has a
uniformly strong tread
and wall bracing. This
gives it better cushioning
properties but causes
some tread deformation
on some surfaces and
while cornering.

Fig. 22.31   Cross-ply tyre construction

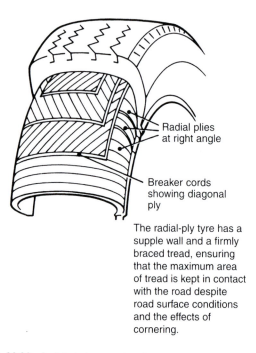

Radial plies
at right angle

Breaker cords
showing diagonal
ply

The radial-ply tyre has a
supple wall and a firmly
braced tread, ensuring
that the maximum area
of tread is kept in contact
with the road despite
road surface conditions
and the effects of
cornering.

Fig. 22.32   Radial-ply tyre construction

Cross-ply versus radial-ply tyres

Radial-ply tyres have a more flexible side wall and this, together with the **braced tread,** ensures that a greater area of tread remains in contact with the road when the vehicle is cornering. Figure 22.33 shows the difference between cross-ply and radial-ply tyres when the vehicle is cornering. In effect, radial-ply tyres produce a better grip.

It is this remarkable difference between the performance of cross-ply and radial-ply tyres that leads to the rules about mixing of cross-ply and radial-ply tyres on a vehicle. Cross-ply tyres on the front axle and radial-ply tyres on the rear axle are the only combination that can be used.

The radial-ply tyre has a great number of advantages over the cross-ply tyre. Cornering stability is greatly improved; as you can see in the Figure the flexible side walls of the radial together with the braced tread ensure that the contact area is held firmly on the road when cornering. For this reason also, tyre life is extended.

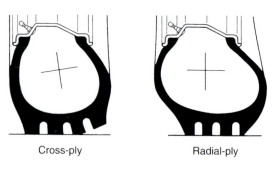

Cross-ply   Radial-ply

Fig. 22.33   Cross-ply and radial-ply tyres under cornering conditions

Tyre sizes

In general tyres have two size markings, one indicating the width of the tyre and other giving the diameter of the wheel rim on which the tyre fits. Thus, a 5.20 by 10 marking indicates a tyre that is normally 5.20 inches wide and fits a 10-inch-diameter wheel rim. Radial-ply tyres have a marking that gives the nominal width in millimetres and a wheel rim diameter in inches; for example, a 145−14 tyre is a tyre with a nominal width of 145 mm with a wheel rim diameter of 14 inches.

Originally vehicle tyres were of circular cross-section and the width of the tyre section was equal to its height. Over the years, developments have led to a considerable change in the shape of the tyre cross-section and it is now common to see tyres where the width of the cross-section is greater than its height. These tyres are known as 'low-profile' tyres. Modern tyres now have an additional number in the size; for example, 185/70−13. The 185 is the tyre width in millimetres, the 13 is the rim diameter in inches, and the 70 refers to the fact that the height of the tyre section is 70% of the width. This 70% figure is also known as the **aspect ratio.** The aspect ratio = (tyre section height/tyre section width) x 100 percent. Tyre size and other information is carried on the side wall and Fig. 22.34 shows how they appear on a typical tyre.

Load and speed ratings

The tyre load and speed ratings are the figures and letters that refer to the maximum load and speed rating of a tyre. For example, 82S marked on a tyre means a load index of 82 and a speed index of S. In this case the 82 means a load capacity of 475 kg per tyre and the S relates to a maximum speed of 180 km/h or 113 mph. One of the main points about these markings

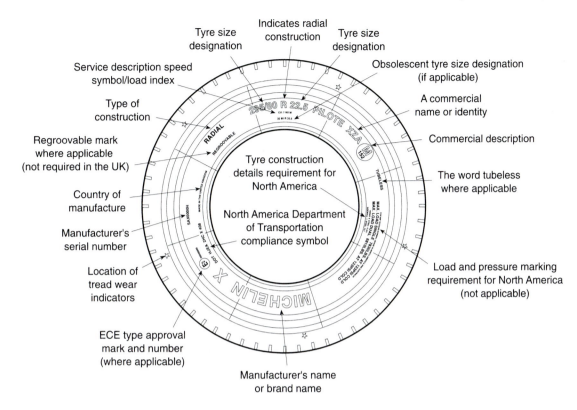

Fig. 22.34 Typical tyre markings

is that they assist in ensuring that only tyres of the correct specification are fitted to a vehicle.

Tread wear indicators (TWI)

Tyres carry lateral ridges (bars) 1.6 mm high in the grooves between the treads at intervals around the tyre. The purpose of these **tread wear indicators** is to show when the tyre tread is reaching its minimum depth. Their position on the tyre is marked by the letters **TWI** on the tyre wall, near the tread.

Tyre regulations

Tyres must be free from any defect that might cause damage to any person or to the surface of the road. There are strict laws about worn tyres. They vary from time to time and it is important that you should be aware of current regulations. The following list is not complete but it does contain three of the most widely known ones. It is illegal to use a tyre that:

- is not inflated to the correct pressure;
- does not have a tread depth of at least 1.6 mm in the grooves of the tread pattern throughout a continuous band measuring at least 3/4 of the breadth of tread and round the entire circumference of the tyre;
- has a cut in excess of 25 mm or 10% of the section width of the tyre, whichever is the greater, which is deep enough to reach the ply or the cord.

There are several other regulations and details can be found in tyre manufacturers' data and Construction and Use Regulations.

In addition to the rules about condition of tyres there are also strict regulations about mixing of cross-ply and radial-ply tyres. For example, it is illegal to have a mixture of cross-ply and radial-ply tyres on the same axle. It is also illegal to have cross-ply tyres on the rear axle of a vehicle that has radial-ply tyres on the front axle. If radial-ply and cross-ply tyres are to be used on the same car, the rule is *radials on the rear.*

Wheel and tyre balance

Unbalanced wheels give rise to vibrations that affect the steering and suspension, they can be dangerous and, if not rectified, can lead to wear and damage to components. Wheel balance can be affected in many ways; for example, a damaged wheel or tyre caused by hitting the kerb. Such damage must be rectified immediately and when the repair has been made the wheel and tyre assembly must be rebalanced.

A **wheel balancer** is a standard item of garage equipment. When new tyres are fitted, or if wheels are being moved around on the vehicle to balance tyre wear, the wheel and tyre assembly should be checked and, if necessary, rebalanced.

Tyre maintenance

Tyre pressures

Tyre pressures have an effect on the steering, braking and general control of a vehicle, and it is important to ensure that pressures are maintained at the correct level for the conditions that the vehicle is operating in. In addition to affecting vehicle handling, tyre pressures have an effect on the useful life of a tyre. The bar chart in Fig. 22.35 shows how various degrees of under-inflation affect tyre life.

Inspecting tyres for wear and damage

Tyres can be damaged in a number of ways, such as running over sharp objects, impact with a kerb, or collision with another vehicle. Tyre damage can also be caused by incorrect tyre pressures, badly adjusted steering track, and accidental damage that has affected steering geometry angles, such as castor and camber.

Tread wear patterns

Inflation pressures

In addition to checking tyre tread depth, the walls and casing of the tyre should be thoroughly examined. Tread wear patterns are a guide to the probable cause of a problem. For example, under-inflated tyres produce more wear at the outer edges because the centre of the tread is pushed up into the tyre. Over-inflated tyres wear at the centre of the tread because that is where most of the load is carried.

Steering track, camber

Toe-in

Figure 22.36 shows an axle and wheel arrangement where the front wheels have toe-in. In this diagram, the outer edges of the tyre are referred to as the shoulders. If the amount of toe-in is too great, the tyre tread will wear in a feathered edge at the outer shoulder.

Toe-out

If the wheels have excessive toe-out, instead of toe-in, the tyre tread will wear in a feathered edge at the inner shoulder. (A feathered edge is a thin layer of rubber that projects outwards away from the tread.). Toe-in and toe-out are checked with the aid of the type of equipment shown in Fig. 22.43.

Camber

Excessive positive camber as shown in Fig. 22.37, where the wheel tilts out at the top, causes the tyre to wear at the outer shoulder. Excessive negative camber causes the tyre to wear at the inner shoulder.

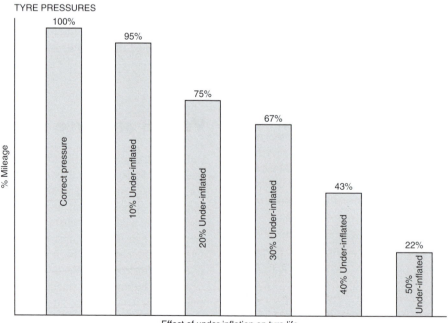

Fig. 22.35 Tyre wear and effect of pressures

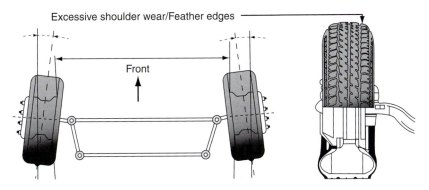

Excessive shoulder wear/Feather edges

Front

Toe-in (positive): faster wear of outside shoulders with feather edges outside to inside.
Toe-out (negative): faster wear of inside shoulders with feather edges inside to outside

Fig. 22.36 Steering toe-in. (Reproduced with the kind permission of Goodyear Tyres.)

Incorrect camber angle also affects the steering, because the vehicle will pull in the direction of the wheel that has the greatest camber angle. Camber angle is checked with the aid of the equipment shown in Fig. 22.41.

Learning tasks

1. Remove the wheels and tyres complete from a vehicle. Check the following and record your results on the job sheet for the customer:
   (a) tread depth
   (b) cuts and bulges
   (c) tyre pressures
   (d) signs of abnormal wear.

   Complete you report with a set of recommendations.
2. Remove a tyre from a wheel, remove all the balance weights, clean the rim, and inspect the tyre bead for damage. Check the inside of the tyre for intrusions and signs of entry of any foreign body such as a nail. Lubricate the bead of the tyre and replace on the wheel, inflate to the correct pressure, rebalance the wheel, and refit to vehicle.
3. Produce a list of the current MOT requirements for wheels and tyres.
4. Describe a procedure that you have used for fitting a new tyre.
5. Make a list of the precautions that must be taken when removing a wheel in order to do some work on the vehicle. Make special note of the steps taken to prevent the vehicle from moving and also the steps taken to ensure that the vehicle cannot slip on the jack.
6. Examine a selection of wheel nuts. State why the conical part of the nut, or set bolt, is necessary. State why some vehicles, especially trucks, have left-hand threaded wheel nuts, on the near side

   (left-hand side of the vehicle when sitting in the driving seat).
7. Make a note of the type of wheel-balancing machine that is used in your workshop. State the safety precautions that must be taken when using it and make a list of the major points that you need to know about when balancing a wheel and tyre assembly.
8. State the type of tyre-thread depth gauge that you use. Describe how to use it and state the minimum legal tread depth in the UK.
9. Examine a number of tyres and locate the tyre-wear indicator bars. State which mark on the tyre wall helps you to locate these wear indicators.

Self-assessment questions

1. Figure 22.38 shows six examples of tyre wear problems. State a probable cause for each case.
2. Tread wear indicators on tyres:
   (a) are located in the grooves between the tyre treads
   (b) only become visible when the tread is worn to the legal limit
   (c) have their positions indicated by an arrow on the wheel rim
   (d) indicate that the tyre is completely legal.
3. A car is to be fitted with two radial-ply tyres and two cross-ply tyres. In this case:
   (a) both radial-ply tyres should be fitted to one side of the vehicle
   (b) the radial-ply tyres should be fitted to the rear wheels
   (c) the tyres should be mixed diagonally
   (d) it does not matter where they are fitted.

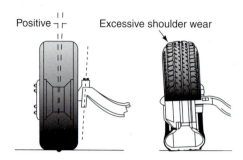

Positive camber (upper wheel part tilted outwards):
faster wear of outside shoulder.
Negative camber (upper wheel end tilted inwards):
faster wear of inside shoulder.

Fig. 22.37 Steering camber angle. (Reproduced with the kind permission of Goodyear Tyres.)

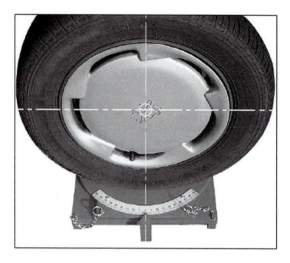

Fig. 22.39   Steering turntable for checking toe-out on turns (Sealey)

4. In a tyre size of 185/70—13, the aspect ratio is:
   (a) 185
   (b) 70%
   (c) 13
   (d) 198.
5. In a steering system that uses a cam and peg steering box, the drop arm:
   (a) controls back lash in the mechanism
   (b) transmits a pull—push motion to the drag link
   (c) connects the steering arms to the stub axle
   (d) transmits steering wheel motion to the cam.
6. When checking the steering geometry on a certain vehicle, it is found that the camber angle on the nearside is 2.6° and on the offside it is only 1.2°. This difference:
   (a) may have caused the vehicle to pull to the right
   (b) may have caused the left front tyre to wear on the inside edge
   (c) may have caused the vehicle to pull to the left
   (d) is not significant.
7. With centre-point steering:
   (a) the centre of tyre contact meets the road surface at the same spot as the centre line of the kingpin
   (b) the vehicle's centre of turn is on a line that passes through the centre of gravity of the vehicle
   (c) the steering gearbox is in the centre of the vehicle
   (d) no kingpin inclination or camber angle is required.

8. Toe out on turns — using turntables
   Figures 22.39 and 22.40 show the equipment and procedure for checking the angles through which the steered wheels are turned through on the left and right locks.
   (a) Why does the inside wheel turn through a greater angle than the outside wheel when a vehicle is turning a corner?
   (b) Why is it important to ensure that the track rod lengths are maintained at the correct length when adjusting the track?
   (c) What is the Ackermann principle as applied to steering.
9. Castor, camber, KPI
   Measuring camber angle
   (a) Check to make sure that the floor is level and that the tyres are correctly inflated.
   (b) Hold the gauge in a true vertical position and adjust the dial marked 'camber' so that the '0' is in line with the index mark.
   (c) Apply the long edge of the gauge to the tyre sidewall holding it in a vertical position — avoid placing the gauge on the bulge where the tyre rests on the floor.
   (d) Turn the camber dial until the bubble in the spirit level is in the central position — note the camber angle as shown on the dial.

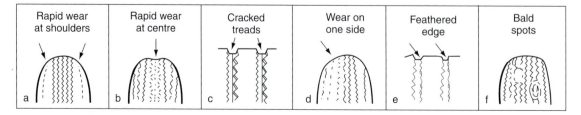

Fig. 22.38   Tyre tread wear patterns

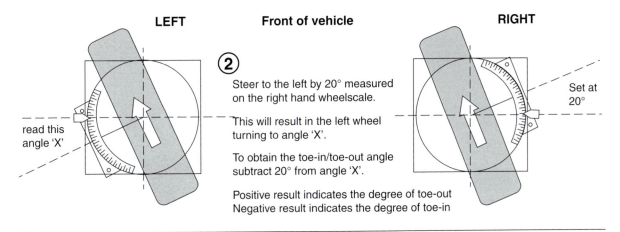

LEFT — Front of vehicle — RIGHT

② Steer to the left by 20° measured on the right hand wheelscale.

This will result in the left wheel turning to angle 'X'.

To obtain the toe-in/toe-out angle subtract 20° from angle 'X'.

Positive result indicates the degree of toe-out
Negative result indicates the degree of toe-in

read this angle 'X'

Set at 20°

4.3 Measure the WHEEL ANGLES and the TOE-IN/OUT ANGLES for both left and right steering as described above and below.
COMPARE YOUR READINGS WITH THE VEHICLE MANUFACTURER'S RECOMMENDATIONS.

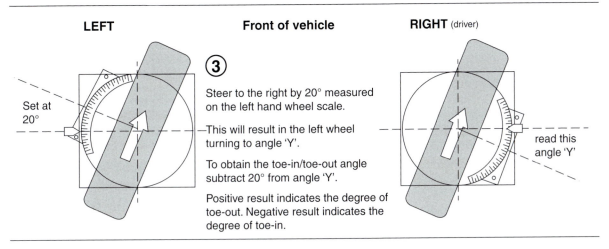

LEFT — Front of vehicle — RIGHT (driver)

③ Steer to the right by 20° measured on the left hand wheel scale.

This will result in the left wheel turning to angle 'Y'.

To obtain the toe-in/toe-out angle subtract 20° from angle 'Y'.

Positive result indicates the degree of toe-out. Negative result indicates the degree of toe-in.

Set at 20°

read this angle 'Y'

4.4 To measure FULL LOCK set the equipment up as described above in sections 4.1 and 4.2 and as shown in diagram 1.
4.5 Remove the turntable locking pins.
4.6 Steer the wheels to FULL LEFT LOCK and take a reading from the right hand wheel.
4.7 Steer the wheels to FULL RIGHT LOCK and take a reading from the left hand wheel.
4.8 COMPARE READINGS AND CHECK WITH THE VEHICLE MANUFACTURER'S RECOMMENDATIONS.

Fig. 22.40 Steering angles

Figure 22.42 shows the additional equipment which, together with the turn tables, is used for measuring castor and kingpin inclination angles.
(a) Refer to the Sealey (www.sealey.co.uk) website and make notes about the procedure for checking castor and KPI angles.
(b) Explain why it is important to ensure that tyres are in good condition and correctly inflated prior to performing steering checks.
(c) Wear in steering and suspension components can affect steering geometry. Write a list of the components that can affect castor and KPI.
10. Laser type tracking equipment

Note on safety

LASER SAFETY

The GA48 utilises a Class II laser that emits low levels of visible radiation (i.e. wavelengths between 400 and 700 nanometres) which are safe for the skin but not inherently safe for the eyes. The Class II emission limit is set at the maximum level for which eye protection is normally afforded by natural aversion responses to bright light. Accidental eye exposure is therefore normally safe, although the natural aversion response can be overridden by deliberately staring into the beam, and can also be influenced by the use of alcohol or drugs.

Fig. 22.41 Camber gauge in position (Sealey garage equipment)

WARNING!

✗ **DO NOT** look or stare into the laser beam as permanent eye damage could result.

✗ **DO NOT** direct the laser beam at any person's (or animal's) eyes as eye damage could result. If the beam is obstructed by a person during use, release the contact switch immediately.

✗ **DO NOT** use the equipment while under the influence of alcohol, drugs or whilst on medication.

✓ Be aware that reflections of the laser beam from mirrors or other shiny surfaces can be as hazardous as direct eye exposure.

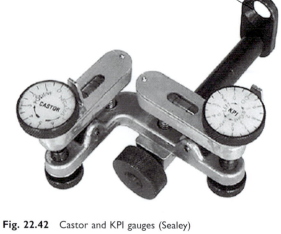

Bracket is attached to a wheel by means of a wheel nut

Fig. 22.42 Castor and KPI gauges (Sealey)

The above safety information is taken from Sealey documentation. For full instructions on calibration and use of this equipment refer to the Sealey website (www.sealey.co.uk).

(a) By reference to several workshop manuals, write down the toe in, or toe out angle for various vehicles.

(b) With the aid of sketches, describe the means by which the track may be adjusted.

11. The type of axle misalignment that is shown in Fig. 22.44 may be the result of:

- Collision damage
- Impact with an obstruction
- Incorrect fitting of components
- Worn or corroded body or suspension parts.

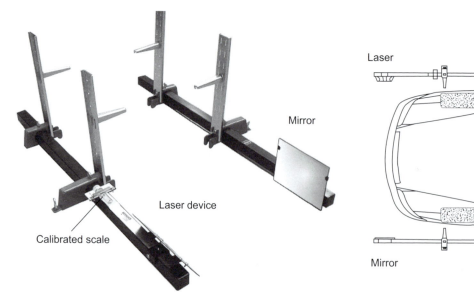

Laser

Mirror

Laser device

Calibrated scale

Mirror

Fig. 22.43 Laser tracking equipment (Sealey)

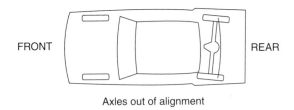

FRONT — REAR

Axles out of alignment

Fig. 22.44 Misaligned axles

If the misalignment is large, the effect on the steering and handling of the vehicle will be pronounced and the defect will require immediate attention. If the misalignment is small it will probably be noticeable in the steering performance of the vehicle, and pronounced wear patterns on the tyres.

Measuring the misalignment

If the misalignment (axle setback) is caused by chassis damage it is possible that a drop test will identify the problem area. A drop test of the type shown in Fig. 22.45 is performed by placing the vehicle on a level surface, on four identical axle stands that are set at the same height. Several fixed points such as suspension fixing points on the chassis are selected, on each side of the underbody of the vehicle. A plumb line is then used to project these points down to make a chalk mark on the floor. By bisecting the distance between the matching points on either side of the underbody, a centre line can be drawn in. When the vehicle has been moved away from the chalked marks on the floor, measurements can be taken and any differences noted. A reasonably accurate check can be performed by measuring the distance between the front and rear axle centres on each side of the vehicle while the

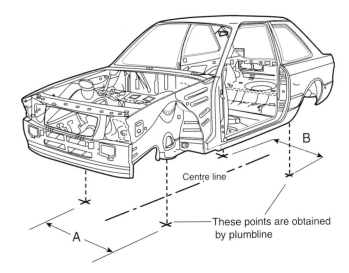

Centre line

B

A

These points are obtained by plumbline

Fig. 22.45 Chassis drop test

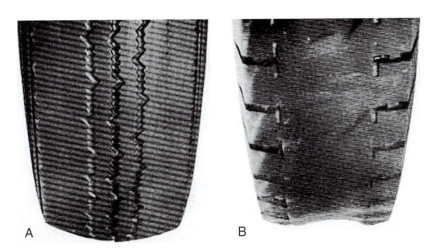

A — B

Fig. 22.46 Examples of tyre wear

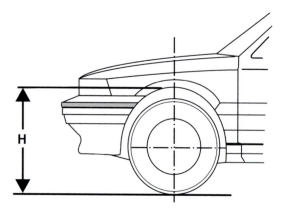

Fig. 22.47 Ride height

vehicle is standing on a smooth, level surface. If there is a difference of more than a couple of millimetres between these two measurements further examination may be required and this can be performed with the aid of four-wheel alignment equipment, which should be available in large body repair depots.

Computerized four-wheel alignment equipment

Various types of computerised alignment equipment are available. Most of them make use of sensors that are attached to each wheel. The vehicle is mounted on a platform that is aligned with a receiving apparatus that is mounted at right angles to the axis of the platform on which the vehicle is mounted. Company websites such as Snapon Tools and others provide descriptions of this type of equipment.

12. Axle misalignment

    Describe some of the factors, other than collision damage, that may cause axles to be misaligned.

13. Tyre wear

    For this exercise refer to Fig. 22.46.

    (a) Approximately what percentage of the tread on tyre A is worn away? Describe what causes a tyre to wear like this.

(b) What do you think caused tyre B to wear like it has? Approximately what percentage of the tread is left on tyre B?

14. Ride height

    In Fig. 22.47 the dimension H is called the ride height.

    (a) Make a list of the factors that can affect ride height.

    (b) Describe some of the steps that can be taken to restore H to its correct level if it is found to be out of limits.

Practical assignment – steering worksheet

Introduction

After this practical exercise you should be able to:

- check and adjust steering alignment
- assess for wear in steering components
- remove and replace track rod ends
- check tyres for misalignment
- make a written report on your findings

Tools and equipment

- vehicle with appropriate steering mechanism
- appropriate workshop manual
- selection of tools and specialist equipment, e.g. alignment gauges
- vehicle lift or jacks and stands

Checklist

Vehicle

Student's signature

Supervisor's signature

Exercise	Date started	Date finished	Serviceable	Unserviceable
Rack and pinion				
Recirculating ball				
Worm and peg				
Worm and roller (hourglass)				
Castor				
Camber				
KPI				
Toe-out on turns				
Optical alignment gauge				
Four-wheel alignment				
Body alignment				

Activity

1. Inspect the steering for wear and security, e.g.
   (a) uneven wear on tyres
   (b) play in track-rod ends, steering rack and column
   (c) play in wheel bearings
   (d) security of shock absorbers/dampers, check for leaks
   (e) wear/play in suspension joints
   (f) security/wear in anti-roll bar mountings.
2. Remove and refit at least one track rod end.
3. Using appropriate equipment check tracking of vehicle's steered wheels.
4. Check steering angles — castor, camber, kingpin inclination, toe-out on turns.
5. Make any adjustments necessary.
6. Investigate where possible the following types of steering gearboxes:
   (a) worm and roller
   (b) cam and peg
   (c) recirculating ball
   (d) rack and pinion.
7. Make any adjustments on the above types of steering gearboxes after stripping and rebuilding them.
8. Produce a report on the condition of the steering including the gearboxes, indicating wear and serviceability.
9. State the meaning of the following terms:
   (a) tracking
   (b) toe-out on turns
   (c) steering angles — castor, camber, KPI, Ackermann angle
   (d) roll radius.

23
Power-assisted steering

Topics covered in this chapter

Types of power-assisted steering
Hydraulic systems
Electronically aided electric power assistance
Four-wheel steering
Electronic sensors and actuators

The two types of power-assisted steering (PAS) that are used in light vehicles can be broadly classified as:

1. Hydraulic power assistance
2. Electrical power assistance.

As with other servo-assisted systems, the amount of power assistance is proportional to the amount of effort applied by the driver, and a major part of a power-assisted steering system is the valve and sensing system that provides this proportional output at the steering gear.

In the hydraulic system the power that helps the driver to operate the steering is provided by a pump that is driven by the engine; in an electric system the assisting power is provided by a reversible electric motor that takes its energy from the main vehicle electrical system.

Hydraulic power-assisted steering

In the hydraulic power assisted steering systems shown in Fig. 23.1 hydraulic pressure is passed from the pump to the control valve. From the control valve, the pressurized fluid is directed to one end of the power cylinder whilst fluid at the other end is released back to the reservoir through the control valve. The power piston is connected to the steering rack and the force that it exerts adds to the force applied by the driver and thus provides the power assistance. The amount of pressure that is exerted on the power piston is determined by the effort that the driver exerts on the steering wheel.

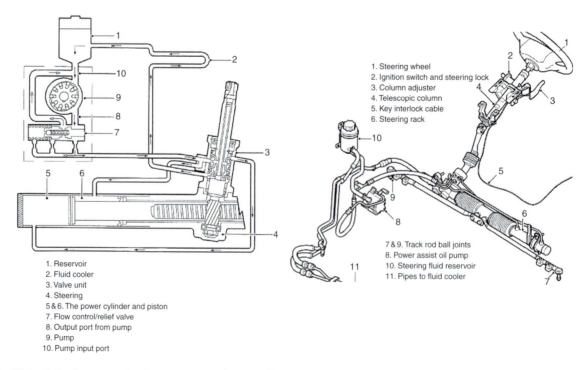

1. Steering wheel
2. Ignition switch and steering lock
3. Column adjuster
4. Telescopic column
5. Key interlock cable
6. Steering rack

7 & 9. Track rod ball joints
8. Power assist oil pump
10. Steering fluid reservoir
11. Pipes to fluid cooler

1. Reservoir
2. Fluid cooler
3. Valve unit
4. Steering
5 & 6. The power cylinder and piston
7. Flow control/relief valve
8. Output port from pump
9. Pump
10. Pump input port

Fig. 23.1 Hydraulic power-assisted steering — neutral position (Land Rover ZF)

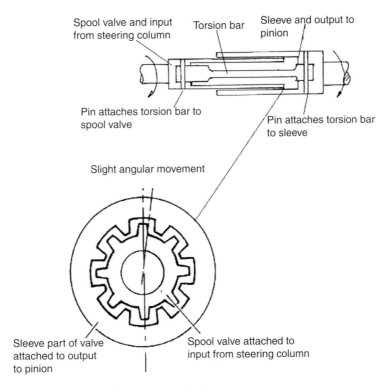

Spool valve and input from steering column

Torsion bar

Sleeve and output to pinion

Pin attaches torsion bar to spool valve

Pin attaches torsion bar to sleeve

Slight angular movement

Sleeve part of valve attached to output to pinion

Spool valve attached to input from steering column

Fig. 23.2 A simplified version of the power-assisted steering control valve

The control valve

The control valve (Fig. 23.2) consists of:

- The outer sleeve
- Input shaft — connected to steering column
- Torsion bar
- Pinion shaft — output to rack.

The input shaft is connected to the pinion shaft and outer sleeve by a small torsion bar and a loose-fitting spline. The torsion bar twists slightly under torque applied at the steering wheel and it is this small amount of twist that determines the flow of fluid to and from the power cylinder. The splined connection between the input and output provides for approximately $7°$ of angular movement and this allows normal steering to be maintained in the event of a failure in the power assistance system.

ECU-controlled power-assisted steering

Item 12 in Fig. 23.3, which is a transducer, operates under the control of the body system's ECU. When conditions require it, the transducer operates on the reaction piston (21) in the reaction chamber (22), to control the input torque required to turn the steering wheel. The hydraulic reaction changes with road speed, with the required steering input increasing as the vehicle moves faster. An advantage claimed for this system is

that fluid pressure and flow through the main valve remains constant and allows full steering pressure to be available in an emergency, where a sudden and unexpected steering correction may be required. The main hydraulic operation is the same as that described in the first part of the section on power-assisted steering.

Servotronic operation

When the vehicle is manoeuvred into or out of a parking space, or similar situation, the body system ECU collects road speed data from the ABS ECU. The body system ECU then outputs an appropriate electric current to the transducer valve. This regulates the pressure in the reaction chamber so that the effort required to turn the steering wheel is kept as low as possible.

As the vehicle is driven away and the road speed increases, the ECU analyses the road speed signals from the ABS ECU and varies the current flowing to the transducer valve so that the steering input effort can be maintained at the correct level for the driving conditions.

Electric power-assisted steering

In electric power-assisted steering systems the power assistance is provided by a reversible electric motor which is built into the steering system as shown in Figs 23.4 and 23.5. The power assistance may be applied to the steering column, or to the steering rack.

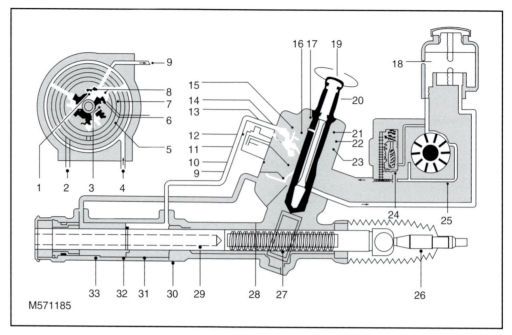

M571185

Circuit shows steering rotary valve in right turn position, at high vehicle speed and with rapid steering corrections. The Servotronic transducer valve is fully open and the maximum hydraulic reaction is limited by the cut-off valve.

1 Return fluid control groove
2 Radial groove
3 Feed fluid control groove
4 Radial groove
5 Axial groove
6 Feed fluid control edge
7 Feed fluid radial groove
8 Retum fluid control edge
9 Retum fluid chamber
10 Cut-off valve
11 Radial groove
12 Servotronic transducer
13 Feed fluid radial groove
14 Radial groove
15 Orifice
16 Balls
17 Compression spring

18 PAS fluid reservoir
19 Torsion bar
20 Valve rotor
21 Reaction piston
22 Reaction chamber
23 Centering piece
24 Pressure relief/flow limiting valve
25 PAS pump
26 Inner track rod
27 Pinion
28 Valve sleeve
29 Steering rack
30 Rack housing
31 Power assist cylinder – right
32 Piston
33 Power assist cylinder – left

Fig. 23.3 An ECU-controlled power-assisted steering system

Modern electric power-assisted steering systems operate under the control of an ECU. The ECU will have inputs from a torque sensor and road speed sensors so the amount of power assistance provided is suitable for the ongoing conditions. Reversible electric motors are used and the direction of rotation occurs instantaneously in response to commands from the ECU so that rapid changes of vehicle direction can be dealt with.

Advantages of electric power assistance are:

• The system is only energized when required, as opposed to the hydraulic system where the engine is driving the hydraulic pump permanently.
• There are no components that may cause fluid leaks.

A possible disadvantage is that quite high current demands may be placed on the vehicle's electrical system. Under some conditions the motor may need to output several hundred watts. Each 100 watts in a 12-volt system represents a current draw of 8.3 amperes.

Four-wheel steering

Some of the advantages claimed for steering the rear wheels in addition to the front wheels are:

• Improved manouevrability when parking or turning tight corners
• Improved vehicle stability when changing direction – in traffic
• Improved high-speed straight-line stability
• Smaller turning circle radius.

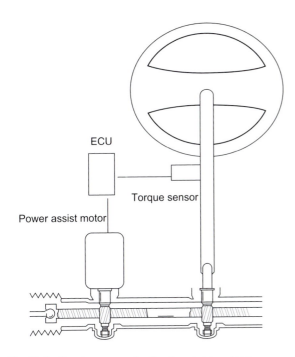

Fig. 23.4 Electric power-assisted rack and pinion steering

The Honda four-wheel steering system that is outlined in the following description gives an indication of the general principles that are employed in this technology.

The Honda electronic four-wheel steering (E4WS) system that is shown in Figs 23.6 and 23.7 performs two distinct operations:

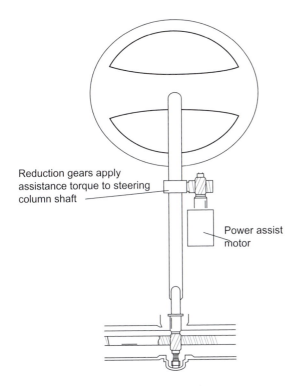

Fig. 23.5 Electric power assistance applied to steering column

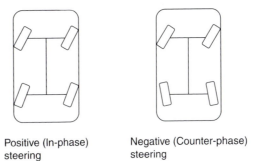

Fig. 23.6 Angle of front and rear wheels in the two phases

1. Positive (in-phase) steering, where the rear wheels are turned in the same direction as the front wheels.
2. Negative (counter-phase) steering, where the rear wheels are turned in the opposite direction.

The two phases (Fig. 23.6) operate as follows:

- Vehicle speed up to 30 km/h. When the front wheels are turned to the right or left, the rear wheels are steered in the opposite direction (counter-phase).
- Vehicle speeds above 30 km/h. When the front wheels are steered to the right or left, the rear wheels are turned in the same direction (in-phase).
- The maximum steered angle at the rear wheels is approximately 7°.

The steering of the front wheels is performed by a rack and pinion system that is controlled by the driver, through the steering wheel. There are two sensors that 'sense' the actions of the front-wheel steering and they relay electrical signals, instantaneously, to the ECU. The memory in the ECU also records electrical signals from the sensors on the rear steering actuator, the vehicle speed sensor, and the rear wheels (Fig. 23.7). These signals are compared with data that is stored in the ECU memory and the ECU then sends an output to the rear steering actuator so that it moves the rear wheels to the required position. The chart in Fig. 23.8 shows how the various sensors and the rear actuator work together to provide rear steering.

Front steering sensors

The main front steering sensor

The front main steering angle sensor (Fig. 23.9) is located on the steering column. It records the angle of the two front wheels, the speed at which the steering angle is changed, and the direction of that change. This data is sent as a digital signal to the ECU of the E4WS system.

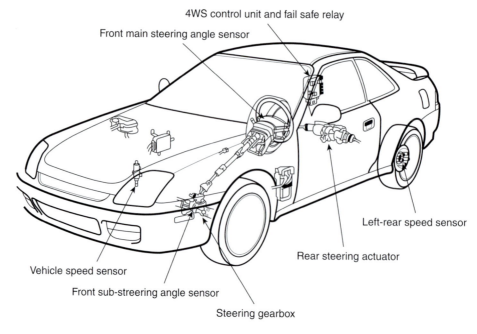

Fig. 23.7   Layout of the components of the Honda four-wheel steering system

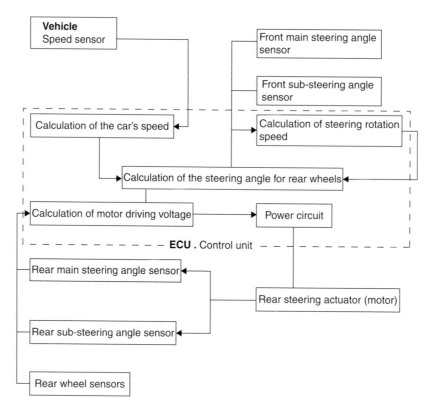

Fig. 23.8   ECU inputs and outputs

The front steering sub-sensor

The sub-steering angle sensor is located on the front steering gearbox. It measures the steering angle on the two front wheels and sends an analogue signal to the E4WS-ECU. Its function is to detect:

1. Steering rack position
2. Neutral position
3. Absolute position of the front main sensor.

The front sub-steering angle sensor (Fig. 23.10) uses a spring-loaded plunger that rides in a sloped groove in

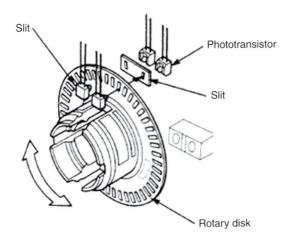

Fig. 23.9   Front steering sensor

the rack. As the rack turns the front wheels, the plunger rides up the slope and causes the sensor element to transmit an analogue voltage signal to the ECU. The plunger's position is used to determine how far the rack is turned.

The rear steering actuator

The rear steering actuator (Fig. 23.11) is a reversible electric motor that causes a nut to rotate around a threaded part of the rack; in order to keep friction to a minimum the threaded portion runs on ball bearings.

Rear main steering angle sensor

The rear main steering angle sensor (Fig. 23.12) is mounted on the rear actuator. It records the steering angle

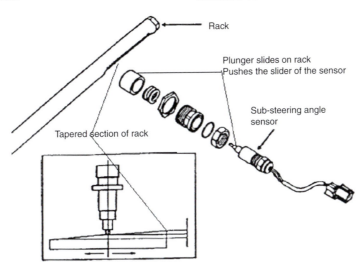

Fig. 23.10   The front steering sub-sensor

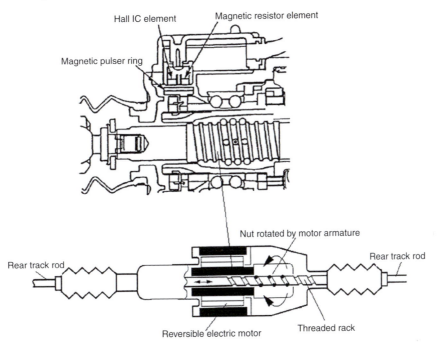

Fig. 23.11   The rear steering actuator and main sensor

Rear Sub-Steering Angle Sensor

The rear sub-steering angle sensor is mounted on the rear actuator spring cover. It detects the position of the rear wheels as a feedback signal for the control unit. It operates in exactly the same way as the front sub-steering angle sensor, and is also adjustable.

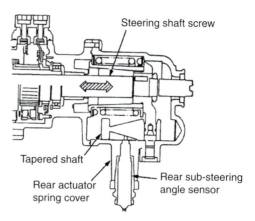

Fig. 23.12 The rear steering sub-sensor

Vehicle Speed Sensor

The vehicle speed sensor is located on the transmission and driven by a ring gear on the differential.

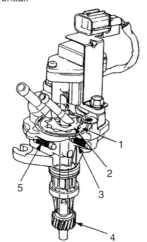

1. Outer rotor
2. Inner rotor
3. Relief valve
4. Speed motor driven gear
5. One way valve

Fig. 23.13 The speed sensor

Table 23.1 Power steering problems

Problem and possible causes	Remedy
Steering feels heavy:	
Incorrect tyre pressures	Check tyre condition and adjust pressures
Loose power steering pump drive belt	Examine belt and adjust or renew
Fluid level in reservoir low	Check for leaks and replenish fluid as necessary
Partly seized steering swivel joints	Check condition and renew if required
Accident damaged steering arms or suspension members	Check visually and check track and steering angles – renew damaged parts
Incorrectly adjusted track	Check track and reset correctly
Partial seizure of steering rack	Inadequate lubrication. Lubricate and try again – if still too heavy, examine rack condition and repair or replace as necessary
Air in PAS hydraulic system	Check for leaks and bleed system – replenish fluid level as required
Low PAS pump pressure	Check pressure – examine pressure hoses. Check the pressure relief valve. Repair or replace pump
Blocked or obstructed pressure hose	Examine hoses – search to see if hose is trapped
PAS fluid leaks:	
Fluid reservoir over-filled	Inspect and adjust to correct level
Damaged or worn hoses	Inspect for cracks or other damage and replace
Incorrectly fitted hoses	Inspect to ensure that there are no crossed threads and that all sealing washers are correctly fitted
Pump seals leaking	Inspect pump condition. Renew seals if possible, otherwise repair or renew the pump

of the rear wheels and sends a digital signal with this information to the ECU. The speed sensor (Fig. 23.13) is used by the ECU to determine when to switch between modes.

Diagnostics

In common with other ECU-controlled systems, the four-wheel steering ECU constantly monitors inputs and outputs; should any of these cease to function correctly the ECU generates a warning and a diagnostic trouble code (DTC). The procedure for retrieving these DTCs is contained in the workshop manual. In the event of a serious fault the rear steering is returned to and held in the straight-ahead position.

Power steering problems, possible causes and remedies are summarized in Table 23.1.

Self-assessment questions

1. Which component of a hydraulic power assisted steering system ensures that the amount of power assistance applied is proportionate to the effort that the driver applies to the steering wheel?
2. In which direction do the rear wheels point when a vehicle with four-wheel steering is moving at slow speed?
3. Which component in a hydraulic power assisted steering system generates the hydraulic pressure that is applied to the power piston?
4. What is the function of the reversible electric motor in an electric power assisted steering system?
5. What is the maximum steered angle of the rear wheels in a four-wheel steering system?
6. What is the function of the front main steering sensor in a four-wheel steering system?
7. What type of fluid is used in hydraulic power assisted steering systems?
8. What is the function of the electronically controlled actuator at the rear wheels of a four-wheel steering system?

24
Computer (ECU)-controlled systems

Topics covered in this chapter

The ECU as a computer
Fault codes
Networking and bus systems
CAN
The CAN bus
High-speed and low-speed CAN
Data transmission rates
Error protection
ISO 9141 system for fault code access
The gateway
Alternative networks
Multiplexing
Power supply and drivers
European on-board diagnostics (EOBD)
Malfunction indicator lamp (MIL)
Engine management system
Feedback system
Oxygen sensor
Exhaust gas recirculation
Traction control
Electronic throttle
Stability control

Electronic control units (ECUs) are computers that are used in many applications on motor vehicles. An ECU consists of a central processor, or microcontroller, and a range of electronic devices that permit the computing unit to interact with vehicle systems. The ECU has a memory in which the program for controlling a system is stored. In operation the ECU receives a constant stream of data from various sensors, which it then processes and compares with data in the memory. The processing takes place almost instantaneously and when it is completed signals are sent to the actuators to make them perform the required action which, in the case of petrol injectors, will be the amount of fuel to be injected.

The ECU memory consists of permanent memory (ROM) in which the control program is stored, and a working memory (RAM) that holds data while it is being worked on. ROM is permanent memory that can only be changed by deliberate action, in which case the memory is called EEPROM, which stands for electrically erasable programmable read-only memory. RAM stands for random access memory and it is said to be 'volatile memory' because it is lost when the electricity supply is removed.

Because the ECU is constantly monitoring many inputs and outputs it is able to detect problems in the circuits and systems that it controls. When a problem occurs, fault codes are flagged up and a record called a freeze frame is held in working memory. In European on-board diagnostics systems (EOBD) any fault in an emissions-related system must cause the malfunction indicator lamp (MIL) to be illuminated.

The fundamental parts of a computer

Figure 24.1 shows the general form of a computer, which consists of the following parts:

1. A central processing unit (CPU)
2. Input and output devices (I/O)
3. Memory
4. A program
5. A clock for timing purposes.

Data processing is one of the main functions that computers perform. Data, in computer terms, is the representation of facts or ideas in a special way that allows it to be used by the computer. In the case of digital computers this usually means binary data, where numbers and letters are represented by codes made up of zeros and ones. The input and output interfaces enable the computer to read inputs and to make the required outputs. Processing is the manipulation and movement of data, and this is controlled by the clock. Memory is required to hold the main operating program and to temporarily hold data while it is being worked on.

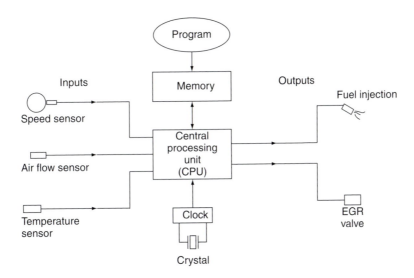

Fig. 24.1 The basic components of a computer system

Computer memory

Read-only memory (ROM)

The ROM is the place where the operating program for the computer is placed. It consists of an electronic circuit that gives certain outputs for predetermined input values. ROMs have large storage capacity. It is a permanent memory that is built into the computer at the production stage.

Read and write, or random access memory (RAM)

The RAM is the place where data is held temporarily while it is being worked on by the processing unit. Placing data in memory is 'writing' and the process of using this data is called 'reading'. RAM is also known as volatile memory because the data is held there electrically and when the source of electricity is removed the contents of the memory are lost.

Keep-alive memory (KAM)

The term keep-alive memory (KAM) refers to the systems where the ECM has a permanent, fused, supply of electricity. Here the fault codes are preserved, but only while there is battery power.

EEPROMs

Electrically erasable programmable read-only memories (EEPROMs) are sometimes used for the storage of fault codes and other data relating to events connected with the vehicle system. This type of memory is sustained even when power is removed.

The clock

The clock is an electronic circuit that utilizes the piezo-electric effect of a quartz crystal to produce accurately timed electrical pulses that are used to control the actions of the computer. Clock speeds are measured in the number of electrical pulses generated in 1 second. One pulse per second is 1 hertz and most computer clocks operate in millions of pulses per second. One million pulses per second is 1 megahertz (1 MHz).

A practical automotive computer system

Figure 24.2 shows a computer controlled transmission system. At the heart of the system is an electronic module, or ECU. This particular module is a self-contained computer that is also known as a microcontroller. Microcontrollers are available in many sizes, e.g. 4-, 8-, 16-, and 32-bit, which refers to the length of the binary code words that they work on. The microcontroller in this transmission is an 8-bit one.

Figure 24.3 shows some of the internal detail of the computer and the following description gives an insight into the way that it operates:

1. The microcomputer. This is an 8-bit microcontroller. In computer language a bit is a 0 or a 1. The 0 normally represents zero or low voltage and the 1 normally represents a higher voltage, probably 1.9 volts. The microcontroller integrated circuit (chip) has an ROM capacity of 2049 bytes. There are 8 bits to 1 byte. There is also an RAM that holds 64 bytes. The microcontroller also has an on-chip capacity to convert four analogue inputs into 9-bit digital codes.

2. The power supply is a circuit that takes its supply from the vehicle battery. It then provides a regulated d.c. supply of 5 volts to the microcontroller and this is its working voltage. The power supply also includes protection against over-voltage and low

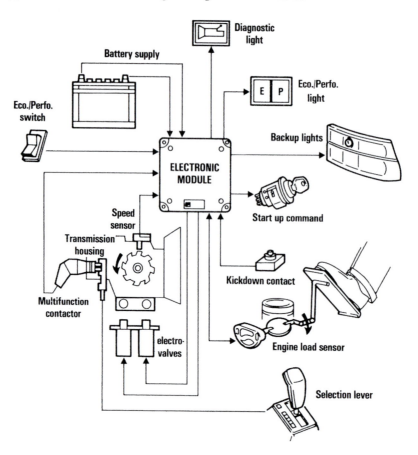

Fig. 24.2   A computer-controlled transmission system

voltage. The low-voltage protection is required if battery voltage is low and it often takes the form of a capacitor.

3. The clock circuit. In this particular application the clock operates at 4 MHz. The clock controls the actions of the computer such as counting sensor pulses to determine speed and timing of output pulses to the electro-valves so that gear changes take place smoothly and at the required time.

4. The input interface. The input interface (4) contains the electronic circuits that provide the electrical power for the sensors and switches that are connected to it. Some of these inputs are in electrical form (analogue) that cannot be read directly into the computer and these inputs must be converted into computer (digital) form at the interface.

5. The output (power) interface. The power driver (5) consists of power transistors that are switched electronically to operate electro-valves that operate the gear change hydraulics.

6. Feedback (at 6). On the diagram the inscription reads 'Reading electrical state'. This means that the computer is being made aware of the positions (on or off) of the electro-valves.

7. The watchdog. The watchdog circuit (7) is a timer circuit that prevents the computer from going into an endless loop that can sometimes happen if false readings occur.

8. The diagnostic interface. The diagnostic interface (8) is a circuit that causes a warning lamp to be illuminated in case of a system malfunction and it can also be used to connect to the diagnostic kit.

Fault codes

When a microcontroller (computer) is controlling the operation of an automotive system such as engine management it is constantly taking readings from a range of sensors. These sensor readings are compared with readings held in the operating program and if the sensor reading accords with the program value in the ROM the microcontroller will make decisions about the required output to actuators, such as injectors. If the sensor reading is not within limits it will be read again and if it continues to be 'out of limits' a fault code will be stored in a section of RAM. It is also probable that the designer will have written the main program so that the microcontroller will cause the system to operate on different criteria until a repair can be made, or until the fault has cleared. The fault codes, or diagnostic trouble codes (DTCs) as they are sometimes called, are of great importance to service technicians and the procedures for gaining access to them need to be understood. It should be clear that, if they are held in ordinary RAM, they will be erased when the ECM power is removed and that is why

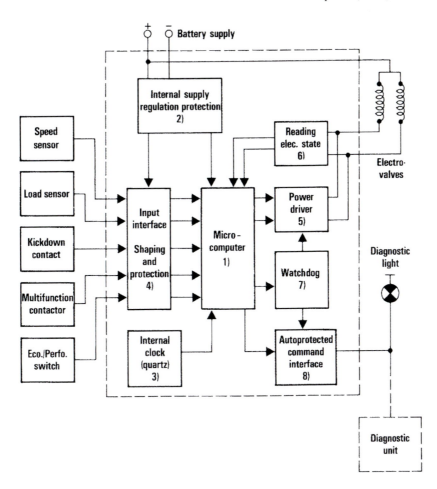

Fig. 24.3 Internal details of the computer

various methods of preserving them, such as KAMs and EEPROMs, are used.

Adaptive operating strategy of the ECM

During the normal lifetime of a vehicle it often happens that compression pressures and other operating factors change. To minimize the effect of these changes, many computer-controlled systems are programmed to generate new settings that are used as references, by the computer, when it is controlling the system. These new (learned) settings are stored in a section of memory, normally RAM. This means that such 'temporary' operating settings can be lost if electrical power is removed from the ECM. In general, when a part is replaced or the electrical power is removed for some reason, the vehicle must be test driven for a specified period in order to permit the ECM to 'learn' the new settings. It is always necessary to refer to the repair instructions for the vehicle in question, because the procedures do vary from vehicle to vehicle.

Networking and bus systems

Computer-controlled vehicle systems such as engine management, traction control, anti-lock braking and stability control, need to communicate with each other in order to obtain the optimum performance from the vehicle. To achieve this end the individual systems are linked together (networked) by a communication bus that permits data to be interchanged between the systems at a very high data transfer speed; systems that operate at high speeds of data transfer are said to operate in '**real time**'.

Bus

Bus is the name that is given to the wires or fibre optics that transmit the data used to exchange information between the computer systems that operate together to control vehicle behaviour. A typical example of a vehicle system that uses a data bus is traction control, where the engine management system and the anti-lock braking system act together to prevent wheel spin. A data bus on a vehicle may be a single wire, a fibre optic or, in the case of CAN, a pair of wires.

CAN – Controller Area Network (Robert Bosch)

The CAN system of data transfer for networking of electronic control units was developed by Robert Bosch GmbH in the 1980s and it has now become widely used in automotive practice. The basic CAN data bus consists of two wires that transmit the data signals in the form of voltages. Two types of CAN are commonly used on vehicles:

1. **Low-speed CAN**, which is used for data transfer speeds of up to 125 kbaud. This is suitable for the control of body systems such as seat adjustment, air-conditioning, and other systems related to driver and passenger comfort.
2. **High-speed CAN**, which is used for data transfer speeds of up to 1 Mbaud. High-speed CAN is used for systems such as engine management, traction control, and transmission control.

Data transmission rates on CAN buses

Data transmission rates are measured in bits/s or baud: 1 baud = 1 bit/s. The maximum bit rate is dependent on the length of the CAN bus. The figures in Table 24.1 are those that are specified by ISO 11898.

CAN voltages

The voltages shown in Fig. 24.4 represent the logic levels, namely binary 0 and 1, that are used for computing purposes are described as dominant and recessive respectively. The **dominant** state represents a binary **0** and the **recessive** state a binary **1**. The voltages that constitute the binary 0 and 1 are made up by the difference between the voltage on the H line and that on the L line (see Table 24.2).

The CAN bus

The two wires that are used to make up the CAN bus are twisted together throughout their length and are normally incorporated into the wiring loom. This method of forming the bus is known as a twisted pair (Fig. 24.5) and it is used to minimize the risk of electrical interference that can occur when data is being transmitted.

Bus layout

For study purposes the CAN bus is represented by a pair of parallel lines as shown in Fig. 24.6. The ECUs that are networked are called nodes. Each node is connected to the two bus wires that are known as CAN_H and CAN_L.

Table 24.1 Data transmission rates

Maximum permitted bit rate	Maximum length of CAN bus (m)
1 Mbit/s	40
500 kbit/s	100
250 kbit/s	250
125 kbit/s	500
40 kbit/s	1000

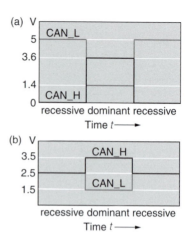

Fig. 24.4 CAN voltages

Table 24.2 CAN voltages

	CAN_H line	CAN_L line	Logic level	Voltage
High-speed CAN-C				
Dominant state	3.5 V	1.5 V	Binary 0	2 V
Recessive state	2.5 V	2.5 V	Binary 1	0 V
Low-speed CAN-B				
Dominant state	3.6 V	1.4 V	Binary 0	2.2 V
Recessive state	0 V	5 V	Binary 1	5 V

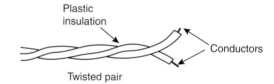

Fig. 24.5 A twisted pair

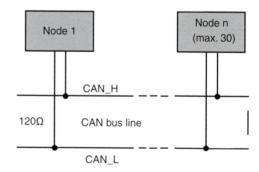

Fig. 24.6 Nodes on the CAN bus

The CAN node

The CAN nodes contain a microcontroller, which is a computer that is able to perform a range of functions. Each node is able to transmit and receive data, and this is achieved by means of a device called a transceiver. Each message on the bus is read by all nodes but the message is only used by those nodes that require it. This is achieved by means of an identifier code that is incorporated into the message.

Should two nodes attempt to connect to the bus simultaneously there is a risk that a collision may occur, resulting in incorrect working of systems. In order to prevent this happening, a system known as 'wired AND' is used; in effect, when the bus has a binary 0 it will only accept an input of the same value. In general, if two messages are trying to access the bus simultaneously the one whose identifier has the lowest binary value will be given access.

Error protection

Because vehicle systems operate under a wide range of hostile conditions, such as varying temperatures — very hot and very cold — and varying voltages, there is a risk of errors. To guard against this, checks are made and when an error in data occurs CAN devices keep a count of the number of times that the error occurs. CAN devices are able to distinguish between temporary errors and permanent ones. When a device is deemed to be permanently defective it disconnects itself electrically from the network. Under such conditions the remainder of the network continues to function.

Cyclic redundancy check (CRC) error detection

Messages that are transmitted to the data bus are checked by the transmitting device and the receiving device using a mathematical process that incorporates a cyclic redundancy code. If the message at the receiving end does not match that sent by the transmitting device the message is rejected.

Types of CAN

High-speed CAN (CAN-C) conforms to ISO Standard 11898-2 and operates at 125 kbit/s up to 1 Mbit/s. It is used for those systems that operate in real time, such as:

- Engine management
- Transmission control, including traction control
- Electronic and anti-lock braking systems
- Vehicle stability systems
- Instrument cluster.

Low-speed CAN (CAN-B) is covered by ISO Standard 11898-3 and it operates at data transmission speeds of 5 kbit/s up to 125 kbit/s. It is used for comfort and convenience systems and others in the body area, such as:

- Air-conditioning

- Lighting system
- Seat adjustment
- Mirrors
- Sliding sun-roof
- Doors.

Number of nodes on a CAN network

Should one system that is connected to the CAN bus fail, the other systems on the bus continue to function. Systems can be added. In most road vehicle applications the number of nodes on a CAN bus is limited to a maximum of 30.

ISO 9141 diagnostic bus system

The International Standards Organization (ISO) develops standards for vehicle bus systems; a significant one that applies to vehicle systems is known as ISO 9141. The ISO 9141 diagnostic bus system is a system that networks vehicle computer control units for diagnostic purposes. The ISO 9141 system may be a single wire known as the K-line (Fig. 24.7), or it may be a two-wire system when the wires are known as the K-line and the L-line. The K-line is connected to a diagnostic socket, which permits certain scan tools and manufacturers' dedicated test equipment to access fault codes and other diagnostic information. The K-line bus operates at a slow speed of 10.4 kbaud.

Under European on-board diagnostics (EOBD) and OBD2 regulations, all systems that contribute to vehicle emissions control must be accessible, for diagnostic purposes, through the diagnostic connector.

Other bus systems used on vehicles

There are several areas of vehicle control where data buses can be used to advantage. Some of these, such as lighting and instrumentation systems, can operate at fairly low speeds of data transfer, say 1000 bits per second, while others such as engine and transmission control require much higher speeds, probably 250 000 bits per second, and these are said to operate in 'real time'. To cater for these differing requirements, the Society of Automotive Engineers (SAE) recommends three classes, known as Class A, Class B, and Class C:

- **Class A.** Low-speed data transmission. Up to 10 kbits per second. Used for body wiring such as exterior lamps, etc.
- **Class B.** Medium-speed data transmission. 10 kbits per second up to 125 kbits per second. Used for vehicle speed controls, instrumentation, emission control, etc.
- **Class C.** High-speed (real-time) data transmission. 125 kbits per second up to 1 Mbits (or more) per second. Used for brake by wire, traction, and stability control, etc.

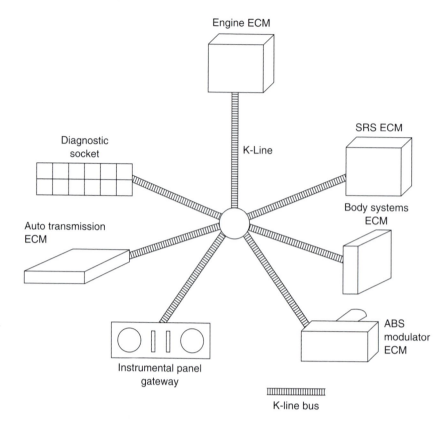

Fig. 24.7   The K-line diagnostic bus

Gateway

The instrument pack acts as a central point, or **gateway**, to which all bus systems are connected. The instrument pack is equipped with sufficient computing power and interfaces to allow the various systems to share sensor data and other information. This means that bus systems such as CAN and the ISO 9141 diagnostic system, which operate at different **baud rates** and use different **protocols**, can communicate effectively for control and diagnostic purposes.

Power supply and drivers

The electrical power supply for the various actuators, such as ABS modulators, fuel injectors, ignition coils, and similar devices, comes from the battery and alternator; on cars this is a 12-volt supply. The actuators are operated by heavy current transistorized circuits (switches) that are operated by the computers (ECUs) that control the respective systems. These transistor switches are known as drivers and a simplified example is shown in Fig. 24.8.

Multiplexing

The 12-volt supply to each injector or other electrically operated unit is carried through a separate cable, which means that there are many cables on a single vehicle.

In order to overcome some of the problems associated with multiple cables, a system known as multiplexing is used. A multiplexer is an electronic device that has several inputs and a single output, and it is used to reduce the number of electrical cables that are needed on some vehicle systems. Some estimates suggest that as much as 15 kg of wire can be eliminated by the use of networking and multiplexing, on a single vehicle.

Figure 24.9 shows the basic concept of multiplexed vehicle wiring – in this case, lights and heated rear window (HRW). In order to keep it as simple as possible, fuses etc. have been omitted. As the legend for the figure states, the broken line represents the data bus. This is the electrical conductor (wire), which conveys messages along the data bus to the respective remote control units. These messages are composed of digital data (zeros and ones). The rectangles numbered 1, 2, 3, and 4 represent the electronic interfaces that permit two-way communication between the ECU and the lamps, or the heated rear window. The dash panel switches are connected to a multiplexer (MUX), which permits binary codes to represent different combinations of switch positions to be transmitted via the ECU on to the data bus. For example, side and tail lamps on and the other switches off could result in a binary code of 1000, plus the other bits (zeros and ones) required by the protocol, to be placed on the data bus so that the side lamps are energized. Operating other switches, e.g. switching on the HRW, would result in a different code and this would be

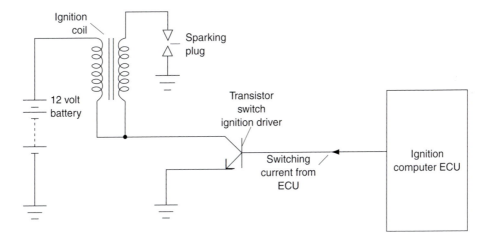

Fig. 24.8 A transistor driver in an ignition circuit

transmitted by the ECU to the data bus, in a similar way. A process known as time-division multiplexing is used and this allows several systems to use the data bus. In effect, the devices are switched on and off many times each second, with the result that there is no effect that is visible to the human eye.

European on-board diagnostics (EOBD) and OBD2

The USA version of on-board diagnostics, known as OBD2, for electronic systems has been in use since the mid-1990s. The European version is much the same as OBD2 and it has been in use since 2000. The regulations for both of these systems require that vehicles should be equipped with a standardized means

of gaining access to defects that may affect the performance of emissions control systems.

European on-board diagnostics (EOBD)

In order to effectively maintain and repair networked and multiplexed systems, it is necessary for diagnostic access to be gained to all systems. Since January 2001, all petrol-engined vehicles sold in Europe must be equipped with a self-diagnostic system that alerts the driver to any fault that may lead to excessive emissions. To comply with EOBD rules, vehicles must be equipped with a malfunction indicator lamp (MIL). The purpose of the MIL is to alert drivers to the presence of a problem and it operates as follows: when the ignition is switched on the MIL illuminates and goes out as soon as the engine is started. If a fault is detected

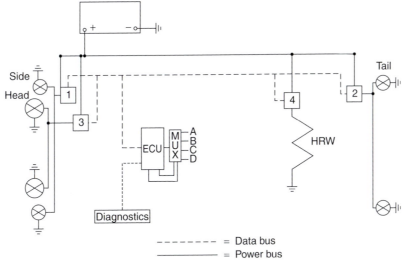

```
--------- = Data bus
————————— = Power bus
```

1 & 2. Electronic switches for side and tail lights.
3. Electronic switch for head lights.
4. Electronic switch for rear window demister.
A, B, C & D. Dash panel switches for lights etc.

Fig. 24.9 The multiplexed wiring concept

while the engine is running, the MIL operates in one of two ways:

1. The MIL comes on and stays on — this indicates a fault that may affect emissions and indicates that the vehicle must be taken to an approved repairer to have the fault investigated, as soon as possible.
2. The MIL flashes at regular intervals, once per second. This indicates an emergency situation and requires the driver to reduce speed because the control computer (ECM) has detected a misfire that has the potential to harm the catalytic converter. Continued flashing of the MIL requires the vehicle to be taken to an approved repairer for early attention.

Additionally, vehicles must be equipped with a standardized diagnostic port that permits approved repairers to gain access to fault codes and other diagnostic data.

Engine management systems

Engine management systems are designed to make the vehicle comply with emissions regulations, as well as to provide improved performance. This means that the number of sensors and actuators is considerably greater than for a simple fuelling or ignition system. The system shown in Fig. 24.10 is fairly typical of modern engine management systems; selected items of technology from the list are picked out for closer attention so that

the aim of 'teasing out' features they have in common with other components is realized. The aim here is to concentrate on the aspects of engine control that were not covered in the sections on fuel and ignition systems.

The intention here is to focus on some of the components in Fig. 24.10. The first component to note is the oxygen sensor at number 20. This is a heated sensor (HEGO) and the purpose of the heating element is to bring the sensor to its working temperature as quickly as possible. The function of the HEGO is to provide a feedback signal that enables the ECM to control the fuelling so that the air—fuel ratio is kept very close to the chemically correct value where lambda = 1, because this is the value that enables the catalytic converter to function at its best. Oxygen sensors are common to virtually all modern petrol-engined vehicles and this is obviously an area of technology that technicians need to know about. The zirconia-type oxygen sensor is most commonly used; it produces a voltage signal that represents oxygen levels in the exhaust gas and is thus a reliable indicator of the air—fuel ratio that is entering the combustion chamber. The voltage signal from this sensor is fed back to the control computer to enable it to hold lambda close to 1.

The downstream oxygen sensor

EOBD and American OBD2 legislation requires vehicles to be fitted with a system that illuminates a warning

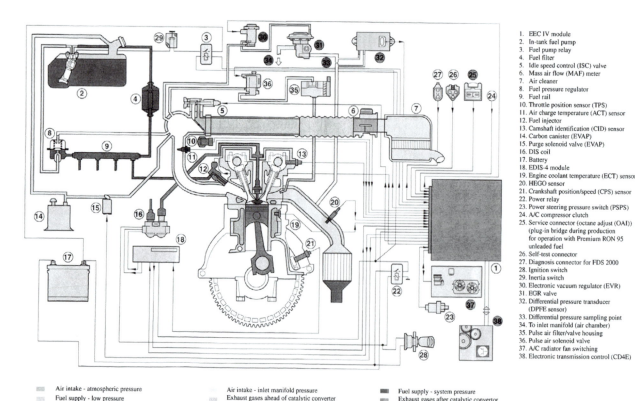

1. EEC IV module
2. In-tank fuel pump
3. Fuel pump relay
4. Fuel filter
5. Idle speed control (ISC) valve
6. Mass air flow (MAF) meter
7. Air cleaner
8. Fuel pressure regulator
9. Fuel rail
10. Throttle position sensor (TPS)
11. Air charge temperature (ACT) sensor
12. Fuel injector
13. Camshaft identification (CID) sensor
14. Carbon canister (EVAP)
15. Purge solenoid valve (EVAP)
16. DIS coil
17. Battery
18. EDIS-4 module
19. Engine coolant temperature (ECT) sensor
20. HEGO sensor
21. Crankshaft position/speed (CPS) sensor
22. Power relay
23. Power steering pressure switch (PSPS)
24. A/C compressor clutch
25. Service connector (octane adjust (OAI)) (plug-in bridge during production for operation with Premium RON 95 unleaded fuel
26. Self-test connector
27. Diagnosis connector for FDS 2000
28. Ignition switch
29. Inertia switch
30. Electronic vacuum regulator (EVR)
31. EGR valve
32. Differential pressure transducer (DPFE sensor)
33. Differential pressure sampling point
34. To inlet manifold (air chamber)
35. Pulse air filter/valve housing
36. Pulse air solenoid valve
37. A/C radiator fan switching
38. Electronic transmission control (CD4E)

Air intake - atmospheric pressure
Fuel supply - low pressure
Fuel vapour

Air intake - inlet manifold pressure
Exhaust gases ahead of catalytic converter

Fuel supply - system pressure
Exhaust gases after catalytic convertor

Fig. 24.10 A typical engine management system

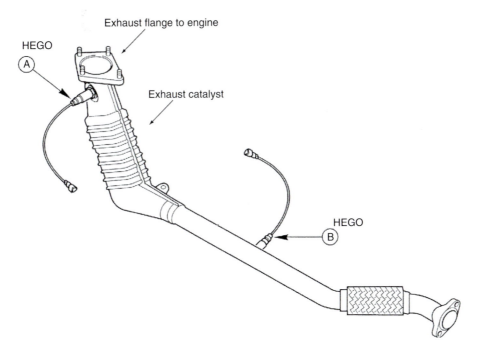

Fig. 24.11 The downstream oxygen sensor

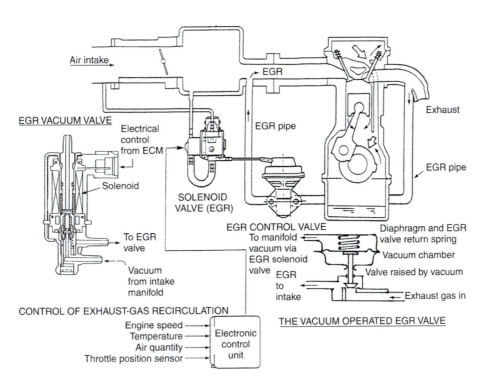

Fig. 24.12 Exhaust gas recirculation system

light when an exhaust catalyst stops working properly. Feedback from the first O_2 sensor is used by the ECU to adjust the amount of fuel injected so that the air–fuel ratio is held at about 15:1; under these conditions the catalytic converter is able to function correctly. The second O_2 sensor that is shown in Fig. 24.11 samples the exhaust gas after it has been processed by the catalyst, and the ECU monitors and compares the second sensor signal with the first; if they are the same, it is taken to mean that the catalyst is not working.

Exhaust gas recirculation

In order to reduce emissions of NO_x it is helpful if combustion chamber temperatures do not rise above approximately $1800°C$ because this is the temperature at which NO_x can be produced. Exhaust gas recirculation (EGR) helps to keep combustion temperatures below this figure by recirculating a limited amount of exhaust gas from the exhaust system, back to the induction system, on the engine side of the throttle valve. Figure 24.12 shows the principle of an EGR system.

Exhaust gas recirculation takes place under the control of the engine management ECU and in order to provide good performance, EGR does not operate when the engine is cold or when it is operating at full load. Figure 24.12 shows the solenoid valve that controls the EGR valve and this type of valve is operated on the duty-cycle principle. Under reasonable operating conditions it is estimated that EGR will reduce NO_x emissions by approximately 30%.

Traction control

The differential gear in the driving axles of a vehicle permits the wheel on the inside of a corner to rotate

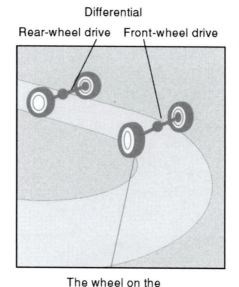

Differential
Rear-wheel drive   Front-wheel drive

The wheel on the
inside does not
travel so far

Fig. 24.13 The need for a differential gear

more slowly than the wheel on the outside of the corner. For example, when the vehicle is turning sharply to the right, the right-hand wheel of the driving axle will rotate very slowly and the wheel on the left-hand side of the same axle will rotate faster. Figure 24.13 illustrates the need for the differential gear.

However, this same differential action can lead to loss of traction (wheel spin). If, for some reason, one driving wheel is on a slippery surface when an attempt is made to drive the vehicle away, this wheel will spin whilst the wheel on the other side of the axle will stand still. This will prevent the vehicle from moving.

The loss of traction (propelling force) arises from the fact that the differential gear only permits transmission of torque equal to that on the weakest side of the axle. It takes very little torque to make a wheel spin on a slippery surface, so the small amount of torque that does reach the non-spinning wheel is not enough to cause the vehicle to move.

Traction control enables the brake to be applied to the wheel on the slippery surface. This prevents the wheel from spinning and allows the drive to be transmitted to the other wheel. As soon as motion is achieved, the brake can be released and normal driving can be continued. The traction control system may also include a facility to close down a secondary throttle to reduce engine power and thus eliminate wheel spin; this action is normally achieved by the use of a secondary throttle, which is operated eletrically. This requires the engine management system computer and the ABS computer to communicate with each other, and the CAN system may be used for that purpose. Figure 24.14 gives an indication of the method of operation of the throttle.

Electric throttle

Drive by wire

The computer-controlled throttle is sometimes referred to as 'drive by wire' because the conventional steel cable or rod connection is discarded. In the type of throttle shown in Fig. 24.14 the throttle butterfly valve

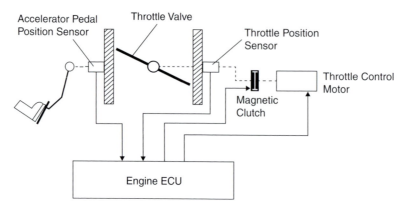

Accelerator Pedal
Position Sensor   Throttle Valve
                                                    Throttle Position
                                                    Sensor

                                                    Throttle Control
                                                    Motor

                                        Magnetic
                                        Clutch

Engine ECU

Fig. 24.14 The electrically operated throttle used with the traction control system

is operated by an electric motor and a magnetic clutch. An analogue signal is sent to the computer from the accelerator pedal sensor and the throttle position sensor; the computer program then decides on the composition of the signal that is sent to the throttle motor and clutch. The throttle valve and engine speed can be varied by the ECU, irrespective of the position of the driver's foot on the accelerator pedal.

An anti-lock braking system (ABS) used in conjunction with an engine control system contains most of the elements necessary for automatic application of the brakes, but it is necessary to provide additional valves and other components to permit individual wheel brakes to be applied. Figure 24.15 shows the layout of a traction control system used on some Volvo vehicles.

In the traction control system shown in Fig. 24.15 the ABS modulator contains extra hydraulic valves, solenoid valves, and bypass valves. These are shown as:

1. The hydraulic valves
2. The solenoid valves
3. The bypass valves.

Figure 24.15 relates to a front-wheel-drive vehicle and for this reason we need to concentrate on the front right (FR) brake and the front left (FL) brake. In this instance, wheel spin is detected at the front right (FR) wheel, which means that application of the FR brake is required.

The solenoid valves (2) are closed and this blocks the connection between the pressure side of the pump (M) and the brake master cylinder. The inlet valve C1 for the front left (FL) brake is closed to prevent that brake from being applied.

The modulator pump starts and runs continuously during the transmission control operation and it takes fluid from the master cylinder, through the hydraulic valve 1, and pumps it to the front right (FR) brake through the inlet valve C4.

When the speed of the front right wheel is equal to that of the front left wheel, the front right brake can be released by computer operation of the valves, and then reapplied until such time as the vehicle is proceeding normally without wheel spin. In the case of spin at the front right wheel that we are considering here, the controlling action takes place by opening and closing of the inlet valve C4 and the outlet valve D4.

When the computer detects that wheel spin has ceased and normal drive is taking place, the modulator pump is switched off, the solenoid valves (2) open, and the valves C4 and D4 return to their positions for normal brake operation. Because the modulator pump is designed to provide more brake fluid than is normally required for operation of the brakes, the bypass valves (3) are designed to open at a certain pressure so that excess brake fluid can be released back, through the master cylinder, to the brake fluid reservoir.

The system is designed so that traction control is stopped if:

1. Wheel spin stops
2. There is a risk of brakes overheating
3. The brakes are applied for any reason
4. Traction control is not selected.

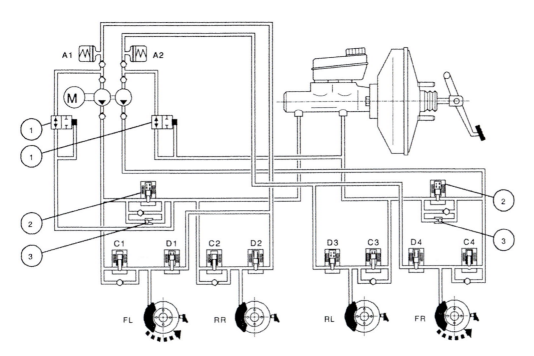

Fig. 24.15 A traction control system

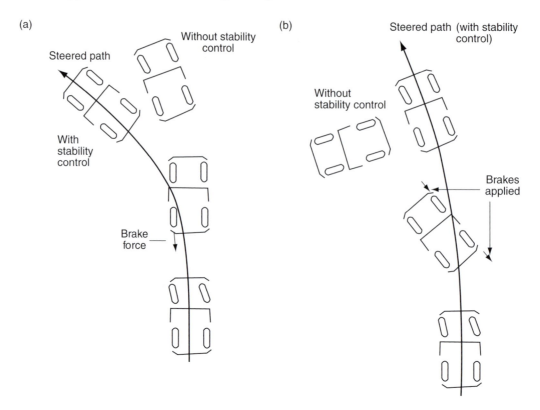

Fig. 24.16 Stability control

Stability control

The capabilities of traction control can be extended to include actions that improve the handling characteristics of a vehicle, particularly when a vehicle is being driven round a bend. The system that results from extending the capabilities of traction control to provide better handling is often referred to as 'stability control'.

Figure 24.16 shows two scenarios. In Fig. 24.16(a), the vehicle is understeering. In effect, it is trying to continue straight ahead and the driver needs to apply more steering effect in order to get round the bend. Stability control can assist here by applying some

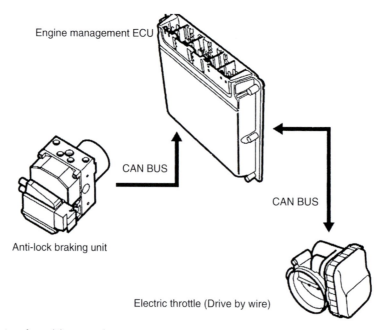

Fig. 24.17 CAN connections for stability control

braking to the wheel on the inside of the bend, at the rear of the vehicle. This produces a correcting action that assists in 'swinging' the vehicle, in a smooth action, back to the intended direction of travel.

In Fig. 24.16(b) oversteer is occurring. The rear of the vehicle tends to move outwards and effectively reduce the radius of turn. It is a condition that worsens as oversteer continues. In order to counter oversteer, the wheel brakes on the outside of the turn can be applied and/or the engine power reduced via the electric throttle, by the computer. In order to achieve the additional actions required for stability control it is necessary to equip the vehicle with additional sensors such as a steering wheel angle sensor and a lateral acceleration sensor that has the ability to provide the control computer with information about the amount of understeer or oversteer.

To achieve stability control it is necessary for the engine control computer and the ABS and traction control computers to communicate; this they do via the CAN network (Fig. 24.17).

Self-assessment questions

1. Serial data transmission:
   (a) requires a separate wire for each bit that is transmitted between the computer and a peripheral such as a fuel injector
   (b) is faster than parallel transmission of the same data
   (c) is transmitted one bit after another along the same wire
   (d) is not used in vehicle systems.
2. In a multiplexed system:
   (a) a data bus is used to carry signals to and from the computer to the remote control units
   (b) each unit in the system uses a separate computer
   (c) more wire is used than in a conventional system
   (d) a high voltage is required on the power supply.
3. In networked systems, messages are divided into smaller packages to:
   (a) prevent problems that may arise because some messages are longer than others

   (b) avoid each computer on the network having to have an interface
   (c) make the system faster
   (d) save battery power.
4. When DTCs are stored in an EEPROM:
   (a) the DTCs are removed when the vehicle battery is disconnected
   (b) an internal circuit must be activated to clear the fault code memory
   (c) they can only be read out by use of a multimeter
   (d) they can only be removed by replacing the memory circuit.
5. A traction control system:
   (a) uses a high-speed bus
   (b) is not networked because it does not need to work with other systems on the vehicle
   (c) is not used on front-wheel-drive vehicles
   (d) can only operate on vehicles equipped with a differential lock.
6. The computer clock is required:
   (a) to permit the time of day to be displayed on the instrument panel
   (b) to allow the time and date to be stored for future reference
   (c) to create the electrical pulses that regulate the flow of data
   (d) to generate the voltage levels required for operation of a data bus.
7. In the CAN system:
   (a) a 'twisted pair' of wires is used so that the correct length of cable can be placed in a small space
   (b) a 'twisted pair' of wires is used to provide the two different voltage levels and minimize electrical interference
   (c) the 'twisted pair' of wires carries the current that drives the ABS modulator
   (d) is used only for the engine management system.
8. The RAM of the ECM computer is:
   (a) the part of memory where sensor data is held while the system is in operation
   (b) only used for storing of fault codes
   (c) held on a floppy disc
   (d) not a volatile memory.

25
Electrical and electronic principles

Topics covered in this chapter

Basic principles of electricity and electronics

Electric current

An electric current is a flow of electrons through a conductor such as a copper wire. It is the movement of free electrons that causes an electric current. Free electrons occur in most metals.

Atoms and electrons

All materials are made up of atoms. An atom has a **nucleus**, around which one or more **electrons** circle in different orbits. An electron has a **negative** electric charge and a nucleus has a **positive** one.

The electrons orbit around the nucleus in rings that are known as **shells**, as shown in Fig. 25.1. The electrons in the outer shells are less strongly attracted to the nucleus than those closer to it. These outer electrons can be moved out of their orbit; they are called **free electrons**.

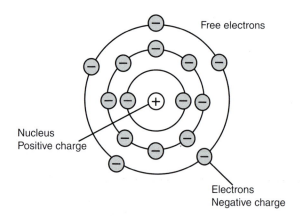

Fig. 25.1 The atom and electron orbits

Conductors and insulators
Conductors

Materials in which free electrons can be made to move fairly easily are known as conductors. Metals such as copper, silver, and aluminium are good conductors of electricity.

Semiconductors

Semiconductors are materials whose free electrons' ability to move falls between that of conductors and insulators. Silicon and germanium are examples of semiconductors.

Insulators

Materials in which free electrons are not readily moved are known as insulators. Ceramics, such as those used on spark plugs, and plastics that are used to insulate cables are examples of insulators that are used in motor vehicle systems.

Electromotive force

The force that moves free electrons is called an electromotive force (e.m.f.). The symbol E is used to denote e.m.f. and it is measured in volts (V). Electromotive forces are produced by power sources. There are three principal types of power source:

1. **Chemical** — batteries. The wet cell shown in Fig. 25.2(a) consists of two metal plates, to which are attached the terminals (poles). The liquid that surrounds the metal plates is called the electrolyte and it conducts electricity. The electric potential (voltage) at the terminals is caused by a chemical reaction between the metal plates and the electrolyte.
2. **Magnetic** — generators (alternators and dynamos). Figure 25.2(b) shows the basic elements of a dynamo. When the loop of wire is moved in a circular motion between the poles of a magnet an electric potential difference (volts) exists at the two ends of the loop.

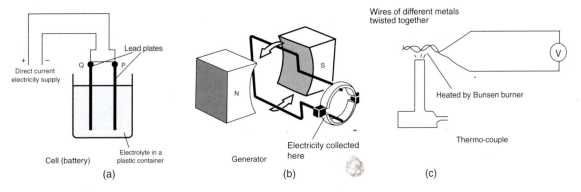

Fig. 25.2 Electrical power sources

3. **Thermal** — thermocouples, used in temperature measurement. Figure 25.2(c) shows two wires of different metals, such as copper and iron, that are twisted together at one end. When the twisted end is heated, a small e.m.f. exists at the opposite ends of the wires. This type of device is called a thermocouple and is used to measure high temperatures such as the temperature of exhaust gas from an engine.

Effects of electric current — using electricity

1. **Heating effect.** When electric current flows through a conductor, the conductor heats up. This effect is seen in electric fire elements, light bulbs (lamps), and similar devices.
2. **Chemical effect.** The main battery on a vehicle makes use of the chemical effect of an electric current to maintain its state of charge, and to provide a source of electric current as required.
3. **Magnetic effect.** The magnetic effect of an electric current is made use of in components such as starter motors, alternators, solenoids, ignition coils, and many other devices.

In order to make use of a power source it must be connected to a circuit. The symbols shown in Fig. 25.3 are used to represent circuit components such as batteries and switches.

The electrical pressure at any point in an electrical system is called the potential. The potential difference (p.d.) is the difference in electrical pressure between two points in an electrical system. When two points of different p.d. are joined together by an electrical conductor, an electric current will flow between them. Potential difference between two points in a circuit can exist without a current. Current cannot exist in a circuit without a potential difference.

The circuit shown in Fig. 25.4 consists of a cell (battery) that is connected to a lamp (bulb) and a switch by means of conducting wires. When the switch is closed (on), the e.m.f. (voltage) at the battery will cause an electric current to flow through the lamp and the switch, back to the cell, and the lamp will illuminate.

By convention, electric current flows from high potential (+; positive) to low potential (−; negative).

Electrical units

Volt

Potential difference and electromotive force are measured in volts (V).

Ampere

Electric current is measured in amperes (A) and it is the number of electrons that pass through a cross-section of the conductor in one second. The symbol used for electric current is I.

Ohm

All materials resist the passage of electricity to some degree. The amount of resistance that a circuit, or electrical component, offers to the passage of electricity is known as the **resistance**. Electrical resistance is measured in ohms and is denoted by the symbol Ω.

Resistors

The component that provides resistance is called a resistor. Several types of resistor are used in vehicle circuits:

- Moulded carbon. These are made from a paste of graphite and resin, which is pressed into shape and then covered with insulating material.
- Wire wound. These are made from a coil of wire that is wrapped round a former. They are sometimes called power resistors because they are capable of dissipating heat.

Name of device	Symbol	Name of device	Symbol
Electric cell		Lamp (bulb)	
Battery		Diode	
Resistor		Transistor (npn)	
Variable resistor		Light emitting diode (LED)	
Potentiometer		Switch	
Capacitor		Conductors (wires) crossing	
Inductor (coil)		Conductors (wires) joining	
Transformer		Zener diode	
Fuse		Light dependent resistor (LDR)	

Fig. 25.3   Circuit symbols

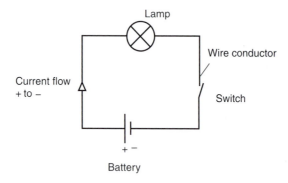

Fig. 25.4   Simple circuit

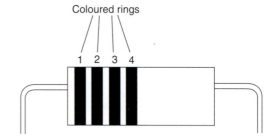

Fig. 25.5   Colour coding of resistors

- Carbon film. These are made by evaporating a thin layer of carbon on to a ceramic insulator.
- Metal film. These are made by evaporating a thin layer of metal on to a ceramic former.

The coloured bands on the resistor (Fig. 25.5) represent the resistance value in ohms. The first two bands give the first two digits of the number. The third band gives the power of 10 by which the first number is multiplied. The fourth band is the tolerance (see Table 25.1).

A resistor with the following bands:

1. Red
2. Red
3. Orange
4. Brown

has a resistance of $22 \times 10^3 = 22\,000 \pm 220\ \Omega$.

Table 25.1 Colour code for resistors

Colour of band	First band	Second band	Third band	Fourth band Colour	Tolerance
Black	0	0	×1	None	20%
Brown	1	1	×10	Silver	10%
Red	2	2	×10^2	Gold	5%
Orange	3	3	×10^3	Red	2%
Yellow	4	4	×10^4	Brown	1%
Green	5	5	×10^5		
Blue	6	6	×10^6		

Watt

Power in a circuit is measured in watts. It is calculated by multiplying the current in the circuit by the voltage applied to it:

$$\text{Power in watts} = \text{volts} \times \text{amperes}$$
$$W = V \times I.$$

Example 25.1. Ohm's law

An electrical component has a current of 5 A flowing through it. If the voltage drop across the component is 10 V, calculate the power used by the component.

$$W = V \times I$$
$$= 10 \times 5$$
$$\text{Power in watts} = 50\ \text{W}.$$

Figure 25.6 shows a circuit that can be used to examine the relation between current, voltage, and resistance in a circuit when the voltage is varied while the resistance and its temperature are held constant.

In this practical exercise, the current in the circuit is varied by changing the setting of the variable resistor. This causes the voltage across the fixed resistor to change. About eight readings of voltage and current are taken and then voltage is plotted against current on a graph. The result should be a straight line of the form shown in Fig. 25.6. This graph shows that the current in this circuit is directly proportional to the voltage producing it, provided that the temperature of the circuit remains constant. This statement is known as Ohm's law and it is normally written in the form:

$$V = I \times R,$$

where V = voltage, I = current in amperes, and R is the resistance in ohms.

Example 25.2. Resistors in series

Figure 25.7(a) shows a circuit that has a 12-volt supply and two resistors that are connected by a conducting cable. When resistors are connected as shown they are in **series** and the resistance of the circuit is found by adding the two resistance values together. In this case,

$$R = R_1 + R_2$$
$$R = 4\ \Omega + 8\ \Omega$$
$$R = 12\ \Omega.$$

Ohm's law can now be applied to find the current in the circuit:

$$I = \frac{V}{R}$$
$$I = \frac{12\ \text{V}}{12\ \Omega}$$
$$I = 1\ \text{A}.$$

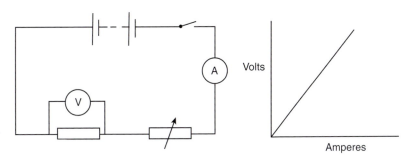

Fig. 25.6 Ohm's law circuit

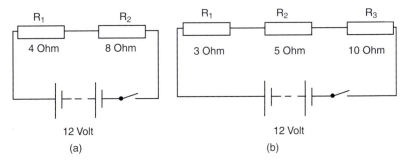

Fig. 25.7 Resistors in series

In Fig. 25.7(b), the total resistance is given by:

$$R = R_1 + R_2 + R_3$$
$$R = 3\,\Omega + 5\,\Omega + 10\,\Omega$$
$$R = 18\,\Omega.$$

The current in this case is:

$$I = \frac{V}{R}$$
$$I = \frac{12}{18}$$
$$I = 0.67\ \text{A}.$$

In a series circuit the current is the same in all parts of the circuit.

Example 25.3. Resistors in parallel

Figure 25.8 shows a circuit in which two resistors are placed in parallel. This ensures that the same p.d. (voltage) is applied to each resistor. Ohm's law can be used to find the current in each resistor as follows:

$$\text{For resistor } R_1, I = \frac{V}{R_1}$$
$$= \frac{12}{6}$$
$$I_1 = 2\ \text{A}.$$
$$\text{For resistor } R_2, I = \frac{V}{R_2}$$
$$= \frac{12}{4}$$
$$I_2 = 3\ \text{A}.$$

The total current drawn from the supply is:

$$I = I_1 + I_2 = 2 + 3 = 5\ \text{A}.$$

There is an alternative method of finding total current in a circuit containing resistors in parallel. The equivalent series resistance of the resistors in parallel can be found as follows:

Let $R =$ the equivalent series resistance, then:

$$\frac{1}{R} = \frac{1}{R_1} + \frac{1}{R_2}.$$

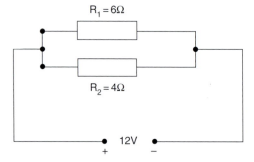

Fig. 25.8   Resistors in parallel

Using the rules for fractions, the equivalent series resistance is:

$$R = \frac{R_1 R_2}{R_1 + R_2}.$$

Example 25.4

The circuit in Fig. 25.8 has the two resistors, $R_1 = 6\,\Omega$ and $R_2 = 4\,\Omega$, in parallel. Using the equation for the equivalent series resistance:

$$R = \frac{6 \times 4}{6 + 4}$$
$$= \frac{24}{10}$$
$$R = 2.4\,\Omega.$$

To find the current in the circuit:

$$I = \frac{V}{R}$$
$$= \frac{12}{2.4}$$
$$I = 5\ \text{A}.$$

Measuring current and voltage

See Fig. 25.9. The ammeter measures the current and it is connected in **series** so that all of the current in the circuit flows through it. The voltmeter is connected in **parallel** and it measures the potential difference between the two points to which it is connected. The p.d. between two points is sometimes referred to as the voltage drop.

Ohmmeter

An ohmmeter is essentially an ammeter that has been adapted to measure resistance (see Fig. 25.10). The ohmmeter operates by applying a small voltage by means of a battery incorporated into the meter. The current flowing is then recorded on a scale that is calibrated in ohms (Ω).

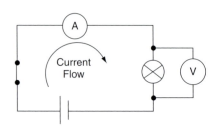

Fig. 25.9   Ammeter and voltmeter connected to a circuit

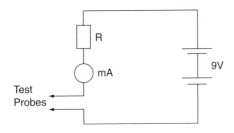

Fig. 25.10 Principle of the ohmmeter

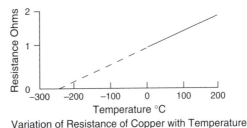

Variation of Resistance of Copper with Temperature

Fig. 25.13 Temperature coefficient of resistance

Open circuit

An open circuit (Fig. 25.11) occurs when there is a break in the circuit; this may be caused by a broken wire or a connection that has worked loose.

Short-circuit

A short-circuit occurs when the conducting part of a wire comes into contact with the return side of the circuit. Because electric current takes the path of least resistance, the short-circuit takes the current back to the battery and the lamp shown in Fig. 25.12 will not function.

Temperature coefficient of resistance

When pure metals such as copper, iron, tungsten, etc. are heated the resistance increases. The amount that the resistance rises for each $1°C$ rise in temperature is known as the temperature coefficient of resistance (see Fig. 25.13).

When some semiconductor ceramic materials are heated their resistance decreases. For each degree rise

in temperature there is a decrease in resistance. These materials are said to have a negative temperature coefficient (Fig. 25.14) and they are used in temperature sensors in various vehicle systems.

Electricity and magnetism

Permanent magnets

Permanent magnets are found in nature; magnetite, which is also known as lodestone, is naturally magnetic and ancient mariners used it as an aid to navigation. Modern permanent magnets are made from alloys. One particular permanent (hard) magnetic material named 'alnico' is an alloy of iron, aluminium, nickel, cobalt, and a small amount of copper. The magnetism that it contains arises from the structure of the atoms of the metals that are alloyed together.

Magnetic force is a natural phenomenon. To deal with magnetism, rules about its behaviour have been established. The rules that are important to our study are related to the behaviour of magnets. The main rules that concern us are:

- Magnets have north and south poles.
- Magnets have magnetic fields.
- Magnetic fields are made up from lines of magnetic force.
- Magnetic fields flow from north to south.

If two bar magnets are placed close to each other, so that the north pole of one is close to the south pole of the

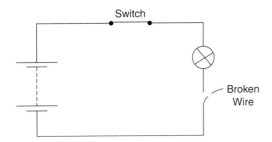

Fig. 25.11 An open circuit

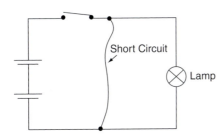

Fig. 25.12 A short-circuit

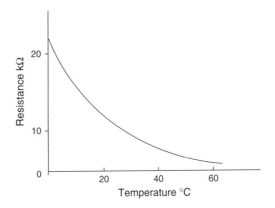

Fig. 25.14 Negative temperature coefficient

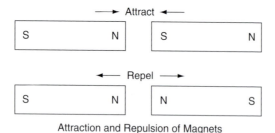

Fig. 25.15 Attraction and repulsion of magnetic poles

other, the magnets will be drawn together. If the north pole of one magnet is placed next to the north pole of the other, the magnets will be pushed apart. This tells us that like poles repel each other and unlike poles attract each other (Fig. 25.15).

The magnetic effect of an electric current

Figure 25.16 shows how a circular magnetic field is set up around a wire (conductor) that is carrying electric current.

Direction of the magnetic field due to an electric current in a straight conductor

Think of screwing a right-hand threaded bolt into a nut, or a screw into a piece of wood. The screw is rotated clockwise (this corresponds to the direction, north to south, of the magnetic field), the screw enters into the wood, and this corresponds to the direction of the electric current in the conductor. The field runs in a clockwise direction and the current flows in towards the wood, as shown in Fig. 25.17. This leads to a convention for representing current flowing into a conductor and current flowing out. At the end of the conductor where current is flowing into the wire, a + is placed; this represents an arrowhead. At the opposite end of the wire, where current is flowing out, a dot is placed; this represents the tip of the arrow (see Fig. 25.18).

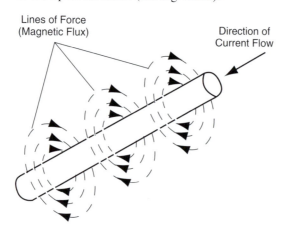

Fig. 25.16 Magnetic field around a straight conductor

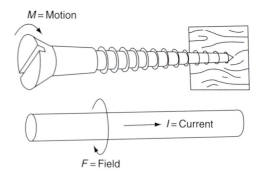

Fig. 25.17 Direction of magnetic field

Fig. 25.18 Current entering and leaving a conductor

Magnetic field caused by a coil of wire

When a conductor (wire) is made into a coil, the magnetic field created is of the form shown in Fig. 25.19; a coil such as this is the basis of a solenoid.

Solenoid and relay

The solenoid principle is used to operate a relay. Relays are used in vehicle systems where a small current is used to control a larger one. When the switch in the solenoid circuit is closed, current flows through the solenoid winding and this causes the magnetic effect to pull the armature towards the soft iron core. This action closes the contacts of the heavy current switch to operate a circuit such as a vehicle headlamp circuit. When the relay solenoid current is switched off, the magnetic effect dies away and the control spring pulls the armature away from the soft iron core to open the heavy current switch contacts.

Electromagnetic induction

Figure 25.20 shows a length of wire that has a voltmeter connected to its ends. The small arrows pointing from the north pole to the south pole of the magnet represent lines of magnetic force (flux) that make a magnetic field. The arrow in the wire (conductor) shows the direction of current flow and the larger arrow, with the curved tail, shows the direction in which the wire is being moved.

Movement of the wire through the magnetic field, so that it cuts across the lines of magnetic force, causes an electromotive force (e.m.f.; voltage) to be produced in the wire. In Fig. 25.20(a), where the wire is being moved upwards the electric current flows in towards the page. In Fig. 25.20(b) the direction of motion of the wire is downwards, across the lines of magnetic force, and the

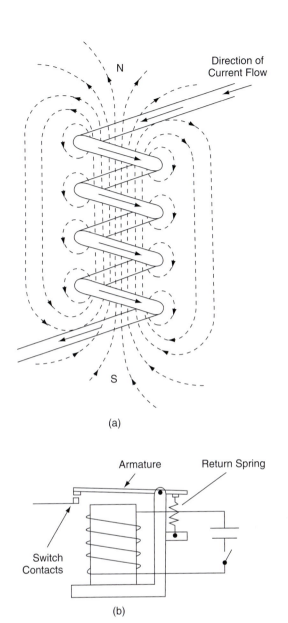

(a)

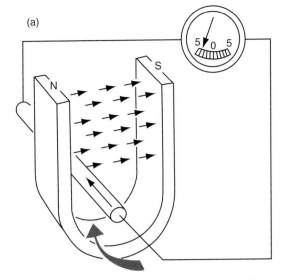

(a)

An EMF (Electromotive Force) is produced in the wire whenever it moves across the lines of magnetic flux.

Note: The reversal of current takes place when direction of movement is reversed.

(b)

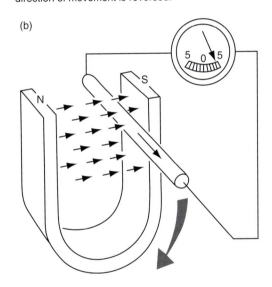

Fig. 25.20 Current flow, magnetic field, and motion

Fig. 25.19 Magnetic field of a coil

Armature   Return Spring

Switch Contacts

(b)

current in the wire is in the opposite direction. Current is also produced in the wire if the magnet is moved up and down while the wire is held stationary. Mechanical energy is being converted into electrical energy, and this is the principle that is used in generators such as alternators and dynamos.

The electric motor effect

If the length of wire shown in Fig. 25.20(a) is made into a loop, as shown in Fig. 25.21, and electric current is fed into the loop, opposing magnetic fields are set up. In Fig. 25.21, the current is fed into the loop via brushes and a split ring. This split ring is a simple commutator. One half of the split ring is connected to one end of the loop and the other half is connected to the other end. The effect of this is that the current in the loop flows in one direction. The coil of wire that is mounted on the pole

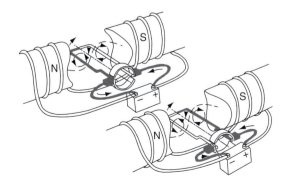

Fig. 25.21 Current flowing through the loop

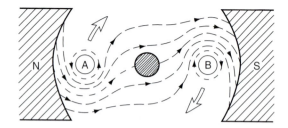

Fig. 25.22 Opposing magnetic fields

pieces, marked N and S, creates the magnetic field. These opposing fields create forces. These forces 'push' against each other, as shown in Fig. 25.22, and they cause the loop to rotate. In Fig. 25.22, the ends of the loop are marked A and B. B is the end that the current is flowing into and A is the end of the loop where the current is leaving the loop.

By these processes, electrical energy is converted into mechanical energy. This is the principle of operation of electric motors, such as the starter motor. Just to convince yourself, try using Fleming's rule to work out the direction of the current, the field, and the motion.

Fleming's rule

Consider the thumb and first two fingers on each hand. Hold them in the manner shown in Fig. 25.23. **Motion** is represented by the direction in which the thumb (M for motion) is pointing. **Field**, the direction (north to south) of the magnetic field, is represented by the first finger (F for field). **Current** − think of the hard **C** in second. The second finger represents the direction of the electric current.

An aid to memory

We think that it is helpful for vehicle technicians to think of the MG car badge. If you imagine yourself standing in front of an MG, looking towards the front of the vehicle, the badge will present you with an M on your left and a G on your right. So you can remember M for Motors and G for Generators − left-hand rule for motors, right-hand rule for generators.

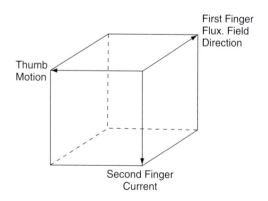

Fig. 25.23 Fleming's rule

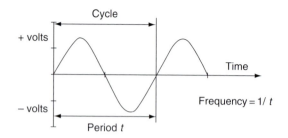

Fig. 25.24 An a.c. waveform

Alternating current

An alternating current produces the type of wave form shown in Fig. 25.24.

- **Cycle.** An alternating voltage rises from zero to a maximum, it then falls back to zero, after that it rises to a maximum in the negative direction before returning to zero. This is called a cycle.
- **Period.** The time taken to complete one cycle is a period. Period is measured in seconds.
- **Frequency.** The frequency is the number of cycles in 1 second. Frequency is measured in hertz (Hz); 1 Hz = one cycle per second.

Applications of alternating current

In a vehicle alternator (Fig. 25.25) the magnet is the rotating member and the coil of wire is the stationary part in which the electrical energy to recharge the battery and operate electrical/electronic systems is generated. The magnetic rotor is driven from the crankshaft pulley and the stationary part − the stator that contains the coils − is attached to the alternator case.

Transformer

A simple transformer consists of two coils of wire that are wound round the limbs of an iron core, as shown

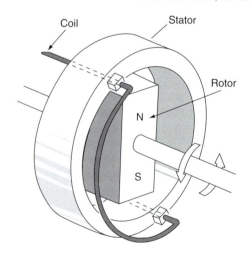

Fig. 25.25 The vehicle alternator

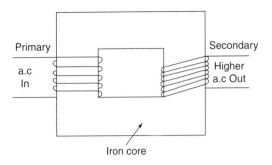

Fig. 25.26 Principle of the transformer

in Fig. 25.26. One coil is the primary winding and the other is the secondary winding.

The iron core concentrates the magnetic field that is created by current in the primary winding. The moving lines of magnetic force that are created by the alternating current in the primary winding cut across the coils in the secondary winding and this creates a current in the secondary winding. The change in voltage that is created by the transformer depends on the number of turns in the coils. For example, if there are 10 turns of coil on the primary winding and 50 on the secondary, the secondary voltage will be five times the primary voltage. This relationship is known as the turns ratio of the transformer.

Lenz's law

In an inductive circuit such as that of the solenoid, there is a reaction to the switching off of the current that creates the magnetic field. As the magnetic field dies away a current is produced that operates in the opposite direction to the magnetizing current. This effect is known as Lenz's law, after the physicist who first explained it in the 1830s.

Lenz's law states: 'The direction of the induced e.m.f. is always such that it tends to set up a current opposing the change responsible for inducing that e.m.f.'.

Inductance

Any circuit in which a change of current is accompanied by a change of magnetic flux, and therefore by an induced e.m.f., is said to possess inductance. The unit of inductance is the henry (H), named after the American scientist Joseph Henry (1797–1878). A circuit has an inductance of $L = 1$ henry (1 H) if an e.m.f. of 1 volt is induced in the circuit when the current changes at a rate of 1 ampere per second. The symbol L is used to denote inductance.

Back e.m.f.

The e.m.f. that is created when a device such as a relay is switched off is known as the back e.m.f. The e.m.f. $= L$ × rate of increase, or decrease, of current:

$$\text{e.m.f.} = L \times \frac{I_2 - I_1}{t} \text{ volts,}$$

where $I_2 - I_1 =$ change of current and $t =$ time in seconds. The back e.m.f. can be quite a high voltage and in circuits where it is likely to be a problem special measures are incorporated to minimize its effect; an example is the surge protection diode in an alternator.

Alternator protection diode

The voltage regulator interrupts the field current through the action of the transistors. When the magnetic field is interrupted a large voltage may be created, which can damage the transistors; to prevent this happening, the protection diode (Fig. 25.27) shorts out the field winding.

Surge protection in a relay

The transistors in computer-controlled circuits may be damaged by the surge voltage that occurs when a relay is switched off; to protect these circuits the relays used are normally equipped with a protection diode of the type shown in Fig. 25.28. The surge voltage may be

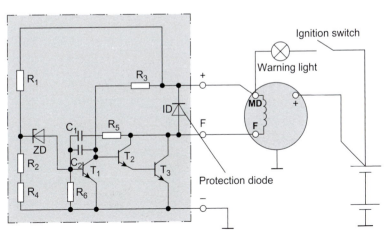

Fig. 25.27 Protection diode in alternator voltage regulator

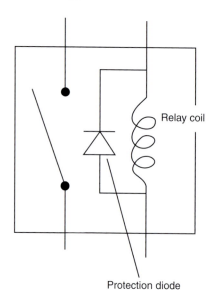

Fig. 25.28   Surge protection diode in a relay

100 volts or so and the diode directs the current back to the relay coil, where it oscillates until its energy is dissipated.

Inductive reactance

Figure 25.29(a) shows that when the switch is on, the rising current causes the coil (inductor) to produce a current that opposes the battery current. In Fig. 25.29(b) the switch is off and the battery current is falling. The current from the inductor keeps a current flowing in the direction of the original battery current. In an a.c. circuit this effect gives rise to inductive reactance that has the same effect as resistance. Inductive reactance is measured in ohms:

$$\text{Inductive reactance } X_1 = 2\pi f L,$$

where f = frequency and L = inductance in henries.

Time constant for an inductive circuit

Figure 25.29(c) shows how the current grows from zero to its maximum value when the switch is on. The time taken for the current to reach two-thirds of its maximum value is known as the time constant:

$$\text{Time constant } t = \frac{L}{R} \text{ seconds},$$

where L = inductance in henries and R = resistance in ohms.

Example 25.5

A circuit containing a coil having an inductance of 2 H is connected in series with a resistance of 4 Ω. Calculate the time constant for the circuit.

$$\text{Time constant } t = \frac{L}{R} = \frac{2}{4} = 0.5 \text{ seconds.}$$

Capacitors

A capacitor is a device that acts as a temporary store of electricity. In its simplest form a capacitor consists of two parallel metal plates separated by a material called a dielectric. The external leads are connected to the two metal plates as shown in Fig. 25.30.

When a voltage is applied to the capacitor electrons are stored on one plate and removed from the other, in equal quantities, until the voltage is the same as that of the supply. The action sets up a potential difference that remains when the external voltage is removed. The capacitor will store an electric charge until it is connected to a circuit; when it discharges into it the voltage may be considerably higher than the charging voltage. The ability of a capacitor to store electricity is called its capacitance. Capacitance is measured in farads after Michael Faraday, who worked in London in the nineteenth century.

Capacitance

Capacitance is determined by the properties of the dielectric and the dimensions of the metal plates. It is calculated from the formula:

$$C = \frac{\varepsilon_0 \varepsilon_r A}{d} \text{ farads.}$$

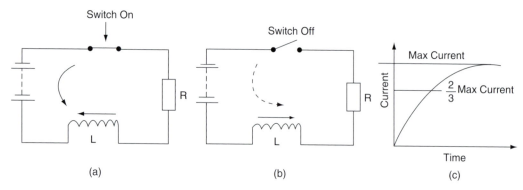

(a)   (b)   (c)

Fig. 25.29   Inductive reactance

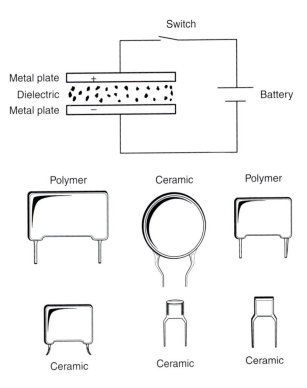

Fig. 25.30 Capacitors

The terms ε_0 and ε_r are properties of a dielectric and they are called permittivity. A is the area of the plates in mm^2 and d is the distance between them in mm. A farad is a very large unit of capacitance and capacitors are normally rated in microfarads, μF.

Capacitance in a circuit

Figure 25.31(a) shows a circuit that can be used to examine the charging of a capacitor. When the switch is closed, the capacitor is discharged. When the switch is opened, the capacitor will recharge. The voltmeter is observed whilst the time is recorded. The result is plotted and a graph of the type shown in Fig. 25.31(b) is produced. This graph shows how the capacitor voltage changes with time. The time taken for the capacitor to reach two-thirds of the supply voltage is known as the time constant for the circuit:

$$\text{Time constant } T = R \times C \text{ seconds,}$$

where R = resistance in ohms and C = capacitance in farads. Capacitance values are normally given in millionths of a farad known as microfarads and denoted by the symbol μF.

Capacitors have the ability to block direct current and pass alternating current.

Example 25.6

A circuit of the type shown in Fig. 25.31(a) has resistance of 10 000 Ω and a capacitance of 10 000 μF. Calculate the time constant for the circuit.

Time constant $T = C \times R = 0.01 \times 10\,000 = 100$ seconds.

Capacitors in circuits

Contact breaker-type ignition

In the contact breaker-type ignition system, a capacitor is placed in parallel with the contact breaker (Fig. 25.32). When the contact points open, a self-induced high-voltage electric charge attempts to arc across the points. The capacitor overcomes this problem by storing the electric charge until the points gap is large enough to prevent arcing.

Capacitive discharge

In the capacitive discharge ignition system (Fig. 25.33) the capacitor stores electrical energy until a spark is required. When a spark is required an electronic circuit releases the energy from the capacitor into the ignition coil circuit to produce a high-voltage spark at the sparking plug.

Capacitors connected in parallel

Figure 25.34(a) shows two capacitors connected in parallel. The total capacitance provided by the two capacitors $= C = C_1 + C_2$. Figure 25.34(b) shows the two capacitors connected in series. In this case the total capacitance is:

$$C = \frac{C_1 C_2}{C_1 + C_2}.$$

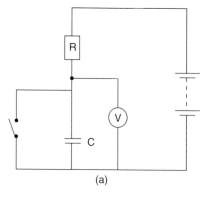

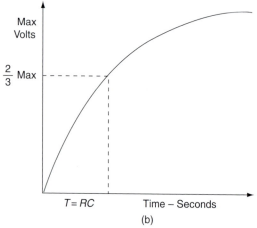

Fig. 25.31   Charging a capacitor

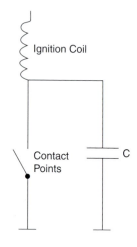

Fig. 25.32   Capacitor in contact breaker ignition circuit

Example 25.7

Two capacitors, one with a capacitance of 8 μF and the other with a capacitance of 4 μF, are connected (a) in parallel, (b) in series. Determine the total capacitance in each case.

(a) Total capacitance $C = 8 + 4 = 12$ μF.

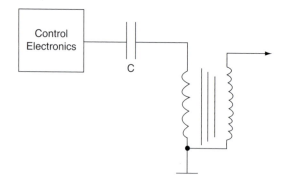

Fig. 25.33   A capacitive discharge ignition system

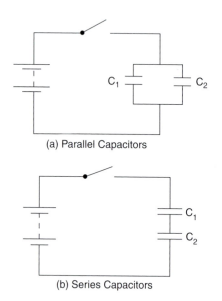

Fig. 25.34   Capacitors in parallel and in series

(b) Total capacitance is:

$$C = \frac{8 \times 4}{8 + 4} = \frac{32}{12} = 2.67 \text{ μF.}$$

Impedance

Figure 25.35 shows a circuit that contains a resistor, an inductor (coil), and a capacitor. The combined effect of these three devices in an a.c. circuit produces an effect known as impedance. Impedance acts like a resistance and is measured in ohms.

Electronic principles

Introduction

Microprocessors are small computers and they are used in many vehicle systems, such as engine management, antilock braking, traction control, etc. In order to function, the microprocessor requires inputs from sensors; these inputs are used to make decisions that produce outputs that operate various actuators. Many of these sensors and actuators

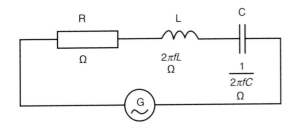

Fig. 25.35 Impedance in a circuit

operate on electronic principles. The subject matter in this chapter is designed to provide an insight into basic electronic principles that will enable readers to take advantage of the many specialized courses that are provided by automobile manufacturers. The basic elements of a microprocessor controlled systems are shown in Fig. 25.36.

Semiconductor materials such as pure silicon have few electrons in their outer shells which makes them poor conductors of electricity. The conductivity of pure silicon is improved by diffusing impurities into it under high temperature. Impurities that are often used for this purpose are boron and phosphorous, and they are called **dopants**.

Effect of dopants

When boron is diffused into silicon the resulting material is known as a **p-type** semiconductor. When phosphorus is diffused into silicon it is known as an **n-type** semiconductor.

Electrons and holes

Phosphorus doping gives **surplus electrons** to the material. These surplus electrons are mobile. Boron doping leads to atoms with insufficient electrons that can be thought of as **holes**; p-type semiconductors have **holes**,

n-type semiconductors have surplus **electrons**. When a p.d. (voltage) is applied, electrons drift from − to +, and holes drift from + to −.

The p—n junction

When a piece of p-type semiconductor is joined to a piece of n-type semiconductor, a junction is formed (Fig. 25.37). At the junction an exchange of electrons and holes takes place until **an electrical p.d., which opposes further diffusion**, is produced between the two sides of the junction. The region where this barrier exists is known as the **depletion layer** and it acts like a capacitor that holds an electric charge.

Forward bias

When a p—n junction is placed in a circuit as shown in Fig. 25.37(b,i), the battery provides a potential difference across the depletion layer. This overcomes the barrier effect and allows electrons to flow through the p—n junction (diode) and the lamp will illuminate. This is known as forward bias.

When the battery connections are reversed, as shown in Fig. 25.37(b,ii), a reverse bias is applied at the p—n junction. This reinforces the barrier effect at the depletion layer and prevents a flow of current through the circuit. The result is that the lamp will not illuminate. The voltage required to overcome the barrier potential is known as the knee voltage and it is approximately 0.7 V for a silicon diode.

A junction diode, or just diode, is an effective one-way valve in an electronic circuit.

Behaviour of p—n junction diode

Figure 25.38(a) shows a circuit in which the resistor can be changed to vary the current in the circuit and hence the voltage across the diode. A series of readings of

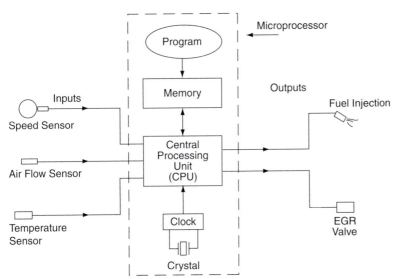

Fig. 25.36 Microprocessor-controlled vehicle system semiconductors

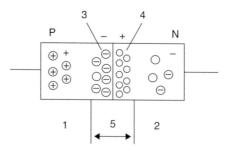

1. Holes
2. Electrons
3. Negative Charges
4. Positive Charges
5. Depletion Layer

(a)

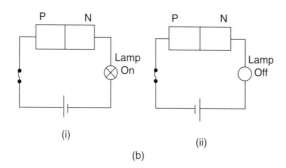

(i)

(ii)

(b)

Fig. 25.37 A p–n junction

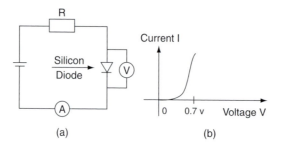

(a)

(b)

Fig. 25.38 Current–voltage in forward-biased diode

current and voltage for various resistance values will produce a graph of the form shown in Fig. 25.38(b). This graph shows that there is very little current below 0.7 V, but as soon as the voltage reaches 0.7 V the current rises rapidly.

Diode protection resistor

If the current in a diode is too large, the heating effect will destroy the diode properties. To prevent this happening a resistor is placed in series with the diode and this restricts the current through the diode to a safe level. The diode current may be calculated from:

$$I = \frac{V_s - V_d}{R},$$

where R = resistance in ohms, V_s = supply voltage, and V_d = diode voltage.

Example 25.8

In the circuit shown in Fig. 25.38(a), the resistor value = 1 kΩ, the supply voltage = 12 V, and the silicon diode voltage = 0.7 V. Calculate the current through the diode.

$$\text{Diode current } I = \frac{12 - 0.7}{1000} = 0.0113 \text{ A} = 11.3 \text{ mA}.$$

Negative temperature coefficient of resistance semiconductor

When the temperature of a semiconductor is increased, thermal activity causes an increase in electrons and holes. These actions increase the conductivity of the semiconductor and the resistance decreases as the temperature rises, as shown in Fig. 25.39.

The Zener diode

The Zener diode is a p–n junction diode that is designed to break down at a specified voltage (Fig. 25.40). The breakdown mechanism is non-destructive provided that the current is limited to prevent overheating. The voltage at which breakdown occurs and current flows in the reverse direction is known as the Zener voltage. Once current flows in the reverse direction, the voltage drop across the Zener diode remains constant at the Zener voltage. Zener voltages normally fall in the range 3–20 V. The actual figure is chosen at the design stage, typical values being 5.6 and 6.8 V. Two common uses of the Zener diode in automotive practice are:

1. Voltage surge protection in an alternator circuit
2. A voltage stabilizer in instrument circuits.

Light-emitting diodes (LEDs)

When electrons have been moved into a larger orbit they may fall back to their original energy level. When this happens, the electrons give up energy in the form of heat, light, and other forms of radiation. The light-emitting diode utilizes this effect and various types of LED emit different colours of light. When elements such as gallium, arsenic, and phosphorus are used, colours such as red, green, blue, etc. are produced.

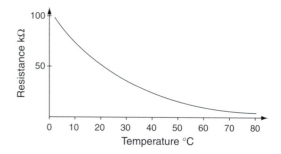

Fig. 25.39 Negative temperature coefficient

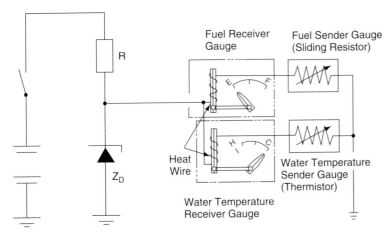

Fuel Receiver Gauge

Fuel Sender Gauge (Sliding Resistor)

Heat Wire

Water Temperature Sender Gauge (Thermistor)

Water Temperature Receiver Gauge

Fig. 25.40 The Zener diode (Z_D) as a voltage regulator

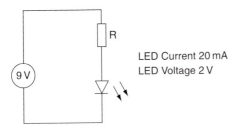

R

9 V

LED Current 20 mA
LED Voltage 2 V

Fig. 25.41 The current-limiting resistor − LED circuit

Voltage and current in an LED

Most LEDs have a voltage drop of 1.5−2.5 V with currents of 10−40 mA. In order to limit the current, a resistor is placed in series with an LED, as shown in Fig. 25.41. The voltage drop across the LED is 2 V and the maximum permitted current is 20 mA. The voltage drop across the resistor is (9 − 2) = 7 V. Using Ohm's law:

$$R = \frac{V}{I} = \frac{7}{0.02} = 350 \ \Omega.$$

Photo-diode

When radiation in the form of light falls on a p−n junction diode some electrons are caused to move, thus creating an electric current. The greater the intensity of the light falling on the diode, the greater the current. This makes photo-diodes suitable for use in a range of automotive systems, such as security alarms and automatic light operation.

Bipolar transistors

Bipolar transistors use both positive and negative charge carriers. This form of transistor is basically two p−n junction diodes placed back to back to give either a p−n−p type or an n−p−n type. In both cases the transistor has three main parts and these are the

collector (c), the emitter (e), and the base (b), as shown in Fig. 25.42(a).

Basic operation of transistor

Figure 25.42(a) shows an n−p−n transistor in a circuit that contains a battery and a lamp. There is no connection to the base − the transistor is not biased. Virtually no current flows in the circuit and the lamp will not illuminate. In Fig. 25.42(b) a voltage source is applied to the base of the transistor and this causes electrons to enter the base from the emitter. The free electrons can flow in either of two directions. That is, they can flow either into the base or into the collector, which is where most of them go. In this phase the lamp begins to light. Once free electrons are in the collector they become influenced by the collector voltage and current flows freely through the main circuit to illuminate the lamp. This phase is shown in Fig. 25.42(c). This switching action takes place at very high speed, which makes transistors suitable for many functions in electronic and computing devices.

Current gain in transistor

Under normal operating conditions the collector current I_c is much larger than the base current I_b. The ratio I_c/I_b is called the gain and it is denoted by the symbol β:

$$\beta = \frac{I_c}{I_b}.$$

Values of current gain vary according to the type of transistor being used; low-power transistors have current gains of approximately 100−300 and high-power transistors have current gains of less than 100.

Current flow in transistors

Figure 25.43 shows the path of current flow in transistors. It is useful to study this when working out how circuits operate.

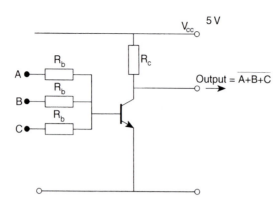

Fig. 25.50   A logic gate circuit

gate is built up from an arrangement of resistors and a transistor. There are three inputs, A, B, and C. If one or more of these inputs is high (logic 1), the output will be low (logic 0). The output is shown as A + B + C with a line, or bar, over the top; the + sign means OR. Thus, A + B + C with the line above means 'not A or B or C' (NOR: NOT OR).

The base resistors R_b have a value that ensures that the base current, even when only one input is high (logic 1), will drive the transistor into saturation to make the output low (logic 0) − RTL stands for resistor transistor logic.

Truth tables

Logic circuits operate on the basis of Boolean logic and terms like NOT, NOR, NAND, etc. derive from Boolean algebra. This need not concern us here, but it is necessary to know that the input−output behaviour of logic devices is expressed in the form of a 'truth table'. The truth table for the NOR gate is given in Fig. 25.51.

In computing and control systems, a system known as TTL (transistor-to-transistor logic) is used. In TTL, logic 0 is a voltage between 0 and 0.8 volts. Logic 1 is a voltage between 2.0 and 5.0 volts.

In the NOR truth table, when the inputs A and B are both 0, the gate output, C, is 1. The other three input combinations each give an output C = 0.

A range of other commonly used logic gates and their truth tables is given in Fig. 25.52.

Self-assessment questions

1. In the circuit shown in Fig. 25.53(a), what is the total series resistance:
   (a) 1.33 Ω
   (b) 6 Ω
   (c) 8 Ω
   (d) 2 Ω.
2. In the circuit in Fig. 25.53(a), what is the current at the ammeter:

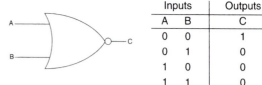

Inputs		Outputs
A	B	C
0	0	1
0	1	0
1	0	0
1	1	0

Fig. 25.51   NOR gate and truth table

   (a) 6 A
   (b) 8 A
   (c) 2 A
   (d) 1 A.
3. In the circuit of Fig. 25.53(a), what is the voltage drop across the 4 Ω resistor:
   (a) 8 V
   (b) 12 V
   (c) 3 V
   (d) 1 V.
4. In the circuit of Fig. 25.53(b), what is the equivalent series resistance of the two resistors in parallel:
   (a) 8 Ω
   (b) 1.5 Ω
   (c) 3 Ω
   (d) 12 Ω.
5. In Fig. 25.53(b), what is current at the ammeter:
   (a) 1 A
   (b) 1.5 A
   (c) 8 A
   (d) 24 A.
6. In Fig. 25.53(c), what is the current in the 2 Ω resistor:
   (a) 2.4 A
   (b) 6 A
   (c) 1.5 A
   (d) 1 A.
7. In the circuit in Fig. 25.53(c), what is the equivalent series resistance of the three resistors:
   (a) 16 Ω
   (b) 12.4 Ω
   (c) 0.55 Ω
   (d) 4 Ω.
8. In the circuit of Fig. 25.53(c), what is the total current flow at the ammeter:
   (a) 48 A
   (b) 3 A
   (c) 0.97 A
   (d) 5 A.
9. In the circuit in Fig. 25.53(c), what is the voltage drop across the 2.4 Ω resistor:
   (a) 9.6 V
   (b) 3 V
   (c) 7.2 V
   (d) 4.8 V.
10. What is the total capacitance of two capacitors in parallel as shown in Fig. 25.53(d):
    (a) 15 000 μF

Type of logic gate	USA symbol	UK symbol	Truth table
AND	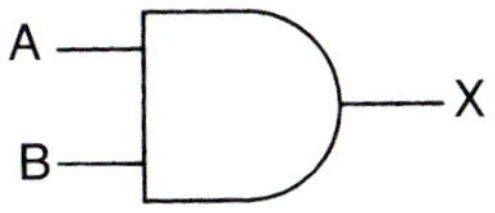	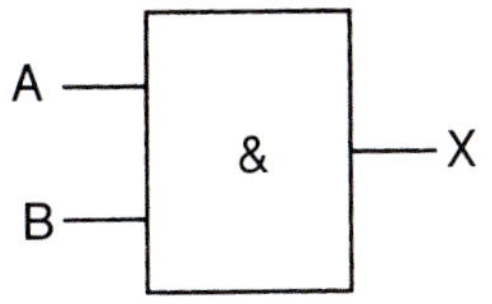	Inputs / Outputs A B | X 0 0 | 0 0 1 | 0 1 0 | 0 1 1 | 1
OR	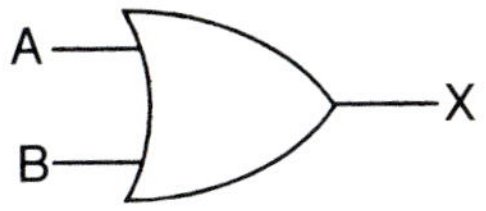	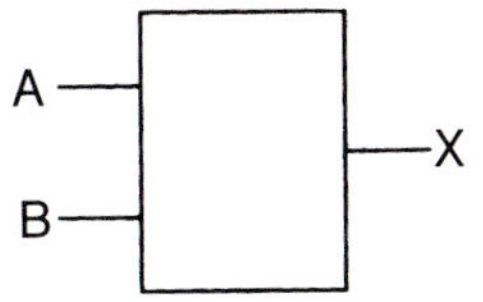	Inputs / Outputs A B | X 0 0 | 0 0 1 | 1 1 0 | 1 1 1 | 1
(NOT AND) NAND	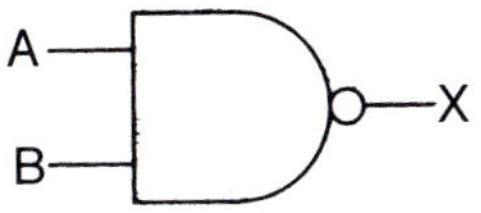	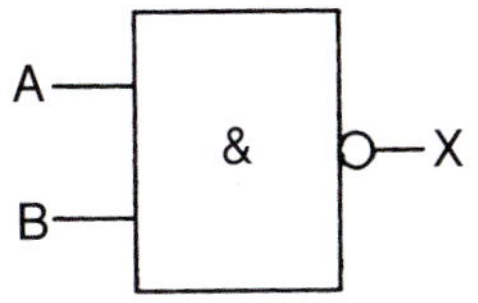	Inputs / Outputs A B | X 0 0 | 1 0 1 | 1 1 0 | 1 1 1 | 0
NOT inverter	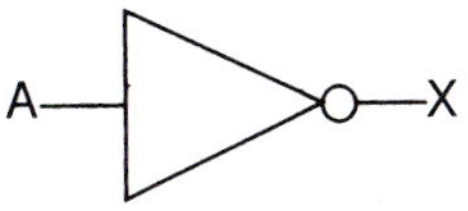	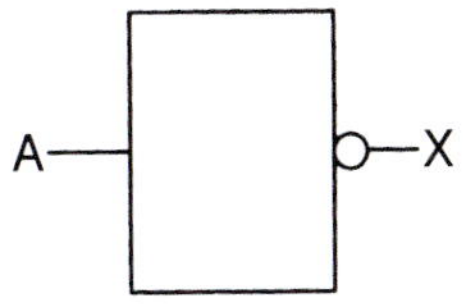	Inputs / Outputs A | X 0 | 1 1 | 0
(NOT OR) NOR	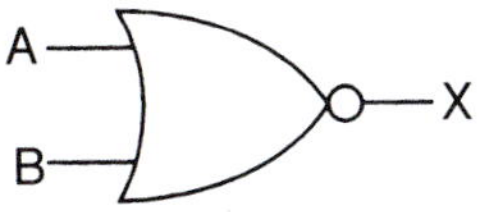	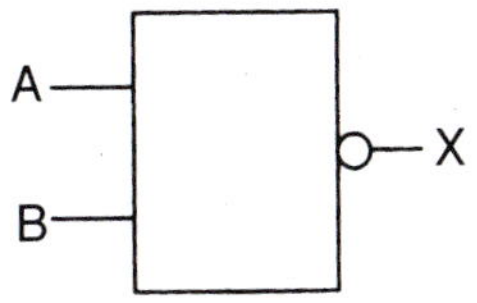	Inputs / Outputs A B | X 0 0 | 1 0 1 | 0 1 0 | 0 1 1 | 0

Fig. 25.52 A table of logic gates and their truth tables

(b) 1.5 μF
(c) 250 μF
(d) 0.67 μF.

11. Does the capacitor in parallel with the ignition points of a coil ignition system, as shown in Fig. 25.53(e):
(a) prevent arcing across the points
(b) act as a suppressor to reduce radio interference
(c) reduce erosion at the spark plug points
(d) prevent overload in the coil primary circuit.

12. The time constant for a capacitor in a circuit that contains resistance is the time that it takes the capacitor to charge to:
(a) 6% of the charging voltage
(b) 67% of the charging voltage
(c) 33% of the charging voltage
(d) the full charging voltage.

13. Fleming's right-hand rule is useful for working out the direction of current flow in:
(a) the armature winding of a starter motor

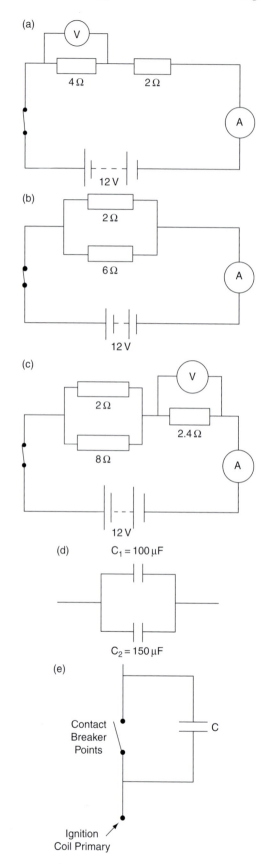

Fig. 25.53   Circuit diagrams

(b) the armature winding of a generator
(c) the winding of a non-inductive resistor
(d) a solenoid.

14. Impedance in an a.c. circuit is the total opposition to current flow that the circuit causes. Is impedance measured in:
   (a) watts
   (b) amperes
   (c) ohms
   (d) coulombs.

15. In a metal conductor the flow of current is caused by the drift of free electrons under the influence of an applied e.m.f. Is the free electron flow:
   (a) from the positive to the negative electrode
   (b) from low electrical potential to higher electrical potential
   (c) from the positive to the negative terminal of the battery
   (d) From high electrical potential to low electrical potential.

16. A 12-volt electric motor takes a current of 20 A. The power of the motor is:
   (a) 1.67 kW
   (b) 0.3 kW
   (c) 240 W
   (d) 32 W.

17. Lenz's law states that:
   (a) The direction of an induced e.m.f. is always such that it tends to set up a current opposing the change that is responsible for inducing the e.m.f.
   (b) $V = IR$
   (c) $T = RC$
   (d) The back e.m.f. is directly proportional to the capacitance.

18. Electrical conductors are materials:
   (a) in which free electrons can be made to move
   (b) that have a high specific resistance
   (c) that do not heat up when electric current flows in them
   (d) that are always physically hard.

19. A circuit that has a resistance of 10 000 Ω and a capacitance of 10 000 µF has a time constant of:
   (a) 10 seconds
   (b) 0.1 seconds
   (c) 100 seconds
   (d) 1000 seconds.

20. Are electronic components such as alternator diodes mounted on a heat sink:
   (a) to avoid damage by vibration
   (b) to conduct surplus electricity to earth
   (c) to allow heat to be conducted away from a diode to prevent damage
   (d) to improve the electrical conductivity of the circuit.

21. A certain transistor has a current gain of 100. When the base current of this transistor is 0.1 mA the collector current is:

(a) 10 A

(b) 10 mA

(c) 99 mA

(d) 1 A.

22. Do active sensors:

(a) process an electrical supply

(b) produce electricity that represents the variable that they are detecting.

23. Do electrical filters:

(a) filter out a.c. frequencies that may interfere with the operation of an electronic system

(b) only work in a d.c. circuit

(c) only find use in radio circuits

(d) never find use in automotive systems.

24. In an OR logic gate:

(a) the output is 1 when both inputs are 0

(b) the output is zero when both inputs are 1

(c) the output is 1 when either or both of the inputs is 1

(d) the output is 0 when any input is 1.

25. Traction control systems have a data transfer rate of:

(a) 10 bits per second

(b) 1000 bits per second

(c) 10 000 bits per second

(d) 250 kbaud to 1 Mbaud.

26. In a computer-controlled automotive system the actuators:

(a) operate devices under instructions from the control computer

(b) sense engine temperature and convert it into a voltage

(c) convert analogue signals into digital ones

(d) store fault codes.

27. The part of the ECU computer that controls the movement of data is:

(a) the A/D interface

(b) the clock

(c) the baud rate

(d) the speed of the vehicle.

28. A petrol injector is:

(a) a sensor

(b) an actuator that operates on the principle of a solenoid

(c) operated from the camshaft

(d) a variable capacitance oscillator.

29. A Zener diode:

(a) is used to convert a.c. to d.c.

(b) is used to provide surge protection

(c) is used as an element in a liquid crystal display

(d) blocks all current flow in the reverse direction.

30. In order to limit the current flow through an LED:

(a) a resistor is placed in parallel with the LED

(b) the supply voltage is limited

(c) a resistor is placed in series with the LED

(d) it is necessary to increase the dopant level.

26
Vehicle electrical systems

Topics covered in this chapter

Batteries
Charging systems
Alternator
Rectifier
Voltage regulator
Starting systems
Charging and starting system diagnostic checks
Circuits and vehicle wiring

Modern vehicles are equipped with many electrical and electronic systems, as shown by the list given in Table 26.1.

We think that it is a good idea to start with the battery. This is where the energy is stored, it operates the starter motor and, as vehicle journeys usually begin by starting up the vehicle engine, this seems like a good point at which to start this study of vehicle electrical and electronic systems.

Table 26.1 The main items of electrical/electronic equipment on vehicles

Vehicle system	Comment
Batteries and charging systems	Alternators, voltage, and current control
Starter motors	Axial engagement
Ignition systems	Distributor type, direct type, distributor-less
Engine management	
Fuel systems	Single-point, multi-point, direct injection
Emission control	Oxygen sensor, catalyst, purge canister
Diagnostics	EOBD
Body electrics	Remote central locking, electrically operated windows, heated screens, screen wash and wipe
Security	Immobilizer, anti-theft alarm
Anti-lock brakes, traction control, stability control	CAN systems
Lights and signals	Headlights, side and tail, stop lights, hazard lights, interior lights

Vehicle batteries

Chemical effect of electric current

If two suitably supported lead plates are placed in a weak solution of sulphuric acid in water (electrolyte) and the plates connected to a suitable low-voltage electricity supply as shown in Fig. 26.1, an electric current will flow through the electrolyte from one plate to the other.

After a few minutes the lead plate P will appear a brownish colour and the plate Q will appear unchanged. The cell has caused the electric current to have a chemical effect.

This electrochemical effect can be reversed, i.e. removing the battery and replacing it with a lamp (bulb) will allow electricity to flow out of the cell and 'light up' the lamp. Electric current will continue to flow out of the battery until the cell has returned to its original state, or until the lamp is disconnected.

As a result of the electrochemical action, the two lead plates develop a difference in electrical potential. One plate becomes electrically positive and the other becomes electrically negative. The difference in electrical potential causes an electric force. This force is known as an electromotive force (e.m.f.). E.m.f.s are

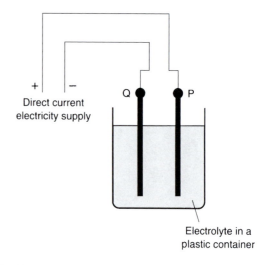

Fig. 26.1 A simple cell

measured in volts and the e.m.f. generated by the two plates, as shown in Fig. 26.1, can be measured by connecting a voltmeter across the metal plates. This simple cell is known as a secondary cell. It can be charged and recharged.

Dry batteries, similar to those used in a hand torch, exhaust their active materials in the process of making an electric current and they cannot be recharged. Such dry batteries are primary cells. These are not to be confused with modern secondary cells of similar appearance.

Because secondary cells can be charged and recharged they are used to make vehicle batteries. There are two types of vehicle batteries: one is lead acid and the other is nickel–iron alkaline. The lead acid battery is the one that is most widely used and it is this type that we propose to concentrate on.

The lead acid vehicle battery

Safety note

Before we proceed with this work on batteries, there are several points that need to be emphasized.

- Batteries are heavy. Proper lifting practices must be used when lifting them.
- Sulphuric acid is corrosive and it can cause serious injury to the person and damage to materials. It must be handled correctly. You must be aware of the actions to take in case of contact with acid. Every workshop must have proper safety provision and the rules that are set must

be observed. In the event of spillage of acid on to a person, the area affected should be washed off with plenty of clean water. Any contaminated clothing should be removed. If acid comes into contact with eyes, they should be carefully washed out with clean water and urgent medical attention obtained.

- Batteries give off explosive gases, e.g. hydrogen. There should be no smoking or use of naked flame near a battery. Sparks must be avoided when connecting or disconnecting batteries. This applies to work on the battery when it is on the vehicle and also when the battery has been removed from the vehicle for recharging.
- When disconnecting the battery on the vehicle, make sure everything is switched off. Remove the earth terminal FIRST.
- When reconnecting the battery, make sure everything is switched off. Connect the earth terminal LAST.

In the case of batteries 'on charge', the battery charger must be switched off before disconnecting the battery from the charger leads.

A typical vehicle battery

Figure 26.2 shows a battery in the engine compartment of a light vehicle. The notes on the diagram indicate factors that require attention.

The type of battery shown in Fig. 26.2 is typical of batteries used on modern vehicles. Most light vehicles use 12-volt batteries. The 12 volts is obtained by

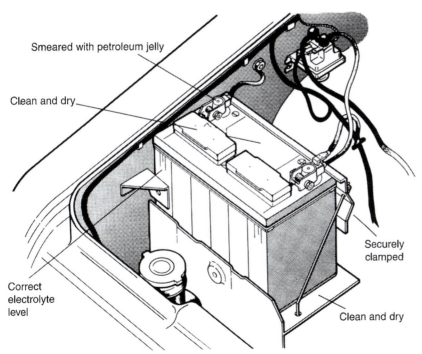

Smeared with petroleum jelly

Clean and dry

Correct electrolyte level

Securely clamped

Clean and dry

Fig. 26.2 The battery installation

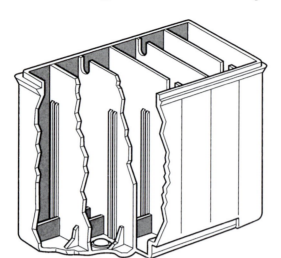

Fig. 26.3   The polypropylene battery case

connecting six cells in series. Each single lead acid cell produces 2 volts.

Some larger trucks and buses use a 24-volt electrical system. In this case, two 12-volt batteries are connected in series to give the 24 volts.

Battery construction

The case

The battery casing needs to be acid resistant and as light and as strong as possible. Plastic material, such as polypropylene, is often used for the purpose. Figure 26.3 shows the construction of a fairly typical battery case.

Each cell of the battery is housed in a separate container and there are ribs at the bottom of the cell compartment that support the plates so that they are clear of the container bottom. Sediment can accumulate in this bottom space without short-circuiting the plates.

The plates are made in the form of a grid, which is cast in lead. In order to facilitate the casting process a small amount of antimony is added to the lead. In maintenance-free batteries, calcium is added to the lead because this reduces losses caused by gassing. Figure 26.4 shows a battery plate.

During the manufacturing process, other materials are pressed into the plate grid. On the positive plate the material added is lead peroxide (a browny-red substance) and on the negative plate it is spongey lead, which is grey. The completed plates are assembled as shown in Fig. 26.5.

The positive plates are connected together with a space in between. The negative plates are also joined together with a space in between. The negative plates are placed in the gaps between the positive plates and the plates are prevented from touching each other by the separators that slide in between them. The separators are made from porous material, which allows the

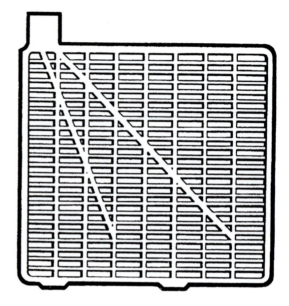

Fig. 26.4   A battery plate

electrolyte to make maximum contact with the surfaces of the plates.

Battery capacity: the ampere-hour (A-H)

The unit of quantity of electricity is the Coulomb. One coulomb is the equivalent of a current of 1 ampere flowing for 1 second. This is a very small amount of electricity in comparison with the amount that is required to drive a starter motor. It is usual, therefore, to use a larger unit of quantity of electricity to describe the

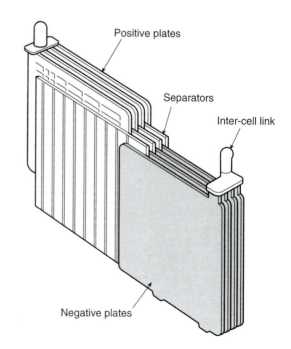

Fig. 26.5   Battery plates

Table 26.2 Relative density of electrolyte and battery charge

State of charge	Relative density
Fully charged	1.290
Half charged	1.200
Completely discharged	1.100

electrical energy capacity of a vehicle battery. The unit that is used is the ampere-hour. One ampere-hour is equivalent to a current of 1 ampere flowing for 1 hour. The size of the battery plates and the number of plates per cell are factors that have a bearing on battery capacity. In general, a 75 ampere-hour battery will be larger and heavier than a 50 ampere-hour battery.

Relative density of electrolyte (specific gravity)

One of the chemical changes that takes place in the battery, during charging and discharging, is that the relative density of the electrolyte (dilute sulphuric acid) changes. There is a link between relative density of the electrolyte and the cell voltage. Relative density of the electrolyte is a good guide to the state of charge of a battery. The following figures in Table 26.2 will apply to most lead acid batteries:

The relative density (specific gravity) varies with temperature, as shown in Fig. 26.6.

Examination of these figures show, that the relative density figure falls by 0.007 for each $10^\circ C$ rise in temperature. This should be taken into account when using a hydrometer.

If the electrolyte (battery acid) is accessible — for example, not a maintenance-free battery — a hydrometer can be used to check battery condition. Figure 26.7 shows a hydrometer together with the readings showing the state of charge.

SG (corrected to 15 °C)	Density (kg/m³) (corrected to 25 °C)	State of charge (%)
1.273	1280	100
1.253	1260	90
1.233	1240	80
1.213	1220	70
1.193	1200	60
1.173	1180	50
1.133	1140	30
1.093	1100	10
1.073	1080	0 (flat)

Fig. 26.6 Variation of electrolyte relative density with temperature

Battery charging

A properly maintained battery should be kept in a charged condition by the vehicle alternator, during normal operation of the vehicle. If it becomes necessary to remove the battery from the vehicle to recharge it, with a battery charger, the ampere-hour capacity of the battery gives a good guide to the current that should be used and the length of time that the battery should stay on charge.

Battery charging rate

The test for battery capacity can be a 10-hour test or a 20-hour test. If, as a guide, we take the 10-hour rating of a 50 A-H battery, in theory this means that the battery will give 5 A for 10 hours, i.e. $5 \times 10 = 50$. If this 50 A-H battery was being completely recharged, it would require a charging current of 5 A to be applied for 10 hours.

Reserve capacity

In recent years it has become the practice to give the reserve capacity of a battery rather than the A-H

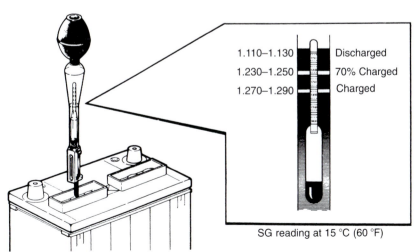

1.110–1.130   Discharged
1.230–1.250   70% Charged
1.270–1.290   Charged

SG reading at 15 °C (60 °F)

Fig. 26.7 The hydrometer

capacity, because reserve capacity is considered to be a better guide to how long the battery will provide sufficient current to operate vehicle systems. Reserve capacity is the time in minutes that it will take a current of 25 A to cause the cell voltage to fall to 1.75 V. That is, the number of minutes that it takes for a 12-volt battery voltage to drop to 10.5 volts, when it is supplying a current of 25 A.

Battery suppliers will provide the data for the batteries that they supply. Manufacturers' charging rates must be used.

Testing batteries

Batteries are normally reliable for a number of years. However, batteries do sometimes fail to provide sufficient voltage to operate the vehicle starting system. Such failure can arise for reasons such as:

(a) Lights, or some other systems, have been left switched on, and the battery has become discharged, etc.
(b) The battery is no longer capable of 'holding' a charge.
(c) The battery is discharged because the alternator charging rate is too low.

If it is (a), and the lights have been left on and caused the battery to 'run down', a quick check of switch positions (i.e. are they in the 'on' position?) will reveal the most likely cause.

A hydrometer test, or a voltage check at the battery terminals, will confirm that the battery is discharged. The remedy will be to recharge the battery, probably by removing it and replacing it, temporarily, with a 'service' battery.

If it is (b) or (c), it will be necessary to check the battery more thoroughly. A high-rate discharge tester that tests the battery's ability to provide a high current for a short period is very effective, but it can only be used if the battery is at least 70% charged. The state of charge of the battery can be determined by using a hydrometer. If the specific gravity (relative density) reading shows that the battery is less than 70% charged, the battery should be recharged at the recommended rate. After charging, and leaving time for the battery to settle and the gases to disperse a high-rate discharge tester of the type shown in Fig. 26.8 can be used to perform a high discharge rate test. Many different forms of high-rate discharge testers are available and it is important to carefully read the instructions for use.

Checking the alternator charging rate is covered in the following section on alternators.

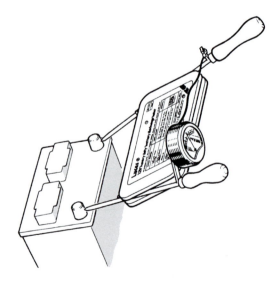

Fig. 26.8 High rate discharge tester

> ### Learning tasks
>
> 1. Ask your supervisor or tutor about the workshop rules that apply to the handling of battery charging in your workshop. Write a summary of the these rules.
> 2. On which types of vehicles are you likely to find two 12-volt batteries connected in series: cars and light commercial vehicles, or heavy trucks and buses?
> 3. On some electronically controlled systems, settings such as idle speed are held in the volatile memory of the ECU. That is, the settings are lost when the battery is discharged or disconnected. Write out a brief description of the procedure that is followed to restore such settings.
> 4. What is a 'rapid' battery charger? What precautions must be observed when using one?
> 5. Examine a vehicle battery installation and make a simple sketch to show how the battery is secured to the vehicle.
> 6. Assuming that you are asked to remove the battery that you have examined for question 5, describe the procedure to be followed. State which battery lead to remove first, and include any steps taken to observe safety rules and protect the vehicle.

The alternator

The alternator supplies the electrical energy to recharge the battery and to operate the electrical/electronic systems on the vehicle, while the engine is running.

The basic vehicle alternator (generator of electricity)

In a vehicle alternator, the magnet is the rotating member and the coil in which current is generated is the stationary part. Figure 26.9 illustrates the general principle. The

(a)

(b)

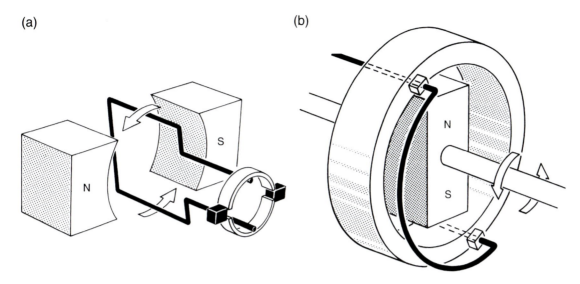

Fig. 26.9 The principle of the generator. (a) Dynamo stationary field. (b) Alternator rotating field

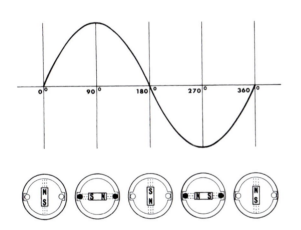

Fig. 26.10 Diagram showing e.m.f. produced in one complete cycle relative to magnet position

rotor shaft is driven round by the pulley and drive belt and the stator is fixed to the body of the alternator.

Figure 26.10 shows how the electromotive force (e.m.f.; voltage) produced by this simple rotating magnet and stator coil varies during one complete revolution (360°) of the magnetic rotor.

Alternating current (a.c.)

Alternating current, as produced by this simple alternator, is not suitable for charging batteries. The alternating current must be changed into direct current (d.c.). The device that changes a.c. to d.c. is the rectifier. Vehicle rectifiers make use of electronic devices, such as the p–n junction diode. The property of the diode that makes it suitable for use in a rectifier is that it will pass current in one direction but not the other. Figure 26.11 demonstrates the principle of the semiconductor diode.

(a)

p   n

(b)

(c)

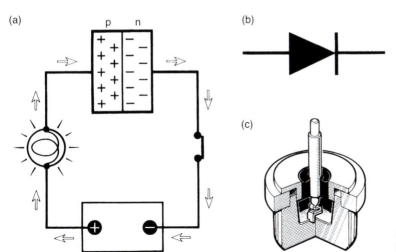

Fig. 26.11 The principle of the diode. (a) Circuit for diode principle. (b) Circuit symbol for a diode. (c) Typical heavy-duty diode

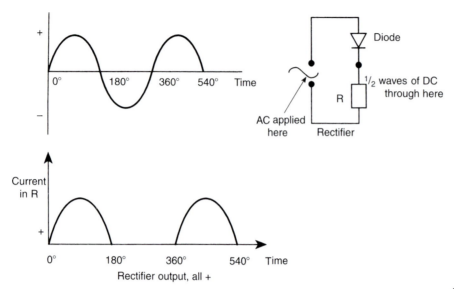

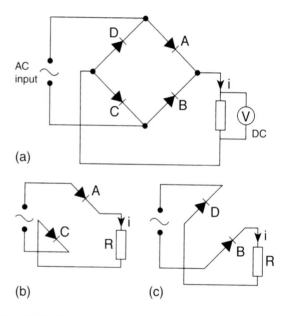

Fig. 26.12   The half-wave rectifier

Using a single diode as a simple half-wave rectifier

As a guide to the principle of operation of the actual rectifier as used in a vehicle alternator, we think that it is useful to briefly consider the operation of the simple half-wave rectifier shown in Fig. 26.12.

In this simple rectifier the negative half of the wave is 'blocked' by the diode. The resulting output is a series of positive half-waves separated by 190°. In order to get a smooth d.c. supply it is necessary to use a circuit that contains several diodes. The bridge circuit rectifier is the basic approach that is used.

A full-wave rectifier: the bridge circuit

Figure 26.13(a) shows a full-wave rectifier circuit. There are four diodes, A, B, C and D, which are connected in bridge form. There is an a.c. input and a voltmeter is connected across the output resistor.

Reference to Fig. 26.13(b) shows how the current flows for one half of the a.c. cycle, and Fig. 26.13(c) shows the current flow in the other half of the a.c. cycle. In both cases, the current flows through the load resistor R in the same direction. Thus, the output is direct current and this is achieved by the 'one-way' action of the diodes.

A typical vehicle alternator

The following description of the parts shown in Fig. 26.14 should help to develop an understanding that will aid the additional study and training that will build the skills necessary for successful work on electrical/electronic systems.

Fig. 26.13   The full-wave rectifier. (a) Complete bridge circuit. (b) Current flow for one half of the voltage waveform. (c) Current flow for the other half of the voltage waveform

The rotor

This is where the magnetic field is generated by the current that flows into the rotor windings through the slip rings and brushes. Figure 26.15 shows three types of rotors that are used in Lucas alternators.

The rotor is made from an iron core around which is wound a coil of wire. The current in this coil of wire creates a magnetic north pole at one end and a magnetic south pole at the other of the stator coil. The iron claws of the rotor are placed on opposite ends of the rotor (field) winding. The 'fingers' of the claws are north or south magnetic poles, according to the end of the field coil that they are attached to. This makes pairs of north and south poles around the circumference of the rotor.

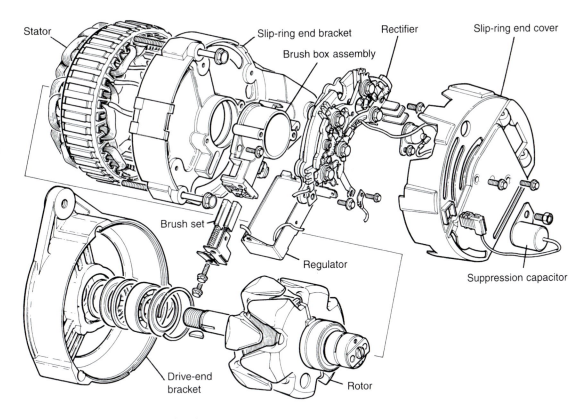

Fig. 26.14 The main component parts of an alternator

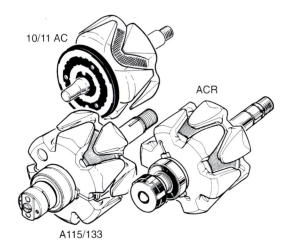

Fig. 26.15 Alternator rotors (electromagnets)

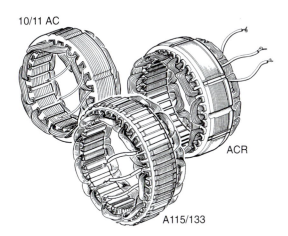

Fig. 26.16 A selection of alternator stators

The whole assembly is mounted on a shaft, which also has the drive pulley fixed to it so that the rotor can be driven by the engine.

The stator

The stators shown in Fig. 26.16 have a laminated iron core. Iron is used because it magnetizes and demagnetizes easily, with a minimum of energy loss. Three separate coils are wound on to the stator core. A separate a.c. wave is generated in each winding, as the rotor revolves inside the stator. The stator windings are either star or

delta connected as shown in Fig. 26.17, according to alternator type.

The rectifier

A six-diode rectifier

A fairly commonly used bridge circuit of an alternator rectifier uses six diodes, as shown in Fig. 26.18. The stator windings are connected to the diodes as shown in this figure.

In operation, a power loss occurs across the diodes and in order to dissipate the heat generated the diodes

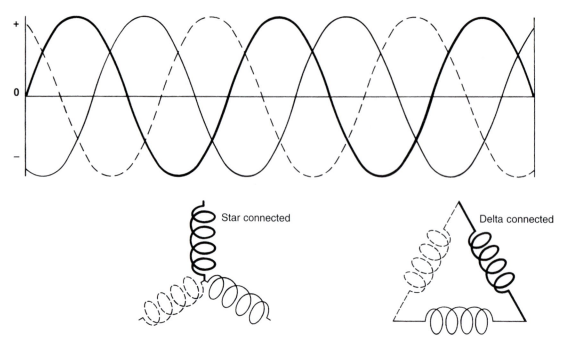

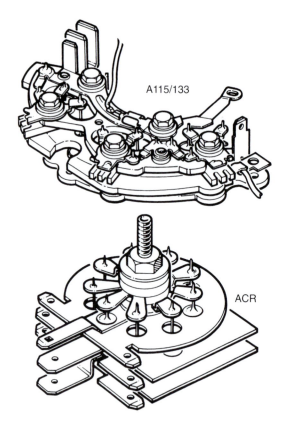

Fig. 26.17 Diagram showing star and delta connected stator windings

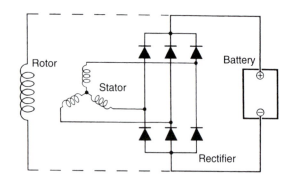

Fig. 26.18 A six-diode rectifier in the alternator circuit

are mounted on a heat sink, which conducts the heat away into the atmosphere. Typical diode packs are shown in Fig. 26.19.

The voltage regulator

The voltage regulator is an electronic circuit that holds the alternator voltage to approximately 14 volts on a 12-volt system, and approximately 28 volts on a 24-volt system. It operates by sensing either the voltage at the alternator (machine sensing), or at the battery (battery sensing). The regulator controls the amount of current that is supplied to the rotor field winding and this alters the strength of the magnetic field and hence the alternator output voltage.

The brushes and brush box

The brushes carry the current to the slip rings, and from there to the rotor winding. Alternator brushes are normally made from soft carbon because it has good

Fig. 26.19 Rectifier diodes and heat sinks

electrical conducting properties and does not cause too much wear on the brass slip rings. In time, the brushes may wear and the brush box is normally placed so that brush checking and replacement can be performed without major dismantling work.

Fig. 26.20 The alternator drive pulley and cooling fan

Bearings

The rotor shaft is mounted on ball races, one at the drive end, the other at the slip-ring end. These bearings are pre-lubricated and are designed to last for long periods without attention. The drive-end bearing is mounted in the drive-end bracket (Fig. 26.14) and the slip-ring end bearing is mounted in the slip-ring-end bracket. These two brackets also provide the supports that allow the alternator to be bolted to the vehicle engine.

Drive pulley and cooling fan

The drive pulley is usually belt driven by another pulley on the engine. The alternator cooling fan of the type shown in Fig. 26.20 is mounted on the rotor shaft and is driven round with it. This fan passes air through the alternator to prevent it from overheating.

Externally, alternators all look very much alike, but they vary greatly in performance and internal construction. The description given here is intended to provide an introduction to the topic. When working on a vehicle, it is essential to have access to the information relating to the actual machine being worked on.

Alternator testing

Before starting on the alternator test procedure, there are a few general points about care of alternators that need to be made:

- Do not run the engine with the alternator leads disconnected. (Note that in the tests that follow the engine is switched off and the ammeter is connected before restarting the engine.)
- Always disconnect the alternator and battery when using an electric welder on the vehicle. Failure to do

this may cause stray currents to harm the alternator electronics.
- Do not disconnect the alternator when the engine is running.
- Take care not to reverse the battery connections.

With these points made we can turn our attention to the alternator tests. As with any test a thorough visual check should first be made. In the case of the alternator this will include:

- Check the drive belt for tightness and condition
- Check leads and connectors for tightness and condition
- Ensure that the battery is properly charged
- Check any fuses in the circuit.

Alternator output tests

The question of whether to repair or replace an alternator is a subject of debate. However, one thing that is certain is the need to be able to accurately determine whether or not the alternator is charging properly. Figure. 26.21 shows the procedure for checking alternator current output under working conditions.

The test link will be a cable with suitable crocodile clips. This link must be securely attached to ensure proper connections. The ammeter connections must also be secure, with the cables and the meter safely positioned to ensure accurate readings and to avoid mishaps.

The drive belt tension must also be checked, as shown in Fig. 26.22.

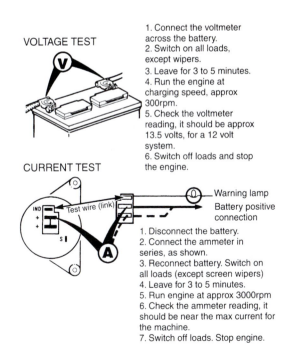

VOLTAGE TEST

1. Connect the voltmeter across the battery.
2. Switch on all loads, except wipers.
3. Leave for 3 to 5 minutes.
4. Run the engine at charging speed, approx 300rpm.
5. Check the voltmeter reading, it should be approx 13.5 volts, for a 12 volt system.
6. Switch off loads and stop the engine.

CURRENT TEST

Test wire (link)

Warning lamp
Battery positive connection

1. Disconnect the battery.
2. Connect the ammeter in series, as shown.
3. Reconnect battery. Switch on all loads (except screen wipers)
4. Leave for 3 to 5 minutes.
5. Run engine at approx 3000rpm
6. Check the ammeter reading, it should be near the max current for the machine.
7. Switch off loads. Stop engine.

Fig. 26.21 Alternator output tests

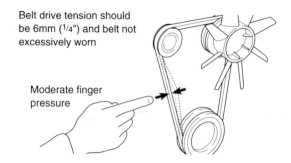

Belt drive tension should
be 6mm (1/4") and belt not
excessively worn

Moderate finger
pressure

Fig. 26.22   Checking the drive belt tension

The belt tension may also be checked by means of a torque wrench, or a Burroughs-type tension gauge, as shown in Fig. 26.23. If the belt is too tight (excessive tension) alternator bearings may be damaged, and if the tension is insufficient the drive belt may slip. This will cause a loss of alternator power and, in time, can lead to worn pulleys, hence the reason for knowing what the tension should be and setting it accurately.

The starter motor

The starter motor shown in Fig. 26.24 is fairly typical of starter motors in modern use.

Most modern starter motors are of the pre-engaged type. This means that the drive pinion on the starter motor is pushed into engagement with the ring gear on the engine flywheel before the starter motor begins to rotate and start the engine.

The amount of torque (turning effect) that the starter motor can generate is quite small and it has to be multiplied to make it strong enough to rotate the engine crankshaft at about 100 revolutions per minute (rpm). This multiplication of torque is achieved by the ratio

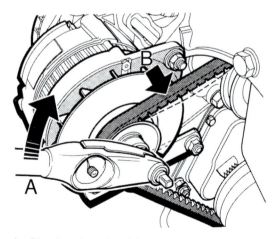

A – Direction of rotation of the gauge
B – Degree of flexure in the drive belt

Fig. 26.23   Using the belt tension gauge

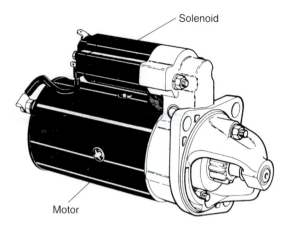

Solenoid

Motor

Fig. 26.24   A starter motor

of the number of teeth on the flywheel divided by the number of teeth on the starter pinion. In the example shown in Fig. 26.25, the ratio is 10:1.

In the starter motor drive shown in Fig. 26.26, the pinion is moved into engagement by the action of the solenoid on the operating lever.

The circuit in Fig. 26.27 shows that there are two solenoid coils that are energized when the starter switch is closed. One coil is the holding coil and the other is the closing coil.

The armature (plunger) of the solenoid is attached to the operating lever. When the pinion is properly engaged, the heavy current contacts of the solenoid will close and the starter motor will rotate; this action bypasses the closing coil and the holding coil remains energized until the starter switch is released. The plunger then returns to the normal position, taking the starter pinion out of engagement with the flywheel.

The starter motor pinion must be taken out of engagement with the flywheel immediately the engine starts. An engine will normally start at approximately 100 rpm. If the starter motor remains in engagement, the engine will drive the starter motor at high speed (because of the gear ratio) and this will destroy the starter motor. To guard against such damage, pre-engaged

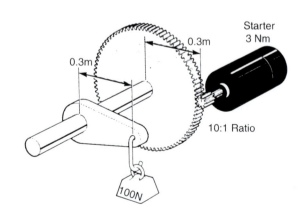

Starter
3 Nm

0.3m

0.3m

10:1 Ratio

100N

Fig. 26.25   Gear ratio of a starter pinion and flywheel gear

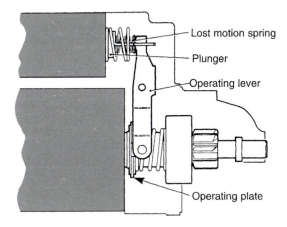

Fig. 26.26 The operating lever

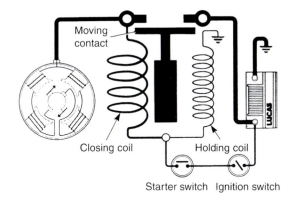

Fig. 26.27 The starter solenoid

starter motors are equipped with a free-wheeling device. A common example of such an over-run protection device is the roller clutch drive assembly shown in Fig. 26.28. The inner member is fixed to the starter motor shaft and the outer member to the starter motor pinion.

This is a convenient point at which to have a look at some tests that can be used to trace faults in starter motor circuits.

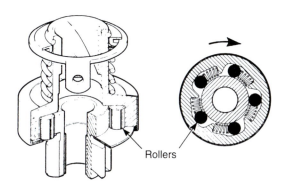

Fig. 26.28 Starter pinion over-run protection

Testing starter motor circuits

As with all diagnostic work, it is advisable to adopt a strategy. One strategy that works well is the six-step approach. The six steps are:

1. Collect evidence
2. Analyse evidence
3. Locate the fault
4. Find the cause of the fault and remedy it
5. Rectify the fault (if different from 4)
6. Test the system to verify that repair is correct.

In the case of a starter motor system and assuming that the starter motor does not operate when the switch is on, step 1 of the six steps will include a visual check to see whether all readily visible parts such as cables, etc. are in place and secure. The next step would be to check the battery. An indication of battery condition can be obtained by switching on the headlights. If the starter switch is operated with the headlights on and the lights go dim, there is a case for a battery test, as described in the section on batteries.

If the battery condition is correct, there are some voltmeter tests that can be performed to check various parts of the starter motor circuit. For these tests, it is best to use an assistant to operate the starter switch because it is unwise to attempt to make the meter connections and observe the meter readings in the vicinity of the starter motor and attempt to operate the starter switch simultaneously. The meter must be set to the correct d.c. range and clips should be used so that secure connections can be made to the various points indicated.

To avoid unwanted start-up, the ignition system should be disabled. On a coil ignition system, the low tension lead between the coil and distributor should be disconnected. On a compression ignition (diesel) engine, the fuel supply should be cut off, probably by operating the stop control. The instructions that relate to the vehicle being worked on must be referred to; this warning is particularly important when working on engine management systems, because fault codes may be generated.

Figure 26.29 shows the test to check voltage at the battery terminals under load.

Here the battery voltage is being checked while the starter switch is being operated. For a 12-volt system on a petrol-engined vehicle, the voltmeter should show a reading of approximately 10 volts. On a diesel-engined vehicle, the load on the battery is higher and the voltage reading will be lower; approximately 9 volts should be obtained.

The starter terminal voltage

The purpose of the test shown in Fig. 26.30, is to check the heavy current circuit of the starter motor and the

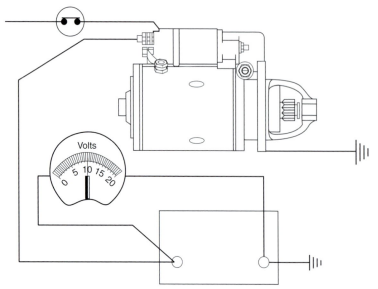

Fig. 26.29   Checking battery voltage under load

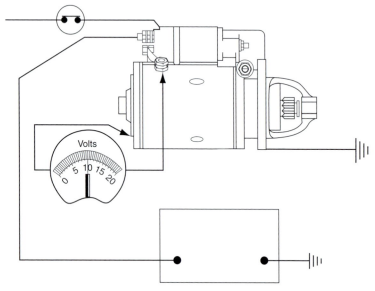

Fig. 26.30   Starter motor voltage test (under load)

condition of the solenoid contacts. When the starter motor switch is in the start position, the voltage recorded between the input terminal and earth should be within 0.5 volts of the battery voltage on load. If that voltage was 10.5 volts, then the voltage recorded here should be at least 10 volts.

Checking the solenoid contacts

Should the starter motor voltage under load be low, it is possible that the heavy current solenoid contacts are defective. The test shown in Fig. 26.31 checks for voltage drop across the solenoid contacts. Good contacts will produce little or no voltage drop. The voltmeter must be connected, as shown, to the two heavy current terminals of the solenoid. When the starter switch is in

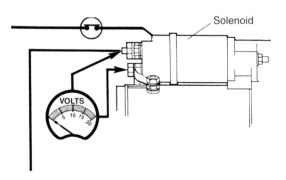

Fig. 26.31   Checking the solenoid contacts

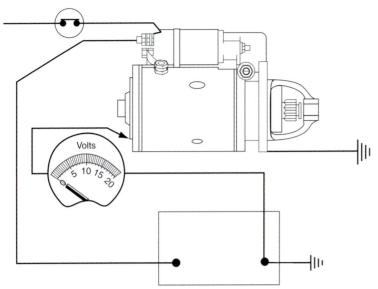

Fig. 26.32 Testing voltage drop on the earth circuit

the off position, the voltmeter should read battery voltage. When the starter switch is in the start position, the voltmeter reading should be zero.

Checking the earth circuit

Because vehicle electrical systems rely on a good earthing circuit, such as metallic contact between the starter and the engine, and other factors such as earthing straps, this part of the circuit needs to be maintained in good order. The test shown in Fig. 26.32 checks this part of the starter circuit. When the starter motor switch is turned to the on position, the voltmeter should read zero. If the reading is 0.5 volts or more, there is a defect in the earth circuit; a good visual examination should help to locate the problem.

Note

These tests are based on Lucas recommendations. Lucas sell a range of inexpensive, pocket-sized test cards. These are recommended for use by those who are learning about vehicle maintenance. The one that relates to 12-volt starting systems has the reference XRB201. We believe that other equipment manufacturers produce similar aids to fault diagnosis.

Learning tasks

Note

In most cases, the procedure for performing the practical tasks is detailed in the text. However, you should always seek the advice of your supervisor before you start.

Working on vehicles can be dangerous so you must not attempt work that you have not been trained for. When you are in training you must

always seek the permission of your supervisor before attempting any of the practical work.

In the case of the first task you will need to have been trained before you can read out the fault codes.

Part of the purpose of these practical tasks is to assist you in building up a portfolio of evidence for your NVQ. You should therefore keep the notes and diagrams that you make neatly in a folder.

1. With the aid of diagrams, where necessary, describe the procedure for obtaining fault codes from an ECU on a vehicle system that you are familiar with.
2. Write down the procedure for removing a battery from a vehicle. Make careful notes of the safety precautions and specify which battery terminal should be disconnected first.
3. Describe the tools and procedures for checking the state of charge of a battery.
4. Carry out a check of the charging rate of an alternator. Make a list of the tools and equipment used and give details of the way in which you connected the test meter. Make a note of the voltage values recorded.
5. Make a sketch of the provision made for tensioning the alternator drive belt on any vehicle that you have worked on. Explain why drive belt tension is important. State the types of problems that may arise if an alternator belt is not correctly adjusted.
6. With the aid of diagrams, explain how to check for voltage at the starter motor terminal when the engine is being cranked.
7. Examine the leads that are connected to the headlights on a vehicle. Make a note of the colours of the cable insulation. Obtain a wiring diagram for the same vehicle and, by reference

to the wiring diagram, and the colours that you have observed the cables to be, locate the headlamp cables on the wiring diagram.

8. Make a note of the headlamp alignment equipment that is used in your workshop. Describe the procedure for carrying out a headlamp alignment check. Make a sketch of the method used to adjust headlamp alignment on any vehicle that you have worked on.

9. Give details of the fuseboxes on any vehicle that you are familiar with. State the type and current rating of the side and tail lamp fuse.

10. Write down the precautions that should be taken to protect vehicle electrical equipment before attempting any electric arc welding on the vehicle.

Self-assessment questions

1. The function of the heat sink part of the rectifier diode pack in an alternator is to:
   (a) conduct heat away from the diodes in order to protect them
   (b) provide a solid base for the transistors
   (c) act as a voltage regulator
   (d) prevent the stator from overheating.

2. Machine sensing, where the alternator output voltage is used to vary the current in the rotor windings, is a method of controlling:
   (a) the phase angle of the a.c. output
   (b) the voltage output of the alternator
   (c) the self-exciting voltage of the alternator
   (d) the brush pressure on the slip rings.

3. A slack alternator drive belt may cause the:
   (a) engine to overheat
   (b) engine to lose speed
   (c) battery to become discharged
   (d) charging rate to be too high.

4. Alternator generated a.c. is converted to d.c. by:
   (a) Zener diode
   (b) the regulator

(c) the battery
(d) the slip rings.

5. The regulated voltage output of the alternator on a 12-volt system should be:
   (a) exactly 12 volts
   (b) 10.9 volts
   (c) approximately 13–15 volts
   (d) 18 volts.

6. The maximum permissible voltage drop on a 12-volt starter motor system as measured across the battery terminals when the starter motor is operating is:
   (a) 0.2 volts
   (b) 2 volts
   (c) 5 volts
   (d) 0.5 volts.

7. If the gear ratio between a starter motor and the flywheel ring gear is 10:1, and the starter motor generates a torque of 5 Nm, the torque at the flywheel will be:
   (a) 50 Nm
   (b) 105 Nm
   (c) 0.5 Nm
   (d) 50 amperes.

8. On a pre-engaged type of starter motor, overspeeding of the armature is prevented by:
   (a) the back e.m.f. in the windings
   (b) Fleming's right-hand rule
   (c) a free-wheeling clutch
   (d) manual withdrawal of the pinion.

9. At normal room temperature, the relative density of the electrolyte in a lead acid battery should be:
   (a) 13.6
   (b) 12.6
   (c) 1.28
   (d) 0.128.

10. A high rate discharge tester:
   (a) should always be used to test a flat battery
   (b) should only be used if the battery is at least 70% charged
   (c) can only be used on nickel alkaline cells
   (d) determines the watt-hour capacity of the battery.

27
Ignition systems

Topics covered in this chapter

Ignition systems — coil, capacitor (condenser)
Electronic ignition systems
Non-electronic ignition systems
Fault tracing and service and maintenance procedures
Sensors and pulse generators
Ignition timing
Diagnostic trouble codes

Ignition systems for spark ignition engines

The purpose of the ignition system is to provide a spark to ignite the compressed air—fuel mixture in the combustion chamber. The spark must be of sufficient strength to cause ignition and it must occur at the correct time in the cycle of operations. The ignition system is the means by which the 12-volt supply from the vehicle battery is converted into many thousands of volts that are required to produce the spark at the sparking plug, as illustrated in Figs 27.1 and 27.2.

Producing the high voltage required to cause ignition

Electromagnetism was discussed in the electrical principles section (Chapter 25). It is these electromagnetic principles that are employed in the induction coil type of ignition system (see Fig. 27.3). In this coil there are two windings, a primary winding and a secondary winding. The primary winding consists of a few hundred turns of lacquered copper wire. The secondary winding consists of several thousand turns of thin, lacquered copper wire. The primary winding is wound around the outside of the secondary winding because the heavier current that it carries generates heat, which needs to be dissipated. The layers of windings are normally electrically insulated from each other by a layer of insulating material.

In the centre of the secondary winding there is a laminated soft iron core. The purpose of this laminated core is to concentrate the magnetic field that is produced by the current in the coil windings. The laminations between the coil's outer casing and the windings serves a similar purpose.

The interior parts of the coil are supported on a porcelain base. In some cases the coil may be filled with a transformer oil to improve electrical insulation and to aid heat dissipation.

Principle of operation

The actual size of the secondary voltage (high tension (HT) voltage) is related to the turns ratio of the coil, e.g. several thousand turns in the secondary winding and a few hundred turns in the primary winding.

In order for the ignition system to operate, the primary winding current must be switched on and off, either by electronic means or, in the case of older systems, the contact breaker.

Figure 27.4 shows a simple ignition coil primary circuit. Current from the vehicle battery flows through the primary winding via the ignition switch and the closed contact breaker points, to earth. The current of about 3 amperes, in the primary winding, causes a strong magnetic field which is concentrated by the soft iron laminations at the coil centre and the outer sheath of iron. It should be remembered that the secondary winding is inside the primary winding so the magnetic field from the primary winding also affects the secondary winding. A primary purpose of the primary winding is to create the strong magnetic field that will induce the high voltage in the secondary circuit.

Method of inducing the high-tension (HT) secondary voltage

Switching off the primary current by opening the contact breaker points has two major effects, which

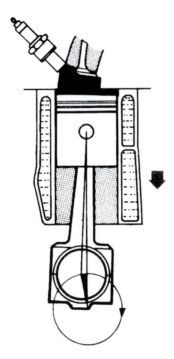

Fig. 27.1 Position of the spark in a reciprocating (piston) engine

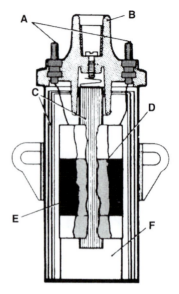

A-Primary terminals
B-High tension terminals
C-Laminations
D-Secondary winding
E-Primary winding
F-Porcelain insulator

Fig. 27.3 A typical ignition coil

are caused by the collapsing inwards of the magnetic field. These two effects are:

1. Self-induction – a voltage of approximately 250 V is generated across the ends of the primary winding. The direction of this induced voltage in the primary winding is such that it opposes the change that produced it, i.e. it attempts to maintain the same direction of current flow.
2. Induction – a high voltage (several thousand volts, kV) is generated across the ends of the secondary winding.

Secondary circuit

Figure 27.5 shows the secondary circuit as well as the primary circuit of the simplified system. One end of the secondary winding (thousands of turns of fine copper wire) is connected to the HT terminal in the coil tower; the other end is connected to the primary winding at one of the low-tension (LT) connections. When the primary winding is connected to the secondary winding it is known as 'auto-transformer' connected. The induced primary voltage is added to the secondary voltage.

The capacitor (condenser)

The simplified ignition system described above would not work very well because the induced voltage in the primary winding would be strong enough to bridge the contact breaker gap and carry current to earth. This would reduce the secondary voltage and also lead to excessive sparking at the contact breaker points; the

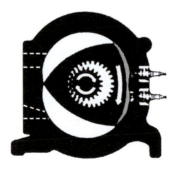

Fig. 27.2 Position of the spark in a rotary engine

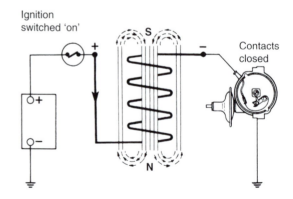

Fig. 27.4 A basic primary circuit of an ignition coil

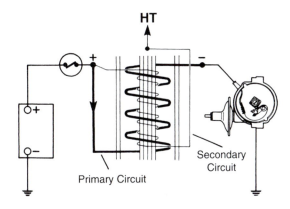

Fig. 27.5 The secondary and primary circuits of a coil

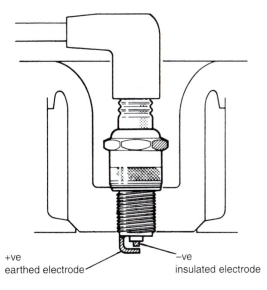

Fig. 27.7 Polarity of spark plug electrodes

points would 'burn' and the efficiency of the system would be seriously impaired. This problem is overcome by fitting a suitable capacitor (approximately 0.2 μF) across (in parallel with) the points.

Figure 27.6 shows the capacitor in a circuit. One terminal of the capacitor is connected to the low-tension connection on the distributor and the other terminal is earthed.

Capacitors store electricity so that when, in the case of the coil ignition system, the contact points begin to open, the self-induced current from the primary winding will flow into the capacitor instead of 'jumping' across the points gap. This flow of current will continue until the capacitor is fully charged. When the capacitor is fully charged it will automatically discharge itself, back into the primary winding. This capacitor discharge current is in the opposite direction to the original flow and it helps to cause a rapid collapse of the magnetic field. This, in turn, leads to a much higher HT voltage from the secondary winding.

Coil polarity

Electrons flow from negative to positive and thermal activity (heat) also aids electron generation; for this

reason the insulated (central) hotter electrode of the spark plug is made negative in relation to the HT winding as shown in Fig. 27.7. The result of this polarity direction is that a lower voltage is required to generate a spark across the plug gap. To ensure that the correct coil polarity is maintained, the LT connections on the coil are normally marked as + and −.

The coil polarity is readily checked by means of the normal garage-type oscilloscope. Figure 27.8 shows the inverted waveform that results from incorrect connection of the coil, or the fitting of the wrong type of coil for the vehicle under test.

Electronic ignition systems

Electronic ignition systems make use of some form of electrical/electronic device to produce the electrical pulse that switches 'on and off' the ignition coil primary current so that a high voltage is induced in the coil secondary winding in order to produce a spark, in the required cylinder, at the correct time.

There are several methods of producing the 'triggering' pulse for the ignition that replaces the 'on and off' action of the contact breaker. The commonly used types are reviewed in the following sections.

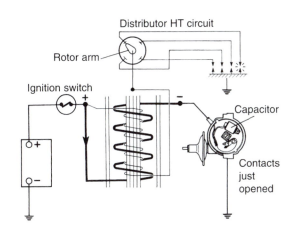

Fig. 27.6 The ignition circuit with the capacitor in place

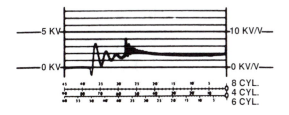

Fig. 27.8 The inverted waveform on the oscilloscope screen

Electromagnetic pulse generators

The constant-energy ignition system

Figure 27.9 shows part of a type of electronic ignition distributor that has been in use for many years. The distributor shaft is driven from the engine camshaft and turns the reluctor rotor that is attached to it. The reluctor thus rotates at half engine speed.

Each time that a lobe on the rotor (reluctor) passes the pick-up probe a pulse of electrical energy is induced in the pick-up winding. The pick-up winding is connected to the electronic ignition module and when the pulse generator voltage has reached a certain level (approximately 1 volt) the electronic circuit of the module will switch on the current to the ignition coil primary winding.

As the reluctor (rotor) continues to rotate, the voltage in the pick-up winding begins to drop and this causes the ignition module to 'switch off' the ignition coil primary current; the high voltage for the ignition spark is then induced in the ignition coil secondary winding. The period between switching on and switching off the ignition coil primary current is the dwell period. The effective increase in dwell angle as the speed increases means that the coil current can build up to its optimum value at all engine speeds. Figure 27.10 shows how the pulse generator voltage varies due to the passage of one lobe of the reluctor past the pick-up probe.

From these graphs it may be seen that the ignition coil primary current is switched on when the pulse generator voltage is approximately 1 volt and is switched off again when the voltage falls back to the same level. At higher engine speeds the pulse generator produces a higher voltage and the switching on voltage (approximately 1 volt) is reached earlier, in terms of crank position, as shown in Fig. 27.10(b). However, the 'switching off' point is not affected by speed and this means that the angle (dwell) between switching the coil primary

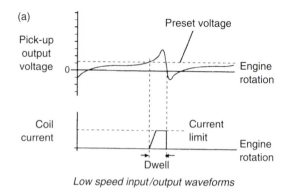

(a)

Low speed input/output waveforms

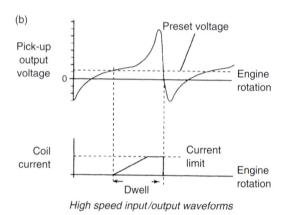

(b)

High speed input/output waveforms

Fig. 27.10 Pick-up output voltage at (a) low and (b) high speeds

current on and off increases as the engine speed increases. This means that the build-up time for the current in the coil primary winding, which is the important factor affecting the spark energy, remains virtually constant at all speeds. It is for this reason that ignition systems of this type are known as 'constant-energy systems'. It should be noted that this 'early' type of electronic ignition still incorporates the centrifugal and vacuum devices for automatic variation of the ignition timing.

Digital (programmed) ignition system — Hall effect sensor

Programmed ignition makes use of computer technology and it permits the mechanical, pneumatic, and other elements of the conventional distributor to be dispensed with. Figure 27.11 shows an early form of digital ignition system.

The control unit (ECU or ECM) is a small dedicated computer that has the ability to read input signals from the engine, such as speed, crank position, and load. These readings are compared with data stored in the computer memory and the computer then sends outputs to the ignition system. It is traditional to represent the data, which is obtained from engine tests, in the form of a three-dimensional map, as shown in Fig. 27.12.

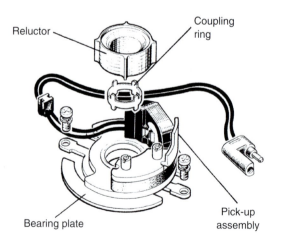

Fig. 27.9 Reluctor and pick-up assembly

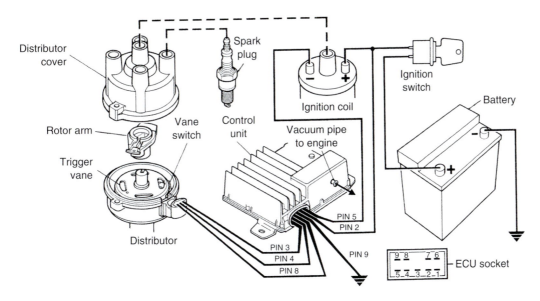

Fig. 27.11 A digital ignition system

Any point on this map may be represented by a number reference. For example, engine speed 1000 rpm, manifold pressure (engine load) 0.5 bar, ignition advance angle 5°. These numbers can be converted into computer (binary) code words, made up of zeros and ones (this is why it is known as digital ignition). The map is then stored in computer memory, where the processor of the control unit can use it to provide the correct ignition setting for all engine operating conditions.

In this early type of digital electronic ignition system the 'triggering' signal is produced by a Hall effect sensor of the type shown in Fig. 27.13. When the metal part of the rotating vane is between the magnet and the Hall element the sensor output is zero and when the gaps in the vane expose the Hall element to the magnetic field a voltage pulse is produced. In this way, a voltage pulse

is produced by the Hall sensor each time that a spark is required. Whilst the adapted form of the older type ignition distributor is widely used for electronic ignition systems, it is probable that the trigger pulse generator driven by the crankshaft and flywheel is more commonly used on modern systems.

Opto-electronic sensing for the ignition system

Figure 27.14 shows the electronic ignition photo-electronic distributor sensor used on a KIA. There are two electronic devices involved in the operation of the basic device. One is a light-emitting diode (LED; it converts electricity into light), the other is a photo diode that

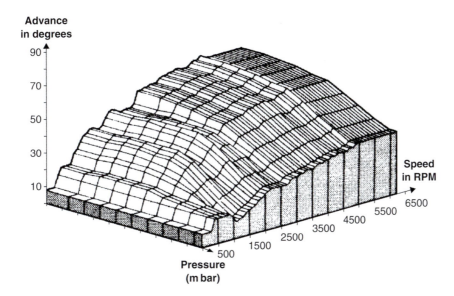

Fig. 27.12 An ignition map that is stored in the ROM of the ECM

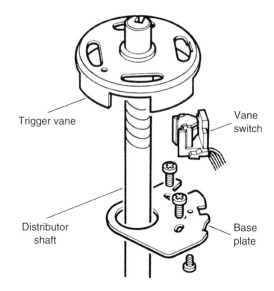

Fig. 27.13 A Hall-type sensor

can be 'switched on' when the light from the LED falls on it.

Another version of this type of sensor is shown in Fig. 27.15. Here the rotor plate has 360 slits placed at 1° intervals, for engine speed sensing, and a series of larger holes for top dead centre (TDC) indication that are placed nearer the centre of the rotor plate. One of these larger slits is wider than the others and it is used to indicate TDC for number 1 cylinder.

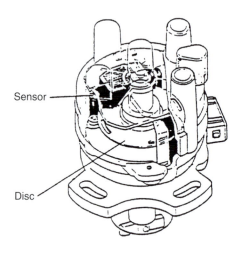

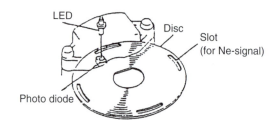

Fig. 27.14 An opto-electronic sensor

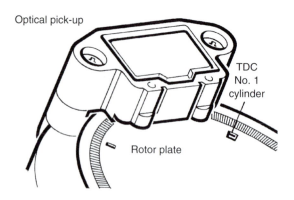

Fig. 27.15 An alternative form of opto-electronic sensor

Distributorless ignition system (DIS)

Figure 27.16 shows an ignition system for a four-cylinder engine. There are two ignition coils, one for cylinders 1 and 4, and another for cylinders 2 and 3. A spark is produced each time a pair of cylinders reaches the firing point, near top dead centre. This means that a spark occurs on the exhaust stroke as well as on the power stroke. For this reason, this type of ignition system is sometimes known as the 'lost spark' system.

Figure 27.16 shows that there are two sensors at the flywheel; one of these sensors registers engine speed and the other is the trigger for the ignition. They are shown in greater detail in Figs 27.17 and 27.18, and they both rely on the variable reluctance principle for their operation.

An alternative method of indicating the TDC position is to use a toothed ring, attached to the flywheel, which has a tooth missing at the TDC positions, as shown in Fig. 27.18.

With this type of sensor, the TDC position is marked by the absence of an electrical pulse; this is also a variable reluctance sensor. The other teeth on the reluctor ring, which are often spaced at 10° intervals, are used to provide pulses for engine speed sensing.

Direct ignition systems

Direct ignition systems are those where each cylinder has a separate ignition coil placed directly on the sparking plug, as shown in Fig. 27.19. The trigger unit is incorporated in the coil housing in order to protect the ECM against high current and temperature.

Ignition system refinements

As the processing power of microprocessors has increased, it is natural to expect that system designers will use the increased power to provide further features such as combustion knock sensing and adaptive ignition control.

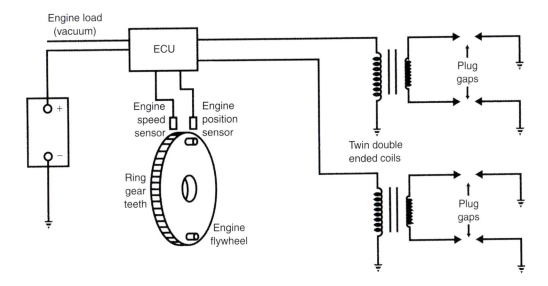

Fig. 27.16 Distributorless ignition system

Knock sensing

Combustion knock is a problem that is associated with engine operation. Early motor vehicles were equipped with a hand control that enabled the driver to retard the ignition when the characteristic 'pinking' sound was heard. After the pinking had ceased the driver could move the control lever back to the advanced position. Electronic controls permit this process to be done automatically and a knock sensor is often included in the make-up of an electronic ignition system.

Combustion knock sensors

A knock sensor that is commonly used in engine control systems utilizes the piezo-electric generator effect. That is, the sensing element produces a small electric charge when it is compressed and then relaxed. Materials such

as quartz and some ceramics like PZT (a mixture of platinum, zirconium, and titanium) are effective in piezo-electric applications. In the application shown in Fig. 27.20, the knock sensor is located on the engine block adjacent to cylinder number 3. In this position it is best able to detect vibrations arising from combustion knock in any of the four cylinders.

Because combustion knock is most likely to occur close to top dead centre in any cylinder, the control program held in the ECM memory enables the processor to use any knock signal generated to alter the ignition timing by an amount that is sufficient to eliminate the knock. When knock has ceased the ECM will advance the ignition, in steps, back to its normal setting. The mechanism by which vibrations arising from knock are converted to electricity is illustrated in Fig. 27.21.

The sensor is accurately designed and the centre bolt that pre-tensions the piezo-electric crystal is accurately

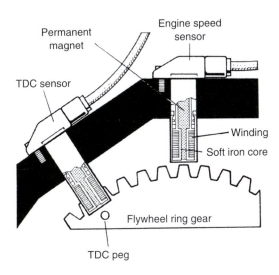

Fig. 27.17 Details of engine speed and crank position sensor

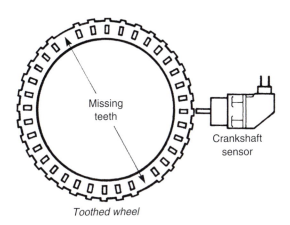

Fig. 27.18 Engine speed and position sensor that uses a detachable reluctor ring

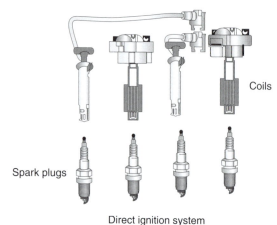

Coils

Spark plugs

Direct ignition system

Fig. 27.19 Direct ignition system

torqued so as to provide the correct setting. The steel washer that makes up the seismic mass has very precise dimensions. When combustion knock occurs, the resulting mechanical vibrations are transmitted by the seismic mass to the piezo-electric crystal. The 'squeezing up' and relaxing of the crystal in response to this action produces a small electrical signal that oscillates at the same frequency as the knock sensor element. The electrical signal is conducted away

from the crystal by wires that are secured to suitable points on the crystal.

The tuning of the sensor is critical because it must be able to distinguish between knock from combustion and other knocks that may arise from the engine mechanism. This is achieved because combustion knock produces vibrations that fall within a known range of frequencies.

Adaptive ignition

The computing power of modern ECMs permits ignition systems to be designed so that the ECM can alter settings to take account of changes in the condition of components such as petrol injectors, as the engine wears. The general principle is that the best engine torque is achieved when combustion produces maximum cylinder pressure just after TDC. The ECM monitors engine acceleration, by means of the crank sensor, to see if changes to the ignition setting produce a better result, as indicated by increased engine speed, as a particular cylinder fires. If a better result is achieved then the ignition memory map can be reset so that the revised setting becomes the one that the ECM uses. This 'adaptive learning strategy' is now used quite extensively on computer-controlled systems and it

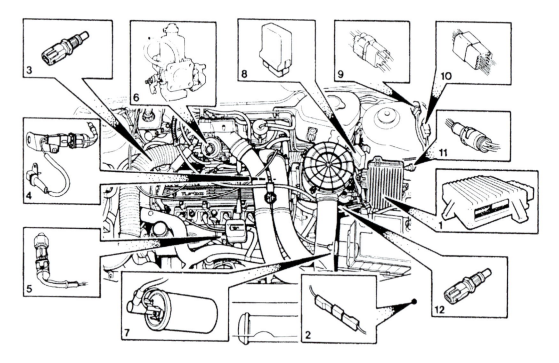

1. ERIC ECU	5. Knock sensor	10. 13-way connector—engine/main
2. Ambient air temperature sensor (located behind horns)	6. Carburetter	11. Serial Diagnostic Link connector
3. Coolant temperature sensor	7. Ignition coil	12. Inlet air sensor
4. Crankshaft sensor	8. Engine MFU	
	9. 4-way connector—engine/main	

Fig. 27.20 The knock sensor on the engine

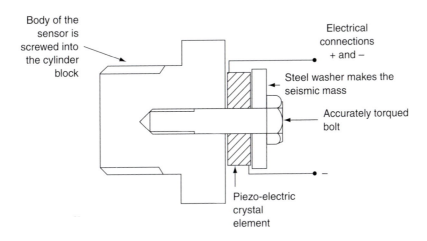

Fig. 27.21 The principle of the piezo-electric combustion knock sensor

requires technicians to run vehicles under normal driving conditions for several minutes after replacement parts and adjustments have been made to a vehicle.

Fault tracing, maintenance, and repair of ignition systems

The majority of the following descriptions are based on the use of a portable oscilloscope. However, for tests on the primary circuit of an ignition system, a good-quality digital multimeter will also serve most purposes.

Performance checks on pulse generators

If preliminary tests on the ignition circuit indicate that the pulse generator may be defective, checks of the following type can provide useful information.

Magnetic (inductive)-type ignition pulse generator

The voltage output of a typical ignition pulse generator of this type varies with engine speed and the actual voltage varies considerably according to manufacturer. However, the voltage waveform is of similar form for most makes and useful test information can be gained from a voltage test. There are two possible states for a voltage test: one is with the pulse generator disconnected from the electronic module when a waveform of the type shown in Fig. 27.22 should be produced, and the other is when the pulse generator is connected to the electronic module, when the waveform shown in Fig. 27.23 should be produced. The difference is accounted for by the fact that the lower part of the

pattern is affected by the loading from the electronic module.

A digital voltmeter set to an appropriate a.c. range may also be used to check the output of this type of pulse generator. The voltages for each pulse should be identical and the maximum value should be at the level given in the vehicle repair data.

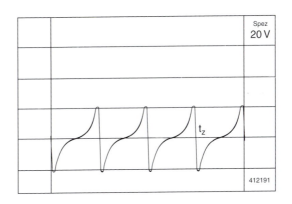

Fig. 27.22 Waveform, pulse generator — electronic module disconnected

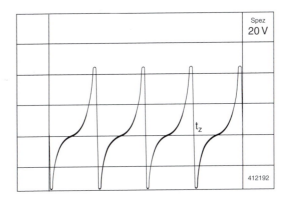

Fig. 27.23 Waveform, pulse generator — electronic module connected

Other checks that can be performed on the inductive-type pulse generator

Resistance check

If the sensor voltage is not correct it is possible to conduct a resistance test that gives an indication of the condition of the winding part of the sensor. Figure 27.24 gives an indication of the connections used for this process.

Hall-type sensor tests

A Hall-type sensor is known as a passive device because it operates by processing electricity that is supplied to it and it can be expected to have at least three cables at its connection. Figure 27.25 shows the approximate position of the oscilloscope test probes on a Hall-type ignition distributor.

If the test is being performed because the engine will not start, the purpose of the test will be to check that there is a signal from the ignition trigger sensor. The oscilloscope, or meter, would be connected as shown and the engine cranked over. The voltage pattern would be observed and, provided the signal is adequate, the fault would be sought elsewhere in the system.

If the engine starts up and the test is being conducted to check for other defects, the test connections remain as for the 'no start' condition but the sensor signal may be observed at a number of engine speeds. The approximate shape of a voltage trace for a Hall-type sensor is shown in Fig. 27.26. The frequency of the voltage pulse will vary with engine speed. The reference voltage that is referred to is the regulated voltage that is supplied to the sensor.

Opto-electronic trigger sensor

This is also a passive device and will have a connection for a supply of electricity. The test procedure is similar to that described for the Hall-type sensor and the type of

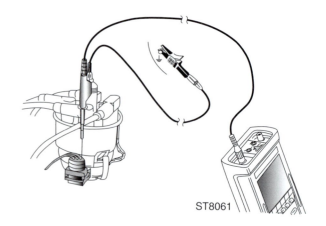

Fig. 27.25 Using a portable oscilloscope to test a Hall-type ignition pulse sensor

voltage output to be expected in the test is shown in Fig. 27.27.

Servicing an electronic-type distributor

The rotor (reluctor) to magnetic pick-up air gap on an electronic ignition distributor is an important setting that requires occasional checking. Figure 27.28(a) shows the rotor and magnetic pick-up and Fig. 27.28(b) shows non-magnetic, i.e. brass or plastic feeler gauges, being used to check the gap. The use of non-magnetic feeler gauges is important because if steel is placed in this air gap the electric charge generated may damage the electronic circuit.

Setting the static timing of an electronic-type distributor

The procedures described here applies to two types of distributors. One is the variable-reluctance pulse generator type and the other is a Hall effect pulse generator. It should be noted that the procedure will vary according

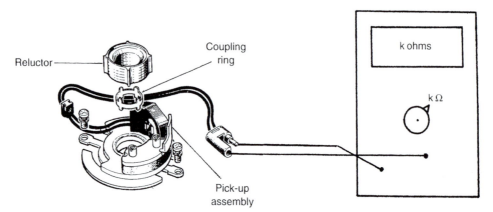

Reluctor and pick-up assembly

Pulse generator winding resistance

Fig. 27.24 Resistance test on sensor winding

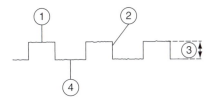

1. The upper horizontal lines should reach reference voltage.
2. Voltage transitions should be straight and vertical.
3. **Peak-Peak** voltages should equal reference voltage.
4. The lower horizontal lines should almost reach ground.

Fig. 27.26 Oscilloscope trace for Hall-type ignition trigger sensor

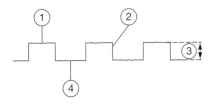

1. The upper horizontal lines should reach reference voltage.
2. Voltage transitions should be straight and vertical.
3. **Peak-Peak** voltages should equal reference voltage.
4. The lower horizontal lines should almost reach ground.

Fig. 27.27 Oscilloscope voltage trace for opto-electronic-type ignition trigger sensor

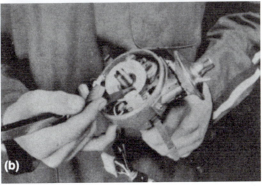

Fig. 27.28 Checking the air gap between pick-up and rotor

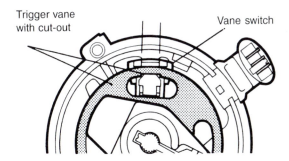

Fig. 27.29 Setting the timing on a Hall effect distributor

to engine type and also the type, and make, of distributor. It is essential to have access to the details relating to a specific application. Of course, if you are very familiar with a particular make of vehicle it is quite possible that you will be able to memorize the procedures.

Static timing — variable-reluctance pulse generator

The cylinder to which the distributor is being timed must be on the firing stroke and the timing marks, on the engine, must be accurately aligned. The distributor rotor must be pointing towards the correct electrode in the distributor cap and the corresponding lobe of the reluctor rotor must be aligned with the pick-up, as in the case shown in Fig. 27.28. When the distributor is securely placed and the electrical connections made, the final setting of the timing is achieved by checking the timing with the strobe light. Final adjustment can then be made by rotating the distributor body, in the required direction, prior to final tightening of the clamp.

Static timing — Hall effect pulse generator

Figure 27.29 shows a Hall effect distributor, for digital electronic ignition, where the distributor itself is provided with timing marks. With the engine timing

marks correctly set, the cut-out in the trigger vane must be aligned with the vane switch, as shown.

Should the setting require adjustment, the distributor clamp is slackened, just sufficient to permit rotation of the distributor body, the distributor is then rotated until the marks are aligned. After tightening the clamp and replacing all leads etc., the engine may be started and the timing checked by means of the stroboscopic light.

Diagnosis — secondary (high-voltage) side of ignition system

Firstly, it is important to **work in a safe manner**. Electric shock from ignition systems must be avoided; not only is the shock dangerous in itself, but it can also cause involuntary muscular actions that can cause limbs to be thrown into contact with moving or hot parts.

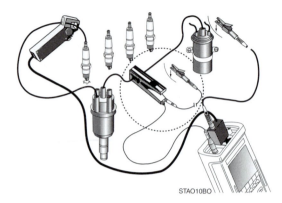

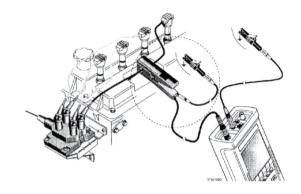

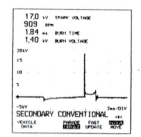

Fig. 27.30   Oscilloscope connections for obtaining ignition trace

Fig. 27.32   The oscilloscope set-up for obtaining a secondary voltage trace from a DIS

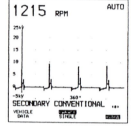

SINGLE cylinder display   PARADE display

Fig. 27.31   Oscilloscope displays of secondary ignition voltage — single pattern and parade

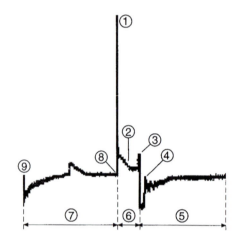

Fig. 27.33   Details of the HT voltage trace for a single cylinder

Tests on distributor-type ignition systems

The king lead and the HT lead to number 1 cylinder are normally connected to the oscilloscope as shown in Fig. 27.30. A typical oscilloscope display for a distributor-type ignition is shown in Fig. 27.31. These displays show the firing voltage for number 1 cylinder. The parade pattern shows the voltage of the remaining cylinders as displayed across the screen, in firing order, from left to right.

Tests on distributorless ignition systems

In this case there is no king lead so the inductive pick-up is clamped to the HT leads individually, as close as possible to the spark plug, as shown in Fig. 27.32. The screen display obtained from such a test shows data about engine speed, burn time, and voltage. These and other details may more easily be seen by taking an enlargement of the trace for a single cylinder, as shown in Fig. 27.33.

The details are:

1. Firing line. This represents the high voltage needed to cause the spark to bridge the plug gap.
2. The spark line.
3. Spark ceases.
4. Coil oscillations.

5. Intermediate section (any remaining energy is dissipated prior to the next spark).
6. Firing section (represents burn time).
7. Dwell section.
8. Primary winding current is interrupted by transistor controlled by the ECM.
9. Primary winding current is switched on to energize the primary. The dwell period is important because of the time required for the current to reach its maximum value.

A comparison of the secondary voltage traces for each of the cylinders should show them to be broadly similar. If there are major differences between the patterns then it is an indication of a defect.

For example, a low firing voltage (1) indicates low resistance in the HT cable or at the spark plug. The low resistance could be attributed to several factors, including oil or carbon fouled spark plug, incorrect plug gap, low cylinder pressure, or defective HT cable. A high voltage at (1) indicates high resistance in the HT cable, or at the spark plug. Factors to consider here include a loose HT lead, wide plug gap, or an excessive amount of resistance has developed in the HT cable. Table 27.1 summarizes the major points.

Table 27.1 Factors affecting firing voltage

Factor	High firing voltage	Low firing voltage
Spark plug gap	Wide	Small
Compression pressure	Good	Low
Air—fuel ratio	Weak	Correct
Ignition point	Late	Early

Sparking plugs can be removed and examined; HT leads can also be examined for tightness in their fittings and their resistance can be checked with an ohmmeter. Resistive HT leads are used for electrical interference purposes and they should have a resistance of approximately **15 000–25 000 Ω per metre length.**

Here, as in all cases, it is important to have to hand the information and data that relates to the system being worked on.

Direct on sparking plug systems

Direct ignition systems of the type shown in Fig. 27.19 are normally controlled by the engine management computer (ECM), which incorporates self-diagnostics that give quite detailed information about ignition faults. Should a misfire be attributed to one of the ignition coils, the condition of the coil may be checked by conducting a resistance test on the primary and secondary windings. The primary winding of the coil will have a resistance of approximately **1 Ω**, whilst the secondary winding resistance is normally **5000–10 000 Ω**.

Diagnostic trouble codes (DTCs)

Modern engine management systems use European or USA on-board diagnostic standards. European on-board diagnostic systems (EOBD) require that emissions-related trouble codes (DTCs) are available through a standard connection that permits an approved repairer to gain access to the DTCs by means of an appropriate scan tool. The form of the EOBD diagnostic plug and the recommended position of it on the vehicle is shown in Fig. 27.34.

Because the vehicle emissions are affected by a number of vehicle systems such as fuelling and ignition, the ECM is programmed to record and report on a range of faults in these systems that may contribute to a vehicle's failure to meet emissions regulation limits.

An example of a DTC is given in Table 27.2. This DTC tells us that there is a misfire. To determine the cause will require further testing of the ignition-related systems in order to trace the cause and effect a repair. Some on-board diagnostics may go deeper into the causes of the misfire, in which case the task of fault tracing should be made simpler. In other cases, the fault-tracing procedure may be expected to involve the procedures described earlier.

Dwell angle

The dwell angle is the period of time, as represented by angular rotation of the cam, for which the contact breaker points are closed. Figure 27.35 shows the dwell angles for the cams of four-, six-, and eight-cylinder engines.

The dwell angle is important because it controls the period of time during which the primary current is energizing the primary winding of the coil. In order for the magnetic field to reach its maximum strength, the primary current must reach its maximum value and this takes time. In an inductance, which is what the primary coil is, it takes several milliseconds for the current to rise to its maximum value, as shown in

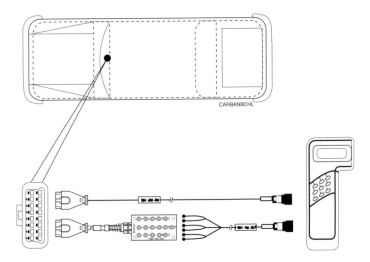

Fig. 27.34 The recommended position and form of the EOBD diagnostic connection

Table 27.2 Example of a DTC

DTC	Function
PO300	Random/multiple cylinder misfire detected

Fig. 27.36. The actual time of current build-up is dependent on the inductance (henry) and the resistance (ohms) of the primary winding; for a typical ignition coil the time constant is of the order of 15 milliseconds.

The dwell angle, which determines current build-up time, is affected by the position of the heel of the moving contact in relation to the cam, as shown in Fig. 27.37. In ECU-controlled ignition systems the dwell angle is determined by the computer program and the angle of advance can be varied to suit ongoing motoring conditions.

Timing of the spark

The process of ensuring that the spark occurs at the right place and at the right time is known as **timing** the ignition. In the first place there is **static timing**. This is the operation of setting up the timing when the engine is not running. Ignition timing settings are normally expressed in degrees of crank rotation. The static timing varies across the range of engines in use today. A reasonable figure for static timing would be approximately $5-10°$. (Please note: this setting is critical and must be verified for any engine that is being worked on – never guess.) When the spark is made to occur before this static setting it is said to be **advanced**, and when the

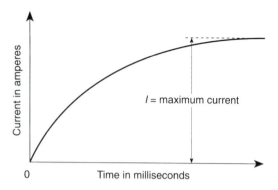

Fig. 27.36 Time and current in a coil

spark occurs after the static setting it is said to be **retarded**.

Because vehicle engines are required to operate under a wide range of varying operating conditions it is necessary to alter the ignition timing while the vehicle and engine are in operation. The devices used for this purpose are often referred to as 'the automatic advance and retard devices'. There are two basic forms of automatic 'advance and retard' mechanisms: one is **speed sensitive**, the other **load sensitive**.

Distributing high-voltage electrical energy

The timing of the spark is achieved by setting the points opening to the correct position in relation to crank and piston position, or the rotor in a rotary engine such as the Wankel engine. Distributing the resultant

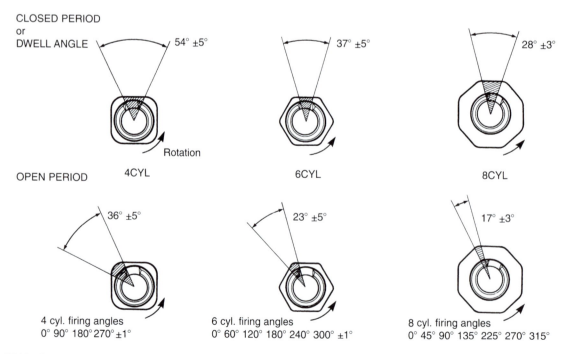

Fig. 27.35 Dwell angle for four-, six-, and eight-cylinder engines

The points gap chosen is a compromise which must be made in order to balance out the advantages with the disadvantages at very high and at very low engine speeds.

Results in small dwell angle, good for ignition at low engine speeds; results in less arcing at contacts with less contact wear but gives poor high speed performance.

Results in large dwell angle, good for ignition performance at high engine speeds; more energy stored in the HT coil, increased arcing and contact wear at low speeds.

POINTS GAP CORRECT

POINTS GAP LARGER THAN SPECIFIED

POINTS GAP SMALLER THAN SPECIFIED

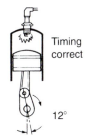 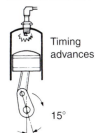 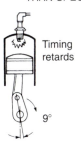

Timing correct

Timing advances

Timing retards

12°

15°

9°

Fig. 27.37 Effect of points gap on dwell angle and ignition timing

high-voltage electricity to produce the spark at the required cylinder is achieved through the medium of the rotor arm and the distributor cap, as shown in Fig. 27.38. In this figure, a spring-loaded carbon brush is situated beneath the king lead (the one from the coil). This carbon brush makes contact with the metal

part of the rotor and the high-voltage current is then passed from the rotor via an air gap to the plug lead electrode and, from there, through an HT lead to the correct sparking plug.

The rotor arm shown in Fig. 27.39 is typical of the type in common usage. It is sometimes the case that the metal nose of the rotor is extended as shown in Fig. 27.40. This is intended as an aid to prevent the engine running backwards by making the spark occur in a cylinder where the piston is near bottom dead centre, should the engine start to rotate backwards.

There is a slot in the shaft above the cam that locates the rotor. Inside the rotor arm, in the insulating plastic, is a protrusion that locates the rotor arm in the slot on the distributor cam. The rotor is thus driven round by the distributor cam. The HT leads are arranged in the cap

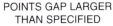

A

B

C

A – Carbon brush
B – Plug lead electrode
C – Rotor gap

Fig. 27.38 The HT circuit inside the distributor cap

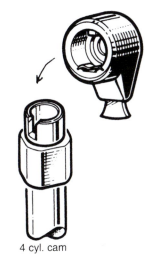 

4 cyl. cam

6 cyl. cam

Fig. 27.39 A rotor arm

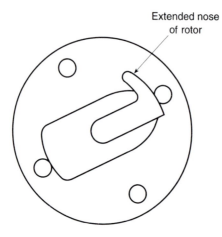

Fig. 27.40   The extended metal nose on the rotor arm

so that they are encountered in the correct firing sequence, e.g. 1—3—4—2 for a four-cylinder and probably 1—5—3—6—2—4 for a six-cylinder, in-line engine.

Cold starting

When an engine is cold the lubricating oil in the engine is more **viscous** than when it is hot. This requires greater torque from the starting motor and this, in turn, places a greater drain on the battery.

Figure 27.41 shows how the voltage required to produce a spark at the spark plug gap varies with temperature; the decimal figures refer to three different sparking plug gaps. **A much higher voltage is required for cold temperatures.**

These are among the factors that sometimes make an engine more difficult to start when it is cold than when it is hot. To ease this problem, coil ignition systems are often designed to give a more powerful spark for starting up than is used for normal running. This process is achieved by the use of a ballast resistor and a coil that gives normal operating voltage at a voltage value lower than the battery voltage, e.g. an 8-volt coil in a 12-volt

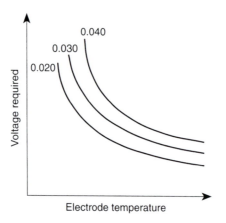

Fig. 27.41   Spark voltage and plug temperature

system. Under running conditions the coil operates on current supplied from the ignition switch, through the ballast resistor, to the primary winding. The size of the ballast resistor, in ohms, is that which produces the desired operating voltage for the coil. Under starting conditions the ballast resistor is bypassed and the coil is supplied with full battery voltage. This gives a higher voltage spark for starting purposes.

Figure 27.42 shows the circuit for a ballasted ignition system. In this system the ballast resistor is a separate component, but some other systems use a length of resistance cable between the ignition switch and the coil; this achieves the same result. Whichever type of ballast resistor is used its function will be the same and that is to produce a voltage drop that will give the coil its correct operating voltage. If the coil operates at 8 volts and the battery voltage is 12 volts, the ballast resistor will provide a voltage drop of 4 volts.

Sparking plugs

The sparking plug is the means of introducing the spark into the combustion chamber of internal combustion engines. Figure 27.43 shows the main details of a widely used make of sparking plug. In order to function effectively, sparking plugs must operate under a wide range of varying conditions. These conditions are considered next, together with some features of sparking plug design.

Compression pressure

The voltage required to provide a spark at the plug gap is affected by the pressure in the combustion chamber. It is approximately a linear relationship, as shown in Fig. 27.44. Note that the higher the compression pressure, the higher the voltage required to produce a spark.

The spark gap (electrode gap)

An optimum gap width must be maintained. If the gap is too small the spark energy may be inadequate for combustion. If the gap is too wide it may cause the spark to fail under pressure. Both the centre electrode and the earth electrode are made to withstand spark erosion and chemical corrosion arising from combustion. To achieve these aims the electrodes are made from a nickel alloy, the composition of which is varied to suit particular plug types. For some special applications the alloys used for the electrodes may contain rarer metals such as silver, platinum, and palladium.

Figure 27.45 shows the spark gap on a new plug; in time the electrodes will wear and this means that spark plugs should be examined at regular intervals. The type of electrode wear that might occur in use is shown in Fig. 27.46. The shape of worn electrodes varies. Small amounts of wear can be remedied by bending the

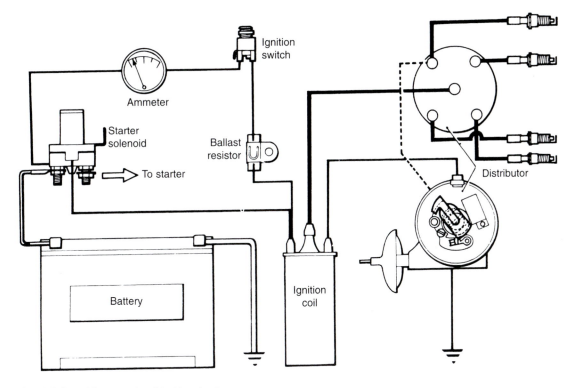

Fig. 27.42 A ballasted (series resistor) ignition circuit

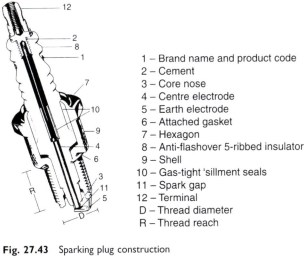

1 – Brand name and product code
2 – Cement
3 – Core nose
4 – Centre electrode
5 – Earth electrode
6 – Attached gasket
7 – Hexagon
8 – Anti-flashover 5-ribbed insulator
9 – Shell
10 – Gas-tight 'sillment seals
11 – Spark gap
12 – Terminal
D – Thread diameter
R – Thread reach

Fig. 27.43 Sparking plug construction

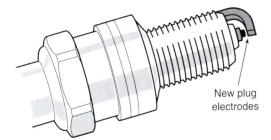

Fig. 27.45 The spark plug electrodes gap

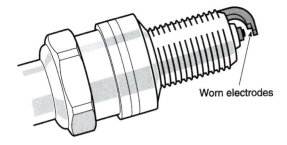

Fig. 27.46 Worn electrodes

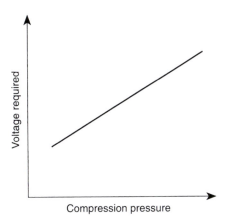

Fig. 27.44 Variation of sparking voltage with compression pressure

side electrode to give the required gap, but in cases of excessive wear the only remedy is to replace the plug.

For good performance the spark plug insulator tip must operate in the range from approximately 350 to 1050°C. If the insulator tip is too hot, rapid electrode wear will occur and pre-ignition may also be caused. If the insulator tip is too cool the plug will 'foul' and misfiring will occur. The heat range of the spark plug is important in this respect.

Heat range of sparking plugs

Figure 27.47 shows the approximate amounts of heat energy that are dissipated through the various parts of the spark plug; by far the greatest amount passes through the threaded portion into the metal of the engine.

A factor that affects the dissipation of heat away from the centre electrode and the insulator nose is the amount of insulator that is in contact with the metallic part of the plug. This is known as the heat path. Spark plugs that 'run' hot have a long heat path and spark plugs that 'run' cold have a short heat path, as shown in Fig. 27.48.

In general, hot running plugs are used in cold running engines, and cold running plugs are used in hot running engines. The type of plug to be used in a particular engine is specified by the manufacturer. This recommendation should be observed. In order not to restrict choice it is possible to obtain charts that give information about other makers' plugs that are equivalent to a particular type.

Plug reach

The length of the thread, shown as dimension R in Fig. 27.43, is a critical dimension. If the correct type of plug is fitted to a particular engine there should be no problem and the plug will seat as shown in

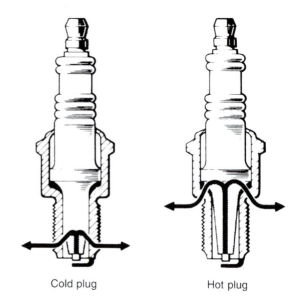

Cold plug   Hot plug

Fig. 27.48   Cold and hot running sparking plugs

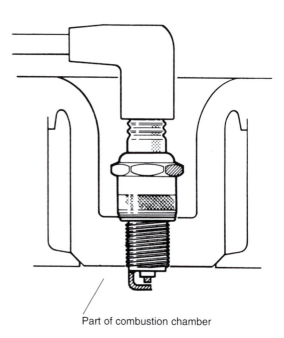

Part of combustion chamber

Fig. 27.49   The importance of plug 'reach'

Fig. 27.49. However, if the reach is too long the electrodes and threaded part of the plug will protrude into the combustion chamber and will probably damage the valves and piston.

Gasket or tapered seat

Spark plugs will be found with a gasket-type seat as shown in Fig. 27.50 or a tapered seat as shown in Fig. 27.51:

- **Gasket seat.** Spark plugs with a gasket-type seat need to be installed 'finger tight' until the gasket is firmly on to its seat; having previously ensured that

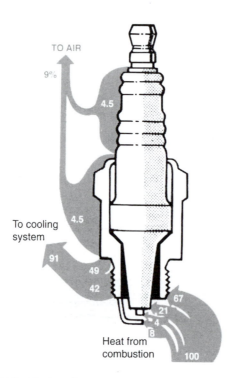

TO AIR

9%

4.5

To cooling system

4.5

91

49

42

67

21

4

8

Heat from combustion

100

Fig. 27.47   The heat flow in a typical sparking plug

Fig. 27.50 A gasket-type plug seat

the seat is clean, the final tightening should be to the engine maker's recommended torque.

- **Tapered seat.** These spark plugs do not require a gasket (commonly known as a plug washer). The tapered faces of both the gasket and the cylinder head need to be clean. The plug must be correctly started in its threaded hole, by hand, and then screwed down, finger tight. The final tightening must be done very carefully because the tapered seat can impose strain on the thread. As with the gasket-type seat, the final tightening must be to the engine maker's recommended torque.

Caution. Alignment of the thread when screwing a spark plug in is critical; always ensure that the plug is not cross-threaded before exerting torque above that which can be applied by the fingers. In aluminium alloy cylinder heads it is very easy to 'tear' the thread out. Repairing such damage is normally very expensive.

Fig. 27.51 A tapered-type plug seat

Maintenance and servicing of sparking plugs

In common with many other features of vehicle servicing, the service intervals for sparking plugs have also lengthened and it is quite common for the servicing requirement to be to keep the exterior of the ceramic insulator clean and to replace sparking plugs at prescribed intervals, perhaps 20 000 miles or so. However, there are still quite a few older vehicles in use and it is useful to have an insight into the methods of spark plug maintenance that may be applicable to such vehicles.

Cleaning spark plugs

Early spark plugs were made so that they could be taken apart to be cleaned. This meant that all parts could be cleaned satisfactorily by the use of a wire brush. When the non-detachable spark plug was introduced it became necessary to clean by abrasive blasting. This requires the use of a special machine. Figure 27.52 shows such a machine in use. It should be noted that protective eyewear must be worn for this operation because it is fairly easy to make a slip and this could lead to injury.

In addition to the abrasive (sand) blast facility, the plug cleaning machine contains an 'air-blast' facility that will remove unwanted abrasive from the spark plug. When the cleaning operation is completed the plug is removed from the machine. The plug thread can then be cleaned and the electrodes inspected. If the electrodes are spark eroded then the surfaces should be filed flat, as shown in Fig. 27.53.

After this operation has been performed, the spark gap electrodes should be adjusted to the correct gap, and this gap is checked by placing a feeler gauge, of the required thickness, between the electrodes. Any

Fig. 27.52 A spark plug cleaning machine

Fig. 27.53 Filing the electrodes

adjustment of the gap is carried out by bending the side contact with the aid of a small 'wringing iron', as shown in Fig. 27.54.

Testing the spark plug

The spark plug servicing machine normally includes a facility for testing the spark plug. This facility is a small pressure chamber into which the spark plug is screwed. A control allows the operator to admit compressed air to pressurize the chamber into which the spark plug is screwed. The degree of compression is registered on a pressure gauge. The machine is also supplied with an HT coil, which is connected by a suitable HT lead to the terminal of the spark plug. When the desired test pressure is reached the 'spark' button is pressed. The pressure chamber also includes a

Fig. 27.54 Bending the side electrode to set the gap

transparent window and it is possible to examine the spark through this window.

Fault tracing in contact breaker ignition systems

Safety note

The HT voltage is of the order of many thousands of volts, perhaps as high as 40 000 volts. The danger of electric shock is ever present and steps must be taken to avoid it. In addition to electric shock the involuntary 'jerking' of limbs, when receiving a shock, may cause parts of the human frame to be thrown into contact with moving parts such as drive belts, pulleys, and fans.

Care must always be taken to avoid receiving an electric shock. It is possible to obtain special electrically insulated tongs for handling HT cables when it is necessary to handle the cables of running engines.

Most of the details about oscilloscope testing of electronic ignition systems also apply to contact breaker systems. The details in this section refer to a more practical type of work that may be required when repairs are required outside a workshop, as may happen in the case of a roadside breakdown.

Fault-finding principles

Before considering ignition system faults and how to deal with them it is wise to remember that much of the skill required to perform diagnosis and repair of electrical and electronic systems is the same as that required for good-quality work of any type. By this I mean that it is important to be methodical, it is unwise to start testing things randomly, or even to try changing parts in the hope that you might hit on the right thing by chance.

Many people do work methodically and they probably employ a method similar to the 'six-step' approach, which is a good, commonsense approach to problem solving in general. The 'six-step' approach provides a good starting reference, although it requires some refinement when used for diagnosis of vehicle systems.

We will briefly consider the 'six steps' and at a later stage take into account the refinements that are considered necessary for vehicle systems.

The 'six-step' approach

This six-step approach may be recognized as an organized approach to problem solving in general. As quoted here it may be seen that certain steps are recursive. That is, it may be necessary to refer back to previous steps as one proceeds to a solution. Nevertheless, it does provide a proven method of ensuring that vital steps are not omitted in the fault-tracing and rectification process. The six steps are:

1. Collect evidence
2. Analyse evidence
3. Locate the fault
4. Find the cause of the fault and remedy it
5. Rectify the fault (if different from 4)
6. Test the system to verify that repair is correct.

Just to illustrate the point, take the case of a vehicle with an engine that fails to start. The six-step approach could be:

1. Is it a flat battery? Has it got fuel, etc.?
2. If it appears to be a flat battery, what checks can be applied, e.g. switch on the headlamps.
3. Assume that it is a flat battery.
4. What caused the battery to become discharged?
5. Assume, in this case, that the side and tail lights had been left on. So, in this case, recharging the battery would probably cure the fault.
6. Testing the system would, in this case, probably amount to ensuring that the vehicle started promptly with the recharged battery. However, further checks might be applied to ensure that there was not some permanent current drain from the battery.

We hope that you will agree that these are good, commonsense steps to take and we feel sure that most readers will recognize that these steps bear some resemblance to their own method of working.

Visual checks

Figure 27.55 shows a preliminary check of the type that is advised in most approaches to fault tracing. Here the technician is looking for any obvious external signs of problems such as loose connections or broken wires, cracked or damaged HT cables, dirty or cracked distributor cap and coil tower insulator, dirty and damaged spark plug insulators.

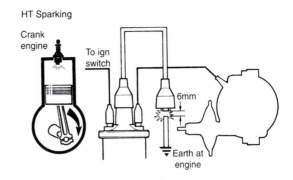

Fig. 27.56 Checking for HT sparking

Checking for HT sparking

A suitable neon test lamp is useful for this purpose. However, Fig. 27.56 shows the king lead removed from the top of the distributor. Here it is held 6 mm from a suitable earth point. The engine is then switched on and the starter motor operated; there should then be a regular healthy spark each time the CB (contact breaker) points open. A regular spark in this test shows that the coil and LT circuits are working properly and that the fault lies in the HT system, i.e. the rotor arm, the distributor cap, the HT leads, or the spark plugs.

Checking the rotor

The rotor arm can be checked by the procedure shown in Fig. 27.57; here the distributor cap end of the king lead is removed from the distributor cap. The end of the king lead is held about 3 mm above the metal part of the rotor. With the CB points closed and the ignition switched on, the points can be flicked open with a suitable insulated screwdriver, or an assistant can be used to turn the engine over on the starter motor. There should be no spark. If there is a spark it means that the rotor insulation has broken down and this means that the rotor must be replaced by a new one.

If there is no spark then this indicates that the rotor is satisfactory. The next step would be to check the HT leads; if, as is likely, they are resistive leads it will be necessary to look up the permitted resistance of the

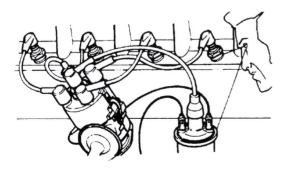

Fig. 27.55 Performing a visual check

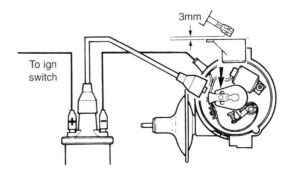

Fig. 27.57 Checking the insulation of the rotor

leads before performing an ohmmeter test. The resistance value must fall within the stated limits. If they are not resistive leads, a straightforward continuity test will prove the conductivity of the HT leads. If these checks prove satisfactory the sparking plugs must be examined. I have already described the procedure for checking sparking plugs; if these facilities are not available the only solution may be to inspect the plugs internally, and the most likely step will be to fit a new set of spark plugs. It is unusual to find that a set of spark plugs has failed completely.

Checking the condition of the contact points

Should the first test (Fig. 27.56) have failed, i.e. no spark, it is advisable to check the contact breaker condition. The contact surfaces should be clean and not badly pitted and piled, and the gap should be checked to ensure that it is within the recommended limits. The securing screws should be checked for tightness and there should be no evidence of fouling of cables inside the distributor. If the points are badly worn and burnt a new set of contacts should be fitted. After this the first check for HT spark should be repeated. If this is successful then there may be no need to proceed further. However, if there is still no spark further tests will be required. These tests are now described. The distributor cap should be removed and the engine rotated, with the ignition off, until the CB points are closed. The voltmeter is then connected between the switch side of the coil and earth. The ignition should then be switched on. The voltmeter should then read battery voltage (very closely) or, if it is a ballasted coil, a lower voltage, probably 6−8 volts.

Checking the voltage on the CB side of the coil

If these tests are satisfactory the next step is to proceed to check the voltage at the contact breaker side of the coil. The voltmeter position for this test is shown in

Fig. 27.59. The CB contacts must be open. The voltmeter is then connected between the CB terminal of the coil and earth. The ignition should now be switched on and the voltmeter reading observed. If the coil primary circuit is satisfactory, a reading that is very close to that obtained in the test shown in Fig. 27.58 should be seen. If the reading is correct, proceed to test the voltage at the coil CB terminal with the contact points closed.

If the voltage reading is zero, disconnect the LT lead to the distributor. The voltmeter should now read the same voltage as on the switch side of the coil. If it still reads zero it indicates that there is a broken circuit in the primary side of the coil and this calls for a new coil.

Voltage at the CB terminal of the coil − point closed

The test shown in Fig. 27.60 checks the condition of the contact breaker points and also checks whether there are any current leakages to earth. The voltmeter should show a zero reading if all is correct, or less than 0.2 volts if it is a very sensitive meter. If the reading is more than zero volts the contacts should be checked to see if they are clean and properly closed; the insulation on the contact to condenser lead and on the LT lead should also be checked. If these are in order it may be that the capacitor itself is providing a leakage path to earth.

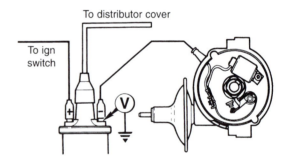

Fig. 27.59 Voltmeter check on CB side of coil

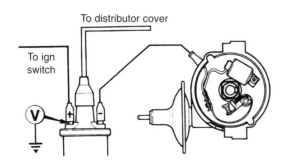

Fig. 27.58 Voltmeter check on coil supply

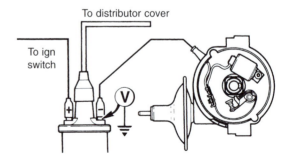

Fig. 27.60 Voltmeter test for voltage drop across the points.

Testing the capacitor

Now that the HT lead has been tested, and if there is still no spark from this test, then the capacitor should be tested. Figure 27.61 shows how a test capacitor can be connected temporarily to perform this test. The original capacitor is unscrewed and lifted away so that its casing does not make contact with earth.

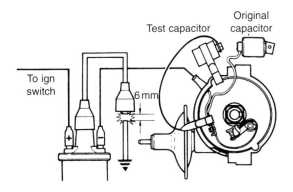

Fig. 27.61 Connecting a test capacitor

A test capacitor with suitable crocodile clips and leads is then connected as shown. The ignition is switched on again and the engine turned over, on the starter motor. If a spark is obtained this is fairly convincing evidence that the capacitor should be renewed. If there is still no spark, after testing the capacitor, the coil should be replaced.

Checking the electrical condition of the ignition coil

The ignition coil can be checked as shown in Fig. 27.62. The secondary winding resistance should be high, probably 10 kΩ. (Note: In some ignition coils the secondary winding is earthed. In such cases the check for secondary winding resistance would be made between the HT connection and the coil earth.) The primary winding resistance should be very low (the meter needs to be set to a very low ohm scale), 0.5−1 Ω.

The tests shown here are of a general nature, i.e. they do not apply to a specific system and, as with most other work on vehicle systems, it is important to have the

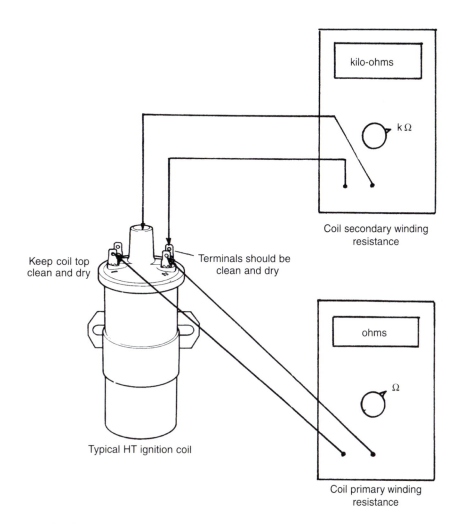

Fig. 27.62 Checking the HT coil

figures that relate to the specific system being worked on. However, the tests do show the practical nature of tests that can be applied with the aid of a good-quality multimeter, sound knowledge of circuits, and the relevant information.

High-tension (HT) leads

The HT leads are the means by which the high-voltage electrical energy that makes the spark is conducted to the sparking plug. The high voltage, perhaps as high as 40 kV, will wish to seek the shortest path to earth. In order to prevent HT leakage from the cables they must be heavily insulated electrically. The insulating material must also be resistant to heat, and to oil, water, and any corrosive agents that they may encounter; PVC is often used for the insulation because it possesses most of these properties. To reduce radio interference, the conducting parts of HT leads are made to have electrical resistance. This resistance can change in use and HT leads should be checked to ensure that the leads have the correct resistance. The resistance of HT leads varies considerably, a rough guide being **18 000 Ω per metre length**. This means that an HT cable of **300 mm length would have a resistance of 5400 Ω.** It is a value that can be checked by means of an ohmmeter and the resistance values and tolerances will be given in the workshop manual.

As with any electric current in a conductor, the HT leads set up a magnetic field. In order to prevent one HT cable inducing current in a neighbouring one, by mutual induction, cable routing as arranged by the vehicle manufacturer should always be adhered to.

Some practical applications, i.e. doing the job

A substantial amount of background theory has been covered and we have shown how this theory is used to make vehicle systems work. When systems break down or cease to function correctly it is the job of the vehicle service and repair technician to find out what is wrong and to put it right, and this is where practical skill, or competence, comes in to play. In a book of this type it is not possible to cover every type of vehicle system and we will not attempt to do that. As with all work, proper safe working practices must be observed. Personal safety and the safety of others nearby must be protected, and the vehicle itself must be protected against damage by ensuring that the correct procedures prescribed by the manufacturer are followed and that wing and upholstery covers etc. are used.

Because of the global (international) nature of the vehicle industry, vehicle technicians will encounter a range of products and it is unwise to attempt even simple work without having good knowledge of the product, or access to the information and data relating to the vehicle and system to be worked on.

The practical work described in this section generally relates to a specific product and that is made clear in the text. However, the devices chosen and the work described have been selected because they highlight types of practical procedures and tests that can be applied across a range of vehicles. It is intended that this book should be used in conjunction with courses of practical instruction and learning on the job.

The main source of information and guidance about repair and maintenance procedures is the manufacturers workshop manual, which is often available on CD-ROM.

Ignition timing

Because ignition timing varies with engine speed, it is necessary to check both the static timing and the dynamic timing.

Static timing

The following list shows the ignition timing details for a four-cylinder in-line engine:

- Firing order 1–3–4–2. Static advance 10°.
- Contact breaker gap 0.40 mm.
- Dwell angle 55°.

For static advance setting, the figures of interest are: static advance angle of 10°, contact breaker gap of 0.40 mm, and the firing order.

When resetting the static timing after performing work on the engine, such as on removing and refitting the distributor, it is necessary to ensure that the piston in the cylinder used for timing (usually number 1) is on the compression stroke. The firing order helps here because in this case the valves on number 4 cylinder will be 'rocking' when number 1 piston is at top dead centre (TDC) on the compression stroke. The 10° static advance tells us that the contact breaker points must start to open at 10° before TDC. Engines normally carry timing marks on the crankshaft pulley and these are made to align with a pointer on the engine block, or timing case, as shown in Fig. 27.63.

To position the piston at 10° before TDC the crank should be rotated, by a spanner on the crankshaft pulley nut, in the direction of rotation of the engine. This ensures that any 'free play' in the timing gears is taken up. To make the task more manageable it is probably wise to remove the sparking plugs. If the alignment is

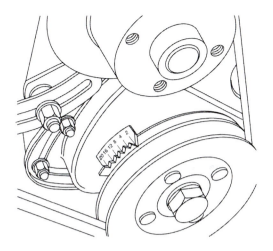

Fig. 27.63 Engine timing marks

missed the first time round, the crank should continue to be rotated in the direction of rotation until the piston is on the correct stroke and the timing marks are correctly aligned.

When the piston position of number 1 cylinder is accurately set the distributor should be replaced with the rotor pointing towards the HT segment, in the distributor cap, that is normally connected to number 1 sparking plug. This process is made easier on those distributors that have an offset coupling.

Setting the timing

The contact breaker gap should be set to the correct value and the distributor clamp loosened. When the low-tension terminal of the distributor has been reconnected to the coil, a timing light may then be connected between the contact breaker terminal of the coil and earth. With the ignition switched on and the contact breaker points closed, the timing lamp will be out. When the points open, the primary current will flow through the timing light, which will light up. An alternative is to disconnect both coil LT connections and then connect the timing light in series, as shown in Fig. 27.64; in this case the timing light will be on when the points are closed and off when they are open.

When the correct setting has been achieved the distributor clamp should be tightened, the timing light removed, and all leads reconnected. The timing light (lamp) can be a light-bulb of suitable voltage (12 V) mounted in a holder and fitted with two leads to which small crocodile clips have been securely attached.

Dynamic timing

As the name suggests, dynamic timing means checking the timing with the engine running. At this present level of study we will restrict ourselves to the use of the stroboscopic timing lamp, often referred to as a strobe lamp. The strobe lamp 'flash' is triggered by the HT pulse from a sparking plug, usually number 1 cylinder. When the strobe light is directed on to the timing marks the impression is given that the mark on the rotating pulley is stationary in relation to the fixed timing mark.

Safety note

This useful stroboscopic effect carries with it potential danger because the impression is given that other rotating parts such as fan blades, drive belts, etc. are also stationary. It is therefore important not

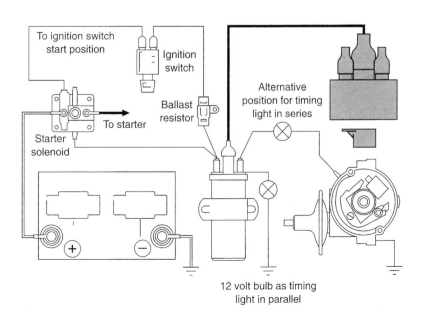

Fig. 27.64 Using a timing light

to allow hair and clothing or any part of the body to come into contact with moving parts. It is also important to avoid electric shock. As with other work that involves running engines, exhaust extraction equipment should be used.

Checking the static timing with the engine running

The static timing setting can be checked by means of the strobe lamp because, at idling speed, the centrifugal advance and the vacuum advance mechanisms should not be adding to the static advance angle. The static timing mark on the engine should be highlighted with white chalk, or some other suitable marking substance. This will enable you to locate the correct marks when conducting the test. Figure 27.65 shows the strobe light; this one incorporates an advance meter, used to check the static timing. For this test the engine will be run at the manufacturer's recommended speed, probably idling speed. If the marks align correctly no further action is required. However, if the marks do not align then the distributor clamp must be slackened and the distributor body rotated until the marks are aligned correctly. When the correct alignment is achieved the clamp must be re-tightened.

Checking the centrifugal advance mechanism

The timing details that we are using as an example gave us a static advance figure of $10°$ and the following for vacuum and centrifugal:

- Automatic centrifugal advance $25°$
- Vacuum advance $12°$.

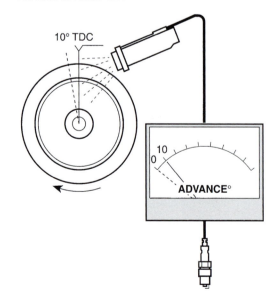

Fig. 27.65   The stroboscopic timing lamp

If the vacuum advance is disabled by disconnecting the pipe at the engine end, and blanking off the hole with a suitable device, the maximum amount of advance obtainable will be $35°$, i.e. $25°$ from the centrifugal device and $10°$ static. In this example this maximum advance occurs at 4300 rev/min of the engine and a tachometer (rev counter) will be needed to record the engine speed. With the vacuum pipe disconnected, the tachometer and strobe light connected, and all leads checked to ensure that nothing is touching moving parts or hot exhausts, the check may proceed. It will probably require another person to operate the accelerator and observe the tachometer. The strobe light should be aimed at the timing marks and when the engine is brought up to the correct speed the control on the advance meter is adjusted until the timing marks align. The number of degrees shown on the advance meter should then show the maximum advance angle. If this is correct the vacuum advance mechanism can be tested.

Checking the vacuum advance mechanism

After removing the blanking device, the vacuum pipe should be reconnected and the engine speed will increase slightly if the vacuum device is working.

Dwell angle

In contact-breaker-type ignition systems the dwell angle is the period during which the contact breaker points are closed. The dwell angle is the period during which the electrical energy builds up in the coil primary winding. If the dwell period is too short the primary current will not reach its maximum value and the HT spark will be accordingly weaker. Figure 27.66 shows dwell angle, which has already been discussed in the context of the operating principles of the ignition system.

On a four-lobed cam as shown here, the dwell angle is $54 \pm 5°$. However, the dwell angle varies according to distributor type and the manufacturer's data should always be checked to ensure that the correct figure is being used. The dwell angle is affected by the points gap (Fig. 27.67). If the gap is too large the heel of the moving contact will be closer to the cam and this will cause the dwell angle to be smaller than it should be.

Adjusting the contact breaker points gap

Not only does the points gap affect the dwell angle, but it also affects the ignition timing. A wide points gap advances the ignition and too small a gap retards the ignition. Setting the contact breaker points to the

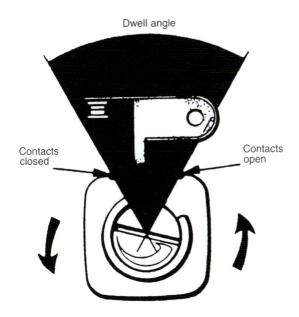

Fig. 27.66   Dwell angle

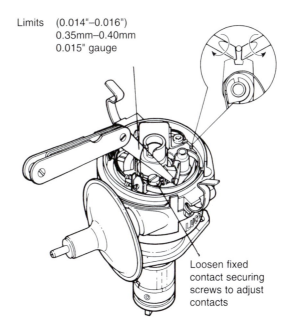

Limits   (0.014"–0.016")
0.35mm–0.40mm
0.015" gauge

Loosen fixed
contact securing
screws to adjust
contacts

Fig. 27.68   Checking points gap

correct gap is therefore a critical engineering measurement in a technician's work. Figure 27.68 shows a set of feeler gauges being used to check the gap, and the small inset shows a method of achieving the fine adjustment.

In performing this task it is essential to use clean feeler gauges and it should be noted that feeler gauges do wear out. For example, if feeler gauges have been used to check valve clearances with the engine running, it is quite possible for the blades to be 'hammered' thin. An occasional check with a micrometer will verify the accuracy of the feeler gauges. Returning to the points gap setting, the points gap is checked with the ignition switched off. The feeler gauge is inserted between the contacts, as shown in Fig. 27.68; very light force should be used to 'feel' the setting and care must be taken to keep the feeler gauge blade in line with the contact face. If the points gap is set accurately, the dwell angle should be correct. However, it is common practice to check the dwell angle by the use of a dwell meter and tachometer. These instruments are normally part of an engine analyser, such as the Crypton CMT 1000 shown in Fig. 27.69.

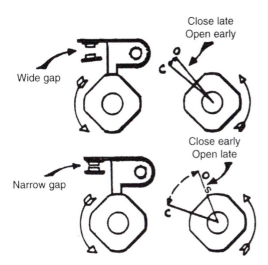

Close late
Open early

Wide gap

Narrow gap

Close early
Open late

Fig. 27.67   The effect of points gap on dwell angle

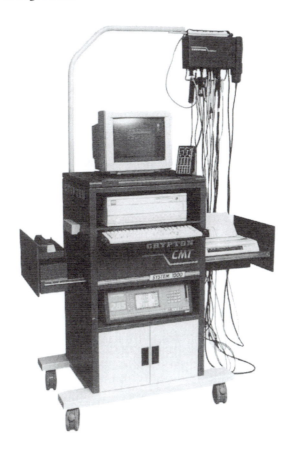

Fig. 27.69   A full-sized diagnostic machine

Servicing a contact-breaker-type distributor

Figure 27.70 shows the main points that require periodic attention. Oil should be applied sparingly and care taken to prevent contamination of the contact points. While the distributor cap is removed it should be wiped clean, inside and out, and checked for signs of damage and tracking (HT leakage). The rotor arm should also be inspected.

Explanation of terms associated with ignition systems

Misfiring

Misfiring is a term that is normally applied to the type of defect that shows up as an occasional loss of spark on one or more cylinders. It can be caused by almost any part of the ignition system and the cause is often difficult to locate. Much depends on the way in which the fault occurs. If it is a regular and constant misfire one could start by checking for HT at each sparking plug. If each plug is receiving HT, then the most likely cause is a spark plug. Plugs should then be removed and tested. In some cases, where the engine has not been running for very long, the insulator of a non-firing spark plug will be significantly cooler to the touch, than the other plug insulators. This may be a help in locating the misfire. Here again, an electronic engine analyser is much more satisfactory, because a power drop test will show which cylinder is not firing.

Cutting out

Cutting out is a term that is used to describe the type of fault where the engine stops completely, perhaps only for a split second. The most likely cause here is that there is a broken or loose connection, which breaks the circuit

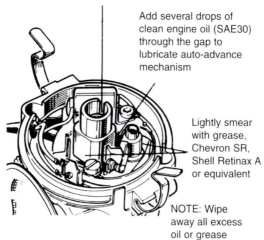

Add 2 or 3 drops of clean engine oil (SAE30)

Add several drops of clean engine oil (SAE30) through the gap to lubricate auto-advance mechanism

Lightly smear with grease, Chevron SR, Shell Retinax A or equivalent

NOTE: Wipe away all excess oil or grease

Fig. 27.70   Service details for a distributor

momentarily. Careful examination of all connections, and 'wiggling' of cables whilst conducting a voltage drop test across the connectors, should help to locate such faults.

Hesitation

This normally happens under acceleration; the symptoms are that the engine does not respond to throttle operation and it can cause problems when overtaking. The voltage required to produce a spark rises with the load placed on the engine as happens under acceleration. Probable causes are wide spark plug gaps, or spark plugs breaking down under load.

Excessive fuel consumption

Incorrect ignition timing is usually associated with high fuel consumption.

Low power

Incorrect ignition timing, weak spark, dirty or badly adjusted plugs are defects associated with low power.

Overheating

The ignition defect most often associated with overheating is the ignition timing, but other defects that cause pre-ignition and detonation can also lead to overheating of the engine.

Running on

This happens when the engine continues to run even after the ignition is switched off. An obvious cause is that parts of the engine interior remain sufficiently hot to cause combustion. This may be caused by a dirty engine, with heavy carbon deposits in the combustion chamber, or by dirty, worn, or incorrect type of sparking plugs. It is overcome, to some extent, by fitting a cut-off valve in the engine idling system. This prevents mixture from entering the combustion chamber through the engine idling system, this being the probable source of fuelling as the throttle is virtually closed.

Detonation and pre-ignition

Both detonation and pre-ignition give rise to a knocking sound that, at times, can be quite violent. On other occasions a high-pitched 'pinking' sound may be heard and, as the term implies, this sound is known as pinking. Pre-ignition arises when combustion happens before the spark occurs. Detonation happens after the spark has occurred.

Pre-ignition

Pre-ignition may be caused by 'hot' spots in the combustion space. These 'hot' spots may be caused by sharp edges and rough metallic surfaces, glowing deposits (carbon), overheated spark plugs, and badly seated valves. As a result of ignition starting before the spark occurs, the pressure and temperature in the cylinder rise to high levels at the wrong time. This gives rise to poor performance, engine knock, and eventual engine damage. Some of the probable causes of pre-ignition are listed below:

- **Sharp edges and rough metallic surfaces.** These are likely to be caused by manufacturing defects and are normally rectified at that stage. However, if an engine has been opened up for repair work it is possible that careless use of tools may damage a surface. Should this be the case, the remedy would be to remove the roughness by means of a scraper, a file, or a burr.
- **Glowing deposits (carbon).** This is most likely to occur in an engine that has covered a high mileage. Decarbonizing the engine (a decoke) would probably be the answer here.
- **Overheated spark plugs.** In this case there are various factors to consider: Are the spark plugs of the correct type and heat range? Are the spark plugs clean? Are the electrodes worn thin?
- **Badly seated valves.** If a poppet-type valve does not seat properly it is possible for part of the valve head to form a hot spot. Failure to seat squarely could be caused by a bent valve, a misaligned valve guide and, in the case of older engines, a badly reconditioned valve. In each case the remedy would be to rectify the fault indicated.

In addition to the above defects, combustion knock can arise from use of the incorrect fuel. The octane rating of a fuel is the critical factor here. Tetra-ethyl lead (leaded fuel) used to be the method of altering the octane rating (anti-knock value) of petroleum spirit. Now that most spark ignition engines are fitted with catalytic converters for emissions control, leaded fuels cannot be used. There are different octane grades of unleaded fuel and many electronically controlled engines are provided with a simple means of altering settings if a driver wishes to use a different octane rating fuel from the one that she/he has been using.

I hope that you will agree that these are good, commonsense steps to take and feel sure that most readers will recognize that these steps bear some resemblance to their own method of working.

In the following cases it is assumed that the above steps have been followed and that the fault has been found to exist in the area of the ignition system.

Poor starting or failure to start

For current purposes we will assume that the fuel and electrical systems are in good order. In particular, that the battery is fully charged and that the starter motor is capable of rotating the engine correctly. We shall also assume that the mechanical condition of the engine is good and that the fault has been identified as being in the area of the ignition system.

Poor starting

If the controls are operated correctly, the engine should start promptly. The time taken to start up may vary with air temperature. For example, when it is very cold the cranking speed of the engine will be affected by the viscosity of the lubricating oil, and the mixture entering the engine is affected by contact with cold surfaces. Generally, an engine will start more promptly when it has warmed up than it will when it is cold.

For 'poor' starting we are therefore looking at cases where, taking account of the above factors, the engine either takes a long time to start or is reluctant to start at all.

The ignition system faults that have a bearing on the problem of poor starting are:

- **Weak spark.** This could be caused by loss of electrical energy in the ignition circuit, or failure to generate sufficient energy for a spark, in the first instance. Loss of electrical energy could be due to defective connection and/or defective insulation in either the low-tension or the high-tension circuit. A range of tests and checks that can be carried out is described in the section on servicing and repair.
- **Spark occurring sometimes but not at other times.** This is known as an intermittent fault. Here again, the problem could lie in either or both the LT and the HT circuits.
- **Spark occurring at the wrong time.** In most cases this means that the ignition timing is incorrectly set.

Failure to start

As it is ignition faults that we are considering, the obvious place to start is the sparking plug gap. Failure to start is probably due to there being no spark, or a spark that is too weak to cause combustion. However, if the ignition timing is not correct, or the HT leads have been wrongly connected (not in accordance with the firing order), the engine may also fail to start.

Learning tasks (ignition)

Do not attempt any practical work until you have been properly trained. You must always seek the permission and advice of your supervisor before you attempt any of the practical tasks.

Much of the NVQ assessment is based on work that you have actually performed. Each time you complete the tasks you should keep a record and note down the tools and equipment that you have used.

1. Make a list of the precautions to be taken when working on ignition systems.
2. Describe the procedure for setting the static ignition timing on an engine that has a contact-breaker-type ignition system.
3. Give details of the procedure and type of feeler gauges used to set the pick-up to rotor air gap in an electronic, contact-breakerless, ignition system.
4. Take a set of ignition HT leads and measure the length in centimetres and the resistance in kΩ of each lead. Compare the figures with those given in this book, or in the workshop manual for the vehicle. Write down the result and state whether or not the leads are in good condition.
5. Examine a set of sparking plugs from an engine. Measure and record:
   (a) the spark gaps
   (b) the reach of each plug
   (c) the thread diameter
   (d) the recommended tightening torque
   (e) note the type of spark plug (not the make) and state whether they are intended for use in a hot running engine or a cold running engine.
6. Examine a distributorless ignition system. Write down the number of ignition coils and the position of the sensor that triggers the ignition pulses.
7. Describe, with the aid of diagrams, a simple practical test that can be used to check the electrical condition of a rotor arm.
8. Perform a test to check the electrical output of a magnetic-pulse-type generator that is used in an electronic ignition system.
9. Obtain an ignition coil and measure the resistance of the primary winding and the secondary winding. Record the figures and compare them with the figures given in this book, then write down your opinion about the condition of the coil. State whether or not the coil is for use in a ballasted ignition system.
10. Describe the procedure for setting up a garage-type oscilloscope to check ignition coil polarity.

Self-assessment questions

1. A Hall effect ignition system sensor:
   (a) requires a supply of electricity
   (b) generates electricity by electromagnetism
   (c) can only be used on ignition systems that have a distributor
   (d) can only be used on four-cylinder engines.

2. In a test on a contact-breaker-type ignition system with the points closed and the ignition switched on, the voltage drop across the points should be:
   (a) 14 volts
   (b) not more than 0.2 volts
   (c) 2.5 volts
   (d) 2 volts.
3. In a 'lost spark' ignition system each cylinder has a spark:
   (a) on the induction stroke as well as at the end of the compression stroke
   (b) on the exhaust stroke as well as at the end of the power stroke
   (c) twice on the power stroke
   (d) on the exhaust stroke as well as at the end of the compression stroke.
4. Ignition-related faults that affect the emissions controls on the vehicle:
   (a) cause the MIL lamp to illuminate on EOBD systems
   (b) cause the vehicle to be disabled
   (c) require the vehicle to be taken off the road immediately
   (d) should be left to rectify themselves.
5. A set of ignition HT leads are required to have a resistance of 18 kΩ per metre of length. One of these leads that is 430 mm long will have a resistance of:
   (a) 41.86 Ω
   (b) 7740 Ω
   (c) 4186 Ω
   (d) 0.7 kΩ.
6. Sparking plugs that 'run' hot have a long heat path, and sparking plugs that 'run' cold have a short heat path. The sparking plugs fitted to an engine that runs 'cool' would have:
   (a) a short heat path
   (b) tapered seats
   (c) a long heat path
   (d) short reach.
7. The 'firing line' on an oscilloscope trace of ignition HT voltage represents:
   (a) the high voltage that is needed to cause the spark to bridge the sparking plug gap
   (b) the current in the primary winding of the ignition coil
   (c) the condition of the condenser
   (d) the air–fuel ratio of the mixture in the cylinder.
8. A knock sensor is used:
   (a) to allow the ECM to alter the ignition timing
   (b) to alert the driver to mechanical noise from wide valve clearances
   (c) to allow catalytic converters to operate on leaded fuel
   (d) on contact-breaker-type ignition systems only.

9. Pre-ignition happens:
   (a) when the static ignition timing is too far advanced
   (b) when combustion occurs before the spark
   (c) when pressure waves cause a region of high pressure, in the combustion chamber, after the spark has occurred
   (d) when the ignition is retarded.

10. In engine management the ignition timing can be altered to suit engine wear. When new parts are fitted to replace worn or damaged ones on this type of ignition system, the vehicle:
    (a) should have a new ECM fitted
    (b) should be operated under normal running conditions for a length of time that allows the ECM to adjust to the new parts
    (c) should be returned to the customer without advising them that it may take time for the vehicle to return to its best working condition
    (d) should have the entire ignition system renewed.

11. The vacuum-operated timing control on distributor-type ignition systems:
    (a) changes the ignition timing to suit engine speed
    (b) changes the ignition timing to suit engine temperature
    (c) changes the ignition timing to suit load on the engine
    (d) changes the air–fuel ratio.

28
Electrical systems and circuits

Topics covered in this chapter

Central door locking — remote
Screen wipers
Screen washers
Vehicle security alarm
Electric window operation
Electrically operated mirrors
Vehicle lighting systems
Supplementary restraint systems — air bags
and seat-belt pre-tensioners
Instruments
Circuit protection — fuses and circuit breakers

Central door-locking systems

Manual operation

A central door-locking system permits all doors, often including the boot lid or tailgate, to be locked centrally from the driver's door and normally includes the front passenger's door as well.

The locking and unlocking action is produced by an electromechanical device, such as a solenoid or an electric motor. The door key operates a switch that controls the electrical supply that actuates the door lock mechanism, and the remote control activates a switching circuit in the electronic control module (ECM) that also actuates the locking mechanism. For safety and convenience, the interior door handles also operate the unlocking mechanism.

Remote operation

The remote control operates in much the same way as the remote control for a domestic television. A unique code is transmitted from the remote control that is normally contained in the ignition key. Two types of transmitter are used, infrared and radio frequency. The electronic control module (ECM) that controls the supply of current to the door lock actuators contains a sensor that detects infrared signals or, in the case of radio frequency systems, an aerial to detect the radio frequency signal. The remote signal is normally effective over a range of several metres. Figure 28.1 shows the general principle of central locking systems.

The remote controller

The encoded signals that are transmitted by the remote controller are produced by electronic circuits that are energized by a small battery that is housed in the controller. A pad on the controller is pressed for unlocking and another pad is pressed for locking purposes. The device may also be provided with an LED that indicates battery level.

A simple central locking circuit

Both the manual central lock system and the remote central lock systems employ similar circuits and actuators inside the doors. A general description of central locking system circuits, of the type shown in Fig. 28.2, should serve to bring about the necessary level of understanding.

Solenoid

The type of solenoid used in door-locking systems is shown in Fig. 28.2(a). The flow of current in one direction causes the doors to lock and when current flows in the opposite direction the doors are unlocked. The reversal of current is performed by the ECM.

Motor

The motors (Fig. 28.2(b)) used for door-locking systems are normally of the permanent magnet type because they are capable of being reversed by changing the direction of current flow through the motor. The motorized operation is considered to be slightly smoother in operation than in the solenoid system. The reversal of current flow is achieved through the switching action provided by the ECM.

Circuit protection

In addition to a fuse, door-locking circuits are often equipped with a thermal circuit breaker.

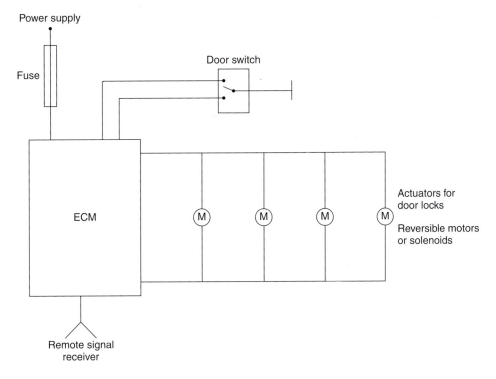

Fig. 28.1 Block circuit diagram for remote central locking

Maintenance and fault tracing

The wiring into the doors is flexible and is protected by a weatherproof sheath. The control rod from the solenoid, or motor, to the mechanical door lock is normally enclosed in a sheath to guard against ingress of water or dust. Periodic visual inspections of the exterior should suffice for these parts. The external mechanical parts of the locks should be lubricated with an approved lubricant whilst ensuring that such lubricant is applied sparingly and carefully so as not to harm the upholstery.

In the event of a locking system failure, every attempt to determine the source of the problem should be made before door trim is removed and work started on the circuit and mechanism inside the door trim. It is recommended that manufacturers' guides are used because of the considerable variations that are to be found throughout the vehicles that are in use. Some of the more obvious general points are covered in Table 28.1.

Screen wipers

Screen wipers are required so that the driver's ability to see out of the vehicle is maintained in all types of conditions. It is a legal requirement that screen wipers must be maintained in good working condition.

Modern wipers are electrically operated and a single electric motor normally drives both front screen wipers through a mechanical linkage. The principle is shown in Fig. 28.3.

Modern styling of vehicles requires the rear screen to be fitted with a separate screen wiper and this is equipped with a separate electric motor. Permanent magnet motors are normally used for screen wiper applications because they are compact and powerful and they are readily equipped with two-speed operation.

Operation

The armature shaft of the motor has a worm gear machined into it and this mates with a worm wheel that provides a gear reduction. The worm wheel carries a crankpin that is connected by a rigid link to the screen wiper linkage and, by this means, the rotary movement of the worm wheel gear is converted into the reciprocating (to and fro) movement of the wiper linkage. The wiper arms are pivoted near the screen and the shaft on which the blade is mounted is connected by a lever to the reciprocating wiper linkage. This results in angular movement of the wiper arm and blade.

Control of the wipers

The control switch for the screen wipers is mounted in a convenient position adjacent to the steering wheel. The switch is multi-functional because it is used to select the options that are available, in addition to the basic on/off operations.

Parking the wiper blades

When the wipers are not in use, the arms and blades are placed (parked) in a position that does not obstruct the

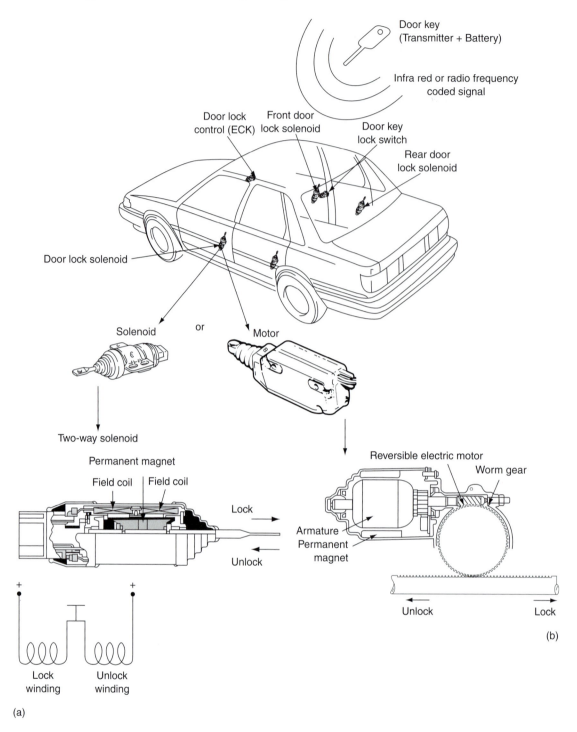

Fig. 28.2 A typical remote central locking system

view from inside the vehicle. This operation is automatically achieved by the use of a cam-operated limit switch that is located inside the wiper motor housing. A simplified circuit for this operation is shown in Fig. 28.4.

Regenerative braking

When the wipers are in action, the mechanism possesses considerable momentum and a form of electric braking is utilized to bring the wiper mechanism to rest in the desired position without placing undue strain on it. The circuit in Fig. 28.5 shows how the limit switch cam activates an additional set of electrical contacts.

When these contacts are closed, an electrical load is placed across the screen wiper motor armature and the generator effect of the moving armature places an effective braking effect on the wiper motor and causes it to stop in the required position.

Table 28.1 Door locking system fault tracing

Defect	Action to be taken
Central locking fails to operate	Check remote control battery
	Make sure that correct key is being used. If key is lost, follow instructions for coding etc. that are normally found in the driver's handbook
	Check vehicle battery condition
	Check fuse, or circuit breaker
	If fuse is blown, check for electrical problems. Replace fuse — if fuse blows again shortly after replacing, start search for the cause of the problem
A single solenoid or motor does not operate	**Electrical:** Check wiring into the door Check operation of solenoid or motor **Mechanical:** Check for broken or seized parts

Two-speed operation

Permanent magnet screen wiper motors are capable of operating at two different speeds to suit different weather conditions. The speed variation is achieved by means of a third brush. This third brush is connected to a smaller number of armature windings than the other main brushes and this causes a higher current to flow in the armature coils; this, in turn, leads to a higher speed of rotation. Figure 28.6 shows the arrangement.

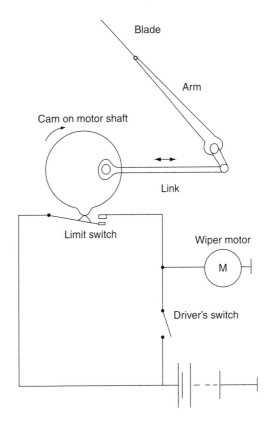

Fig. 28.4 Limit switch circuit for 'Off' position of wiper arm

Intermittent operation

Most modern wiper systems make provision for the driver to select a switch position that allows the wipers to provide a short spell of wiper action at set intervals of

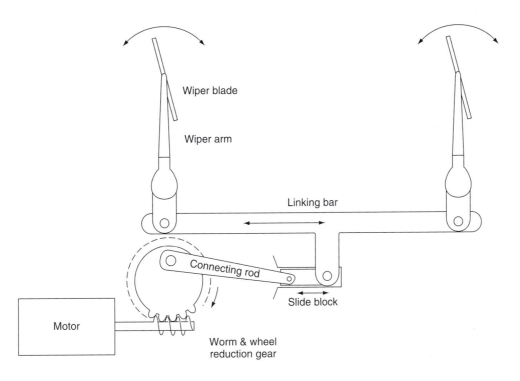

Fig. 28.3 Screen wiper system

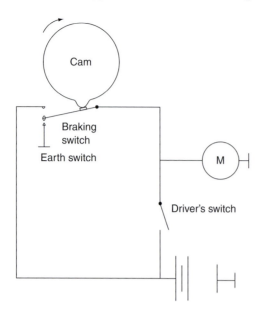

Fig. 28.5   Regenerative braking

time. This is achieved by means of a timer circuit that controls a relay. The wipe action usually occurs at intervals of 15–20 seconds and is designed to help the driver cope with conditions where light rain or showers are encountered. In some designs the delay period may be adjusted at the wiper switch. In such cases, the timer circuit will contain a resistive and capacitive device in which the resistance may be altered to vary the interval between wipe actions. An outline circuit is shown in Fig. 28.6.

Screen washers

A windscreen washer on the driver's side of the vehicle is an essential piece of equipment and it must be maintained in working order. In addition to this basic requirement, most modern vehicles are equipped with screen wash facilities on the passenger's side and the rear

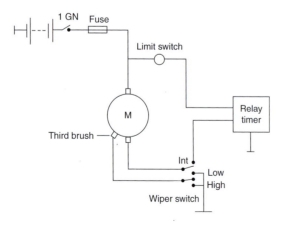

Fig. 28.6   Two-speed and intermittent wiper control (simplified circuit)

screen, and many vehicles are now equipped with headlamp washers. Figure 28.7 shows the essential parts of a screen washer system.

The pump is electrically operated from the screen wiper switch and it is normally mounted at the base of washer fluid reservoir, where it is gravity fed. The inlet to the pump is provided with a filter in order to exclude foreign matter and thus prevent blockages and damage to the pump. The fluid, under a pressure of approximately 0.7 bar from the pump, is conveyed to the washer jets through plastic piping. The jets are arranged so that a maximum washing effect is gained from the wash/wipe action. The washer should only be operated for a few seconds at a time as long periods of continuous use may result in damage to the pump motor. A few seconds of use normally suffices for effective cleaning of the screen. The fluid used is water to which is added an antifreeze agent, a detergent, and an anti-corrosion agent. The strength of the washer fluid that is added to the water in the washer reservoir is varied according to the season of the year.

Rear screen washer

Figure 28.7 shows how the screen washer reservoir and pump may be connected so that it supplies both front and rear screens.

A solenoid-operated washer valve is placed in the pipeline and operation of the washer switch permits the solenoid valve to direct the fluid to the required washer jets.

Maintenance and repair

It should be evident that preventive maintenance of wipers and washers is important because there may be quite long intervals when they are not in use and, if it only becomes clear that the wipe/wash system is not working properly during a journey, an inconvenient and possibly dangerous situation may arise. Maintenance of screen wipers is largely a matter of taking preventive measures, such as examining the condition of wiper blades to ensure that they are not worn, perished or hardened, and replacing them as necessary. The action of wipers is readily observed by the simple expedient of operating them in their various modes of operation, for example slow speed, intermittent, etc.

Screen washer fluid levels must be checked at regular intervals and the strength of the washer fluid mixture that is added to the system must be adjusted to suit weather conditions. The wash action may readily be examined by operating the system and observing the pattern of fluid sprayed on to the screen through the washer jets. The condition of the fluid reservoir and its fixings, and the condition of the plastic

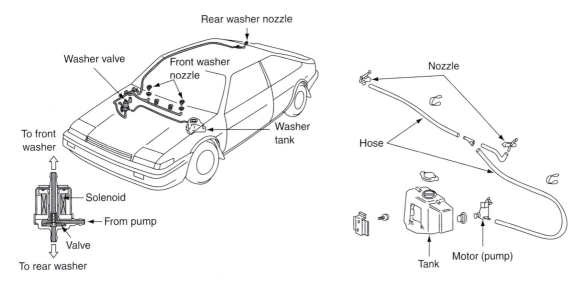

Fig. 28.7 Front and rear screen washers

tubes that connect the pump outlet to the jets, should be checked to ensure that they are secure and in good condition.

Repairs and fault tracing

For current purposes, screen wiper defects may usefully be grouped into two categories, namely (1) electrical and (2) mechanical.

Electrical defects

Problem — electrical overload

Possible effect — fuse blown or circuit breaker triggered

Some thermal circuit breakers operate by breaking the circuit at intervals, in which case the sporadic behaviour of the wipers will indicate that there is something wrong. Persistent operation of the circuit breaker means that there is a problem that requires attention because the wiper circuit is being overloaded. In the case of a blown fuse, it may be that the fuse has blown because of a temporary overload such as ice on the screen or excessively dirty conditions. In this case it would probably be in order to replace the fuse and perform a full test on the wipers, ensuring a clean and wet screen before attempting any test. Should the fuse blow again, or the circuit breaker come into operation again, the system must be examined to see whether it is an electrical fault or a mechanical fault.

Electrical overload may arise from a short-circuit where some part of the wiring has become damaged, or it may arise from friction in the motor bearing, damage in the linkage, or excessive pressure on the wiper blades. Most manufacturers' dedicated test equipment makes provision for electrical tests on wiper circuits.

In general, the wiper current may be checked by placing an ammeter in series with the motor and then conducting a test to observe the current flow. On a 12-volt system, the wiper motor operating current is approximately 1.5–4 A. The diagrams in Fig. 28.8, show the positions of the ammeter for conducting current tests on Lucas-type wiper motors in the low- and high-speed modes. The procedure for performing such tests varies for different makes of vehicle and electrical equipment, and they should only be attempted by skilled personnel and in accordance with the manufacturer's instructions.

Motor does not operate or motion is weak

- **Motor** Weak motion of the wipers may arise from a voltage drop in the circuit. Causes may be dirty or loose connections, which may usually be detected by means of voltage tests at the motor and the switch and suitable intermediate points. The brushes in the motor may be worn and, in some cases, these can be replaced. However, the overall efficacy and cost of such work must be compared with the alternative of a replacement motor.

- **Switch** The control switch may fail completely, or it may fail to operate some of the wipe/wash functions. In the case of complete failure, circuit tests should be carried out in order to ascertain that it is switch failure and not some other electrical problem, such as a broken or disconnected wire or a loose connection. Once the cause of the malfunction has been confirmed as a defective switch, the remedy will probably be to replace the switch.

Mechanical defects

- **Blades** Poor wiping action may result from hardened, cracked or perished blades. Before

TEST PROCEDURE FOR SCREEN WIPER MOTOR

Note: All tests with screen wet.

CONNECTIONS		TYPE OF MOTOR	RESULT
BATT. '+'	BATT. '−'		
*TEST 1 Red/Green	Brown/Green	All types	Motor should run at normal speed
*TEST 2 Blue/Green	Brown/Green	2-speed motors only	Motor should run at high speed
TEST 3 Red/green	Green	Self-switching types only	Motor should run to park position then stop
TEST 4 Green	Red/Green	Self-parking types only	Motor should run to extended park position then stop

*Do not disconnect battery supply from plug while the wiper blades are in the parked position.

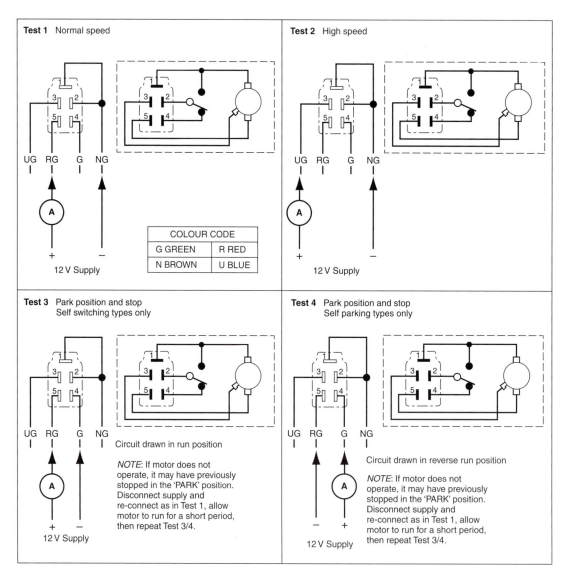

Fig. 28.8 Using an ammeter to test the operating current of a screen wiper system

condemning wiper blades the screen should be thoroughly cleaned to remove any traffic film that may have stuck to the screen. The wipers should then be retested with the screen wet. Once satisfied that the wiper blades are at fault, it is usually possible to effect a cure by replacing the rubber part of the blades.

- **Arms** Wiper arms are spring loaded and are designed to exert sufficient blade pressure on the screen to ensure effective wiping action. If the spring tension becomes weak, the blade pressure will be lower and wiping action will be poor with the possibility of blade chattering. Too great a spring pressure may cause scratching of the glass and poor wiper action due to extra load on the motor. Most wiper arm problems may be detected by visual examination and comparison with known good examples.

- **Linkage** Wear in the linkage will result in lost motion at the wiper blades and excessive noise when the wipers are in operation. Looseness in the wiper mechanism may result from the blades not being properly secured to the spindle.

Anti-theft devices

The coded ignition key and engine immobilizer system

The engine immobilizer system is a theft deterrent system that is designed to prevent the engine of the vehicle from being started by any other means than the uniquely coded ignition key. The principal parts of a system are shown in Fig. 28.9.

The transponder chip in the handle of the ignition key is an integrated circuit that is programmed to communicate with the antenna (transponder key coil) that surrounds the barrel of the steering lock. When electromagnetic communication is established between

the chip in the key and the transponder key coil, the immobilizer section of the ECM recognizes the key code signal that is produced and this effectively 'unlocks' the immobilizer section so that starting can proceed. Vehicle owners are provided with a back-up key that can be used to reset the system should a key be lost.

Burglar alarm

Modern vehicles are normally equipped with some form of alarm to alert people within hearing range to the fact that the vehicle is being forcibly entered. In most cases, the alarm system sounds the vehicle horn and flashes the vehicle lights for a period of time. The theft alarm system shown in Fig. 28.10 is armed (set) whenever the ignition is switched off and the doors and boot lid are locked. If an unauthorized entry is attempted through doors, bonnet or boot, an earth path will be made via the alarm ECM and the alarm will be triggered.

Electrically operated door mirrors

Many vehicles are equipped with door mirrors that can be adjusted by the driver, from inside the vehicle, in order to achieve the optimum rearward view. Adjustment of mirror position is achieved by means of an electric motor that is housed inside the mirror's protective moulding, as shown in Fig. 28.11, and the driver's control is situated in a suitable position close to the driving position. In many cases, the electrically operated mirrors are fitted with an electric demister facility.

Accurate setting of the mirrors, to suit an individual driver, is vital if the mirrors are to be used to full effect. Because the door mirrors protrude from the side of the vehicle they are normally retractable so

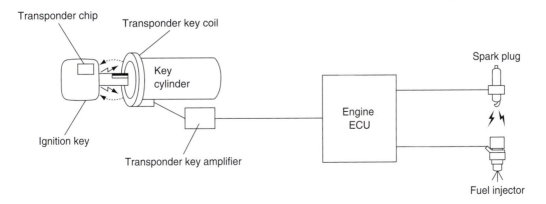

Fig. 28.9 The engine immobilizer system

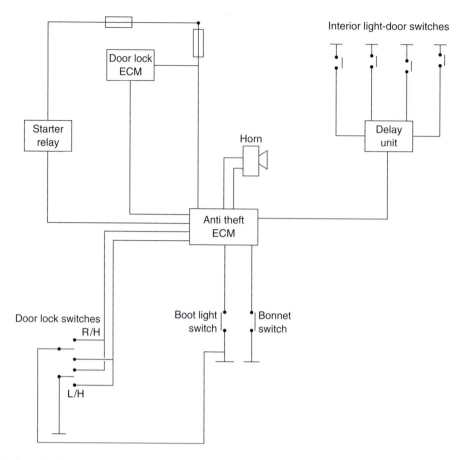

Fig. 28.10 Theft alarm circuit

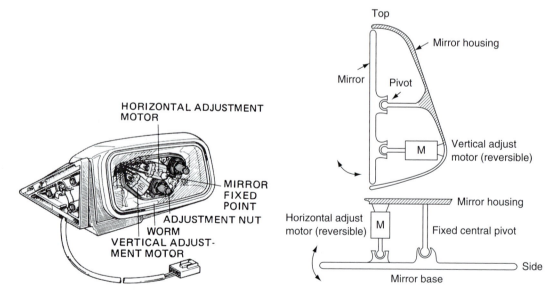

Fig. 28.11 Electrically operated door mirrors

that they can be folded away when necessary. Cracked mirrors can normally be replaced because the glass part of the mirror is normally attached to a mirror holder by means of adhesive pads. After fitting a replacement glass part of the mirror, it is necessary to ensure that the mirror is readjusted to suit the driver.

Electric horns

Motor vehicles must be equipped with an audible warning of approach and this normally takes the form of an electrically operated horn. The basic structure of a simple electric horn is shown in

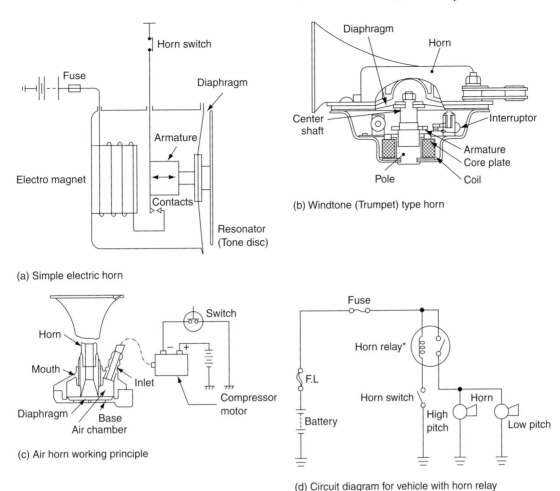

(a) Simple electric horn

(b) Windtone (Trumpet) type horn

(c) Air horn working principle

(d) Circuit diagram for vehicle with horn relay

Fig. 28.12 Basic principles of an electric horn

Fig. 28.12(a). Pressing the horn switch completes the circuit that energizes the electromagnet and this action draws the armature towards the electromagnet, causing the metal diaphragm to deflect. At the same time, the contacts open to break the circuit and this de-energizes the electromagnet so that the diaphragm springs back. Making and breaking the circuit causes the diaphragm to vibrate, thus setting up the sound. The resonator that is attached to the diaphragm acts as a tone disc in order to give the horn a suitable sound. The diaphragm is normally made from high-quality carbon or alloy steel, the thickness and hardness being factors that give the required sound properties and ensure a long working life. The contact points that make and break the circuit are made from tungsten alloy and are designed to be maintenance-free. The simple horn takes a current of approximately 4 amperes, but the wind tone and the air horn require a current of 10 amperes or more and it is common practice for the circuits for these horns to use a relay, as shown in the circuit in Fig. 28.12(d).

Wind tone horns

This type of horn also operates through the action of a vibrating diaphragm. The action is slightly different from that of the simple horn because the diaphragm controls the movement of air in a trumpet-shaped tube. The tube is often spiral in form so that a relatively long 'trumpet' can be contained in a short space.

Air horns

This type of horn is operated by means of a supply of compressed air that is provided by a small electrically operated compressor controlled by the horn switch. Figure 28.12(c) shows the principle of operation.

Repair and maintenance

Horns are normally quite reliable. Problems are most likely to arise in the electrical circuit and the procedure for fault tracing should probably start with a visual inspection followed by checking the fuse. Horns are normally secured to the vehicle structure by means of a flexible bracket, the purpose of which is to prevent vehicle vibrations from distorting the quality of the sound emitted by the horn. This is a point to note when considering the fitment of a different type of horn.

> ### Learning task
>
> Inspect the horn installations on several different vehicles. Pay particular attention to the fixing bracket and the type of horn. With the aid of the notes in this book, identify the type of horn in each case and then look up the vehicle details to compare your result with the horn details provided by the vehicle maker.

Electrically operated windows

Electrically operated windows are commonly used in many types of vehicle. The two systems, illustrated in Fig. 28.13, make use of reversible electric motors. The mechanism that is used to raise and lower the windows is known as the regulator. In Fig. 28.13(a), the regulator is operated by a small gear on the motor shaft that engages with a quadrant gear on the arm of the regulator. Rotation of the motor armature causes a semi-rotary action of the quadrant. This semi-rotary action is converted into linear motion by the regulator mechanism. The system shown in Fig. 28.13(b) makes use of cables that are wound on a drum that is driven by the operating motor. In both cases, the electric motors are of the permanent magnet type. The permanent magnet motor is reversible via the switching system and this permits the windows to be operated as required. An electrically operated window circuit normally contains a thermally operated switch to protect the circuit in the event of severe icing or other condition

that may lead to an excessive load on the motors and winding mechanism.

Lighting

The main purpose of lights on a vehicle is to enable the driver to see and for other road users to be able to see the vehicle after dark and in other conditions of poor visibility. In the UK there are legal requirements for the lights that must be fitted. In summary, these requirements are:

1. Headlamps (minimum of two, one each side)
2. Side, rear, and number plate lamps
3. Direction indicator lamps (flashers)
4. Stop lamps (brake lights)
5. Rear fog lamp (at least one).

All of these lamps must be maintained in working condition, including proper alignment.

Lighting regulations

The law relating to vehicle lights is quite complicated and you are advised to check the regulations to ensure that you understand them. This is particularly important when fitting extra lights to a vehicle.

Bulbs

The main source of light for the lamps listed above is the traditional bulb. Electric current in the bulb (lamp) filament causes the filament to heat up and give out whitish light. Where other colours are required, for example for stop and tail lights, the lamp lens is made from coloured material. Figure 28.14 shows two commonly used types of bulbs.

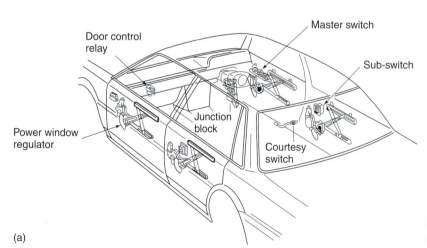

(a)

Fig. 28.13 Electrically operated windows. *(Reproduced with the kind permission of Ford Motor Company Limited.)*

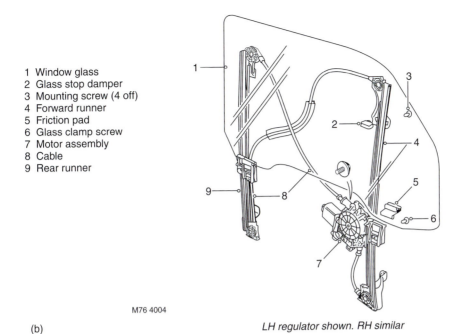

1 Window glass
2 Glass stop damper
3 Mounting screw (4 off)
4 Forward runner
5 Friction pad
6 Glass clamp screw
7 Motor assembly
8 Cable
9 Rear runner

M76 4004

(b)

LH regulator shown. RH similar

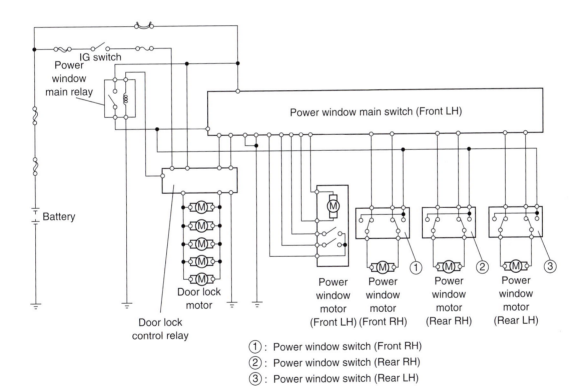

(1) : Power window switch (Front RH)
(2) : Power window switch (Rear RH)
(3) : Power window switch (Rear LH)

(c)

Ordinary twin-filament headlamp bulb

The filaments are made from tungsten wire and the glass bulb is often filled with an inert gas such as argon. This permits the filament to operate at a higher temperature and increases the reliability. One of the filaments provides the main beam and the other filament, which is placed a little above centre, provides the dipped beam. The effect is shown in Fig. 28.15.

In order to locate the bulb accurately in the lamp reflector, the metal base of the bulb is equipped with a notch, as shown in Fig. 28.16.

(a)

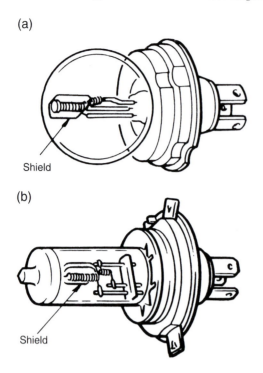

Shield

(b)

Shield

Fig. 28.14 Typical vehicle light bulbs. (a) Ordinary bulb. (b) Quartz halogen bulb

Quartz halogen twin-filament headlamp bulb

A problem with ordinary tungsten lamp bulbs is that, over a long period of time, the filament deteriorates (evaporates) and discolours the glass. The rather more elaborate quartz halogen bulb is designed to give brighter light and to prevent the evaporated tungsten from being deposited on the inside of the bulb. Halogens are gases, such as iodine, chlorine, etc., which react chemically inside the bulb to provide the 'halogen cycle'. The 'halogen cycle' preserves the life of the tungsten filament.

Oil, grease and salt from perspiration can damage the quartz, and it is recommended that these bulbs are

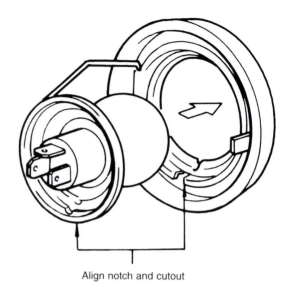

Align notch and cutout

Fig. 28.16 The location notch in the bulb base — regular type

handled by the metal part to prevent damage. Figure 28.17 shows a typical method of locating a quartz halogen-type bulb.

Stop and tail lamps

The stop/tail bulb has two filaments. They are normally 21-watt and 5-watt filaments. The 21-watt filament is for the stop (brake) lights and the 5-watt filament is for the tail lights. The metal base of the bulb is provided with offset pins so that it cannot be fitted incorrectly.

Sealed beam units

The construction of a sealed beam unit is shown in Fig. 28.18.

There is no separate bulb. The assembly — lighting filaments, reflector, and lamp lens — is a single unit. During manufacture, the inside of the unit is evacuated

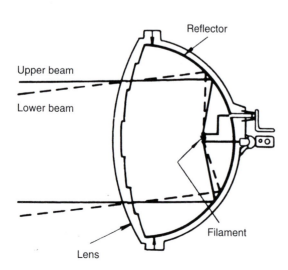

Reflector
Upper beam
Lower beam
Lens
Filament

Fig. 28.15 The dipped beam principle

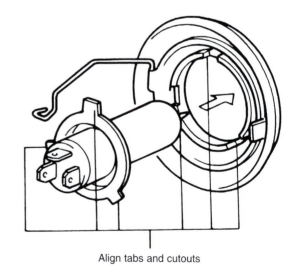

Align tabs and cutouts

Fig. 28.17 Locating a quartz halogen bulb

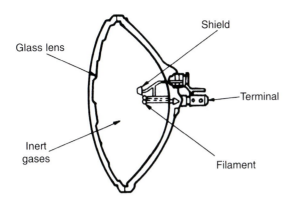

Fig. 28.18 A sealed beam headlamp unit

and filled with inert gas and the unit is then hermetically sealed. The idea is that the dust and other contaminants cannot enter and the accuracy of the setting of the filament in relation to the focal point of the reflector cannot be altered. A disadvantage is that, in the event of filament failure, the whole unit must be replaced.

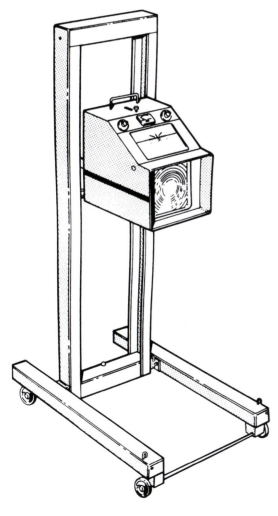

Lucas Beam Tester Mk. II

Fig. 28.19 Lucas headlamp alignment machine

Headlamp dipping

The reflector concentrates the light produced by the bulb and projects it in the required direction. The lens is specially patterned to 'shape' the light beam so that the beam is brighter in the centre and less bright on both sides. The intention is to reduce the risk of 'dazzling' oncoming drivers and pedestrians. Headlamps must also be provided with the means to deflect the beam downwards and, in a two-headlamp system, this is achieved by switching the lamps from the main beam filament to the dip beam filament.

Headlamp alignment

To ensure that headlamps are correctly adjusted, it is necessary to check the alignment. Special machines of the type shown in Fig. 28.19 are often used for this purpose.

If such a machine is not available it is acceptable to use a flat, vertical surface such as a wall or a door.

The wall should be marked out, as shown in Fig. 28.20.

The vehicle is placed so that the headlamps are parallel to the wall and 8 metres from the wall. The centre line of the vehicle must line up with the centre line marked on the wall and, with the headlamps switched on to main beam, the area of concentrated light should be very close to that shown in Fig. 28.20. Should the settings be incorrect, it will be necessary to inspect for the cause of the inaccuracy. It may be a case of incorrect adjustment. If adjustment is required, it is probable that the headlamp unit will be equipped with screws to alter the horizontal and vertical settings, as shown in Fig. 28.21.

Gas discharge lamps

Xenon lamps operate on the gas discharge principal. The light source (bulb) contains a pair of electrodes that are encased in a special type of quartz glass bulb. The bulb is filled with xenon gas, traces of mercury and other

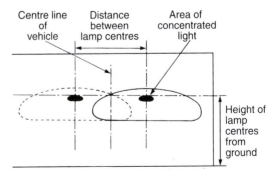

Front of vehicle to be square with screen
Vehicle to be loaded and standing on level ground
Recommended distance for setting is at least 8m
For ease of setting, one headlamp should be covered

Fig. 28.20 Using a marked wall for headlamp alignment

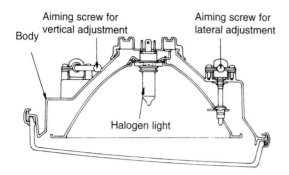

Fig. 28.21 Headlamp alignment screws

elements, under pressure. When a high voltage of approximately 20 kV is applied to the electrodes, the xenon gas emits a bright white light and evaporates the mercury and other elements. Once illuminated, the light output of the bulb is maintained by a lower voltage. Gas discharge (xenon) lamps produce a greater amount of light than conventional bulbs and are designed to last longer. The main features of a xenon headlamp are shown in Fig. 28.22.

Safety note

The electronic control unit for the xenon lamp produces a very high voltage and the following precautions must be observed:

- Ensure that the lights are switched off before attempting to work on the system. This is to guard against electric shock that can arise from the high voltage.
- Gloves and protective eyewear must be used when handling xenon bulbs and the glass part of the bulb must not be touched.
- Discarded xenon bulbs must be treated as hazardous waste.

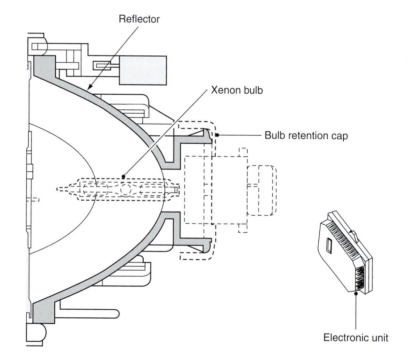

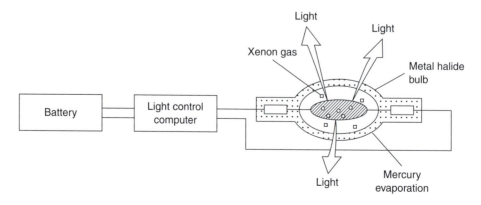

Fig. 28.22 The gas discharge (xenon) lamp. *(Reproduced with the kind permission of Ford Motor Company Limited.)*

Vehicles that are equipped with xenon lamps must also be equipped with a means of automatically adjusting the lamps to prevent dazzling of oncoming drivers.

Headlight and rear fog light circuit

In the circuit shown in Fig. 28.23, the rear fog lamp circuit is fed from the main light switch. The purpose of the rear fog lamp is to make the presence of the vehicle more visible in difficult driving conditions. The headlamps are connected in parallel so that failure of one lamp does not lead to failure of the others.

Direction indicators

Direction indicator lights (flashers) are required so that the driver can indicate any intended manoeuvre. Figure 28.24 shows a typical indicator lamp circuit.

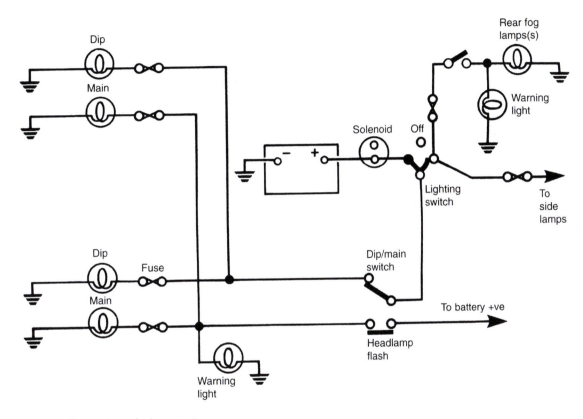

Fig. 28.23 Headlamp and rear fog lamp circuit

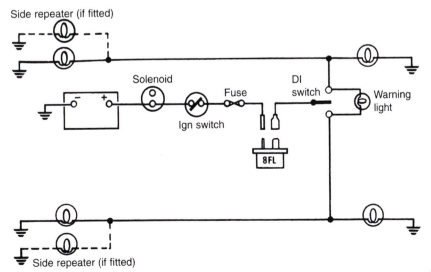

Fig. 28.24 Indicator lamp circuit

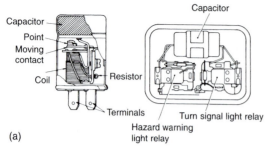

Capacitor
Point
Moving contact
Coil
Resistor
Terminals
Turn signal light relay
Hazard warning light relay
Capacitor

(a)

Current type

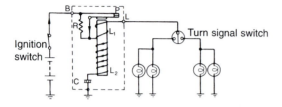

B P L
R
L₁
Ignition switch
Turn signal switch
L₂
C

Voltage type

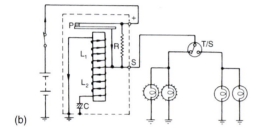

P +
L₁
R
L₂
S
T/S
C

(b)

Fig. 28.25 Capacitor and relay flasher unit and circuit. (a) Construction of the flasher units. (b) Circuits for the two types of flasher unit

The 'flasher' unit is designed so that the frequency of flashes is not less than 60 per minute and not more than 120 per minute. The circuit is also designed so that failure of an indicator lamp will lead to an increased frequency of flashing and the driver will notice this at the warning light.

The 'flasher' unit

The unit shown in Fig. 28.24 is marked 8FL. This is a Lucas unit that operates on the basis of thermal expansion and contraction of a metal strip. The flasher unit shown in Fig. 28.25 employs a capacitor and a relay to provide the flashing action. Examination of the circuit diagram shown in Fig. 28.25 shows that the contact points are normally at rest in the closed position. When the indicator switch is moved to indicate a turn, to left or right, current flows to the indicator lamps, through the upper winding of the relay. This current energizes the relay and opens the contact points, the current flow is interrupted, and the lamps 'go out'. As the lamps go out, the contact points close and the lights 'come on' again. This happens at a frequency of approximately 90 flashes per minute and the timing of the flash rate is controlled by the capacitor.

Circuits and circuit principles

Circuit (wiring) diagrams

A complete circuit is needed for the flow of electric current. Figure 28.26 shows two diagrams of a motor in a circuit. Should the fuse be blown, the circuit is incomplete, current will not flow, and the motor will not run. In the right-hand diagram, the fuse has been replaced. The circuit is now complete and the motor will run. Basic electrical principles such as this are fundamental to good work on electronic systems because much of the testing of electronic systems requires checking of circuits to ensure that they are complete.

It is essential to be able to understand and follow circuit diagrams and this requires a knowledge of circuit symbols. There is a set of standard circuit symbols, some of which are shown in Fig. 28.27.

However, non-standard symbols of the type shown in Fig. 28.28 are sometimes used and this can cause confusion. Fortunately, when non-standard symbols

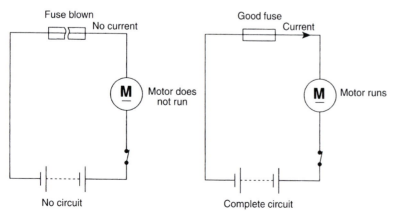

Fuse blown — No current
Motor does not run M
No circuit

Good fuse — Current
Motor runs M
Complete circuit

Fig. 28.26 An electric circuit

Name of device	Symbol	Name of device	Symbol
Electric cell	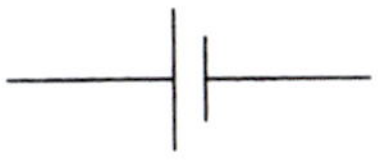	Lamp (bulb)	
Battery		Diode	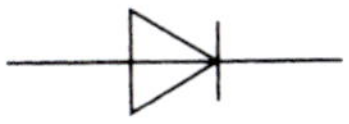
Resistor		Transistor (npn)	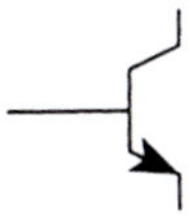
Variable resistor	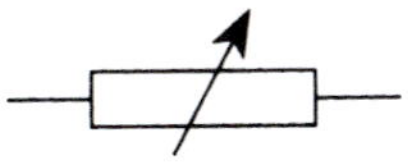	Light emitting diode (LED)	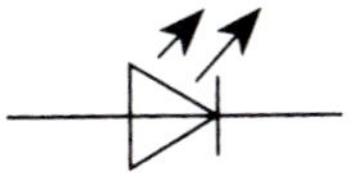
Potentiometer	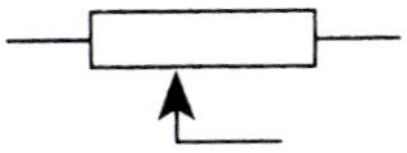	Switch	
Capacitor		Conductors (wires) crossing	
Inductor (coil)		Conductors (wires) joining	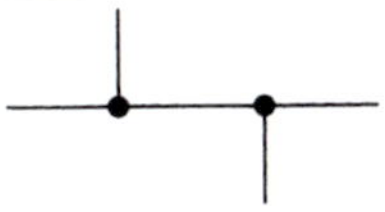
Transformer	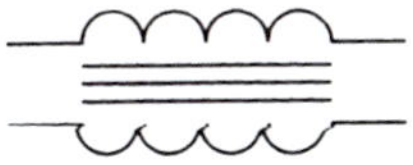	Zener diode	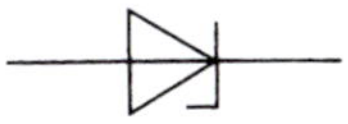
Fuse	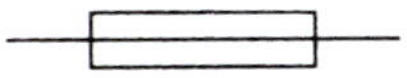	Light dependent resistor (LDR)	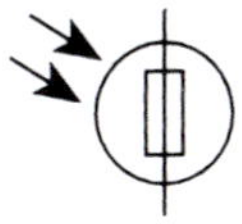

Fig. 28.27   A selection of circuit symbols

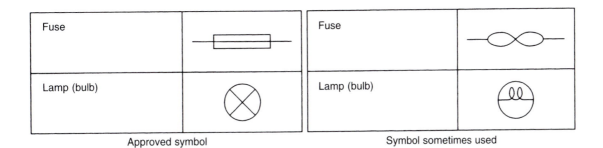

Fuse		Fuse	
Lamp (bulb)		Lamp (bulb)	
Approved symbol		Symbol sometimes used	

Fig. 28.28   Non-standard circuit symbols

are used, they are normally accompanied by a descriptive list.

Figure 28.29 gives an example of the use of a descriptive list to support the wiring diagram; the injector resistors are shown as a sawtooth line, number 21 on the list, and on the diagram.

This solves the problem of deciphering the diagram.

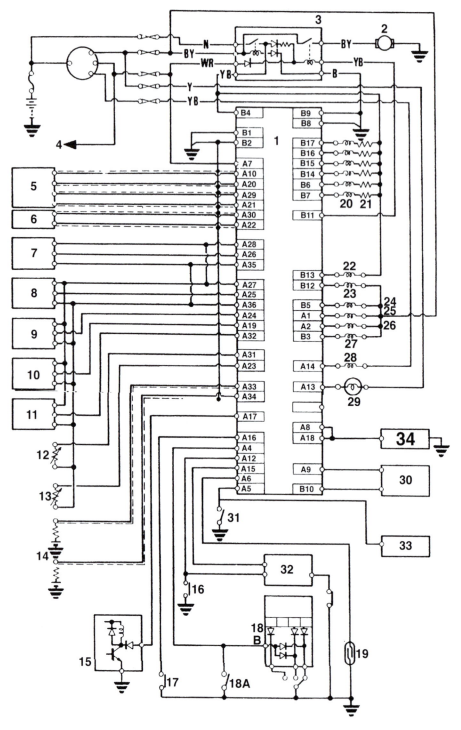

1 – ECU	13 – Air intake temperature sensor	24 – EGR solenoid
2 – Fuel pump	14 – Oxygen sensors*	25 – Air suction control solenoid*
3 – Main relay	15 – Alternator	26 – By-pass solenoid B
4 – To starter	16 – Cooling fan switch	27 – By-pass solenoid A
5 – Cyl/crank sensor	17 – Power steering switch	28 – Air-con clutch relay
6 – TDC sensor	18a – Neutral switch (MT)	29 – Check engine light
7 – MAP sensor	18b – A/T position switch	30 – EAT ECU
8 – Atmospheric pressure sensor	19 – Vehicle speed sensor	31 – Clutch switch M/T
9 – Throttle angle sensor	20 – Injectors	32 – Radiator fan control unit
10 – Ignition timing adjuster	21 – Injector resistors	33 – Cruise control
11 – EGR lift sensor	22 – EICV	34 – Igniter unit
12 – Water temperature sensor	23 – Pressure regulator solenoid	*Emission vehicles only

Fig. 28.29 System diagram with list describing circuit elements

Colour code

The wiring diagram is an essential aid to fault tracing. The colour code for wires is an important aid to fault tracing and most wiring diagrams make use of it.

A commonly used colour code is:

N	= brown	Y	= yellow
P	= purple	K	= pink
W	= white	R	= red
O	= orange	LG	= light green
U	= blue	B	= black
G	= green	S	= slate

In order to assist in tracing cables, they are often provided with a second colour tracer stripe. The wiring diagram shows this by means of letters; for example, a cable on the wiring diagram, with GB written on it, is a green cable with a black tracer stripe. The first letter is the predominant colour and the second is the tracer stripe.

Note

This is not a universal colour code and, as with many other factors, it is always wise to have accurate information to hand, that relates to the product being worked on.

The predominant (main) colours frequently relate to particular circuits as follows:

- Brown (N) = Main battery feeds
- White (W) = Essential ignition circuits (not fused)
- Light green and also Green (LG) (G) = Auxiliary ignition circuits (fused)
- Blue (U) = Headlamp circuits
- Red (R) = Side and tail lamp circuits
- Black (B) = Earth connections
- Purple (P) = Auxiliary, non-ignition circuits, probably fused.

A complete wiring diagram

Figure 28.30 shows a full wiring diagram for a vehicle. In order to make it more intelligible, this diagram uses a grid system: numbers 1–4 across the page, and letters A, B, C, at the sides. This means that an area of diagram can be located. For example, the vehicle battery is in the grid area 1A.

The wiring loom, or harness

It is common practice to bind cables together to facilitate positioning them in the vehicle structure. It used to be the practice to fabricate the loom, or harness, from a woven fabric. Modern loom sheathing materials are usually a PVC type of material. Figure 28.31 shows the engine compartment part of a typical wiring loom.

Cable sizes

The resistance of a cable (wire) is affected by, among other factors, its diameter and its length. For a given material, the resistance increases with length and decreases with diameter. Doubling a given length of cable will double its resistance and doubling the diameter of the same length of cable will decrease the resistance to one-quarter of the original value. Cable sizes are therefore important, and only those sizes specified for a given application should be used. Most cables used on vehicles need to be quite flexible. This flexibility is provided by making the cable from a number of strands of wire and it is common practice to specify cable sizes by the number of strands and the diameter of each strand; for example, 14/0.30 means 14 strands of wire and each strand has a diameter of 0.30 mm.

Some typical current-carrying capacities of cables for vehicle use are given in Table 28.2.

The choice of cable is a factor that is decided at the design stage of a vehicle. However, it sometimes affects vehicle repair work when, for example, an extra accessory is being fitted to a vehicle, or a new cable is being fitted to replace a damaged one. In such cases, the manufacturers' instructions must be observed.

Circuit protection — fuses

The purpose of the fuse is to provide a 'weak' link in the circuit that will fail (blow) if the current exceeds a certain value and, in so doing, protect the circuit elements and the vehicle from the damage that could result from excess current.

The fuse is probably the best known circuit protection device. There are several different types of fuses and three of these are shown in Fig. 28.32. Fuses have different current ratings and this accounts for the range of types available. Care must be taken to select a correct replacement and larger rated fuses must never be used in an attempt to 'get round' a problem. Many modern vehicles are equipped with a 'fusible link' that is fitted in the main battery lead as an added safety precaution.

It is common practice to place fuses together in a reasonably accessible place on the vehicle. Another feature of the increased use of electrical/electronic circuits on vehicles is an increase in the number of fuses to be found on a vehicle. Figure 28.33 shows an engine compartment fuse box that carries fusible links in addition to 'normal' fuses.

The same vehicle also has a dashboard fuse box. This also carries 'spare' fuses that appear to the right of the other fuses (Fig. 28.34).

Whilst, in the event of circuit failure, it is common practice to check fuses and replace any 'blown' ones, it should be remembered that something caused the fuse to blow. Recurrent fuse 'blowing' requires that

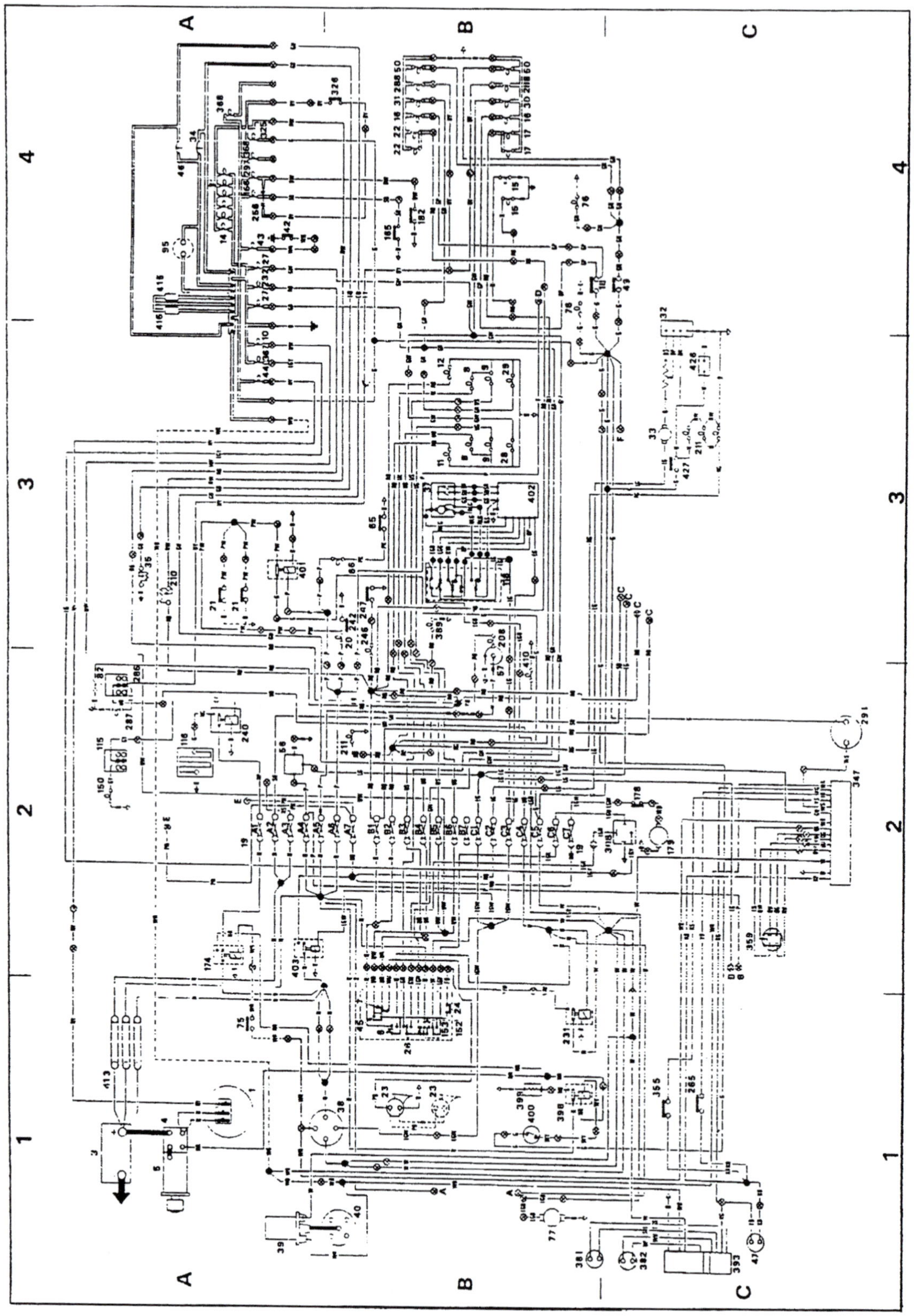

Fig. 28.30   A typical wiring diagram

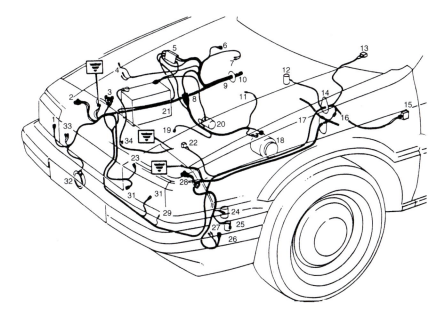

1 – RH direction indicator lamp connector
2 – RH headlamp connector
3 – Relays – air-conditioning
4 – Ignition coil connector
5 – Fusible link box
6 – Windscreen wiper motor plug
7 – Handbrake warning lamp switch
    connector
8 – Main/engine harness connector
9 – Main harness
10 – Bulkhead grommet
11 – Fuel cut-off solenoid
12 – Throttle solenoid – air-conditioning
13 – Switch – air-conditioning
14 – Bulkhead grommets
15 – Relay – air-conditioning – 'A' post
16 – Main/air-conditioning harness connector
17 – Harness – air-conditioning
18 – Alternator

19 – Reverse light switch
20 – Starter solenoid
21 – Battery
22 – Charging pressure switch –
    air-conditioning
23 – Cooling fan connector
24 – Headlamp washer motor
25 – Windscreen washer motor
26 – LH direction indicator lamp
    connector
27 – LH side lamp
28 – LH headlamp connector
29 – Compressor clutch –
    air-conditioning
30 – Fan – air-conditioning
31 – Thermostatic switch
32 – Horn
33 – RH side lamp
34 – Cable clip

Fig. 28.31 Engine compartment wiring loom

circuits should be checked to ascertain the cause of the excess current that is causing the failure.

The circuit breaker

The thermal circuit breaker relies for its operation on the principle of the 'bimetallic strip'. In Fig. 28.35, the bimetal strip carries the current between the terminals of the circuit breaker. Excess current, above that for which the circuit is designed, will cause the temperature of the bimetal strip to rise to a level where it will curve and cause the contacts to separate. This will open the circuit and current will cease to flow. When the

temperature of the bimetal strip falls the circuit will be remade. This action leads to intermittent functioning of the circuit, which will continue until the fault is rectified. An application may be of a 7.5 ampere circuit breaker to protect a door lock circuit. The advantage of the circuit breaker, over a fuse, is that the circuit breaker can be reused.

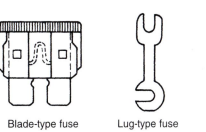

Blade-type fuse   Lug-type fuse   Cartridge-type fuse

Fig. 28.32 Fuse types

Table 28.2 Cable current-carrying capacities and use

Size of cable	Current rating	Typical use
14/0.30	8.75 amps	Side and tail lamps
28/0.30	17.5 amps	Headlamps
120/0.30	60 amps	Alternator to battery

No.	Rating	Function
G	50 amp	Radio, power amplifier, electric seats
H	50 amp	Ignition switch circuit
I	80 amp	Battery output
J	50 amp	Window lift
K	50 amp	ABS brake system
L	50 amp	Supply to fuses 4, 5 and 6 and sidelight relay

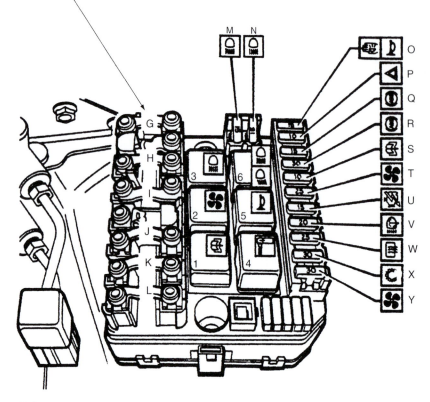

Relays
1 – Cooling fan changeover or manifold heater
2 – Cooling fan
3 – Lighting
4 – Starter
5 – Horns
6 – Main/dip beams
7 – Air conditioning changeover

Fig. 28.33 Engine compartment fusebox (Rover 800)

Other circuit protection

Vehicle circuits are subject to 'transient' voltages that arise from several sources. Those that interest us here are load dump, alternator field decay voltage, switching of an inductive device (coil, relay, etc.), and over-voltage arising from incorrect use of batteries when 'jump starting'.

Load dump occurs when an alternator becomes disconnected from the vehicle battery while the alternator is charging, i.e. when the engine is running.

Figure 28.36 shows a Zener diode, as used for surge protection in an alternator circuit. The breakdown (Zener) voltage of the diode is 10–15 volts above the normal system voltage. Such voltages can occur if an open circuit occurs in the main alternator output lead when the engine is running. Other vehicle circuits,

such as coil ignition, can also create inductive surges. Should such voltage surges occur, they could damage the alternator circuits but, with the Zener diode connected as shown, the excess voltage is 'dumped' to earth via the Zener diode. Should such a voltage surge occur, it may destroy the protection diode. The alternator would then cease charging. In such a case, the surge protection diode would need to be replaced, after the cause of the surge had been remedied.

Figure 28.37 shows another form of circuit protection where a diode is built into a cable connector. This reduces the risk of damage from reversed connections and it is evident that one should be aware of such uses because a continuity test on such a connector will require correct polarity at the meter leads.

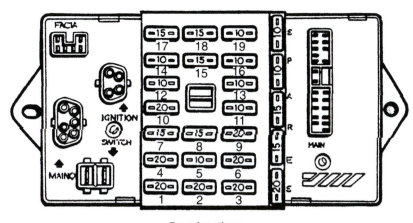

Fuse functions

Fuse No.	Rating	Wire colour	Function
1	20 amp	N/O	Sunroof, driver's seat heater
2	20 amp	S/U	RH front door window lift
3	20 amp	S/R	LH front door window lift
4	20 amp	P/N	Front and rear cigar lighters Footwell lamps
5	10 amp	P	Interior lights, boot light, interior light delay unit, map light, door open lights, trip computer memory, radio station memory, clock memory, headlight delay unit
6	20 amp	P/O	Burglar alarm ECU (optional) Central locking ECU
7	15 amp	R/O	RH number plate lamp RH side, tail and marker lights Trailer plug
8	10 amp	R/B	Cigar lighter illumination, LH licence plate lamp, sidelight warning light, glovebox light, trailer plug, dashboard illumination, LH side tail lights
9	20 amp	S/O	LH rear window lift
10	20 amp	S/G	RH rear window lift

Fig. 28.34   Dashboard fusebox (Rover 800)

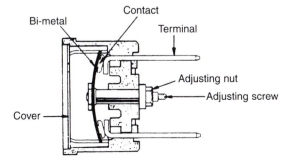

Fig. 28.35   A circuit breaker (Toyota)

Connectors

Connectors are used to connect cables to components and to join cables together. They are made in a range of different shapes and sizes for a variety of purposes, as shown in Fig. 28.38. In Fig. 28.38(b), a pair of cables are joined together by a connector and in Fig. 28.38(c), some cables are connected to a component. Figure 28.38(a) shows the principal parts of a connector. These parts are designed to be resistant to vibration and moisture. In some cases, such as the

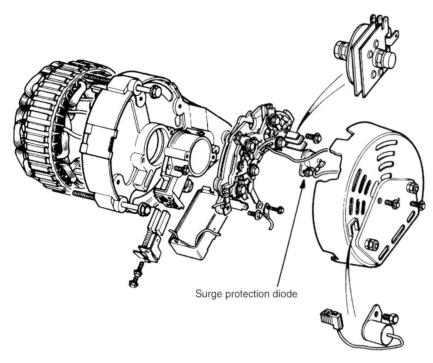

Surge protection diode

Fig. 28.36   Voltage surge protection by Zener diode

multiple pin connector for an ECU shown in Fig. 28.38(d), the pins are coated with gold to ensure the type of electrical connection needed for very small electric current.

Repair and maintenance

Some service and repair operations may require components and cables to be disconnected. In such cases, the procedures for taking connectors apart should be observed and care taken to avoid damage to the pins. Damage to pins may cause the connection to develop a high resistance or to fail completely. Should it be necessary to probe connections, in an endeavour to trace a fault, every effort should be made to prevent damage to waterproofing and electrical insulation.

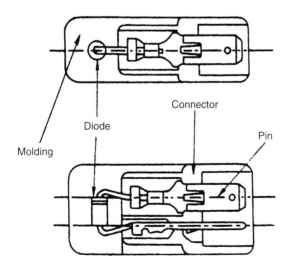

Fig. 28.37   A protection diode in a cable connector

Instrumentation

Vehicles are equipped with instruments and other warning devices, in order to inform the driver about the current state of various systems on the vehicle. Some instruments, such as the speedometer, the fuel tank contents gauge, the oil pressure gauge, and the ignition warning, have been features of the instrument display on vehicles for many years. Several of the more commonly used instruments are examined below, in order to assess the technology involved and to examine some approaches to fault tracing.

Thermal type gauges

These instruments make use of a bimetal strip to provide the movement that gives an indication of the quantity that is being measured. In the example of the fuel tank contents gauge, shown in Fig. 28.39, the current flowing in the meter circuit is dependent on the resistance of the sensing unit at the fuel tank. As the float of the fuel tank sensing unit rises and falls with fuel level, so the resistance that it places in the meter circuit changes. Changes in resistance cause the current flowing in the heating coil that surrounds the bimetal strip to change and this leads to changes in the temperature of the bimetal strip. Thus, the heating effect of the current that deflects the bimetal strip varies with fuel level and, by careful calibration, the fuel gauge is made to give a reliable guide to the quantity of fuel in the tank.

The engine coolant temperature gauge, also shown in Fig. 28.39, operates on the same principle as the fuel

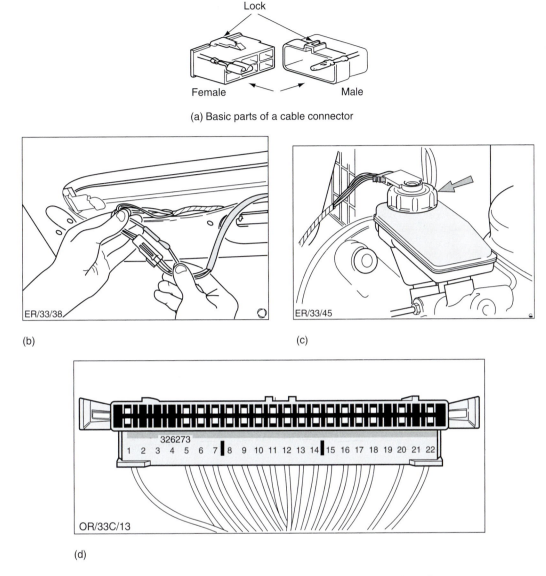

Lock

Female   Male

(a) Basic parts of a cable connector

ER/33/38

(b)

ER/33/45

(c)

326273
1 2 3 4 5 6 7 8 9 10 11 12 13 14 15 16 17 18 19 20 21 22

OR/33C/13

(d)

Fig. 28.38 Types of cable connectors. *(Reproduced with the kind permission of Ford Motor Company Limited.)*

contents gauge, except that the variable resistance is provided by the thermistor. A thermistor has a negative temperature coefficient, which means that the resistance of the thermistor decreases as its temperature rises. It is this variation of resistance that causes the meter current to change and thus indicate the coolant temperature.

The voltage stabilizer

Reliable operation of thermal-type meters is dependent on a reasonably constant voltage supply and the voltage stabilizer (regulator) is the means of providing a stable voltage. In the type of voltage stabilizer shown in Fig. 28.39, the current in the meter circuit causes the temperature of the voltage stabilizer bimetal strip to heat up and bend. The heating and bending action causes the voltage stabilizer contact to open, thus stopping the current flow. The temperature of the bimetal

strip then drops and the contacts close. This process proceeds at high frequency and provides a reasonably constant voltage level at the meter. An alternative method of voltage stabilization is shown in Fig. 28.39(c). This type of stabilizer makes use of the breakdown voltage of an avalanche (Zener) diode. When the voltage at the terminals exceeds the breakdown value, the diode conducts and effectively 'absorbs' the excess voltage.

A vacuum fluorescent display gauge

The vacuum fluorescent display (VFD) makes use of a number of segments. Each segment of the display is activated electronically through a circuit similar to that shown in Fig. 28.40(a). Figure 28.40(b) shows the main features of a single element of a VFD. The fluorescent material, the grid, and the filament are encased in a vacuum compartment. When a segment is activated

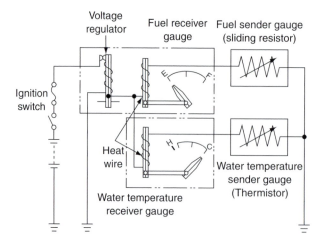

(a) Thermal-type gauge circuits

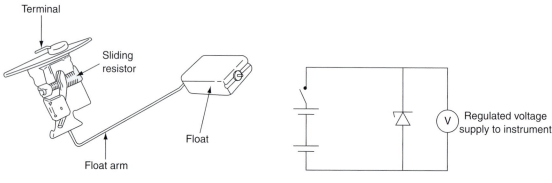

(b) Variable resistor fuel tank sender unit

(c) Zener diode type voltage regulator

Fig. 28.39 A thermal type fuel gauge and variable resistor sender unit. *(Reproduced with the kind permission of Ford Motor Company Limited.)*

by the electronic circuit, a stream of electrons bombards the fluorescent material, which causes it to emit light. The filament is negatively charged and heated to approximately 600 °C. At this temperature electrons are released to pass through the grid towards the positively charged segment, where they cause the display to emit light.

Liquid crystal displays (LCDs) and light-emitting diode displays (LEDs) are also used in many similar applications. They are also driven by electronic circuits similar to that shown in Fig. 28.40(a).

Vehicle condition monitoring (VCM)

Gauges, such as the oil pressure gauge, the ammeter, and warning lights (i.e. those that light up when the brake pads are worn), are all examples of instruments that continuously monitor the condition of various parts of the vehicle. The extensive use of computer-controlled systems on vehicles means that many of the factors that affect the use of a vehicle are constantly monitored because it is the monitoring that provides the data that allows the systems to operate. Vehicle condition monitoring has thus evolved so that it has become a feature on all modern vehicles. The most striking example is probably that provided by on-board diagnostics. Modern

computer-controlled systems have sufficient memory and processing capacity to constantly check a large number of variables and to take action if any variable (measurement) is wrong. This action may be to store a fault code in the memory, or it may be to illuminate a warning lamp to alert the driver. The European (EOBD) version of condition monitoring requires the use of a malfunction indicator lamp (MIL). The MIL is illuminated when any item of equipment that affects vehicle emissions develops a fault.

Supplementary restraint systems (SRS)

Air bags of the type shown in Fig. 28.41, and seat-belt pre-tensioners, such as that shown in Fig. 28.42, are features of a basic supplementary restraint system. In the event of a frontal impact of some severity, the air bags and seat-belt pre-tensioners are deployed. The air bags are inflated to protect those provided with them from impact with parts of the vehicle. The seat-belt pre-tensioners are made to operate just before the air bags are inflated and they operate by pulling about 70 mm of seat-belt on to the inertia

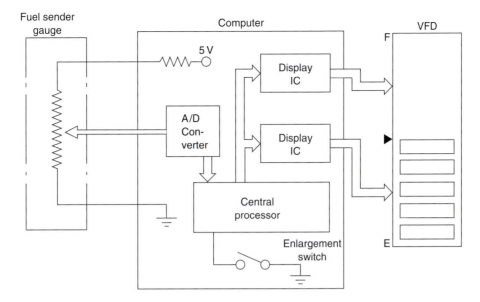

(a) A vacuum fluorescent gauge circuit

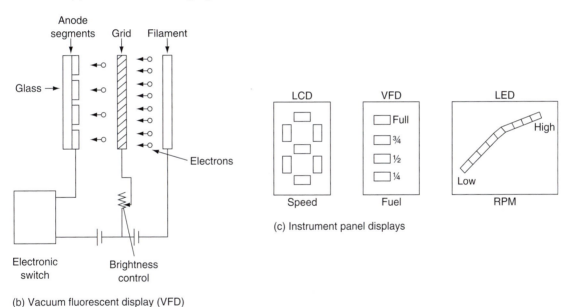

(b) Vacuum fluorescent display (VFD)

(c) Instrument panel displays

Fig. 28.40 Types of instrument displays. (Fig. 28.40(a) — *Reproduced with the kind permission of Ford Motor Company Limited.*)

reel of the belt. This serves to pull the seat occupant back on to the seat.

The deployment of these supplementary restraint devices is initiated by the action of the collision detection sensing system. A collision detection sensing system normally utilizes signals from two sensors. One sensor is a 'crash sensor' and the other is a 'safing sensor'. The safing sensor is activated at a lower deceleration than the crash sensor (about 1.5 g less) and both sensors must have been activated in order to trigger the supplementary restraint system. The safing sensor is fitted to reduce the risk of a simple error bringing the air bag into operation. Both of these sensors may be fitted inside the electronic control unit which, in some

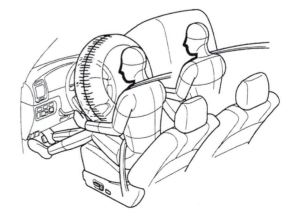

Fig. 28.41 Air bags for driver and front seat passenger

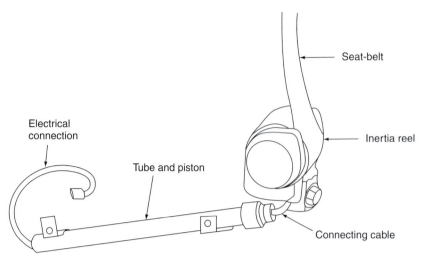

Fig. 28.42   Operation of pre-tensioner

cases, is known as a diagnostic and control unit (DCU) because it contains the essential self-diagnosis circuits, in addition to the circuits that operate the SRS. Figure 28.43 shows the layout of a supplementary restraint system on a Rover Mini.

Air bags and seat-belt pre-tensioners

Air bags are made from a durable lightweight material, such as nylon, and in Europe they have a capacity of approximately 40 litres. The pyrotechnic device

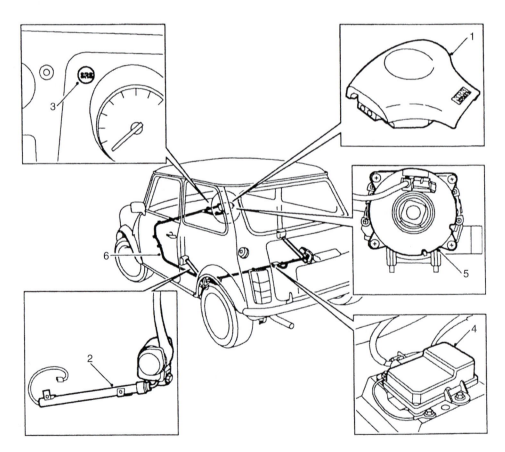

1. Driver's air-bag module	4. Diagnostic and control unit
2. Seat-belt pre-tensioner	5. Rotary coupler
3. SRS warning light	6. SRS wiring harness

Fig. 28.43   The elements of a supplementary restraint system (Rover Mini)

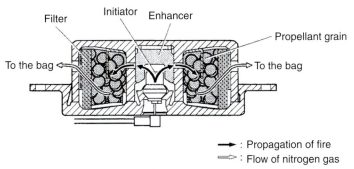

Filter Initiator Enhancer

Propellant grain

To the bag ←   → To the bag

→ : Propagation of fire
⇒ : Flow of nitrogen gas

Fig. 28.44 A pyrotechnic device for inflating air bags

that provides the inert gas to inflate the air bag contains a combustion chamber filled with fuel pellets, an electronic igniter and a filter, as shown in Fig. 28.44. Combustion of the fuel pellets produces the supply of nitrogen that inflates the air bag. The plastic cover that retains the folded air bag in place at the centre of the steering wheel is designed with built-in break lines. When the air bag is inflated, the plastic cover separates at the break lines and the two flaps open out to permit unhindered inflation of the air bag.

The seat-belt pre-tensioners are activated by a similar pyrotechnic device. In this case the gas is released into the cylinder of the pre-tensioner, where it drives a piston along the cylinder. The piston is attached to a strong flexible cable that then rotates the inertia reel of the seat-belt by a sufficient amount to 'reel in' the seat-belt by approximately 70 mm.

The rotary coupler

This device is fitted beneath the steering wheel to provide a reliable electrical connection between the rotating steering wheel and air bag, and the static parts of the steering column. The positioning of the rotary coupler is a critical element of the air-bag system and it should not be tampered with. When working on supplementary restraint systems, it is important that a technician is fully acquainted with the system and procedures for working on it.

Handling SRS components

The following notes are provided to Rover trained technicians and they are included here because they contain some valuable advice for all vehicle technicians.

Safety precautions, storage, and handling

Air bags and seat-belt pre-tensioners are capable of causing serious injury if abused or mishandled. The following precautions must be adhered to:

In vehicle

- ALWAYS fit genuine new parts when replacing SRS components.
- ALWAYS refer to the relevant workshop manual before commencing work on a supplementary restraint system.
- Remove the ignition key and disconnect both battery leads, earth lead first, and wait 10 minutes to allow the DCU back-up power circuits to discharge before commencing work on the SRS.
- DO NOT probe SRS components or harness with multi-meter probes, unless following a manufacturer's approved diagnostic routine.
- ALWAYS use the manufacturer's approved equipment when diagnosing SRS faults.
- Avoid working directly in line with the air bag when connecting or disconnecting multi-plug wiring connectors.
- NEVER fit an SRS component that shows signs of damage or you suspect has been abused.

Handling

- ALWAYS carry air-bag modules with the cover facing upwards.
- DO NOT carry more than one air-bag module at a time.
- DO NOT drop SRS components.
- DO NOT carry air-bag modules or seat belt pre-tensioners by their wires.
- DO NOT tamper with, dismantle, attempt to repair or cut any components used in the supplementary restraint system.
- DO NOT immerse SRS components in fluid.
- DO NOT attach anything to the air-bag module cover.
- DO NOT transport an air-bag module or seat-belt pre-tensioner in the passenger compartment of a vehicle. ALWAYS use the luggage compartment.

Storage

- ALWAYS keep SRS components dry.
- ALWAYS store a removed air-bag module with the cover facing upwards.

- DO NOT allow anything to rest on the air-bag module.
- ALWAYS place the air-bag module or pre-tensioner in the designated storage area.
- ALWAYS store the air-bag module on a flat, secure surface well away from electrical equipment or sources of high temperature.

Note

Air-bag modules and pyrotechnic seat-belt pre-tensioners are classed as explosive articles and, as such, must be stored overnight in an approved, secure steel cabinet that is registered with the Local Authority.

Disposal of air-bag modules and pyrotechnic seat-belt pre-tensioners

If a vehicle that contains air bags and seat-belt pre-tensioners that have not been deployed (activated) is to be scrapped, the air bags and seat-belt pre-tensioners must be rendered inoperable by activating them manually prior to disposal. **This procedure may only be performed in accordance with the manufacturer's instruction manual.**

Learning tasks

1. Identify the headlamp alignment equipment that is used in your workplace. Study the instructions for use and consult with your training supervisor to arrange sessions when you will learn how to use the equipment.
2. Examine a number of vehicles and note the procedures that must be followed in order to replace headlamp bulbs and bulbs in any other external lamps on the vehicles.
3. Examine a number of screen wiper blades and, with the aid of sketches and a few notes, explain the procedure for replacing worn squeegees.
4. With the aid of workshop manuals, locate the fuse boxes on a number of vehicles. Make a note of the different types of fuses that you have seen and discuss with your supervisor the procedure for replacing blown fuses.
5. Locate the heated rear screen relay on a vehicle and write down an explanation as to why it is placed where it is. Explain this to your training supervisor.
6. Examine a number of vehicles to locate the position of the horn. Use the manual to find the position of the horn relay and fuse.
7. Describe the procedure for dealing with air bags when a repair operation entails removal of a steering wheel.

Self-assessment questions

1. The approximate electric current used by a 12-volt wiper motor, when it is in proper operating condition, is:
   (a) 1.5—4 A
   (b) 15—40 A
   (c) 0.15—0.4 A
   (d) 6 watts.
2. Remote controllers for central locking are infrared or radio frequency devices. The source of energy that provides the power for the remote control device is:
   (a) an electromagnet in the key
   (b) a small battery in the key
   (c) static electricity
   (d) a thermionic valve.
3. Two additives that are added to water to make screen washer fluid are:
   (a) liquid paraffin and salt
   (b) antifreeze and detergent
   (c) boiling point depressant and descaler
   (d) distilled water and rubber lubricant.
4. Regenerative braking of a screen wiper motor is achieved by:
   (a) a clamping action on the wiper mechanism
   (b) electrically connecting the two main brushes of a permanent magnet motor
   (c) a mechanical brake on the wiper motor driving gears
   (d) the back e.m.f. from the wiper control switch.
5. By what means is the self-parking action of wiper blades achieved:
   (a) a cam-operated trip switch at the motor switches the motor off when the park position is reached
   (b) a mechanism that permits the driver to stop the wiper arms at any point on the screen
   (c) a rheostat that is built into the switch circuit
   (d) a sensor built into the screen.
6. A wind tone horn circuit contains a relay:
   (a) to reduce power loss
   (b) because of the high current
   (c) to vibrate the diaphragm
   (d) to reduce arcing at the horn switch.
7. Instrument circuits require a steady voltage that is provided by:
   (a) a relay
   (b) a voltage stabilizer
   (c) a field effect transistor
   (d) a separate battery.
8. Electrically operated windows are operated by:
   (a) linear motors
   (b) semi-rotary induction motors
   (c) reversible permanent magnet motors
   (d) powerful solenoids.

9. Headlamps are connected:
   (a) in series
   (b) in parallel
   (c) by a Wheatstone bridge
   (d) by a failure mode analysis circuit.
10. Hazard warning lights:
    (a) make use of the direction indicator lamps circuit
    (b) operate in conjunction with the fuel circuit inertia switch
    (c) operate by causing the stop lamps to flash
    (d) use 5-watt MES bulbs.
11. List three factors that may cause scratching of the windscreen.
12. Make a list of the possible causes of a weak spray of washer fluid from the jets on to the screen.
13. Describe a simple procedure to test a theft alarm system.

29
Electrical and electronic fault diagnosis

Topics covered in this chapter

Systematic fault diagnosis in electronically controlled systems

Diagnostics on ECU-controlled systems

Whilst a computer-controlled system is operating normally, the processor is constantly monitoring the electrical state of input and output connections at the various interfaces of the ECU. This monitoring (reading) of the inputs and outputs occurs so that the instructions that the computer processor has to perform, such as to compare an input value with a programmed value stored in the ROM, means that the ECU is ideally placed to 'know' what is happening at many parts of the system that is connected to it. If, for example, a throttle position sensor is producing a reading that does not match the engine speed and load signals that the ECU is reading, the software (program) in the ROM may cause the following to happen:

- A 'diagnostic trouble code' (DTC) to be flagged up
- The malfunction indicator lamp to be illuminated
- A limited operating strategy (limp home mode) to be followed.

On-board diagnostics (OBD)

The term 'on-board diagnostics' refers to the self-diagnosis that is built into the ECU. Modern ECU-controlled systems have a standard type of connection (Fig. 29.1) that permits diagnostic equipment such as a scan tool to be connected to the system so that DTCs can be read out and other diagnostic functions performed.

Scan tool

A scan tool is a small computer that incorporates a display on which DTCs and other information can be displayed.

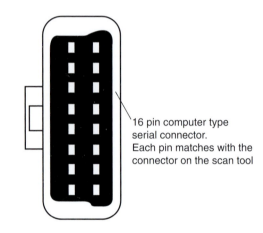

16 pin computer type serial connector. Each pin matches with the connector on the scan tool

Fig. 29.1 The standard diagnostic connection

Fault codes (DTCs)

Under USA OBD2 and European EOBD rules, the DTCs associated with emissions systems must be accessible to any qualified person who is equipped with suitable test instruments. To support this rule a system of standard DTCs has been produced to cover all emissions-related faults. The system of standard DTCs has also been extended to cover other ECU-controlled systems and some manufacturers have adopted them.

Structure of DTCs

The EOBD diagnostic codes comprise five digits, for example **P0291**, which refers to a petrol injector. The first character shows which area of the vehicle generated the code:

- P: Power train
- B: Body
- C: Chassis
- U: Network.

In the example (P0291) it is P because the petrol injector is part of the power train (engine).

The next character can be a 0 or a 1 or a 2:

- 0: standard (SAE) OBD code
- 1: manufacturer's own code
- 2: manufacturer's own code

In the example (P0291) the next character is 0, which shows that it is an OBD code.

If the first character was P (Power train), then the next will be a number, which identifies the specific power train system concerned:

- 0: Fuel and air metering
- 1: Fuel and air metering
- 2: Fuel and air metering, specifically injector circuit
- 3: Ignition system and misfire detection
- 4: Auxilliary emission controls
- 5: Vehicle speed control and idle control system
- 6: Computer output circuit
- 7: Transmission-related faults
- 8: Transmission-related faults.

In the example (P0291) the next character is 2, which shows that it is a fault connected with fuel and air metering — specifically an injector circuit.

The last two characters are numbers that will identify the specific fault as seen by the on-board systems.

Most scan tools provide additional information about the exact nature of the fault, as shown in Table 29.1.

Limitations of DTCs

The DTC shows that a particular sensor or actuator is not functioning correctly, but it does not always identify exactly what is wrong. Most sensors and actuators form part of a circuit and the problem may lie elsewhere in a circuit; if this is the case, it then becomes necessary to perform tests to 'pinpoint' the actual fault. This searching for the cause of the fault is an important part of the process of fault diagnosis and it is in this area that simple electrical testing plays an important part.

Take the example of an air-flow sensor fault (Table 29.2). This DTC is based on the voltage signal that the ECU reads at its input terminal. The reason for the voltage being out of limits may be due to several factors associated with the air-flow sensor circuit so,

Table 29.1 Example fault descriptions for DTCs

Malfunction code DTC	Description of fault
P0301	Misfire in cylinder 1 detected
P0741	Torque converter clutch malfunction
P0303	Misfire in cylinder 3
P0421	Catalyst not warming up efficiently

Table 29.2 Example air-flow sensor fault

Diagnostic trouble code	Description
P0104	Mass or volume air-flow circuit intermittent

before condemning the sensor, it is advisable to conduct tests to ascertain the cause of the problem.

The types of tests that can be performed on the air-flow sensor circuit include the following:

- Checking the air-flow meter
- Checking the circuit between the sensor and the ECU.

The various types of air-flow sensors used in automobile systems produce an electrical signal that is transmitted to the ECU to represent the amount of air that is flowing to the combustion chambers. As an example to demonstrate the principles involved in validating a DTC, I have chosen a potentiometer-type sensor. These sensors are normally supplied with a voltage of 5 V d.c. The potentiometer part of the sensor then outputs a voltage between zero and 5 V, which can be examined on an oscilloscope trace, as shown in Fig. 29.2. If the test shows the sensor to be functioning correctly and the fault persists, it becomes necessary to

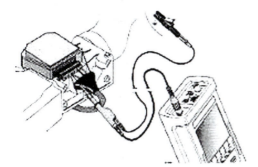

Oscilloscope connection

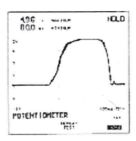

Sensor voltage output
Closed throttle to fully
open.

Fig. 29.2 An oscilloscope trace for a potentiometer-type air-flow sensor

check the circuit between the sensor and the ECU input connections.

Using the same procedure at the ECU end of the air sensor cable (Fig. 29.3) should produce an identical result to that obtained at the sensor. A material difference between the two results indicates that there is a defect in the circuit between the sensor and the ECU. The types of test that can be deployed from there on depend on the nature of the circuit, and it will be necessary to examine circuit diagrams and other workshop manual data.

The example shown here indicates that there is a general procedure for fault diagnosis that can be applied to any ECU-controlled system that provides a range of DTCs.

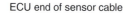

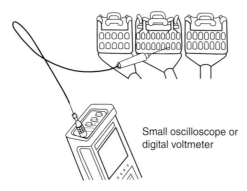

ECU end of sensor cable

Small oscilloscope or digital voltmeter

Fig. 29.3   Checking air sensor output at ECU end of cable

Fault tracing in electrical circuits

Various forms of equipment are available for circuit and component testing. Some of the commonly used ones are listed below:

- Digital multimeter — for voltage, current, and resistance
- Oscilloscope
- Scan tools and code readers
- Test lamps.

The digital multimeter

Digital multimeters are relatively inexpensive instruments that are often used in tests to assess the condition of a circuit. Digital instruments are used on ECU-controlled systems because they have a high internal resistance (impedance) that makes them suitable for testing components that incorporate electronic circuits. Digital multimeters such as the Fluke instruments, shown in Fig. 29.4, are specially designed for automotive work and can be used for the following measurements:

- Voltage
- Current — amperes
- Resistance — ohms for continuity
- Duty cycle — on/off time of electronic pulse
- Frequency — Hz
- Temperature — via probe in $^{\circ}$C or $^{\circ}$F.

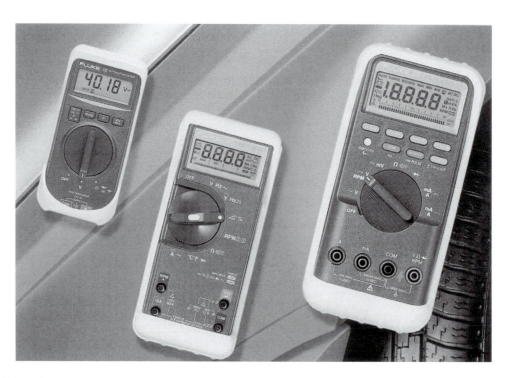

Fig. 29.4   Fluke multimeters

Sensors

It is important to recognize that two types of sensors are used in vehicle systems:

1. **Active sensors.** These normally generate electricity by electromagnetic means. A common use is in the crank position sensor.
2. **Passive sensors.** These receive an electrical supply, which they process to produce an output that represents some variable such as engine coolant temperature.

Passive sensors only work when the system is switched on, whereas active ones will produce a measurable output when they are activated, usually by rotating a component such as the crankshaft.

Voltage drop tests

In order for an electrical component to function correctly it must form part of a complete circuit. On most light vehicles, electrical circuits are completed by the 'earth return' principle, which means that the metal parts of the vehicle act as part of a circuit. The connection between a component and the vehicle metalwork is normally completed through an earth connection, and the electrical joint between the earth lead and the vehicle metalwork is a vital link in the circuit. Because the engine is normally mounted on rubber, the electrical connection between the engine and the chassis is an important part of the circuit of the electrical components that are attached to it.

Faults in automotive electrical circuits may be due to several factors, including the following:

- **Open circuit.** An open circuit occurs when there is a break in the circuit, which may be caused by:
  - A cable connector that has worked loose
  - A broken cable
  - A blown fuse
  - A broken wire in a device such as an injector
  - A broken switch.
- **Short-circuit.** Short-circuits are caused when a cable, or some part of a component, makes contact with the chassis.
- **High resistance.** This may be caused by dirty or corroded connections, or badly fitted components.
- **Intermittent breaks in a circuit.** These may be caused by loose cable connectors or insecure fixings of components.

The following examples give an indication of the basic procedures that can be used to trace the causes of these faults.

Engine to chassis earth lead

Because the engine is mounted on flexible rubber blocks of the type shown in Fig. 29.5 it is necessary to provide

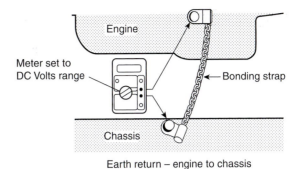

Earth return – engine to chassis

Fig. 29.5   Voltage drop test on earthing strap

a good electrical connection between it and the vehicle frame. Systems such as fuel and ignition are normally earthed to the metal of the engine and their circuits are then completed through the earth strap. If the earth strap electrical connections are not sound, the operation of components will be affected – this is particularly so in the case of the starter motor, which draws a heavy current from the battery.

Voltage drop test on earth strap

1. The battery state of charge should be checked to ensure that it is at least 70% charged. A voltmeter is then connected across the battery terminals while the starter is operated. The voltage should then be recorded.
2. The voltage between the engine and the chassis should then be checked while the starter switch is operated. The voltage should be zero or, at the most, 0.50 volts on a 12-volt system.

Lighting circuit tests

The examples of lighting circuit tests are carried out with the lights switched on. In checking lighting circuit connections, the battery voltage should first be checked. In the example shown in Fig. 29.6, the test is applied to the number plate light, but the same test can be applied to any of the lamps. The voltage drop as measured at these points should be zero, or not more that 10% of battery voltage.

Continuity

Multimeters normally contain a system that enables checks to be made to see if there is a break in a cable or other element of a circuit. This test is carried out while the circuit is switched off. The continuity tester applies a small voltage to the circuit section and, if the circuit element is complete, a current will flow and a buzzer will sound. In some circumstances it may be difficult to perform the test because the test leads are applied to both ends of the cable.

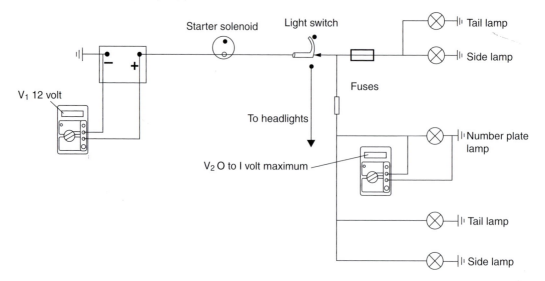

Fig. 29.6   Voltage drop at lamp connections

Resistance tests

The amount of resistance in a circuit is normally quoted in service manuals and valuable information about the condition of circuits can be obtained by performing a resistance check. The ohmmeter section of the digital meter is energized by the battery that powers the meter. When a resistance check is being performed, the meter applies a small voltage to the component under test and the current flowing through the meter is then recorded in ohms — some resistances are very high in terms of the number of ohms and the digital meter has a number of scales to provide for small resistances, up to large ones in megaohms ($M\Omega$). It must be understood that resistance and continuity tests are performed when the circuit, or unit, is switched off.

Figure 29.7 shows a resistance test as applied to a petrol injector. It is known that the resistance should be between 5 and 7 Ω; if it is less than this it indicates a short-circuit, and if it is greater than 7 Ω there is probably a poor connection inside the injector.

Resistance checks at control unit and sensor

Figure 29.8 shows some detail of a sensor that is used to detect wear in brake pads. The sensor contains two resistors; when the sensing element is not broken only the 180 Ω resistor is active and the resistance at the sensor terminals will be 180 Ω. When the sensing wire wears through, the 1200 Ω resistor becomes active and the sensor terminal resistance will be 1380 Ω. By checking the resistance at the control unit and at the sensor, it is possible to assess the condition of the entire circuit.

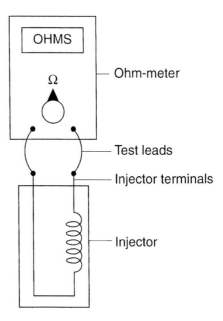

Resistance should be 1.5 to 2.5 ohms

Fig. 29.7   Resistance test

Breakout box

On ECU-controlled systems a large number of cables connect various sensors and actuators to the input side of the ECU, through a multiple-pin connector. There may be 60 or more pins on the connector and in workshop manuals they are identified by a numbering system. In order to check components their circuits are often accessed from the ECU end of the circuit. The breakout box (Fig. 29.9) provides a means of accessing the circuits without the need to break through cable insulation. The input side of the breakout box has an adaptor to which the ECU input, multiple-pin connector is applied. Each pin connects to an easily accessible, numbered pin on the breakout box so that a multimeter

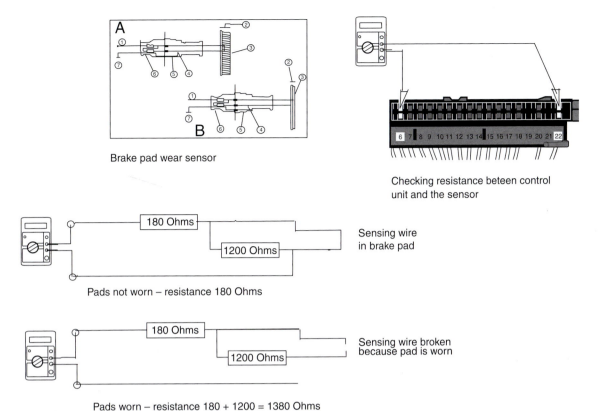

Checking resistance beteen control
unit and the sensor

Brake pad wear sensor

180 Ohms

1200 Ohms

Sensing wire
in brake pad

Pads not worn – resistance 180 Ohms

180 Ohms

1200 Ohms

Sensing wire broken
because pad is worn

Pads worn – resistance 180 + 1200 = 1380 Ohms

Fig. 29.8 Checking sensor resistance at the multi-pin connector

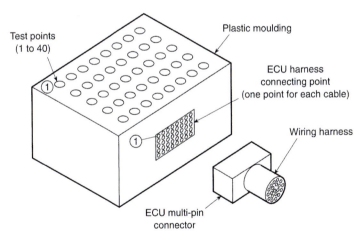

Test points
(1 to 40)

Plastic moulding

ECU harness
connecting point
(one point for each cable)

Wiring harness

ECU multi-pin
connector

Fig. 29.9 A breakout box

can be connected. It is common practice for vehicle manufacturers to give resistance values for components that can be accessed through a breakout box.

A breakout box can be used to check circuits when:

1. The ECU is disconnected. These checks are similar to the one described in the brake pad sensor test.
2. When the ECU is connected and systems are operating. When systems are active it is possible to connect test equipment such as meters and oscilloscopes to the breakout box pins so that their performance can be assessed.

Oscilloscope tests at the breakout box

Checking wiring between sensor and ECU

If the ECU records a fault in an injector and a check of the injector shows that it is in good order, the fault probably lies in the part of the circuit between the injector and the ECU. A defect in the wiring may be due to:

- A bad connection at the injector
- A bad connection at the ECU
- A badly fitting or corroded cable connector.

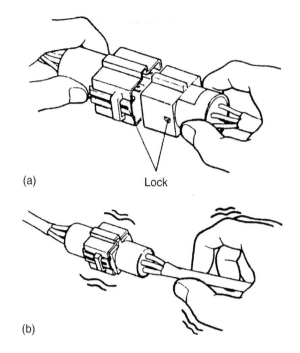

(a)   Lock

(b)

Fig. 29.10 The wiggle test

Wiggle test

Defects in cable connectors may cause intermittent circuit problems. The type of test shown in Fig. 29.10 can sometimes identify problems in this area, particularly when a system is in operation. If a connector is at fault, the wiggle test should affect the operation of the system — voltage and other performance tests should then confirm whether or not the connector is defective.

Sensor and actuator tests

When connected to the diagnostic port, a scan tool will normally permit the operator to:

1. Perform a check, through the ECU, of the various sensors in a given system
2. Activate petrol injectors and other actuators to gain an assessment of their performance.

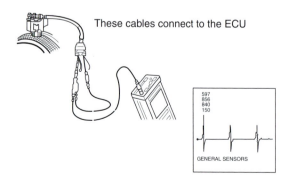

These cables connect to the ECU

Checking output at magnetic type Crankshaft sensor

Fig. 29.11 Checking a crankshaft sensor

The results of these tests are then compared with the data provided by the vehicle manufacturer. If the figures indicate that performance of a particular sensor or actuator is outside the limits, it may be necessary to perform further tests of the type shown in Figs 29.11 and 29.12.

Actuator tests

Actuators are those electromechanical devices that operate under the control of the ECU to operate some system, or part thereof. A relatively simple actuator is the valve that controls the recirculation of exhaust gas back to the induction system. Exhaust gas recirculation (EGR) occurs at certain engine and load speeds, and emissions and engine performance is badly affected if the system is not working correctly. The test is shown in Fig. 29.12.

● MultiScope key sequence

3. Connect the test leads as displayed by the MultiScope's **Connection Help** as shown below.

4. [F1] Starts the EGR test.

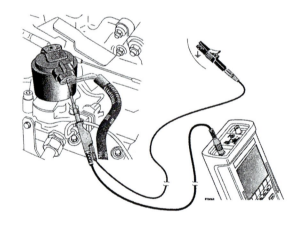

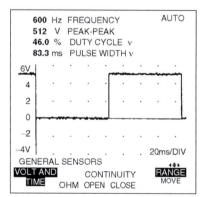

Fig. 29.12 Testing the EGR actuator

Because there are many different types of actuators and sensors, it is important to be able to compare test results with the model scope patterns and test data provided by the vehicle manufacturer; most diagnostic equipment contains a database that is frequently updated by the equipment manufacturer.

Logic probe

A logic probe of the type shown in Fig. 29.13 may be used to check a logic circuit where the transistorized element operates by pulsing a voltage on and off. The hand-held probe is applied to the point at which the voltage level is to be tested. The display is via the LEDs, of which there are normally three:

- Red usually indicates the high (logic) voltage level
- Green usually indicates the low (logic) voltage level
- Amber indicates that a pulse is present.

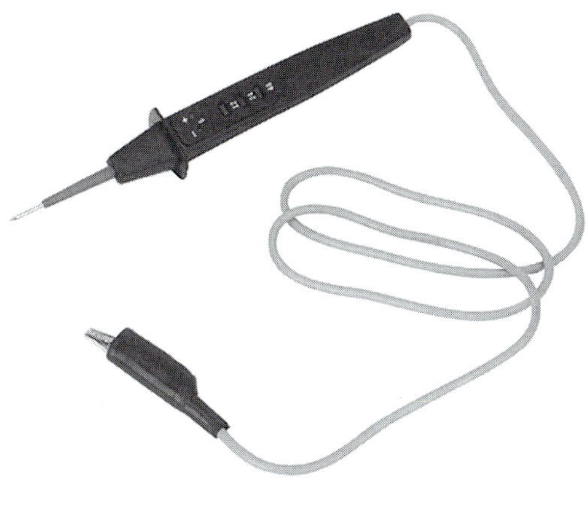

Fig. 29.13 A circuit probe

Intermittent faults

Intermittent faults can be particularly difficult to trace; 'freeze frames' and flight-recorder-type diagnostic equipment can be useful in this area of diagnostic work. Diagnostic scan tools have a data logger capability that is similar to, but of smaller capacity than, the flight recorder on aircraft. Because of this similarity, the data logger function of the scan tool is often known as the 'flight recorder' function.

Flight recorder (data logger) function

The data logger aspect of test equipment capability permits the test equipment to store selected data that the test equipment 'reads' through the serial data diagnostic connector of the ECM. It is particularly useful for aiding the diagnosis of faults such as an unexpected drop in power that occurs during the acceleration phase. When the test equipment is connected, and proper preparations have been made for a road test, the vehicle is driven by a person who should be accompanied by an assistant to operate the test equipment (for safety reasons). With the test equipment in

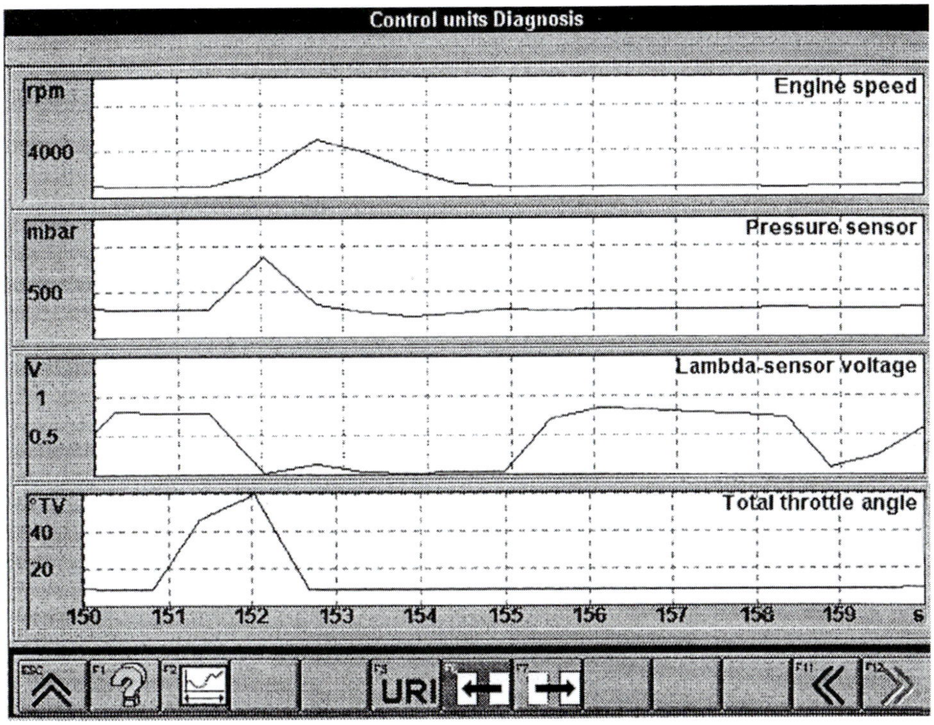

Fig. 29.14 A display of live data as 'read' out from the diagnostic link

'record' mode, the vehicle is driven in an attempt to recreate the default condition, and when the 'fault' occurs the test equipment control button is pressed. From this point, data from just before the incident and for a period after is recorded. The stored data can then be played back on an oscilloscope screen, or printed out later for analysis in the workshop. Figure 29.14 shows an example of live data that was obtained from a test using the Bosch KTS 500 equipment on a Peugeot vehicle.

In this case the signals all appear to be in order. However, should a defect occur it will show up in displays such as these. Figure 29.15 gives an impression of the type of information that might be seen. The reasoning here is that the vehicle is accelerating. This is borne out by the increased throttle opening, increased engine speed, and increased fuel injection. One would expect these events to be accompanied by an increase in air flow. But the air-flow sensor shows a drop in air flow — maybe there is a fault at the air-flow sensor!

Alternative data loggers

The large memory capacity of modern memory sticks allows them to be incorporated into a small data logger that can be inserted into the diagnostic port of an OBD system (see Fig. 29.16). The insertion of the data logger into the OBD socket does not interfere with the operation of the vehicle. Once energized, the device can be left operating while the vehicle is in normal use. Any defect in an ECU-controlled system will be recorded together with other information about time, mileage, and other operational details. After a period of recording, the data logger can be connected to a PC, via a suitable interface and computer program, and the data downloaded for analysis.

Fig. 29.16 The Texa OBD data logger

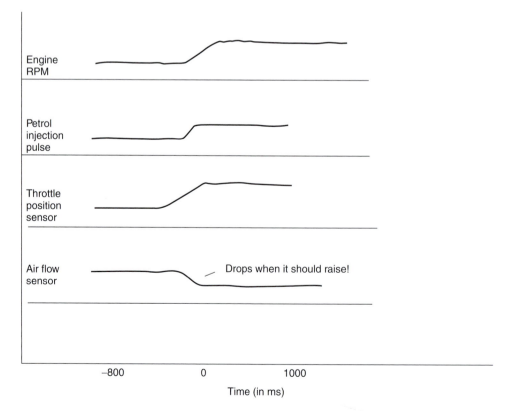

Fig. 29.15 How a probable fault shows up in a live data display

Self-assessment questions

1. Diagnostic trouble codes are:
   (a) computer codes that can be displayed at a fault code reader
   (b) only readable through the serial connector of the ECM
   (c) generated at random whenever there is a fault on any part of the vehicle
   (d) information that tells the user exactly what the fault is.

2. A limited operating strategy:
   (a) permits the vehicle to be driven until a repair can be made
   (b) cuts out fuel injection above a certain engine speed
   (c) retards ignition timing to stop combustion knock
   (d) refers to the limited nature of fault codes for diagnostic purposes.

3. Microprocessor-based diagnostic testers can:
   (a) only read fault codes
   (b) read fault codes and perform actuator tests via the serial port of the vehicle
   (c) reset the values stored in a mask programmable ROM
   (d) only read out diagnostic data from CAN systems.

4. The standardized serial port for diagnostics that is used with OBD2 has:
   (a) a three-pin connector
   (b) no specified number of pins but its position on the vehicle is specified
   (c) a 16-pin connector
   (d) no pins specifically allocated to the OBD2 emissions systems.

5. In order to read out diagnostic trouble codes (fault codes) it is necessary to:
   (a) earth the K-line and read the flashing light
   (b) carry out the manufacturer's recommended procedure
   (c) have the engine running
   (d) take the vehicle for a road test first.

6. An ABS ECM should have good self-diagnostics because:
   (a) the sensor output signals cannot be measured independently
   (b) it is difficult to simulate actual anti-lock conditions with a stationary vehicle
   (c) if the ABS warning light comes on it will stop the vehicle
   (d) the fault codes are always stored in an EEPROM.

7. A 'freeze frame' is:
   (a) a set of ROM data that is used in very cold weather
   (b) data that is used by the ECM when there is an emergency
   (c) a set of data about operating conditions that is placed in the fault code memory when the self-diagnostics detects a fault
   (d) a diagnostic feature of very early types of electronic control only.

8. In standardized fault codes:
   (a) digit 1, at the left-hand end, identifies the vehicle system
   (b) the digit at the far right-hand end identifies the system
   (c) all computer-controlled vehicle systems must use them
   (d) the identifying digits can appear in any order.

9. Pulse width modulation (PWM) is a system that uses pulses of electric current to operate actuators. Make a list of ECU-controlled actuators that use this method of operation. How is the duty cycle measurement related to PWM?

10. Why is it that diagnostic fault codes are not always the complete answer to tracing a fault in an OBD system?

11. Make a list of the diagnostic equipment in your workshop. Describe the opportunities that are available for you to keep abreast of developments in vehicle technology.

12. Describe any electronics-related diagnostic problem that you have experienced. What do you think that you learned from the experience? How much do think that underpinning knowledge of vehicle systems helped you and your colleagues in making a successful diagnosis?

30
Nuts, bolts, and spanners

Topics covered in this chapter

The origin of standards in sizes and thread forms
Lock nuts
Spanner sizes
Thread restoration

Dimensions of nuts, bolts, and spanners

In the early 1800s, Joseph Whitworth devised a set of standards for screw threads and their associated nuts and bolts. Under the Whitworth system the size of the bolt head and nut is determined by the diameter of the threaded part of the bolt. For engineering drawing purposes, the distance across the flats of the bolt head (*S*) was set at:

$$S = 1.25D + \frac{1}{8}\ \text{inch},$$

where D = diameter of the threaded part of the bolt.

The size of the spanner required for these nuts and bolts is marked on the spanner, near the jaws (see Fig. 30.2). A spanner for a half-inch Whitworth bolt has a distance across the jaws of 0.75 inch.

A similar system that is used in modern automobile work makes use of metric dimensions. In this case the size marked on the spanner is the distance across the flats of the nut as shown in Fig. 30.1. Table 30.1 shows the relation between thread diameter and spanner size.

Screw threads

The pitch of a screw thread is the distance between corresponding points on a thread, as measured form peak to peak. The angle of the thread is determined by the type of thread that is being used (see Fig. 30.3). A number of different thread types may be found in automobile practice and it is useful to have a thread gauge to hand whenever there is a question about the exact type of thread that is in use (see Fig. 30.4).

Studs

Studs are used in many applications, such as that shown in Fig. 30.5. The threaded part that screws into the main component is normally shorter than the thread at the other end of the stud; this is an important point to note because damage may be caused if a stud is fitted the wrong way round. Special tools called stud removers are required for removing and fitting studs.

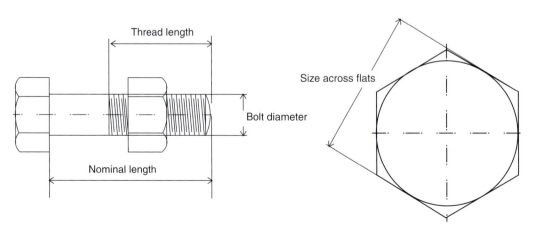

Fig. 30.1 Dimensions of a bolt

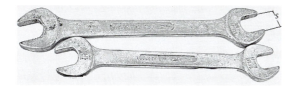

Fig. 30.2 Spanner size

Table 30.1 Thread and spanner sizes

Thread size M = mm	Spanner size (mm)
4	7
5	8
6	10
7	11
8	13
9	14
10	17
12	19
14	22
16	24
18	27

Fig. 30.4 Screw thread pitch gauges

Lock nuts and other securing devices

Locking nuts such as the castle nut and the Nyloc self-locking nut (Fig. 30.6) are used in applications where extra security of fixing is required in steering and other systems. In other applications a lock nut or spring washers of the type shown in Fig. 30.7 may be used.

Taps and dies

Sometimes, during the course of repair work, threads become damaged and it then becomes necessary to cut a new thread (see Fig. 30.8 for thread dimensions). Taps are used to cut new, or restore old, internal threads. Three types of tap are available and they are used in the sequence shown in Fig. 30.9. External threads are cut by means of a die. The tool into which the die fits is called a die stock and it contains screws that allow the cutting surfaces to be adjusted so that the thread can be cut in stages. Die nuts are also available and they are useful for refurbishing damaged threads.

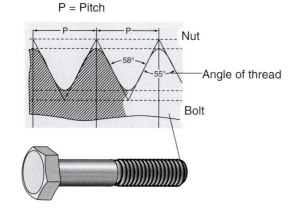

Fig. 30.3 Screw thread pitch and angle

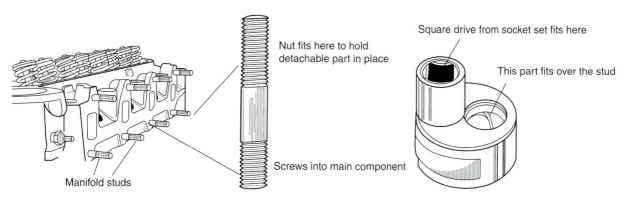

Fig. 30.5 Manifold stud and stud remover

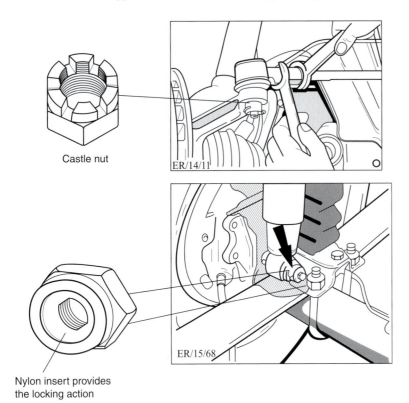

Castle nut

ER/14/11

ER/15/68

Nylon insert provides
the locking action

Fig. 30.6   Castle nut and Nyloc self-locking nuts

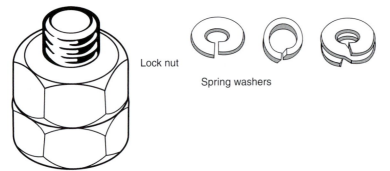

Lock nut

Spring washers

Fig. 30.7   Lock nut and spring washers

Drilling and tapping sizes

When a hole is being made for a bolt to pass through, the hole is slightly larger than the major diameter of the bolt; the hole is then called a clearance hole and the drill required is called the clearance drill. When a hole is being made to accommodate an internal thread, the drill size required is called the tapping drill. Table 30.2 shows a range of tapping and clearance sizes.

Removing broken studs

When a stud, or bolt, breaks off in a component it is useful to have a procedure for removing the broken piece. The screw extractors shown in Fig. 30.10 can

be used for this procedure. The broken surface should be cleaned so that its centre can be accurately marked; the centre should then be marked with a centre punch. A small pilot hole, followed by a larger hole, is then drilled into the broken portion. A suitable diameter of screw extractor is then inserted into the hole and the extractor is turned by a tap wrench. The design of the spiral part of the extractor should then cause the broken portion to be screwed out of the component.

Thread restoration

File

When a thread is slightly damaged it may be possible to repair it with the aid of a thread restoration file of the

UNC Major diam Inches	TPI	UNF Major diam Inches	TPI	ISO Metric Coarse	Pitch mm	ISO Metric Fine	Pitch mm
$\frac{1}{4}$	20	$\frac{1}{4}$	28	6	1.00	6	0.75
$\frac{5}{16}$	18	$\frac{5}{16}$	24	7	1.00	7	0.75
$\frac{3}{8}$	16	$\frac{3}{8}$	24	8	1.25	8	1.00
$\frac{7}{16}$	14	$\frac{7}{16}$	20	9	1.25	10	1.25
$\frac{1}{2}$	13	$\frac{1}{2}$	20	10	1.50	12	1.25
$\frac{9}{16}$	12	$\frac{9}{16}$	18	11	1.50	14	1.50
$\frac{5}{8}$	11	$\frac{5}{8}$	18	12	1.75	16	1.50
$\frac{3}{4}$	10	$\frac{3}{4}$	16	14	2.00	18	1.50
$\frac{7}{8}$	9	$\frac{7}{8}$	14	16	2.00	20	1.50
1	8	1	12	18	2.50	22	1.50
				20	2.50	24	2.00
				22	2.50		
				24	3.00		

Fig. 30.8 Table of thread dimensions (TPI = threads per inch)

Table 30.2 Tapping and clearance drill sizes

Diameter ISO metric coarse threads	Tapping size drill (mm)	Clearance size drill (mm)
M8	6.9	8.4
M10	8.6	10.5
M12	10.4	13.0
M14	12.2	15.0
M16	14.25	17.0
M18	15.75	19.0
M20	17.75	21.0
M22	19.75	23.0
M24	21.25	25.0

type shown in Fig. 30.11. These files have thread-shaped sections of various thread sizes on each of the four faces and when they are used carefully they may help to restore a slightly damaged thread.

Die nut

A damaged external thread can often be refurbished by means of a die nut (Fig. 30.12). Die nuts are made from tool steel; the thread is designed to follow the original thread and remove burrs and other damage. After preparing the damaged thread, the die nut is applied and slowly rotated by means of a spanner — it is helpful if a cutting oil is used in the process.

Helicoil inserts

When an internal thread is damaged it may be possible to repair the damage with the aid of a thread insert. The Helicoil thread inserts shown in Fig. 30.13 are manufactured for this purpose. The damaged thread portion is removed and a thread the size of the outside of the insert is made in its place; the insert is then screwed into the new thread, and the internal thread in the insert then accommodates the bolt or stud.

Screws

Screws are used in a range of automobile applications; they are available in several forms, as shown in Fig. 30.14. The round-headed screw is used where the

1 First tap – taper. Starts the thread
2 Second tap – the plug tap. Completes most of the thread
3 Parallel tap. Completes the thread cutting

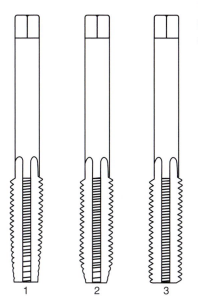

Die that cuts the external thread

Fig. 30.9 Taps and dies

The spiral part is entered into the prepared hole in the broken stud. The rotation of the tap wrench forces the spiral into the stud and should cause the broken section to screw out

Wrench

Fig. 30.10   Screw extractors

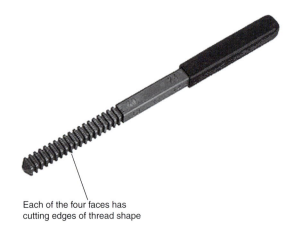

Each of the four faces has cutting edges of thread shape

Fig. 30.11   A thread restoration file

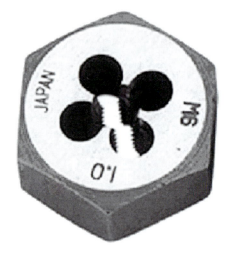

Fig. 30.12   A die nut

HELICOIL INSERTS

Fig. 30.13   Thread restoration inserts

head can protrude above the surface of the object being secured; the cheese head and the countersunk head are used where the top part of the head is flush with the surface of the object. The four types of screwdriver that are used to fit and remove screws are also shown in the diagram. The Torx screw and the Allen key type of head are also used on bolts; they permit greater torque to be applied than could be applied via a slot.

Bolt, stud, and screw strength

High tensile steel is used in applications such as cylinder head studs. The strength of the steel that is used in a bolt is marked on the head (see Fig. 30.15).

Tightening torques

The tightening torque is an important factor to consider when assembling automobile components. If the torque is excessive it may lead to a broken stud or bolt, and if the torque is not sufficient a broken gasket and leakage of gas, or liquid, may ensue. **The figures provided by the vehicle manufacturer should always be referred to, particularly where a nut is being used to pre-load a bearing.**

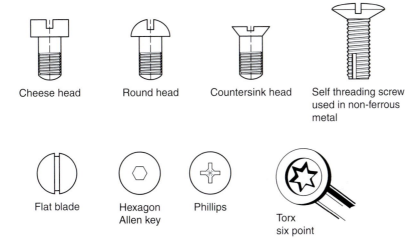

Fig. 30.14 Screws and screw heads

Metric Class	Marks on Head	Material	Tensile Strength N/mm²	psi	Yield Strength N/mm²	psi
8.8	8.8	Steel	800	116,000	640	93,000
10.9	10.9	Steel	1040	151,000	940	136,000
12.9	12.9	Alloy Steel	1220	177,000	1100	160,000

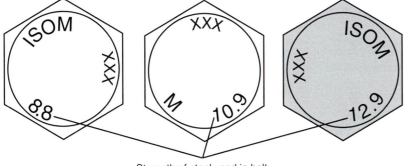

Strength of steel used in bolt

Fig. 30.15 Strength of a bolt

Table 30.3 Tightening torque (bolt strength 8.8)

Bolt size	Tightening torque (Nm)
M4	1.75–3
M6	8.5–10
M8	20–25
M10	40–51
M12	70–90
M14	114–146
M16	175–220
M18	252–317
M20	345–430

Table 30.3 gives an indication of a range of tightening torque settings.

Self-assessment questions

1. What type of screw fixing would the ring spanner shown in Fig. 30.16 be used for?
2. How big is the width across the jaws of an open-ended spanner for a 14-mm-diameter bolt?
3. Describe some of the uses for the type of puller shown in Fig. 30.17.

Fig. 30.16   A slotted ring spanner

Fig. 30.17   A puller

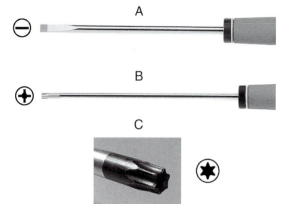

Fig. 30.18   Screwdriver types

4. Name the three types of screwdriver shown in Fig. 30.18. Which of the screws is sometimes called a tamper-proof screw?

5. What size clearance drill would be used for an M18 bolt?

6. Owing to an error in setting a torque wrench, a cylinder head stud is broken off in the cylinder block. The part of the stud that remains in the cylinder block is flush with the face of the block. State which tool would be useful in making a repair and outline the procedure involved.

7. Make a sketch of a wheel nut that is used to attach a steel wheel. Describe the procedure for removing and replacing a wheel. State the sequence that would be used for tightening the nuts. What precautions must be taken when using an air wrench for this operation?

8. On what type of vehicle are left-hand threaded wheel nuts likely to be found?

31
Measuring instruments and measurement

Topics covered in this chapter

Micrometers
Vernier calipers
Feeler gauges
Number approximations — typical motor vehicle measurements
Electrical test meters

Accurate measurement plays an important part in vehicle maintenance and repair. For example, measurements that are routinely taken are:

- valve clearances
- steering angles and track alignment
- exhaust emissions
- compression pressures
- alternator charging rate
- tyre pressures
- brake disc 'run-out'
- brake disc thickness, brake pad thickness
- shaft and bearing clearances.

Figure 31.1 shows a number of examples of the types of measurements and the measuring instruments that are used in motor vehicle maintenance and repair work.

Measuring instruments

Feeler gauges

Feeler gauges are used in motor vehicle repair and maintenance for such purposes as checking valve clearances, piston ring gaps, and many other clearances. The feeler gauges, shown in Fig. 31.2, are typical of the type that form a part of a technician's tool kit. They are made from high-grade steel that is tempered for hard wear and durability and the size of each blade is etched into the surface.

Figure 31.1(a) shows a feeler gauge in use, measuring valve clearance. In this example, the feeler gauge is being inserted between the heel of the cam and the cam follower, with the valve in the fully closed position. When ready to proceed with the measurement, the gauges of the required thickness are inserted in the clearance between the heel of the cam and the cam follower, and a light-hand force, sufficient to slide the feeler gauge through the gap, is applied.

Non-magnetic feeler gauges

Feeler gauges made from brass or some other non-magnetic material, such as a suitable plastic, are used to check the air gap on some types of electronic ignition distributors.

Vernier calipers

Figure 31.1(b) shows the depth gauge section of a vernier caliper being used to check the drive train end-float in a final drive assembly. In this application the vernier gauge is being used in conjunction with an engineer's straight edge. Figure 31.3 shows a vernier gauge in greater detail. This instrument measures to an accuracy of 0.02 mm, which is quite adequate for most vehicle repair and maintenance work. As indicated in the diagram, the instrument is versatile as it may be used for internal, external, and depth measurements. The instrument should be treated carefully and kept in clean, dry conditions when not in use.

Reading a vernier scale

For the following explanation please refer to the metric reading at the bottom of Fig. 31.3.

The **main scale** is graduated in millimetres and are numbered at every 10 divisions. The **vernier scale** is divided into 50 divisions over a distance of 49 mm and each division is equal to 49/50ths of a mm — that is, 0.98 mm. The difference between the divisions on the two scales is $1.00 - 0.98 = 0.02$ mm.

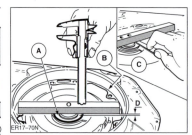

Check valve clearances.

(a) Feeler gauges

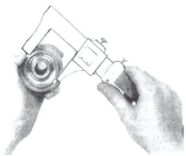

Measure drive train end float with measuring tool 17–025 and depth gauge (or feeler gauge)
A – Needle bearing No.11
B – Oil pump gasket
C – Measuring with feeler gauge
D – Distance between lower edge of measuring tool and needle bearing

(b) Straight edge & vernier gauge

(c) Micrometer – crank dimensions

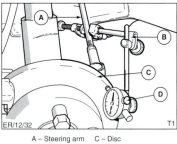

A – Steering arm   C – Disc
B – Holding fixture   D – Dial indicator

(d) Dial test indicator

(e) Vernier caliper – outside diameter of a bush

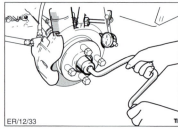

Rotating the disc using a socket and wrench.

(f) Rotating a brake disc to check run-out

Fig. 31.1 Measuring instruments and their uses

Fig. 31.2 A set of feeler gauges. *(Reproduced with the kind permission of Bowers Metrology Group.)*

To read the measurement, note the **main scale** reading up to the zero on the **vernier scale**. In this example, that is a reading of 21 mm. To this reading must be added the decimal reading on the **vernier scale**. This is obtained by noting the line on the **vernier scale** that exactly lines up with a line on the **main scale**. In this case the scales align at seven large divisions and three small divisions of the **vernier scale**. The seven large divisions represent 0.7 mm and the three small divisions represent 3 × 0.02 mm = 0.06 mm.

This gives a total reading = 21 mm + 0.7 mm + 0.06 mm = 21.76 mm

Micrometers

Figure 31.1(c) shows an external micrometer being used to check the diameter of a crankpin. Crankshaft measurement is normally only performed when an engine has operated for many thousands of miles, or when some failure has occurred as a result of running the engine short of oil or a related problem. The reason for checking the crank diameter is to assess its condition and thus ensure that the crank is suitable for further use.

Micrometers are available in a range of sizes, for example 0 inch to 1 inch (0 mm to 25 mm), 1 inch to 2 inch (25 mm to 50 mm), etc.

A metric external micrometer

Figure 31.4 shows the main features of a micrometer and reference should be made to this diagram when reading the following description.

The screw thread on the spindle of the metric micrometer has pitch of 0.5 mm so that one complete turn of the thimble will move the spindle by 0.5 mm. The thimble is marked in 50 equal divisions, each of which is equal to 0.01 mm.

The sleeve is graduated in two sets of lines, one set on each side of the datum line. The set of divisions below the datum line read in 1 mm and the set above the datum

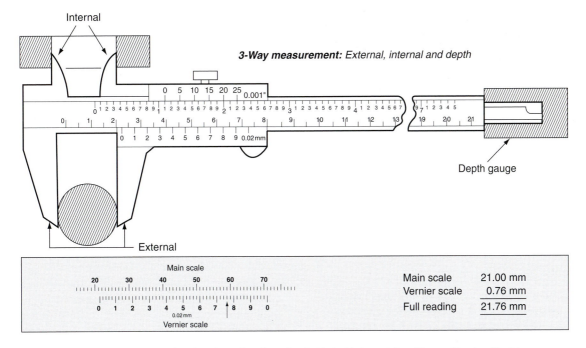

Fig. 31.3 A vernier gauge and method of reading the scale. *(Reproduced with the kind permission of Bowers Metrology Group.)*

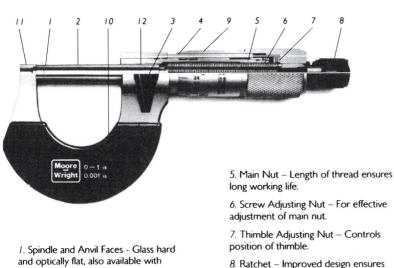

1. Spindle and Anvil Faces - Glass hard and optically flat, also available with TUNGSTEN CARBIDE faces.

2. Spindle – Thread ground, and made from alloy steel, hardened throughout, and stabilised.

3. Locking Lever – Effective at any position. Spindle retained in perfect alignment.

4. Sleeve – Adjustment for zero setting. Accurately divided and clearly marked. Pearl chrome plated.

5. Main Nut – Length of thread ensures long working life.

6. Screw Adjusting Nut – For effective adjustment of main nut.

7. Thimble Adjusting Nut – Controls position of thimble.

8. Ratchet – Improved design ensures even pressure.

9. Thimble – Accurately divided and every graduation clearly numbered. Pearl Chrome plated.

10. Steel Frame - Precision formed and Pearl Chrome plated with insulation pad.

11. Anvil End – Cutaway frame facilitates usage in narrow slots.

12. Larger spindle bearing for longer life.

Fig. 31.4 An external micrometer. *(Reproduced with the kind permission of Bowers Metrology Group.)*

line are in 0.5 mm divisions. (NOTE − on some micrometers, these positions are reversed.)

Reading a micrometer

When taking measurements, the force exerted on the thimble must not be excessive and to assist in achieving this, most micrometers are equipped with a ratchet that ensures that a constant force is applied when measurements are being taken. For measuring small diameters, the micrometer should be held approximately as shown in Fig. 31.5. Larger micrometers require the use of both hands.

Method of holding the micrometer

Fig. 31.5 Using the micrometer on a small diameter. *(Reproduced with the kind permission of Bowers Metrology Group.)*

Taking the reading

Please refer to Fig. 31.6.

The procedure is as follows:

- First note the whole number of millimetre divisions on the sleeve. These are called **major** divisions.
- Next observe whether there is a 0.5 mm division visible. These are called **minor** divisions.
- Finally read the thimble for 0.01 mm divisions. These are called **thimble** divisions.
- Taking the metric example shown in Fig. 31.6, the reading is:

$$\text{Major divisions} = 10 \times 1.00 = 10 \text{ mm}$$
$$\text{Minor divisions} = 1 \times 0.50 = 0.50 \text{ mm}$$
$$\text{Thimble divisions} = 16 \times 0.01 = 0.16 \text{ mm}.$$

- These are then added together to give a reading = 10.66 mm.

> *Learning activity*
>
> Figure 31.7 shows a range of micrometer scales. Write down the dimensions represented by these scale readings and compare your answers with those given at the end of this chapter.

Micrometers for larger diameters

A wide range of diameters may be encountered in vehicle repair work and an adjustable micrometer is

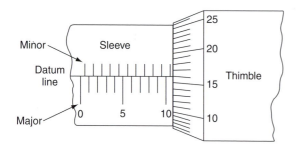

Fig. 31.6 The micrometer scales

useful for covering several ranges of diameters. The micrometer shown in Fig. 31.8(a) has a number of interchangeable anvils that may be screwed into place to suit the diameter being measured.

Figure 31.8(b) shows the range of diameters, from a large piston to a small shaft, that can be measured with an instrument of this type. Larger micrometers of the type shown here should be held in one hand while the thimble is turned with the other.

Internal micrometer

Internal micrometers are used to measure dimensions, such as cylinder bore diameter, internal diameter of a hole into which a bearing is to be inserted, and other machined holes where the dimensions of the internal diameter are critical. Figure 31.9(a) shows an internal micrometer being used for such measurement.

Internal micrometers consist of a measuring head and extension pieces suitable for a range of measurements, and the handle permits measurements to be taken in deep holes such as the cylinder of an engine. The principal features of an internal micrometer are shown in Fig. 31.9(b). The micrometer scale is read in a similar way to that shown in the external micrometer section of this chapter.

Mercer cylinder gauge

Because piston rings do not reach the top edge of the cylinder bore, when the piston is at the top of its stroke, an unworn ridge develops at the top of the cylinder and further cylinder bore wear occurs on the side where most thrust is exerted. This means that the cylinders of an engine wear unevenly from top to bottom of the stroke. In order to assess cylinder bore wear, it is necessary to measure the diameter at several points and the Mercer gauge has been developed for this purpose. Figure 31.10 shows a Mercer gauge. The gauge can be zeroed by setting it to the diameter of the unworn section at the top of the cylinder bore or, more accurately, by means of an outside micrometer.

Once zeroed, the bar gauge is inserted into the cylinder, as shown in Fig. 31.10. Measurements are taken in line with, and at right angles to, the crankshaft. The maximum difference between the standard size and gauge reading is the wear. The difference between the in-line reading and the one at right angles is the ovality. The measurements are repeated at points in the cylinder, as shown in Fig. 31.11, and differences between the diameter at the top and bottom of the cylinder are referred to as taper.

The measurements taken may then be compared with the quoted bore size and excessive wear can be remedied by reboring or, in the case of minor wear, new piston rings.

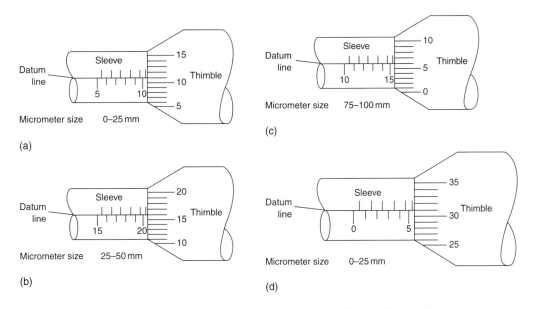

(a) Datum line, Sleeve, Thimble, Micrometer size 0–25 mm

(b) Datum line, Sleeve, Thimble, Micrometer size 25–50 mm

(c) Datum line, Sleeve, Thimble, Micrometer size 75–100 mm

(d) Datum line, Sleeve, Thimble, Micrometer size 0–25 mm

Fig. 31.7 An adjustable micrometer and its uses. *(Reproduced with the kind permission of Bowers Metrology Group.)*

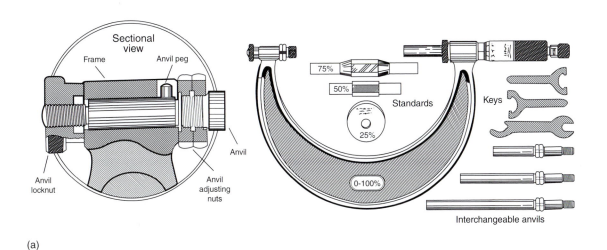

(a) Sectional view — Frame, Anvil peg, Thimble, Anvil, Anvil locknut, Anvil adjusting nuts, Standards 75%, 50%, 25%, 0-100%, Keys, Interchangeable anvils

Typical applications

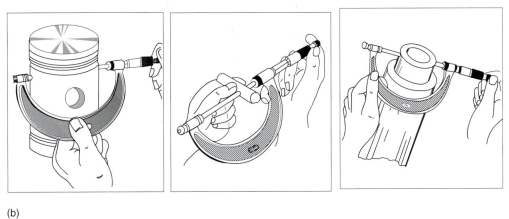

(b)

Fig. 31.8 Exercises on reading micrometer scales

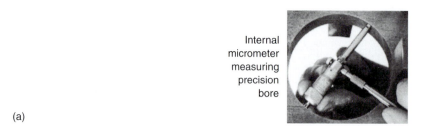

(a)

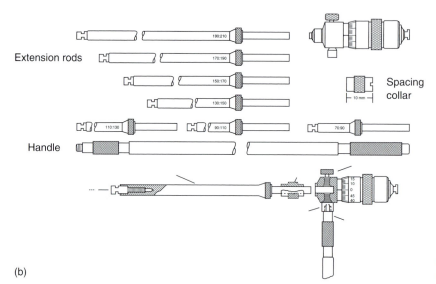

Extension rods

Spacing collar

Handle

(b)

Fig. 31.9 An internal micrometer. *(Reproduced with the kind permission of Bowers Metrology Group.)*

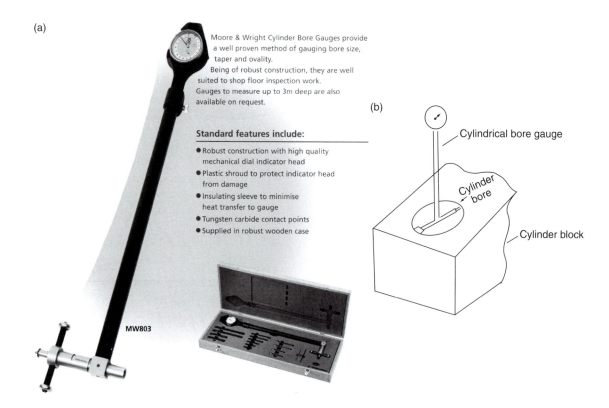

(a)

Moore & Wright Cylinder Bore Gauges provide a well proven method of gauging bore size, taper and ovality.

Being of robust construction, they are well suited to shop floor inspection work.

Gauges to measure up to 3m deep are also available on request.

Standard features include:

- Robust construction with high quality mechanical dial indicator head
- Plastic shroud to protect indicator head from damage
- Insulating sleeve to minimise heat transfer to gauge
- Tungsten carbide contact points
- Supplied in robust wooden case

(b)

Cylindrical bore gauge

Cylinder bore

Cylinder block

Fig. 31.10 A Mercer cylinder gauge. *(Reproduced with the kind permission of Bowers Metrology Group.)*

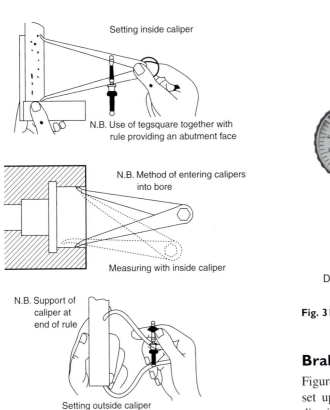

Setting inside caliper

N.B. Use of tegsquare together with
rule providing an abutment face

N.B. Method of entering calipers
into bore

Measuring with inside caliper

N.B. Support of
caliper at
end of rule

Setting outside caliper

N.B. Movement of caliper
giving "feet" at
measurement

Measuring with
outside caliper

Fig. 31.11 Engineer's calipers—checking diameters. *(Reproduced with the kind permission of Bowers Metrology Group.)*

Calipers, inside and out

Figure 31.11 show internal and external calipers being set against a steel rule prior to making measurements of inside and outside diameters. These types of measurements may be found useful in assembly work, such as fitting new parts where it may be necessary to check one diameter against another.

Dial test indicators

Dial test indicators (DTIs) of the type shown in Fig. 31.12 are used for measurements, such as checking run-out on brake discs and checking end float on a crankshaft.

There are many other instances where a DTI would be used, but the following two examples will suffice to explain the principles involved.

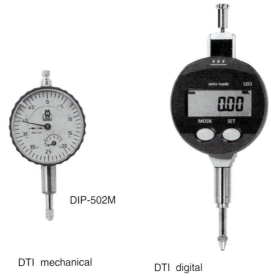

DIP-502M

DTI mechanical DTI digital

Fig. 31.12 Dial test indicators (DTIs)

Brake disc run-out

Figure 31.1(d) shows a dial test indicator that has been set up to test the axial run-out that occurs when the disc is rotated, as shown in Fig. 31.1(f). This measurement is important because of the effect that excessive run-out may have on the operation of the brakes. Among the brake problems that may be caused are defects, such as excessive pad wear and low brake efficiency.

Excessive disc run-out may be caused by a distorted disc, dirt between the disc and the flange that it is fitted to, or other problems in the wheel hub. Run-out should not exceed 0.15 mm (0.006 in) when measured in a position near the edge of the disc, but clear of any corrosion that may exist on the edge of the disc.

Crankshaft end float

End thrust on the crankshaft occurs when the clutch is operated and from the weight of the crankshaft when the vehicle is operating on an incline. The end thrust is normally taken by thrust bearings that are fitted to the centre main bearing of the crankshaft, as shown in Fig. 31.13.

There must be sufficient clearance between the machined faces on the crankshaft and the thrust faces of the thrust bearing to permit proper lubrication and allow for thermal expansion. If the clearance is too great, lubrication may fail because of loss of oil pressure; if the clearance is too small, seizure may occur. The clearance is checked by means of a DTI that measures the end float on the crankshaft, as shown in Fig. 31.14. A lever, suitably positioned, is used to move the crankshaft to and fro in order to check the end float.

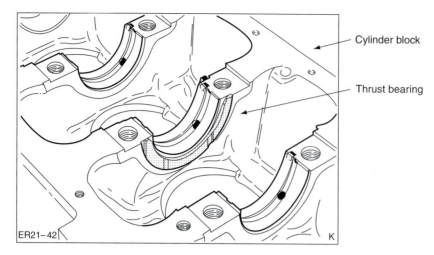

Installation of bearing shells and thrust half washers

Fig. 31.13   Crankshaft thrust bearing

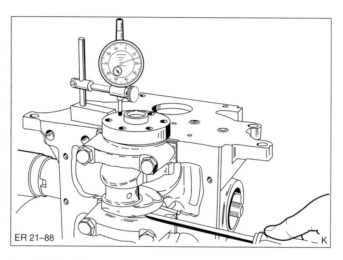

Crankshaft end float measurement

Fig. 31.14   Checking crankshaft end float

Pressure gauges

Pressure checks on various vehicle systems and components are performed as part of routine service procedures or as part of a series of fault-tracing tests. The following examples show the principles involved.

Engine compression test

The compression test is performed in order to ascertain compression pressure because this is a reliable guide to engine condition. The test is performed with all sparking plugs disconnected and removed from the engine. When the compression tester adaptor is firmly pressed on to the sparking plug hole seat, as shown in Fig. 31.15, the engine throttle is opened fully and the engine is cranked over a prescribed number of times, until the maximum pressure is recorded.

In the Ford tester shown here, the pressure is recorded on the graph paper in the instrument, in other cases a note should be made of the maximum pressure recorded. This procedure is repeated for each of the cylinders. The figures are then compared with those given in the manufacturer's data. The pressure readings for each cylinder should also be compared because this information can inform the user about the possible source of a problem. With worn engines, small variations of pressure are to be expected although variations of more than 10% should be investigated.

Oil pressure checks

In addition to its function of lubricating and helping to cool the engine, the engine lubrication system plays a part in operating hydraulic tappets and variable valve-timing mechanisms, and possibly other devices

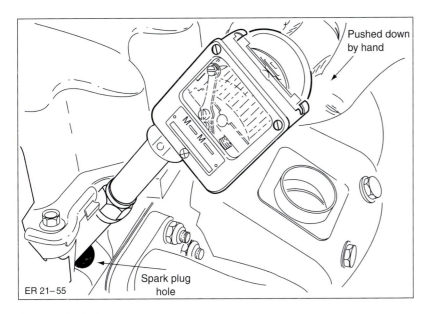

Compression tester

Fig. 31.15 Using the compression test gauge

on the engine. In order for the engine to function properly, the lubrication system must be able to sustain the correct oil pressure at all engine speeds. The oil pressure check that is shown in Fig. 13.16 is carried out by

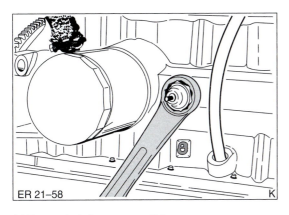

(a) Removal of oil pressure switch

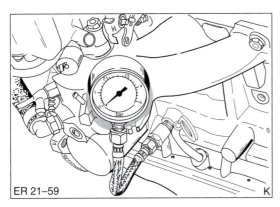

(b) Oil pressure gauge connected to engine

Fig. 31.16 An oil pressure check

temporarily removing the oil pressure detection switch and replacing it by the test type pressure gauge. The engine is then run under the recommended conditions of speed and temperature, etc. The readings should be noted and compared with those provided by the manufacturer.

Radiator pressure cap test

Figure 31.17 shows a piece of equipment that consists of a small pump, a pressure gauge, and an adaptor to which the radiator cap is secured. The radiator cap is removed from the vehicle and then attached to the pressure tester. When the cap has been secured to the tester, the pump is operated and the maximum pressure reached is noted. This figure should be compared with operating pressure for the cap, which is often stamped on the cap.

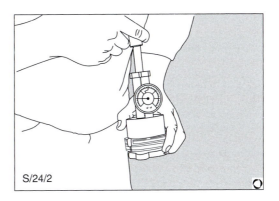

Testing pressure cap

Fig. 31.17 Pressure testing a radiator cap

Belt tension gauge

In order to function correctly, the cam drive belt must be set to the correct tension. The belt tension is set at the assembly stage by means of a special gauge that is attached to the 'tight' side of the belt, as shown in Fig. 31.18. This operation is carried out whenever a new cam belt is fitted or the adjustment is disturbed for some repair operation to be performed.

Cathode ray oscilloscope

The cathode ray tube (oscilloscope), as used in domestic television sets and flat screen displays, is used for a wide range of vehicle tests. With suitable interfaces, a range of voltages, from very small voltage values to very high ones, such as ignition secondary voltages, can be measured. Oscilloscope-based test equipment for vehicle systems normally includes a range of cables and adaptors that contain the interfaces that permit electrical signals to be collected and transferred to the oscilloscope. Because oscilloscopes measure voltage against time, the shape of the image shown on the screen enables a technician to observe the changes that are taking place while the system under test is in operation. Figure 31.19 shows a portable oscilloscope that is connected to the HT lead of a modern ignition system.

Figure 31.20 shows the oscilloscope pattern that is displayed on the screen during an ignition test.

Explanation of HT trace shown in Fig. 31.20

1. Firing line − this represents the high voltage that is required to cause the spark to bridge the spark plug gap.
2. This is the spark line.

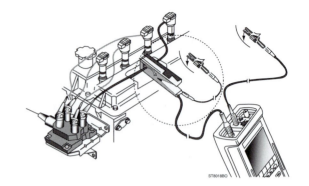

Fig. 31.19 Oscilloscope testing on an ignition system

3. Spark ceases at this point.
4. Coil oscillations − caused by collapse of the magnetic field.
5. Any remaining electromagnetic energy is dissipated.
6. Firing section (represents burn time).
7. Dwell section.
8. Primary winding current of the coil is interrupted by the transistor switch that is controlled by the engine control computer (ECM).
9. Primary winding current is switched on by the ECM and this energizes the primary winding of the ignition coil. The dwell period, between 9 and 8 on the trace, is important because it is the period in which primary current builds up to its maximum value.

Figure 31.20, and the accompanying explanation, give an indication of the amount of information that can be obtained from an oscilloscope test. Engine management systems on modern vehicles rely on the accuracy of factors such as burn time, etc.

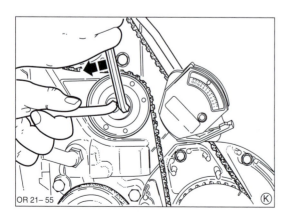

Tension toothed belt

Fig. 31.18 Cam belt tension gauge

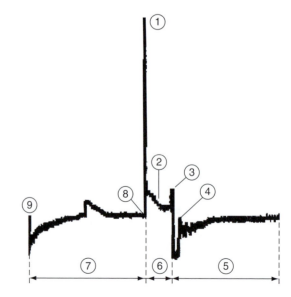

Fig. 31.20 Details of HT voltage pattern for a single cylinder

Sensors

Many vehicle systems are now controlled by a computer on the vehicle.

Such control computers (ECMs) operate on low-voltage electrical pulses that are interpreted as the zeros (0) and ones (1) of digital codes. The electrical pulses originate from devices such as the crank position sensor, the coolant temperature sensor, the engine air-flow sensor, and many others. The sensors on a vehicle measure some quantity, such as speed or position, and then convert the quantity into an electrical signal. The electrical signal that is produced is an accurate representation of the variable being measured. Sensors play a vital part in the operation of vehicle systems and the maintenance and testing of them is a major element of the work of a vehicle technician.

Figure 31.21 shows an oscilloscope screen display of the voltage pattern obtained at the output terminals of a throttle position sensor. A separate window on the screen shows an ideal pattern that is stored in test equipment memory. This is the lower trace in Fig. 31.21. The upper trace shows the voltage pattern obtained from a single throttle operation. This trace is obtained by moving the throttle from the closed position to the fully open position. Note from Fig. 31.21:

1. The actual test pattern showing a defective throttle position sensor trace.
2. The voltage spikes in a downward direction indicate a short-circuit to earth or a break in the resistive strip in the sensor.

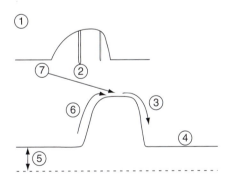

1. Defective TPS pattern.
2. Spikes in a downward direction indicate a short to ground or an intermittent open in the resistive carbon strips.
3. Voltage decrease identifies enleanment (throttle plate closing).
4. Minimum voltage indicates closed throttle plate.
5. DC offset indicates voltage at key on, throttle closed.
6. Voltage increase identifies enrichment.
7. Peak voltage indicates wide open throttle (WOT).

Analysis of voltage trace for a potentiometer-type throttle position sensor

Fig. 31.21   Testing a throttle position sensor

The lower trace shows the pattern for a sensor in good condition. The trace between 5 and 4 is obtained by a full operation of the throttle from closed to fully open and back to closed. The voltage at 5 represents the throttle closed position, the section marked 6 shows the voltage increasing smoothly to the throttle fully open position, and the section marked 3 shows the voltage decreasing smoothly back to the throttle closed position at 4.

Digital multimeter

Multimeters are electrical measuring instruments (meters) that can be set to read voltage, current, resistance, frequency, etc. Digital instruments are most suited to testing modern vehicle systems because they have the high impedance (internal resistance) that is required for testing circuits that contain sensitive electronic components.

Using a multimeter

Figure 31.22 shows a sensor that is built into an engine oil dipstick. The sensor, marked 3, is a small length of fine wire that has a high temperature coefficient of resistance, which means that small changes in temperature produce large changes in resistance. The resistance of the sensing element is $7\ \Omega$ at room temperature.

When the control unit passes a current of 200 mA through the sensing element, for a period of approximately 2 seconds, the sensing wire is heated and this produces an increase in resistance. This increased resistance leads to a voltage increase across the sensor terminals.

If the oil is level is correct, the sensor is in oil and the heat is dissipated. Any voltage increase caused by the

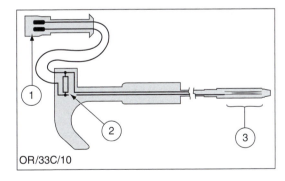

OR/33C/10

Low oil sensor circuit diagram
1 – Connector (to loom)
2 – Trimming resistor
3 – Sensor (fine wire with a high temperature coefficient of resistance)

Fig. 31.22   Low oil sensor

electrical pulse is small. If the oil level is low, the sensor wire will be in air. The temperature increase from the electrical pulse will be large and the voltage across the sensor terminals will be high.

The voltage across the sensor terminals is measured by the control unit. If the voltage increase is greater than the set value that is kept in the control unit memory, the 'low oil' warning light is switched on.

Testing a low oil sensor by means of a digital multimeter

A typical use of the multimeter is to measure the resistance of a sensor because, in many cases, this gives a guide to its electrical condition.

In Fig. 31.23(b), the ignition has been switched off and the cable connector has been detached from the sensor. The multimeter has been set to measure resistance in ohms and the meter probes are connected to the two sensor pins. In this case, the resistance should lie between 6.4 Ω and 7.6 Ω. The test should be done quickly, because the current from the ohmmeter will heat the sensing element and change the resistance. If this test is satisfactory, the sensor should be reconnected. The multi-plug is then disconnected at the ECM and the ohmmeter probes are applied to the appropriate pins on the multi-plug; in this case, pins 10 and 11, as shown in Fig. 31.23(c). The resistance reading at this point should be between 6.4 Ω and 8.6 Ω, the higher figure being attributable to the length of cable between the sensor and the multi-plug.

Applications of number

Approximations and degrees of accuracy

Significant figures Measurements and calculations often produce numbers that are more useful when they are rounded to a certain number of significant figures. For example, the swept volume of a certain single cylinder engine is 249.94 cc. For many purposes, this figure would be rounded to three significant figures, to give 250 cc.

The rule for rounding to a certain number of significant figures is: *If the next digit is 5 or more, the last digit is increased by 1. A zero in the middle of a number is counted as a significant figure, for example 3.062 to 3 significant figures is 3.06.*

Decimal places Numbers can be rounded to a certain number of decimal places, dependent on the required degree of accuracy. For example, 3.56492, rounded to two decimal places, becomes 3.56.

The rule for decimal place approximations is: *If the next number (digit) is 5 or more, the previous digit is increased by 1.*

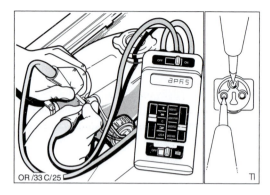

(a)

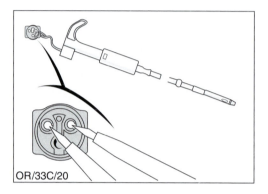

(b) Resistance at sensor 6.4 Ω to 7.6 Ω

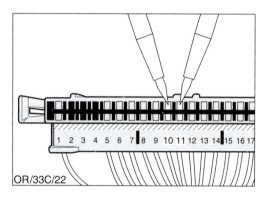

(c) Resistance at ECM 6.4 Ω to 8.6 Ω

Fig. 31.23 Resistance check of a sensor

Examples

1. The test data for a throttle position sensor specifies a voltage of 2.50 volts for the half-open throttle position. A test with a digital meter gives a voltage reading of 2.496 volts for the half-open throttle position. Does this mean that the sensor is faulty?

The answer is probably not, because the test data is given to two places of decimals. When corrected to two decimal places, the digital meter reading becomes 2.50 volts, which is as it should be.

2. The maximum permitted diameter of a certain crankpin is given as 43.01 mm. When measured with a micrometer, the diameter is found to be 43.009 mm. Is this acceptable?

Yes. The test data is given to two decimal places. When the measured reading of 43.009 mm is rounded to two decimal places, a figure of 43.01 mm is produced and this is exactly the same as the quoted figure.

Limits

Dimensions of components are normally given in the following form:

- Crankshaft main bearing journals: 53.970−53.990 mm
- Crankshaft end float: 0.093−0.306 mm
- Valve length: 107.25−107.35 mm.

In these cases, the larger figure is known as the **upper limit** and the smaller figure as the **lower limit.** The difference between the upper limit and the lower limit is known as the **tolerance.**

To calculate the tolerance, the smaller dimension is subtracted from the larger. For example, in the case of the main bearing journal, the tolerance is: 53.990 mm−53.970 mm = 0.020 mm.

These tolerances allow for slight variations in measurements at the manufacturing stage, and also variations in dimensions that arise from wear in use.

Fits

The terminology of engineering fits is based on the dimensions of a shaft and a hole into which the shaft is inserted.

When the shaft is slightly smaller than the hole to which it is to be fitted, the result is known as a **clearance fit.** The clearance is required for lubrication and thermal expansion purposes.

When the shaft is slightly larger than the hole to which it is to be fitted the result is known as an **interference fit.** Interference fits are used where the shaft is to be firmly secured by the material surrounding the hole.

Examples of the two types of fit are shown in Fig. 31.24.

With a **floating** gudgeon pin, the pin is a **clearance fit** in the small-end bush. The gudgeon pin is located in the piston by means of circlips that fit into grooves on each side of the piston.

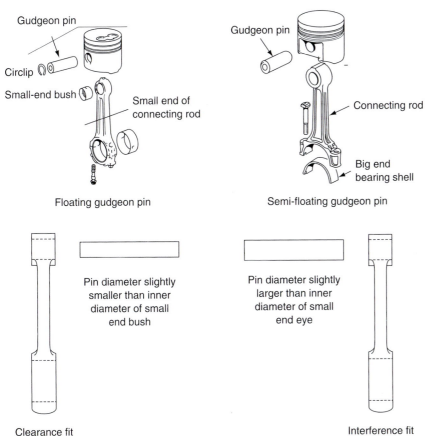

Gudgeon pins − clearance & interference

Fig. 31.24 Floating and semi-floating small ends − clearance and interference fits. *(Reproduced with the kind permission of Bowers Metrology Group.)*

In the **semi-floating** gudgeon pin arrangement, the gudgeon pin is an **interference fit** in the small-end eye of the connecting rod. The gudgeon pin is held in place by the strain energy in the materials. At the fitting stage, the gudgeon pin is either pressed into the connecting rod using a hand-operated screw press or the connecting rod is heated to make the small-end eye expand so that the gudgeon pin may be inserted. When the assembly cools down, the gudgeon pin is firmly held in the small-end eye.

Angular measurement

Examples of angles that occur in motor vehicle dimensions are shown in Table 31.1.

Specification of angles   The degree is the basic measurement of angle. As there are 360 degrees in a complete circle, for more accurate measurement, a degree is subdivided into 60 minutes.

A degree is denoted by the small zero above the number, i.e. $1°$.

A minute is denoted by a small dash above the number, i.e. $5'$.

The camber angle given in the table is $1° 26'$; this reads as one degree, twenty-six minutes.

Figure 31.25 shows two examples of angular measuring devices in use. In Fig. 31.25(a) a $360°$ protractor is attached to the timing gear of a diesel in order to set the injection timing. In Fig. 31.25(b), the angle is read from the protractor on the track alignment gauges.

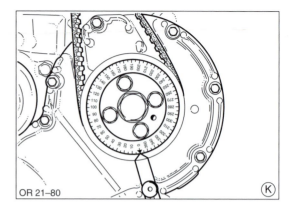

(a)

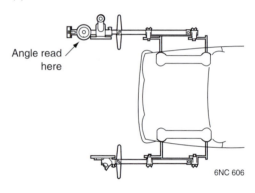

(b)

Fig. 31.25   Angular measurement on motor vehicles

Table 31.1   Examples of angles in motor vehicle dimensions

	Angle (approximate)
Steering	
Camber angle	$1° 26'$
Castor angle	$2° 11'$
Kingpin inclination	$3°$
Toe-out on turns	Outer wheel $20°$
	Inner wheel $25° 30'$
Engine	
Spark ignition — static advance	$12°$ before TDC
Valve timing — inlet valve opens	$8°$ before TDC
Diesel fuel pump — start of injection	$15°$ before TDC

Learning activity answer, Figure 31.7.

Answer (a) 10.61 mm
Micrometer size 0–25 mm
Reading:

- Major divisions = 10 × 1 = 10 mm
- Minor divisions = 1 × 0.5 = 0.5 mm
- Thimble divisions = 11 × 0.01 = 0.11 mm
- Total = 10.61 mm

Answer (b) 45.66 mm
Micrometer size 25–50 mm
Reading:

- Major divisions 20 = 20 mm
- Minor divisions 1 = 0.50 mm
- Thimble divisions 6 = 0.06 mm
- Total = 45.66 mm

Answer (c) 90.56 mm
Micrometer size 75–100 mm
Reading:

- Major divisions 15 = 15 mm
- Minor divisions 1 = 0.50 mm
- Thimble divisions 6 = 0.06 mm
- Total 90.56 mm

Answer (d) 5.81 mm
Micrometer size 0–25 mm
Reading:

- Major divisions 5 = 5.00 mm
- Minor divisions 1 = 0.50 mm
- Thimble divisions 31 = 0.31 mm
- Total 5.81 mm

Self-assessment questions

1. Correct the following figures to two decimal places:
   (a) 43.001
   (b) 0.768
   (c) 5.887.
2. Correct the following figures to the specified number of significant figures:
   (a) 19.28 to three significant figures
   (b) 2.034 to three significant figures
   (c) 337.9 to three significant figures.
3. The capacity of a certain engine is calculated to be 1998 cm^3. For descriptive purposes, this figure is expressed in litres. Calculate the engine capacity in litres, correct to two significant figures.
4. 43 bolts, priced at 45.6 pence each, are to be added to an account. Calculate the total amount for the 43 bolts, correct to the nearest 1 penny. Give the answer in pounds and pence.
5. Figure 31.26 shows a feeler gauge being used to check the gap in a piston ring. In order to perform this check:
   (a) the piston ring should just be pushed into the cylinder
   (b) the piston ring should be gently pushed into the cylinder using a piston as a guide to ensure that the ring is at right angles to the axis of the cylinder
   (c) the ring gap must not exceed 0.001 mm
   (d) it does not matter if the ring is tilted in the cylinder.
6. Figure 31.27 shows a dial gauge that is set up to measure the end float on an engine crankshaft. End float is important because:
   (a) it allows space for lubrication of the crankshaft thrust bearings
   (b) it provides free travel at the clutch pedal
   (c) it helps to keep the timing gears in alignment
   (d) it allows oil to be sprayed on to the cylinder walls to lubricate them.

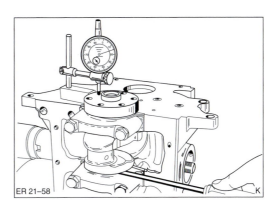

Fig. 31.27 Checking crankshaft end float

7. A crankpin has a diameter of 58.20 mm and the big-end bearing that is fitted to it has an internal diameter of 58.28 mm. The clearance between the crankpin and the big-end bearing is:
   (a) 1.0013 mm
   (b) 0.008 mm
   (c) 0.08 mm
   (d) 0.48 mm.
8. Figure 31.28 shows a feeler gauge marked 5 being used to check the clearance between the cam marked 4 and the cam follower. This clearance is known as the valve clearance and it is checked when:
   (a) the cam is in the fully closed position
   (b) when the engine is at the recommended temperature and the valve is in the fully closed position
   (c) the valve is fully open
   (d) the cam is at any position between valve fully open and valve fully closed.
9. The crankpin shown in Fig. 31.29 is measured at A, B, C and D. Dimensions A and B are taken to

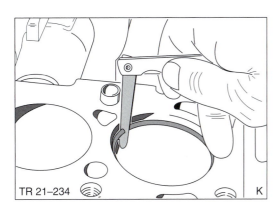

Fig. 31.26 Checking a piston ring gap

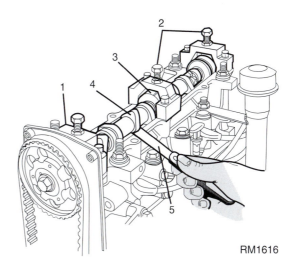

Fig. 31.28 Checking valve clearances

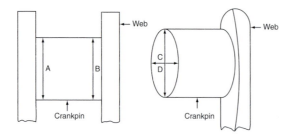

Fig. 31.29 Crankshaft measurement

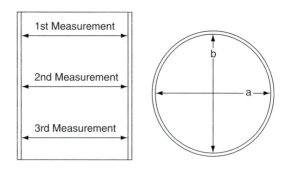

Fig. 31.30 Cylinder bore measurement

determine taper, and dimensions C and D to determine ovality. In this case the dimensions are:

- A = 42.99 mm
- B = 42.75 mm
- C = 42.99 mm
- D = 42.60 mm.

The taper on this crankpin is:

(a) 0.39 mm
(b) 0.24 mm
(c) 0.15 mm
(d) 0.01 mm.

10. The ovality on this crankpin is:
(a) 0.24 mm
(b) 0.00 mm
(c) 0.39 mm
(d) 0.15 mm.

11. Figure 31.30 shows the positions in an engine cylinder at which measurements of diameter are taken in order to assess wear. The measurements are taken in line and at right angles to the crankshaft, at the three levels shown. The third measurement is taken at the lowest extent of piston ring travel. The terms ovality and taper are used to describe the condition of the cylinder bore.

Ovality is the difference betweeen *a* and *b*, and taper is the difference between the largest diameter at the first measurement and the smallest diameter at the third measurement. In an example, the following dimensions were recorded:

- First measurement, a = 74.22 mm, b = 73.95 mm
- Second measurement, a = 74.19 mm, b = 74.00 mm
- Third measurement, a = 74.00 mm, b = 73.97 mm.

The maximum ovality is:

(a) 0.22 mm
(b) 0.19 mm
(c) 0.27 mm
(d) 0.03 mm.

12. The taper is:
(a) 0.25 mm
(b) 0.03 mm
(c) 0.19 mm
(d) 0.27 mm.

32
Motor industry organizations and career paths

Topics covered in this chapter

Qualifications and careers
Organizations in the motor trade

Qualifications

It is unfortunate that training and education tend to be treated as separate activities because this leads to difficulty when one is attempting to choose a course of study or training. However, there are many fundamentals of motor vehicle engineering where the distinction between training and education cannot sensibly be made. With valve timing of an engine, it is essential to understand cycles of engine operation, for example the four-stroke cycle. The knowledge required is known as engine technology and is generally considered to be education and therefore is part of the syllabus for courses such as City and Guilds Progression Award, BTEC First Diploma and Certificate, and similar qualifications. The actual setting of the valve timing is a practical task and learning how to do it is considered to be training. The instruction and practice that leads to becoming proficient at the task is also considered to be training and is considered to be part of the National Vocational Qualification (NVQ). The outcome of this rather artificial distinction between education and training is that educational courses contain much that contributes to one's ability to be a vehicle technician, while practical training courses for NVQ also contain much that is of educational value and, as a result, there are several routes that people can follow in order to qualify as a motor vehicle technician.

Modern motor vehicle qualifications

National Vocational Qualification (NVQ), Scottish Vocational Qualification (SVQ)

A list of skills that are required for competent performance of the tasks involved in maintenance and repair of vehicles has been devised and these skills are known as National Occupational Standards, a typical example being: *Carry out routine vehicle maintenance*. These Occupational Standards are used to design qualifications such as the NVQ. Because the work of a motor car technician differs in some respects from that of a heavy vehicle technician, the training programme and course content is not the same and NVQs have been designed to cater for these differences. The same is true of other motor vehicle occupations, such as Auto Electrical Technician, Vehicle Body Repair Technician, Parts Department Technician, etc.

Levels

In the school years, levels of attainment tend to be expressed in terms of year 2, year 6, etc., but in post-school training and education, the use of years as a description of a degree of difficulty involved in some learning step is not appropriate, and the training and education courses are classified in Levels. The result is that NVQs are available at several Levels, the main ones being at Levels 1, 2, and 3.

Studying for NVQ

To qualify for NVQ, it is necessary to demonstrate competence, namely ability to perform a given task in a normal work situation. However, as already explained in the example of valve timing, the performance of a practical task often requires an underpinning knowledge of technology. The result is that courses of study for NVQ consist of a balanced programme that takes place on the job (in a workshop) for the practical and off the job (in a classroom) for the underpinning knowledge. NVQs in Motor Vehicle Maintenance and Repair certainly require a good deal of underpinning technical knowledge together with practical training, and courses of education and training are designed to cater for these factors.

Modern apprenticeships

Modern apprenticeships provide paid employment, practical training, and further education. During the apprenticeship, trainees must study for NVQ and Key Skills, and a Technical Certificate, such as City and Guilds or BTEC.

There are two levels of apprenticeships: foundation for beginners – the training and education programme normally lasts for approximately 1 year, and the advanced apprenticeship, which lasts for approximately 2 years.

Foundation modern apprenticeship

This first level of apprenticeship trains apprentices for Level 2 NVQ and the associated Key Skills. In addition, a Technical Certificate is studied and this may be a BTEC First, City and Guilds, or Institute of Motor Industry Certificate.

Advanced modern apprenticeship

Advanced Apprenticeships are designed to last for 2 years. They prepare trainees for Level 3 NVQ and Key Skills, and a Technical Certificate, such as BTEC National or similar educational qualification, is studied. Apprenticeships should, in future, provide the most attractive and efficient route to qualification and employment in the vehicle maintenance and repair industry.

Courses in colleges

Many Further Education Colleges provide courses of Motor Vehicle Studies. The courses vary from those for beginners, such as the City and Guilds Progression Award, to mainstream courses for BTEC National Certificates and Diplomas. Many of these courses are linked to employers so that the students gain work experience and this practical work, plus work done at college, may be taken into account when making a later claim for NVQ. For students who opt for study in a Further Education College, as opposed to school sixth form, the BTEC National and Higher National programmes may provide a route into university education.

Courses in secondary schools, technology colleges, etc.

Over the past few years, a number of secondary schools have offered motor vehicle courses and, with the renewed emphasis on vocational education, such courses are likely to become more widely available.

This book is designed for Levels 1 and 2 of current training and education programmes but there are elements that extend into the Level 3 region of some topics. The text shows a large number of engineering concepts and it should help readers to make links between practical work and other subjects, such as science and number work, and thus assist in providing students with a balanced educational experience.

Career prospects

When embarking on a career in the vehicle repair and maintenance industry, thoughts about future jobs may not be uppermost in one's mind. As one's career progresses this may change and it is useful to have an insight into the types of jobs that may be open to people who have trained as technicians. The following list gives an indication of the types of jobs that are often taken by people who started work as trainees or apprentices and who, through gaining experience and qualifications, equipped themselves for promotion:

- Foreman
- Technical receptionist
- Service manager
- Branch manager
- Training instructor
- Teacher/lecturer
- Insurance engineers and assessors
- Transport managers.

How to get there

There are approximately 30 000 businesses, large and small, in the UK that carry out vehicle repair and maintenance. There are therefore many opportunities for trainee technicians to advance to more senior positions, if they so desire. Holders of the jobs in the above list need management skills and these are often taught through specialized management courses. Management studies are also included in higher-level qualifications such as Higher National Certificates and Diplomas. A well-presented curriculum vitae (CV) is likely to be an asset when applying for more senior positions.

Apprenticeships and other opportunities for beginners are likely to be advertised and promoted through the career service. College courses that contain an element of work experience often lead to employment — it is probably a good idea to shop around in your area to see what courses are on offer.

Motor industry organizations

The Institute of the Motor Industry (IMI) and the Institute of Road Transport Engineers (IRTE) are two of the engineering/management institutes that many people engaged in motor vehicle maintenance and repair belong to. These institutes have regional centres that arrange lectures and other events relating to the industry. These events provide an opportunity to meet people from many different companies in an informal setting, in addition to allowing an opportunity to help people keep abreast of developments. Two magazines, *Motor Industry Management* published by the IMI and *Road Transport Engineer* published by the IRTE, advertise posts nationally, and the local press normally carries advertisements about job vacancies in their areas.

Appendix A

A selection of answers to the numerical exercises

Chapter 2

Q1. Exhaust.
Q2. (c).
Q3. (a) $11°$; (b) $35°$; (c) $38°$; (d) (i) $224°$; (ii) $220°$.
Q4. Transfer port.
Q5. Compression stroke.
Q6. 720 degrees.
Q8. 10.8:1.
Q9. 60 mm.
Q10. 679 cc.
Q13. (a).
Q15. (c).
Q16. (c).

Chapter 3

Q1. (a). 6.03 mm.
Q3. (b).
Q5. (c).
Q6. (b).
Q7. 0.16 mm.
Q8. (b).
Q19. L.
Q21. 0.040 mm.

Chapter 4

Q1. (c).
Q2. (a).
Q4. 0.040 mm.

Chapter 6

Q1. 0.24 mm.
Q9. 440 N.
Q10. (a).

Chapter 8

Q2. (a) 33.89 mpg; (b) 8.3 litres/100 km.
Q3. (a) 6250 gallons; (b) £7500.
Q4. 6135 gallons. $6250 - 6135 = 115$ gallons.
    Saving £138.

Chapter 10

Q5. Once.
Q6. 0.48 ms.
Q7. Weaker.

Chapter 13

Q2. 40.80 kJ.
Q10. 25 A.

Chapter 14

Q1. 31.2 kW.
Q2. 134 kW; 89.6%.
Q3. bp = 95 kW; ip = 118.8 kW.
Q4. 38.4 kg/h.
Q5. 6.7%.
Q6. 91 mm.
Q7. 28%.
Q8. ip = 33 kW; ME = 76%.
Q9. 5.08 m^3/h.
Q10. BTE = 25.6%, 28.2%, 30%, 29.2%, 27.3%.

Chapter 16

Q1. 1600 W.
Q6. (b).
Q10. (a).

Q17.

Speed	Difference in volume	% Difference
300	1.6	18.2
600	0.8	7.5
900	1.4	12.3

Q18. $60°$.
Q19. Element 2 minus $1°$.
    Element 4 plus $2°$.

Q20.

Injector number	Error in bar	% Error
2	10	6.7
3	12	8
4	9	6
5	8	5.3
6	1	0.7

Q21.
(a)

Cylinder number	Pressure lbf/in^2	% Error
1	460	6% below
2	395	20% below
3	380	22% below
4	495	1% above

(d)

Cylinder	Pressure in bar
1	31.83
2	27.33
3	26.30
4	34.25

Chapter 17

Q1.

Cylinder	1	2	3	4
Pressure bar	11.2	11.6	12	9
% difference	6.7%	3.3%	0	25%

Q6.

Main bearing caps	88—102 lbf.ft	119.7—138.7 Nm
Inlet manifold bolts	16—20 lbf.ft	21.8—27.2 Nm

Q10. (c).

Q14.

Component	Torque lbf.ft	Torque NM
Main bearing cap	65—75	88.4—102
Connecting rod bolts	26—33	35.4—44.90
Oil pump set screws	16—20	21.76—27.20
Exhaust manifold bolts	14—17	19.04—23.12
Spark plugs	15—20	20.4—27.20
Oil pressure switch	13—15	17.68—20.4

Q16. 0.298 mm.

Chapter 18

Q1. (c).
Q2. (d).
Q3. (c).
Q4. (b).
Q5. (a).
Q6. (a).
Q7. (b).
Q8. (c).
Q9. (c).
Q10. (c).

Chapter 19

Q12. (c).
Q15. (a).

Chapter 21

Q4. (a).

Chapter 31

Q1. (a) 43.00; (b) 0.77; (c) 5.89.
Q2. (a) 19.3; (b) 17.03; (c) 338.
Q3. 17.0 litres.
Q4. £19.61.
Q5. (b).
Q6. (a).
Q7. (c).
Q8. (b).
Q9. (b).
Q10. (c).
Q11. Ovality = (c).
Q12. (a) taper = 0.25 mm.

Appendix B

Conversion factors

Conversion factors and units

Quantity and symbol	SI unit	Imperial unit	Conversion factor
Mass, m	Kilogram, kg Tonne (1000 kg)	Pound, lb (Ton 2240 lb)	1 lb = 0.454 kg
Length, l	Metre, m Millimetre, mm	Foot, ft Inch, in	1 ft = 0.3048 m 1 in = 25.4 mm
Area, A	Square metre, m^2	Square ft, ft^2	1 ft^2 = 0.093 m^2
Volume, V	Cubic metre, m^3	Cubic foot, ft^3	
Capacity, V	Litre, L	Gallon	1 gallon $\approx$ 4.5 litres
Force, F	Newton, N	Pound force, lbf	1 lbf = 4.448 N
Power, P	Kilowatt, kW	Horse power, hp	1 hp = 0.746 kW
Speed	Kilometres per hour, km/h	Miles per hour, mph	1 km/h = 0.622 mile/hour
Velocity, v	Metre/second, m/s	Foot per second, ft/s	1 ft/s = 0.3048 m/s
Torque, T	Newton metre, Nm	Pound foot, lbf.ft	1 lbf.ft = 1.4 Nm
Pressure, P	Pascal, Pa Kilo pascal, kPa Bar, bar	Lbf/in^2 Atmosphere	1 Lbf/in^2 = 6.89 kPa 1 atmosphere $\approx$ 1 bar

Some other units:

The short, or USA, ton = 2000 English pounds.
The USA gallon = 0.83 English gallons.

List of acronyms

ABS	anti-lock braking system
a.c.	alternating current
ACEA	European Automobile Manufacturers' Association
API	American Petroleum Institute
ATDC	after top dead centre
ATF	Authorized Treatment Facility
ATF	automatic transmission fluid
BDC	bottom dead centre
bmep	brake mean effective pressure
bsfc	brake specific fuel consumption
BTDC	before top dead centre
CAN	controller area network
CB	contact breaker
CFPP	cold filter plugging point
CGLI	City and Guilds of London Institute
CI/CIE	compression ignition engine
CRC	cyclic redundancy check
CV	constant velocity
d.c.	direct current
DCU	diagnostic and control unit
DIN	
DIS	distributorless ignition system
DPA	distributor pump application
DPF	diesel particulate filter
DTC	diagnostic trouble code
DTI	dial test indicator
E4WS	electronic four-wheel steering
EA	Environment Agency
ECM	electronic control module
ECU	electronic control unit
EEPROM	electrically erasable programmable read-only memory
EGR	exhaust gas recirculation
ELV	End-of-Life Vehicle
e.m.f.	electromotive force
EOBD	European on-board diagnostics
EP	extreme pressure
EU	European Union
EVAP	

FL	front left
FR	front right
FWD	front-wheel drive
GEA	Garage Equipment Association
HASAW	health and safety at work
HC	hydrocarbons
HFC	hydrofluorocarbons
HRW	heated rear window
HSE	Health and Safety Executive
HT	high tension
IFS	independent front suspension
imep	indicated mean effective pressure
IMI	Institute of the Motor Industry
ip	indicated power
IRS	independent rear suspension
IRTE	Institute of Road Transport Engineers
ISC	idle speed control
ISO	International Standards Organization
KAM	keep-alive memory
KPI	kingpin inclination
LCD	liquid crystal display
LED	light-emitting diode
LGV	large goods vehicle
LPG	liquid petroleum gas
LT	low tension
MACS	Mobile air-conditioning systems
MAP	manifold absolute pressure sensor
MIL	malfunction indicator light/lamp
MON	motor octane number
MOT	Ministry of Transport
MUX	multiplexer
NVQ	National Vocational Qualification
OBD	on-board diagnostics
OBD2	on-board diagnostics (USA)
OFT	Office of Fair Trading
OHC	overhead camshaft
OHV	overhead valve

PAS	power-assisted steering	sfc	specific fuel consumption
PCV	positive crankcase ventilation	SHC	specific heat capacity
p.d.	potential difference	SRS	supplementary restraint system
ps	pferdestärke (pulling power of a horse)		
PWM	pulse width modulation	TDC	top dead centre
PZT	platinum, zirconium, titanium	TRL	British Transport Research Laboratory
		TTL	transistor-to-transistor logic
RAM	random-access memory (read and write memory)	TWI	tread wear indicators
ROM	read-only memory (permanent memory)	UJ	universal joint
RON	research octane number		
rpm	revolutions per minute	VCM	vehicle condition monitoring
RTL	resistor transistor logic	VFD	vacuum fluorescent display
RWD	rear-wheel drive	VOSA	Vehicle and Operator Services Agency
SAE	Society of Automotive Engineers		
SCR	selective catalyst reduction	ZEV	zero emissions vehicle

Index